Plant Biotechnology

This book explores our knowledge of biotechnology and its application to improving the quality of medicinal plants. With its unique and sustained focus on medicinal plant biotechnology, it offers an essential guide and a systematic reference for the development of medicinal products with the help of biotechnology from natural sources. With contributions from world-renowned experts in the fields of biotechnology, pharmaceutical biology, pharmacognosy, chemistry, and pharmaceutical biotechnology, *Plant Biotechnology* was written while keeping in mind the requirements of botanists, the pharmaceutical industry, biotechnologists, microbiologists, and specialists working on plant biotechnology. It can serve as either a textbook or a reference work for students, teachers, or scientists working in the field of medicinal plant biotechnology, and its readership also includes natural product chemists, biotechnologists, pharmacognosists, and pharmacologists, as well as academic and industry researchers.

Features:

- Provides essential evidence for all specialists overseeing supportive biotechnology on its utility.
- Discusses the fundamental techniques in biotechnology and their implementation with medicinal plants.

Plant Biotechnology

The Medicinal and Phytopharmaceutical Aspects

Edited by
Deependra Singh, Durgesh Nandini Chauhan,
Nagendra Singh Chauhan, and Manju Rawat Singh

CRC Press
Taylor & Francis Group
Boca Raton London New York

CRC Press is an imprint of the
Taylor & Francis Group, an **informa** business

Designed cover image: Shutterstock

First edition published 2026
by CRC Press
2385 NW Executive Center Drive, Suite 320, Boca Raton FL 33431

and by CRC Press
4 Park Square, Milton Park, Abingdon, Oxon, OX14 4RN

CRC Press is an imprint of Taylor & Francis Group, LLC

ISBN: 978-1-032-35446-0 (hbk)
ISBN: 978-1-032-35447-7 (pbk)
ISBN: 978-1-003-32693-9 (ebk)

DOI: 10.1201/9781003326939

Typeset in Times
by Apex CoVantage, LLC

Contents

Tissue Culture Techniques

Biotransformation and Transgenesis

Fermentation Technology

Newer Paradigm in Medicinal Plant Biotechnology

Preface

Plant biotechnology has revolutionized drug discovery and production, with advancements in genetic engineering, tissue culture, and metabolic engineering. Ethical, safety, and regulatory considerations are crucial in biotech applications in medicine. Plant biotechnology can be combined with conventional breeding techniques, transgenic approaches, and integrated crop management to address specific agronomic challenges. Techniques like cellular culture, molecular genome analysis, and tissue culture have been used to cope with genetic erosion, strengthen ex situ collections, and accomplish in situ conservation. Advances in genetics and molecular biology have led to improved agricultural products worldwide. Tissue culture techniques, including micropropagation, embryo rescue, and microspore culture, facilitate large-scale plant propagation. DNA-based approaches, such as genetic engineering and recombinant DNA techniques, are essential for developing transgenic individuals and regenerating plants from precursor cells.

Chapter 1 describes the prospects of how integrating plant biotechnology into pharmacy could make biotechnologically derived plant medicines increasingly crucial in health care. Chapter 2 describes the requirements for a laboratory setup for plant biotechnology. Chapter 3 relates genetic engineering or recombinant DNA (rDNA), a method that uses scientific research on DNA in creating practical applications. It has revolutionized medical genetics and biomedicine by transforming microbes, animals, and plants to produce medically beneficial chemicals, and by improving health outcomes through innovative treatments. Chapter 4 gives details on protoplast culture isolation and culture methods. Tissue culture techniques are crucial in preserving rare, endangered medicinal and aromatic plants (MAPs) and in producing secondary metabolites. However, excessive collection threatens these plants' extinction. Modern methods, such as callus induction, cell suspension culture, and organ culture, have gained industrial importance in rapid propagation, resource conservation, and secondary metabolite production. Chapter 5 reviews the improved biotechnological methods for micropropagation strategies, direct organogenesis, and somatic embryogenesis. In Chapter 6, some common sterilization methods and protocols for different plant parts are discussed. Chapter 7 explains the advancements of gene transfer methodologies in the development, modification, and protection of crops and the overall enhancement in boosting agricultural productivity. Chapter 8 covers the immobilization techniques of plant cells, including entrapment, solid support, embedding, bioreactors, and secondary metabolite formation. Chapter 9 explores the importance of the cloning process used to create identical or nearly identical organisms in animals, humans, and plants. It is crucial in genetic engineering, gene research, and plant science to improve yield, insect and herbicide resistance, drought and salt tolerance, and other traits.

Chapter 10 provides an overview of secondary metabolism in tissue cultures, its importance in generating medicinal agents, the role of precursors and elicitors in secondary metabolism, and its future directions. Chapter 11 provides a comprehensive overview of biotransformation using plant cell culture systems. It involves enzymatic catalysis in cells, allowing for the synthesis of novel chemicals and an increased product accumulation. Chapter 12 explores the design, operation, and optimization of bioreactors for scaling up plant cell cultures while preserving their biosynthetic potential. Chapter 13 describes the methods used in gene identification, localization, and sequencing. Gene identification/localization is one of the key steps involved in understanding the genetic bases of economic traits, disease and stress response, and the underlying mechanisms of multiple physiochemical reactions. Chapter 14 explains fermentation technology for the production of ergot alkaloids. The fermentation-based saprophytic generation of ergot alkaloids is more advantageous than synthetic routes. Enzymes produced through fermentation have a high pharmaceutical interest, targeting specific biological reactions to treat diseases. Chapter 15 gives details on computer-aided drug design discovery (CADDD). It involves molecular design, molecular modeling, computational chemistry, and rational drug design. CADDD uses virtual screening, molecular docking, in silico prediction of ADMET properties, and lead optimization to identify and predict lead structures or pharmacophores of novel drugs from secondary metabolites of medicinal plants. Chapter 16 covers metabolic engineering in optimizing microbial pathways for specific plant chemicals, introducing novel biochemical pathways, route optimization, and gene expression modulation. Synthetic biology techniques like CRISPR-Cas9 enable the development of customized microbial hosts for a variety of plant-based chemicals. Microorganisms like *Saccharomyces cerevisiae*, *Escherichia coli*, and filamentous fungus are commonly used. Chapter 17 explains how DNA-based methods have emerged to validate therapeutic plants and separate them from adulterants. Identifying species adulteration visually is challenging due to the plants' dry state. Chapter 18 defines the biotechnological approach in medicinal plants and their chemical compounds in ruminant nutrition and antimethanogenic effects. Essential oils, extracts, and plant components can inhibit rumen methanogenesis and promote protozoa removal. Chemical classes like saponins and tannins reduce environmental pollution without impairing feed digestion. Chapter 19 explores the potential of natural products derived from remedial plants, including phytoconstituents and extracts, for new drug molecules. The chapter also discusses the methodologies in extracting and isolating phytoconstituents from medicinal plants. Chapter 20, meanwhile, explores metabolomics as a powerful tool in analyzing and quantifying metabolites in medicinal plants using techniques like mass spectrometry and NMR spectroscopy. It highlights how metabolomics accelerates drug discovery, quality control, and integration with other omics to understand plant metabolism. Additionally, the chapter discusses its applications in optimizing cultivation, genetic engineering, and enhancing plant resilience.

This book encourages scientists, medical professionals, and/or industrial persons to work on medicinal plant biotechnology.

Last but not the least, we would like to express our earnest gratitude to all the authors who have taken time from their busy timetables to be part of this endeavor and to back perfect chapters that add both gravity and value to this book. We welcome suggestions and criticisms from our readers too. Special thanks to our families for their sustenance and inspiration. We express our acknowledgment to the publishing and production team, especially Renu Upadhya and Jyotsna, for their substantial, skillful, and motivating management.

The Editors: Deependra Singh, Durgesh Nandini Chauhan,
Nagendra Singh Chauhan, and Manju Rawat Singh
February 3, 2025

Editors' Biographies

Prof. Deependra Singh (M.Pharm, Ph.D, Post-doc (USA)) is presently working at University Institute of Pharmacy, Pt. Ravishankar Shukla University, Raipur (C.G.) India. He is chairman of the education regulation committee and central council member of the pharmacy council of India (PCI). He has around 20 years of teaching and research experience. He has 145 research and review papers published in national and international journals. He has written 30 book chapters and two books with Elsevier. He has successfully completed projects funded by ICMR, UGC, AICTE, DST, CGCOST, and several industrial projects. He has citations more than 7943 (h-index of 36 and i10 index of 123-Google scholar) and scopus h-index 30; 3514 citations. He has 2 international, 3 Indian patents and 2 filed. He has been among the top 2% of scientists across the world in the field of pharmacy and pharmacology in a survey by Stanford University, USA for four consecutive years (2021, 2022, 2023, 2024). He was a visiting scientist at the National Center for Natural Products Research, School of Pharmacy, University of Mississippi, USA. He was a recipient of the DHR-ICMR award for young biomedical scientist in 2019, UGC RAMAN international fellow in 2014, UGC research award in 2012, and a number of awards at national and state level for excellence in research. He is a life member of various professional bodies such as the Indian Pharmaceutical Association (IPA), Association of Pharmacy Teachers of India (APTI), Indian Society of Pharmacognosy (ISP), and serving as Vice-President, Central zone APTI 2022–2027 (Elected), Vice-President, Central zone IPGA 2022–2025 (Elected), Secretary, Chhattisgarh state branch IPA.

Durgesh Nandini Chauhan, M.Pharm., has several years of academic (teaching) experience at institutes in India in pharmaceutical sciences, and currently works in Columbia College of Pharmacy, Raipur CG India. She has taught subjects such as pharmaceutics, pharmacognosy, traditional concepts of medicinal plants, drug delivery, phytochemistry, cosmetic technology, pharmaceutical engineering, pharmaceutical packaging, quality assurance, dosage form designing, and anatomy and physiology. She is a member of the Association of Pharmaceutical Teachers of India, SILAE: Società Italo-Latinoamericana di Etnomedicina (The Scientific Network on Ethnomedicine, Italy), and others. She has attended several workshops, conferences, and symposiums, including the AICTE-Sponsored Staff Development Program on "Effects of Teaching and Learning Skills in Pharmacy: Tool for Improvement of Young Pharmacy Teachers." She has written more than 10 articles published in national and international journals, 15 book chapters, and edited 10 books. She is also active as a reviewer for several international scientific journals and is an active participant at national and international conferences, including Bhartiya Vigyan Sammelan and the International Convention of Society of Pharmacognosy. Presently she has joined Ishita Research Organization, Raipur, India as a freelance writer and guide for students in pharmacy, Ayurvedic, and science in their research projects. She is interested in data analysis, proposal writing, questionnaire design, designing and developing research questionnaires and policy. She is actively engaged with accurately monitoring, investigating, and analyzing all statistical information, professionally presenting information and reports, and investigating and documenting key data and information at the request of senior researchers.

Dr. Nagendra Singh Chauhan obtained his M.Pharm, Ph.D. from the Department of Pharmaceutical Sciences, Dr. H.S. Gour University, Sagar, MP, India. He has around 20 years of research experience. He is presently working as a Senior Scientific Officer Grade-I and Government Analyst at Drugs Testing Laboratory Avam Anusandhan Kendra, Raipur, Chhattisgarh, India. He has professional expertise in Natural Product Isolation and phytopharmacology. He has written more than 70 articles published in national and international journals, and 32 book chapters, and edited 7 books. He is the recipient of the Dr. P. D. Sethi Memorial National Award for the best HPTLC publication from India in the years 2021 and 2022. He is a member of various professional and academic bodies such as the Society of Pharmacognosy India, the International Natural Product Sciences Taskforce (INPST), SILAE: Società Italo-Latinoamericana di Etnomedicina (The Scientific Network on Ethnomedicine, Italy), Institutional Human Ethical Committee, Association of Pharmaceutical Teachers of India (APTI), and Indian Pharmacy Graduates' Association.

Prof. Manju Rawat Singh (M.Pharm, Ph.D) is presently working at University Institute of Pharmacy, Pt. Ravishankar Shukla University, Raipur (C.G.) India. She has around 20 years of teaching and research experience. She is among the top 2% of scientists across the world in the field of Pharmacy and Pharmacology surveyed by Stanford University, USA. She is engaged in research projects focused on studying the role of plant bioactives in devising new treatments mainly for autoimmune disorders and inflammatory diseases in collaboration with various universities in India and abroad. She has around 140 research and review papers published in national and international journals. She has written 30 book chapters and two books with Elsevier. She has citations more than 6867 with h-index 37, i10 index of 94 (Google scholar) and scopus h-index 30; 3525 citations. She has 2 international, 3 Indian patents and 2 filed. She has successfully completed several projects funded by the Indian Council of Medical Research (ICMR), University Grants Commission (UGC), C.G. State Council of Science and Technology (CGCOST), and the Department of Science and Technology (DST). She was a visiting scientist at the Department of Pharmaceutics, University of Mississippi, USA in 2015. She is the recipient of the DHR-ICMR international young Indian biochemist award 2014–15 and the APTI Prof. C.J. Shishoo Award for Research in Pharm Sciences 2024. She is a life member of professional bodies such as the Indian Pharmaceutical association (IPA), Association of Pharmacy Teachers of India (APTI), Indian Pharmacy Graduates Association (IPGA), and serving as President, IPGA state branch Chhattisgarh 2022–2025 (Elected).

Contributors

Ahmed Abdel-Megeed
Department of Plant Protection
Faculty of Agriculture (Saba Basha)
Alexandria University
Alexandria, Egypt

Brijesh Ajjappala
Department of Plant Biotechnology
University of Agricultural Sciences
GKVK, Bangalore, Karnataka, India

Aqsa Akram
Department of Plant Biotechnology
Atta-ur-Rahman School of Applied Biosciences
National University of Sciences and Technology
Islamabad, Pakistan

Edwin Rafael Alvarado-Ramírez
Faculty of Veterinary Medicine and Zootechnics
Autonomous University of Tamaulipas
Victoria City, Tamaulipas, Mexico

Santosh Anand
Department of Biotechnology
School of Applied Sciences
REVA University
Bangalore, Karnataka, India

Ram Baran Singh
Department of Biotechnology
School of Applied Sciences
REVA University
Bangalore, Karnataka, India

Manisha Bhatia
MM College of Pharmacy
Maharishi Markandeshwar (Deemed to be University)
Mullana, Ambala, Haryana, India

Hemavathi Brijesh
Department of Biotechnology
School of Applied Sciences
REVA University
Bangalore, Karnataka, India

Subhash Chandra
Computational Biology and Biotechnology Laboratory
Department of Botany
SSJ University
Almora, Uttarakhand, India

Jasmine Chaudhary
MM College of Pharmacy
Maharishi Markandeshwar
Ambala, Haryana, India

Durgesh Nandini Chauhan
Columbia Institute of Pharmacy
Raipur, Chhattisgarh, India

Nagendra Singh Chauhan
Drugs Testing Laboratory
Avam Anusandhan Kendra
Raipur, Chhattisgarh, India

Mayur Chaurey
School of Pharmacy
Devi Ahilya Vishwavidyalaya
Indore, MP, India

Pankaj Dagur
Department of Pharmaceutical Sciences and Technology
Birla Institute of Technology
Ranchi, Jharkhand, India

G. N. Darwhekar
Acropolis Institute of Pharmaceutical Education and Research
Indore, Madhya Pradesh, India

Rohit Dutt
Gandhi Memorial National College
Ambala, Haryana, India

Akanksha Dwivedi
Acropolis Institute of Pharmaceutical Education and Research
Indore, Madhya Pradesh, India

Sumitha Elayaperumal
Department of Biotechnology and Bioinformatics
JSS Academy of Higher Education and Research
Mysore, Karnataka, India

Gajanand Engla
School of Pharmacy
Devi Ahilya Vishwavidyalaya
Indore, MP, India

Dipak D. Gadade
Department of Pharmaceutical Sciences
Delhi Skill and Entrepreneurship University
Dwarka Campus, New Delhi, India

Preeti Garg
Hindu College of Pharmacy
Sonipat, Haryana, India

Vandana Garg
Department of Pharmaceutical Sciences
Maharshi Dayanand University
Rohtak, Haryana, India

Rupesh Kumar Gautam
IES Institute of Pharmacy
IES University
Bhopal, India

Manik Ghosh
Department of Pharmaceutical Sciences and Technology
Birla Institute of Technology
Ranchi, Jharkhand, India

Ekta Gurnany
Arihant School of Pharmacy and Bio-Research Institute
Gandhinagar, Gujarat, India

Pedro Enrique Hernández-Ruiz
Higher School of Veterinary Medicine and Zootechnics
Autonomous University of Guerrero
Tecpán de Galeana, Guerrero, Mexico

Muhammad Irfan
Department of Healthcare Biotechnology
Atta-ur-Rahman School of Applied Biosciences
National University of Sciences and Technology
Islamabad, Pakistan

Aakanchha Jain
National Institute of Pharmaceutical Education and Research (NIPER Ahmedabad)
Gandhinagar, Gujrat, India

Akash Jain
MM College of Pharmacy
Ambala, Haryana, India

C. P. Jain
Department of Pharmaceutical Sciences
Mohanlal Sukhadia University
Udaipur, Rajasthan, India

Sourabh Jain
Arihant School of Pharmacy and Bio-Research Institute
Gandhinagar, Gujarat, India

N. Joshi
School of Biotechnology and Bioinformatics
D. Y. Patil Deemed to Be University
Navi Mumbai, Maharashtra, India

Tushar Joshi
Department of Biotechnology
Kumaun University
Bhimtal, Uttarakhand, India
and
Computational Biology and Biotechnology Laboratory
Department of Botany
SSJ University
Almora, Uttarakhand, India

V. Joshi
Center for Basic Sciences
Pt. Ravishankar Shukla University
Raipur, Chhattisgarh, India

Jyoti
Department of Pharmacognosy
Hindu College of Pharmacy
Haryana, India

G. Karri
School of Biotechnology and Bioinformatics
D. Y. Patil Deemed to Be University
Navi Mumbai, Maharashtra, India

Jeevan Jyoti Kaushik
Department of Microbiology
School of Basic and Applied Sciences
Shri Guru Ram Rai University
Dehradun, Uttarakhand, India

Vichitra Kaushik
Shanti Niketan College of Pharmacy
Mandi, Himachal Pradesh, India

Rakhi Khabiya
Acropolis Institute of Pharmaceutical Education and Research
Indore, Madhya Pradesh, India

Khushbukhat Khan
Department of Healthcare Biotechnology
Atta-ur-Rahman School of Applied Biosciences
National University of Sciences and Technology
Islamabad, Pakistan

Zeeshan Khan
Department of Plant Biotechnology
Atta-ur-Rahman School of Applied Biosciences
National University of Sciences and Technology
Islamabad, Pakistan

Kiran
Department of Pharmaceutical Sciences
Maharshi Dayanand University
Rohtak, Haryana, India

Afrah Kounin P.
Department of Biotechnology and Bioinformatics
JSS Academy of Higher Education and Research
Mysore, Karnataka, India

Love Kumar Soni
School of Pharmacy
Devi Ahilya Vishwavidyalaya
Indore, MP, India

Manish Kumar
Department of Pharmaceutics
ISF College of Pharmacy
Moga, Punjab, India

Vimal Kumar
School of Pharmacy
ITM SLS Baroda University
Vadodara, Gujarat, India

Renuka Madhu
Department of Biotechnology
School of Applied Sciences
REVA University
Bangalore, Karnataka, India

Priyanka Maiti
Centre for Environmental Assessment and Climate Change
G. B. Pant National Institute of Himalayan Environment (GBP-NIHE)
Almora, Uttarakhand, India

Shilpa Malakar
Department of Microbiology
Kalinga University
Raipur, Chhattisgarh, India

Ramya Manjunath
Department of Biotechnology
School of Applied Sciences
REVA University
Bangalore, Karnataka, India

Shalini Mathpal
Department of Biotechnology
Kumaun University–Bhimtal Campus
Bhimtal, Uttarakhand, India
and
Computational Biology Biotechnology Laboratory
Department of Botany
SSJ University
Almora, Uttarakhand, India

Anuj Modi
Department of Pharmaceutical Sciences
Delhi Skill and Entrepreneurship University
Dwarka Campus, New Delhi, India

Mahesha Nand
ENVIS Centre on Himalayan Ecology
G. B. Pant National Institute of Himalayan Environment (GBP-NIHE)
Almora, Uttarakhand, India

Sunita Panchawat
Department of Pharmaceutical Sciences
Mohanlal Sukhadia University
Udaipur, Rajasthan, India

A. Pandurangan
Swami Vivekanand College of Pharmacy
Patiala District, Punjab, India

Janmejay Pant
University Institute of Pharma Sciences
Chandigarh University
Mohali, Punjab, India

Arun Patidar
School of Pharmacy
Devi Ahilya Vishwavidyalaya
Indore, MP, India

Joohee Pradhan
Department of Pharmaceutical Sciences
Mohanlal Sukhadia University
Udaipur, Rajasthan, India

Devyani Rajput
Department of Pharmacognosy
Amity Institute of Pharmacy
Amity University–Madhya Pradesh
Gwalior, Madhya Pradesh, India

Ramachandra Ramanjinappa
Department of Biotechnology
School of Applied Sciences
REVA University
Bangalore, Karnataka, India

Lokendra Singh Rathor
University Institute of Pharmacy
Pt. Ravishankar Shukla University
Raipur, Chhattisgarh, India

Alakh Niranjan Sahu
Department of Pharmaceutical Engineering & Technology
Indian Institute of Technology (Banaras Hindu University)
Varanasi, Uttar Pradesh, India

Divya Sahu
University Institute of Pharmacy
Pt. Ravishankar Shukla University
Raipur, Chhattisgarh, India

Vipin Saini
MM College of Pharmacy
Maharishi Markandeshwar (Deemed to be University)
Mullana, Ambala, Haryana, India

Mohamed Z. M. Salem
Forestry and Wood Technology Department
Faculty of Agriculture (EL-Shatby)
Alexandria University
Alexandria, Egypt

Jashanpreet Singh Sandhu
Alpha MD Pvt Ltd
Mumbai, Maharashtra, India

Kamal Shah
Institute of Pharmaceutical Research
GLA University
Mathura, Uttar Pradesh, India

Priyanka Sharma
Department of Botany
Kumaun University
Nainital, India

Ajay Kumar Shukla
Institute of Pharmacy
Dr Ram Manohar Lohia Avadh University
Ayodhya, Uttar Pradesh, India

Sampat Singh Tanwar
Department of Pharmacy, Shri Vaishnav Vidyapeeth Vishwavidyalaya
Indore, M.P., India

Deependra Singh
University Institute of Pharmacy
Pt. Ravishankar Shukla University
Raipur, Chhattisgarh, India

Manju Rawat Singh
University Institute of Pharmacy
Pt. Ravishankar Shukla University
Raipur, Chhattisgarh, India

Rahul Pratap Singh
School of Medical and Allied Sciences
G. D. Goenka University
Sohna, Gurugram, India

Neha Sisodiya
Acropolis Institute of Pharmaceutical Education and Research
Indore, Madhya Pradesh, India

Yuvaraj Sivamani
Crescent School of Pharmacy
B.S. Abdur Rahman Crescent Institute of Science and Technology
Chennai, Tamilnadu, India

Priya Sutaoney
School of Studies in Life Science
Pt. Ravishankar Shukla University
Raipur, Chhattisgarh, India

Hayagriba Swain
Department of Biotechnology and Bioinformatics
JSS Academy of Higher Education and Research
Mysore, Karnataka, India

Rahul Singh Thakur
Adina College of Pharmacy
Sagar, Madhya Pradesh, India

Abdelfattah Z. M. Salem
Dipartimento di Scienze del Suolo, della Pianta e degli Alimenti (Di.S.S.P.A.)
Università degli Studi di Bari, Via Giovanni Amendola, Bari BA, Italy

1 Historical Perspectives, Prospects for the Development of Plant Biotechnology as a Source of Medicinal Agents

Applications in Pharmacy and Allied Fields

Divya Sahu, Lokendra Singh Rathor,
Manju Rawat Singh, and Deependra Singh*
* Corresponding author: manjursu@gmail.com

1.1 INTRODUCTION

"Biotechnology" was first introduced by Karl Ereky in 1919. It refers to the use of living organisms to create products from raw materials. Historically, humans have relied on plants as a primary source of medicine. Traditional medicine systems across the world, such as Ayurveda, traditional Chinese medicine, and Native American herbalism, have extensively utilized plant-based compounds. The modern era of plant biotechnology began with the isolation of active ingredients from plants (Alamgir 2018). For example, the discovery of salicin from willow bark, which led to the development of aspirin, marked a significant milestone. The 20th century saw rapid advancements in genetics and molecular biology, which have been applied to plants. Genetic engineering techniques have enabled the modification of plants to enhance their medicinal properties. The goal is to enhance the value of plants, improve agricultural productivity, or create novel products (An et al. 2024). It also means the collection of abilities needed to use living systems or manipulate natural processes to create settings, systems, or products that support human progress. At the moment, biotechnology is more focused on creating hybrid genes and then transferring them into creatures where the gene is normally absent in part or as a whole. Agriculturalists used rudimentary biotechnology in the prehistoric era to create superior plant and animal species through cross-pollination and cross-breeding techniques. Earlier applications of biotechnology included raising crops, teaching and selectively breeding animals, and using microorganisms to make cheese, yogurt, bread, beer, and wine (Bhatia 2018). The primary goal of early agriculture was food production. The most archaic forms of biotechnology include plant cultivation and animal training, especially domestication. Our ancestors began domesticating plants as a dependable source of food more than ten thousand years ago, at the same time that they began domesticating animals. Rice, barley, and wheat were among the first examples of these domesticated plants. Additionally, wild animals were tamed to yield meat or milk. There are additional reports of the traditional use of microorganisms to produce bread, yogurt, and cheese. Around this time, when fermentation was first discovered, a variety of alcoholic beverages, including wine and beer, were invented. Products were created as a result of this procedure (food and drink). Because no one knew how fermentation worked in the past, it is possible that it was discovered by accident. Louis Pasteur

DOI: 10.1201/9781003326939-1

was the first scientist to describe fermentation in the late 1800s (Mustafa et al. 2023). He gave a demonstration of the germ hypothesis, which describes how microorganisms survive and how this affects the fermentation process. Pasteur's work advanced a number of scientific fields. Several traditional remedies were utilized as biotechnology products in the past. One example is honey, which was used as an ointment for wounds and to treat a variety of respiratory conditions. Honey is used to treat wounds and is regarded as a natural antibiotic due to its several antibacterial components. Similarly, soybean curds were used to heal boils as early as 600 BC in China. Moldy cheese was once used as a wound treatment by Ukrainian farmers. Later research revealed that the antibiotics in these molds eliminated bacteria and stopped infection from spreading. The year 1928 saw the extraction of penicillin, the first antibiotic, from mold by Alexander Fleming. Since antibiotics are more potent and effective than older medications, this discovery completely changed the range of treatments that are currently available. The late 18th and early 19th centuries saw the development of biotechnology in the fields of vaccines, animal-drawn technology, and crop rotation (particularly leguminous crops). There was a known turning point in biology around the late 19th century (Malik et al. 2021).

1.1.1 Prospects for Development

Genetic Engineering for Enhanced Production. There is ongoing research to increase the yield of medicinal compounds from plants using genetic engineering. For example, modifying pathways in plants to boost the production of specific compounds (Kowalczyk 2022).

Synthetic Biology. Synthetic biology holds great potential in plant biotechnology. It entails creating whole new biological components, tools, and systems, as well as redesigning naturally occurring biological systems for practical uses (Garner 2021).

Pharmacogenomics and Personalized Medicine. The field of pharmacogenomics, which involves tailoring medical treatment to the individual characteristics of each patient, could greatly benefit from plant biotechnology. Plants could be engineered to produce a range of compounds to suit individual genetic profiles (Babu 2024).

1.1.2 Types of Biotechnology

These are the major branches of biotechnology:

1.1.2.1 Medical Biotechnology (Red Biotechnology)

Medical biotechnology, also known as red biotechnology, emphasizes on developing medical therapeutics, diagnostics, and treatments. This includes the production of vaccines and antibiotics, the development of genetic testing, and gene therapy (Elsayed, Kanwugu, and Ivantsova 2019).

1.1.2.2 Agricultural Biotechnology (Green Biotechnology)

Agriculture biotechnology, referred to as green biotechnology, involves the genetic modification of plants and animals to improve their yield, nutritional value, and resistance to pests and diseases. It also includes the development of biopesticides and biofertilizers (Barkha and Bharti 2016).

1.1.2.3 Industrial Biotechnology (White Biotechnology)

Industrial biotechnology (white biotechnology) applies for industrial purposes, including the production of chemicals, biofuels, and materials. It aims to create more sustainable and eco-friendly industrial processes. Plant tissue culture is one example for white biotechnology (Haas et al. 2009).

1.1.2.4 Marine Biotechnology (Blue Biotechnology)

In marine biotechnology, the marine and aquatic environments are explored to develop health-care products, new resources for agriculture, and industrial materials. It includes the discovery of marine biomolecules with potential application in medicine, cosmetics, and nutrition. Therefore, it is also called blue biotechnology (Daniotti and Re 2021).

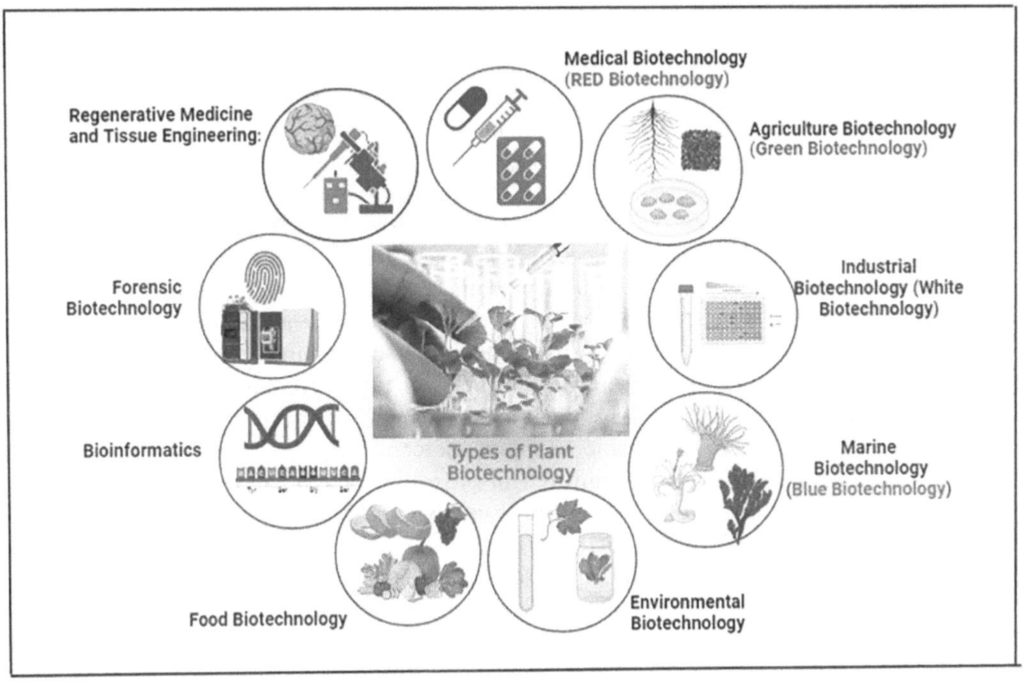

FIGURE 1.1 Diagrammatic representation of types of plant biotechnology.

1.1.2.5 Environmental Biotechnology

Environmental biotechnology is directly connected with nature. This branch is focused on the application of biotechnology in solving environmental problems, including the treatment of waste, bioremediation (using organisms to clean up contaminated environments), and the production of renewable energy from biological sources (Ezeonu et al. 2012).

1.1.2.6 Food Biotechnology

In biotechnology, food production and processing are involved in the area of food biotechnology. This includes the development of genetically modified foods, fermentation processes, and food additives.

1.1.2.7 Bioinformatics

Bioinformatics combines biology, computer science, and information technology to analyze and interpret biological data, such as DNA and protein sequences. It plays a crucial role in identifying genes, understanding diseases, and developing drugs (Bayat 2002).

1.1.2.8 Forensic Biotechnology

DNA and other biological markers are powerful tools in legal and criminal investigations, known as forensic biotechnology. It includes DNA fingerprinting in identifying individuals, solving crimes, and exonerating the wrongly accused (Althobaiti et al. n.d.).

1.1.2.9 Regenerative Medicine and Tissue Engineering

This is the most prominent field in biotechnology, which always converges on repairing, replacing, maintaining, or enhancing tissue function or the whole organ using cells, engineering methods, and materials. This includes stem cell therapy and the development of artificial organs (Han et al. 2020).

1.2 APPROACHES IN PLANT BIOTECHNOLOGY AS A SOURCE OF MEDICINE

Plant biotechnology is a key area of biotechnology that utilizes genetic engineering, tissue culture, and molecular markers to improve plants and their products. This field has been particularly influential in the development of medicinal agents. Genetic engineering is one of its most promising fields, which explains gene cloning and transformation. In gene cloning and transformation, specific genes responsible for the synthesis of medicinal compounds are identified, cloned, and then inserted into plants (Cecchin and Stocco 2020). This genetic transformation enables plants to produce higher levels of specific medicinal compounds, or even novel compounds they do not naturally produce. Metabolic engineering is also a part of genetic engineering, which applies for the modification of the metabolic pathways within plants to increase the production of desired medicinal compounds. By overexpressing or silencing certain genes, the flux toward the synthesis of target compounds is enhanced. The second approach of plant biotechnology includes tissue culture techniques. Various types of tissue culture techniques are applicable for plant biotechnology, such as: (1) *Callus culture and organ culture.* Plant tissues are cultured in vitro to produce masses of cells (callus) or organs (Dong and Ronald 2019). These cultures can be manipulated to produce higher levels of specific compounds than the plants themselves. (2) *Hairy root culture. Agrobacterium rhizogenes*–mediated transformation is used to induce the production of hairy roots in plants, which often produce higher levels of secondary metabolites than the parent plant. (3) *Cell suspension cultures.* Plant cells are cultured in liquid media for the production of valuable phytochemicals. This method allows for the controlled, sterile production of plant compounds on a large scale. (4) *Molecular farming* plays a fourth type of approach, where plants genetically produce proteins, enzymes, and vaccines (Efferth 2019). This approach, also known as "pharming," offers a cost-effective, scalable, and safe alternative to traditional methods of producing pharmaceuticals. Some edible vaccines are developed by transgenic plants that can produce antigens for use as vaccines. When consumed, these antigens stimulate an immune response without the need for traditional injection-based vaccination. Marker-assisted selection (MAS) uses molecular markers to select plants with desirable traits, including enhanced medicinal properties. This accelerates the breeding process by allowing for the selection of individuals carrying genes of interest without having to wait for the plant to grow and exhibit the desired trait. CRISPR/Cas9 genome editing allows for the precise editing of plant genomes to either enhance the production of beneficial compounds or eliminate the production of undesirable ones. This method offers a high degree of specificity and efficiency in modifying plant metabolic pathways (Gavrilescu 2010).

TABLE 1.1
Various Approaches of Plant Biotechnology

S. No.	Approaches	Field	References
1	Genetic engineering	Gene cloning and transformation Metabolic engineering	(Ormandy, Dale, and Griffin 2011)
2	Tissue culture technique	Callus culture and organ culture Hairy root culture Cell suspension cultures	(Wijerathna-Yapa and Hiti-Bandaralage 2023)
3	Molecular farming	Pharmaceutical proteins and vaccines Edible vaccines	(Chung et al. 2022)
4	Marker-assisted selection (MAS)	Breeding for traits	(Collard and Mackill 2008)
5	CRISPR/Cas9 genome editing	Precision editing	(Li et al. 2023)

1.3 MODERN BIOTECHNOLOGY TECHNIQUES

1.3.1 Plant Tissue Culture

Plant tissue culture (Figure 1.2) is an in vitro technique that develops plant tissue fragments (explants) in an artificial medium under aseptic conditions. In order to promote cell proliferation and plant regeneration, it entails cultivating explants—such as branch tips, root tips, calluses, seeds, embryos, pollen grains, ovules, or even a single cell—that have been separated from the mother plant on a sterile nutritional medium. The Murashige and Skoog (MS) basal medium is the most common medium for this technique (Deckers et al. 2020). The various stages involved in this technique are the following: (1) stage 0: selection of explants; (2) stage I: initiation of aseptic culture; (3) stage II: proliferation of axillary shoots; (4) stage III: hardening. The plant part used to initiate tissue culture is called the explant. The advantage of explant depends upon its location, age, and phase of development. Disinfectants used include calcium hypochlorite ($Ca(ClO)_2$), ethanol (CH_2OH), and mercuric chloride ($HgCl_2$). The cultured plant is incubated in an incubation chamber under dark and light conditions. The addition of cytokinin helps in the proliferation of axillary shoots to the culture medium (Garner 2021). Within four weeks, the new shoot may be subcultured, and root formation can then be done using auxin. The induced root is attached to the shoot to form a small plantlet and then transferred to soil. The cultured media responsible for the production of multiple explants can simultaneously occur in the rooting stage. Otherwise, in prominent conditions, the culture media may be changed to induce rooting or the formation of strong roots. At last, the produced plant is transferred to a suitable substrate (compost, fertilizer, dung, etc.), followed by exposure to low humidity and light in a greenhouse, a process known as hardening (Han et al. 2020).

1.3.2 CRISPR-Cas9 Genome Editing

The most suitable genome editing tool in all living cells is called CRISPR (clustered regularly interspaced short palindromic repeat), and the associated protein is known as Cas-9 (Figure 1.3). The CRISPR/Cas-9 system has two essential components: (1) CRISPR-associated Cas-9 protein and (2) guide RNA (gRNA). Recognition, cleavage, and repair are the three major steps in these methods.

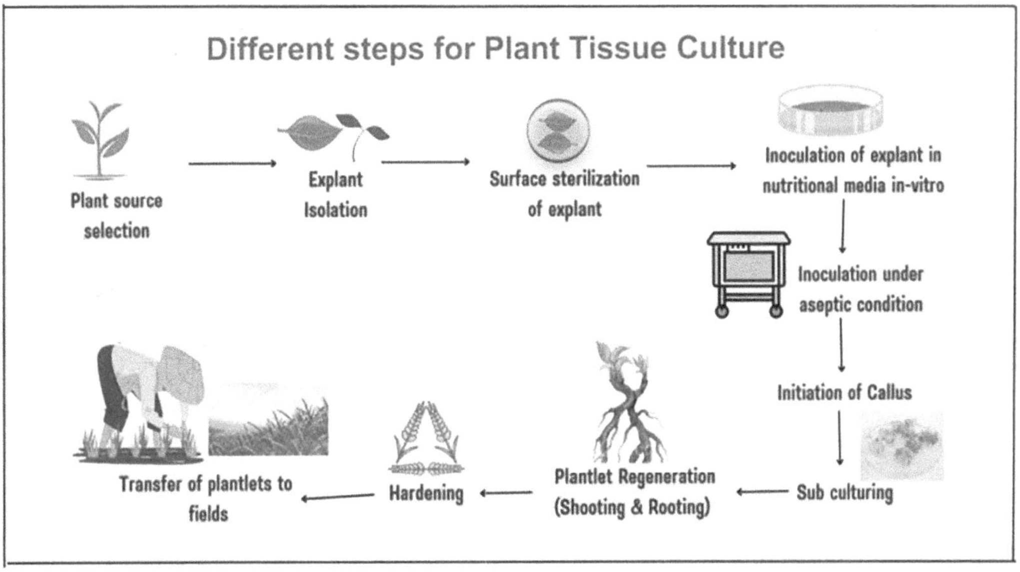

FIGURE 1.2 Schematic representation of the different steps involved in the plant tissue culture process.

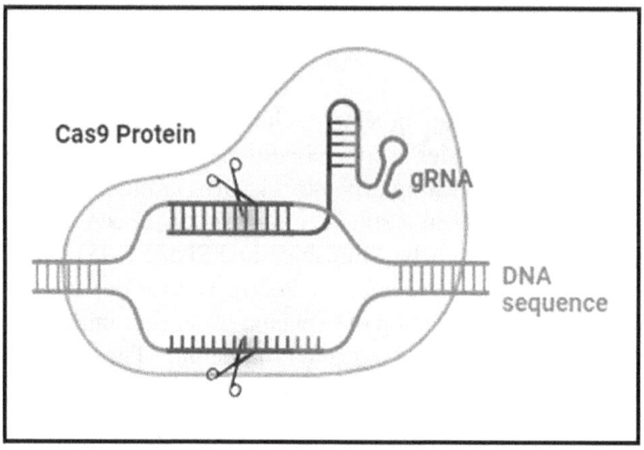

FIGURE 1.3 Schematic representation of CRISPR-Cas9 genome editing.

The technique utilizes the Cas9 enzyme as a molecular scissor to make cuts at specific DNA sequences, while the CRISPR sequence serves as a guide that directs Cas9 to the exact location in the genome that needs editing (Figure 1.3). On the basis of function and structure of Cas protein, the CRISPR/Cas system is further classified into the following categories of classes: class I (types I, III, and IV) and class II (types II, V, and VI). The first class of systems is of multi-subunit Cas proteins, and the second system is to utilize single Cas proteins. This process begins with the formation of a guide RNA (gRNA) designed to match a specific DNA sequence in the genome that is targeted for editing (Zahmanova et al. 2023). This gRNA is complexed with the Cas9 enzyme (molecular scissors). The gRNA guides Cas9 to the exact location of the target DNA sequence, ensuring high specificity. Once the Cas9–gRNA complex finds its target sequence, Cas9 cuts the DNA at the precise location, creating a double-strand break. The cell then attempts to repair this break using its native DNA repair mechanisms, namely, non-homologous end joining (NHEJ) or homology-directed repair (HDR). NHEJ can lead to the insertion or deletion of small DNA sequences at the repair site, potentially disrupting the gene's function. Alternatively, HDR can be harnessed to introduce specific changes or insert new genetic material by providing a repair template with the desired DNA sequence. This template is integrated into the genome at the cut site during the repair process, allowing for the precise editing of the genetic code (Kordi et al. 2022). The CRISPR/Cas-9 genome editing tool is used in the areas of medicine, agriculture, and biotechnology.

1.3.3 RNA Interference (RNAi)

RNA interference (RNAi) is a key method of gene regulation and in mounting defense against viral infection in plants. Either the siRNA or miRNA is involved in controlling the activity of specific genes. RNAi has thus become an invaluable tool in modifying plant traits by either suppressing undesirable gene functions or enhancing the expression of beneficial genes (Figure 1.4) (Kushanov et al. 2021). The mechanism involved in the RNAi process begins with the introduction of dsRNA into the cell. This dsRNA is recognized and processed by an enzyme called Dicer, which cleaves it into small fragments, typically 20–25 base pairs in length. These fragments are known as small interfering RNAs (siRNAs). Small interfering RNAs are loaded into a multifaceted complex known as RISC (RNA-induced silencing complex) (Kowalczyk 2022). Within RISC, one of the siRNA strands is selected as the guide strand, while the other is discarded. The guide strand within RISC binds to messenger RNA (mRNA) molecules that have a complementary sequence. Once bound, RISC cleaves the target mRNA, preventing it from being translated into a functional protein.

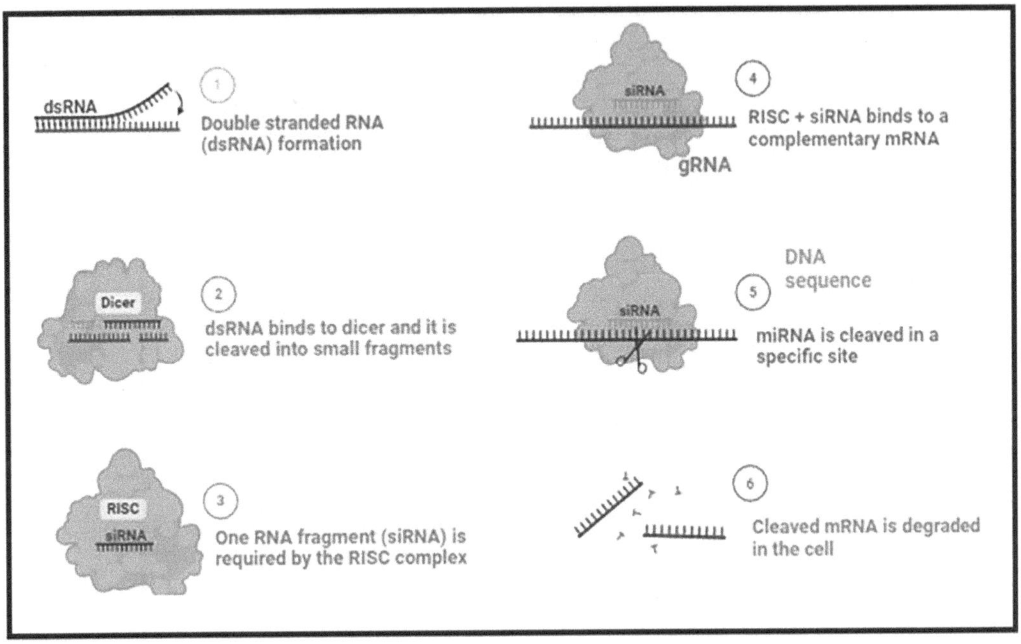

FIGURE 1.4 Schematic representation of RNA interference (RNAi).

This process effectively reduces the expression of the targeted gene. Either the cleaved mRNA is degraded or its translation is inhibited, leading to post-transcriptional gene silencing (Najmi et al. 2022). This process is helpful in plant pest resistance, in disease resistance, in improving agronomic traits, and in functional genomics in plant biotechnology.

1.3.4 Gene Synthesis

Gene synthesis is a powerful molecular biology technique that involves the artificial assembly of DNA sequences to create custom-designed genes (Figure 1.5) (Tyczewska, Twardowski, and Woźniak-Gientka 2023). In the context of plant biotechnology, gene synthesis plays a crucial role in engineering plants with desired traits, as it allows researchers to design and construct synthetic genes tailored to specific purposes, such as improving crop yield, resistance to pests, or tolerance to environmental stress (Hughes and Ellington 2017).

1.3.5 Molecular Cloning

Molecular cloning is an innovator biotechnology technique used to create identical copies of a specific DNA sequence. The process begins with the extraction of DNA from an organism. The DNA segment of interest is then identified and excised using restriction enzymes, which act like molecular scissors to cut DNA at specific sequences. This isolated DNA fragment is then ligated into a plasmid vector—a small circular DNA molecule that can replicate independently within a cell. Once the DNA fragment is inserted into the plasmid, the recombinant DNA molecule is introduced into a host cell, typically a bacterium, like *Escherichia coli*, through a process called transformation. Inside the host cell, the plasmid is replicated whenever the cell divides, creating numerous copies of the DNA fragment. The host cells grow on a selective medium that only allows the cells containing the plasmid to survive and proliferate, enabling the isolation of the cloned DNA (Figure. 1.6). Molecular cloning has vast applications in genetics, medicine, and agriculture (Haque et al. 2023).

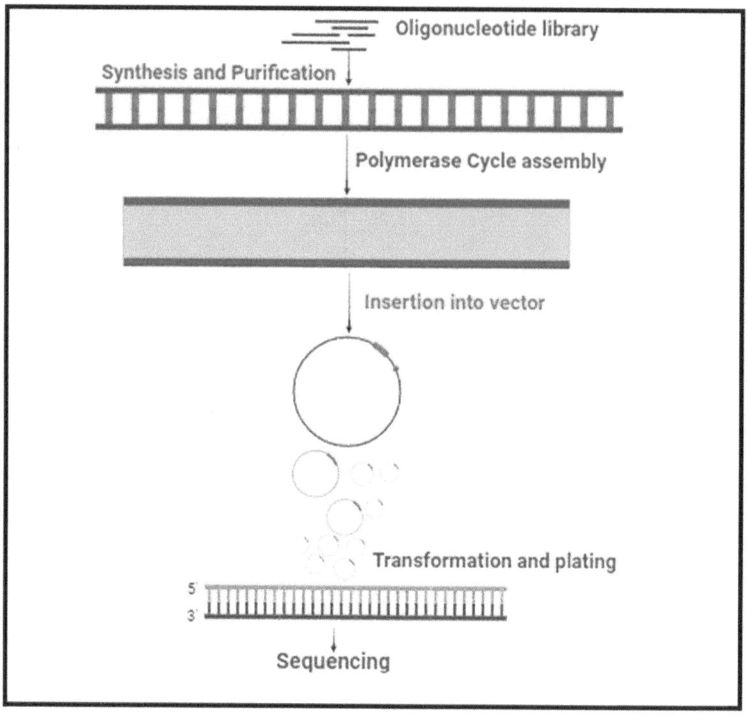

FIGURE 1.5 Schematic representation of gene synthesis.

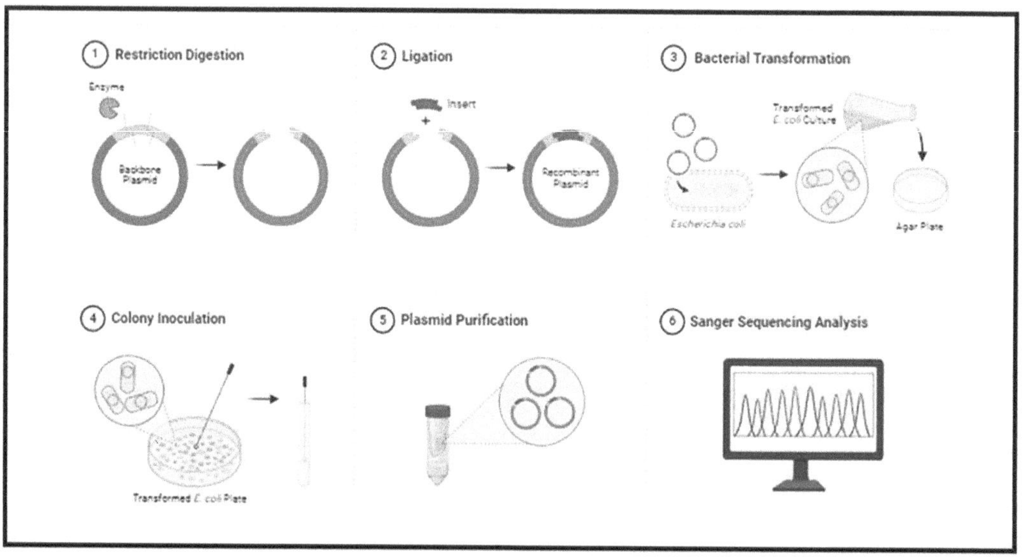

FIGURE 1.6 Schematic representation of molecular cloning.

1.4 APPLICATIONS IN PHARMACY AND ALLIED FIELDS

1.4.1 Drug Discovery and Development

Plants remain a significant source for the discovery of new drugs. Biotechnological tools have accelerated the process of identifying, extracting, and synthesizing plant compounds (Rotter et al. 2021).

Example: artemisinin, a therapeutic compound derived from the plant *Artemisia annua*. This process began with the selection of *Artemisia annua* based on its historical use in traditional Chinese medicine in treating fevers indicative of malaria. The active component, artemisinin, was isolated through solvent extraction and identified as a sesquiterpene lactone with a unique peroxide bridge. Its efficacy against malaria was confirmed through comprehensive preclinical and clinical testing, leading to the development of more soluble derivatives, like artesunate and artemether (Trendafilova et al. 2021).

1.4.2 Nutraceuticals

The use of plant-derived compounds in dietary supplements and functional foods, known as nutraceuticals, is a growing area. Biotechnology plays a key role in enhancing the health benefits of these products (Nasim, Sandeep, and Mohanty 2022). Example: Omega-3 fatty acid is well-known for cardiovascular health. It reduces triglyceride levels, which can decrease the risk of heart disease. A regular intake of omega-3 supplements is recommended for patients with a high cardiovascular risk, and these are often used in conjunction with other lipid-lowering medications to enhance cardiovascular outcome (Tadic et al. 2021).

1.4.3 Cosmeceuticals

This involves the use of bioactive plant compounds in cosmetic products. Biotechnology helps in the efficient extraction and stabilization of these compounds for use in skin care and other cosmetic products (Halder et al. 2022). Example: Topical retinoids, such as tretinoin, adapalene, and tazarotene, are derivatives of vitamin A and are first-line treatments for acne. They work by promoting cell turnover and preventing the clogging of hair follicles, which reduces the formation of acne lesions. These retinoids are also the first choice in skin aging, pigmentation disorder, psoriasis, photoaging protection, and skin barrier restoration (Latter et al. 2019).

1.4.4 Sustainable Production

Plant biotechnology also contributes to sustainable practices in pharmaceutical production. Cultivating medicinal plants using biotech methods can reduce environmental impact and ensure a steady supply of medicinal compounds (Niewiadomski and Stachowiak 2024). Example: The production of atorvastatin, commonly known as Lipitor, traditionally involves multiple steps that can be inefficient and can generate significant chemical waste. Employing green chemistry principles, such as optimizing chemical reactions to increase yields and reduce waste products, can significantly impact sustainability (Zarganes-Tzitzikas and Neochoritis 2019).

1.4.5 Metal Nanoparticle Preparation

Metal nanoparticles like silver, zinc, gold, copper, and other nanoparticles are prepared from natural plants and plant extracts. They have their own antimicrobial and antioxidant properties. Metal nanoparticles are useful in widespread diseases, such as wounds, cancer, and rheumatoid arthritis (Burlec et al. 2023). Green tea is rich in polyphenols (particularly epigallocatechin gallate, or EGCG), which serve as both reducing and stabilizing agents in the synthesis of gold nanoparticles. Metal nanoparticles are natural and biodegradable materials with major biological effects, such as antioxidant and antibacterial properties (Singh et al. 2021).

1.5 RESULTS AND DISCUSSION

Recent advancements illustrate a significant progress in the application of plant biotechnology in medicine. Research has shown that genetically modified plants can produce phytochemicals at higher rates than do their natural counterparts, with some studies demonstrating a doubling in yield. Furthermore, the ability to grow medicinal plants in controlled environments has led to a more sustainable and environmentally friendly production process. Critically, the integration of plant biotechnology into pharmaceutical production poses ethical and regulatory challenges. The genetic modification of plants raises concerns about biosafety, ecological impacts, and the potential for genetic cross-contamination. Additionally, there is a need for stringent regulatory frameworks to ensure that these innovations benefit public health without compromising safety.

1.6 CONCLUSION

The use of plants as sources of medicinal compounds dates back thousands of years, with ancient civilizations relying on plant-based remedies for health care. The scientific study of these plants began in earnest in the 19th and early 20th centuries, with the isolation of active ingredients such as morphine, quinine, and aspirin. However, the development of plant biotechnology as a field began in the late 20th century, with advances in genetic engineering, tissue culture, and molecular biology techniques. These advances have made it possible to enhance the production of desired compounds, understand the biosynthetic pathways, and even engineer plants to produce novel compounds or those typically produced by other organisms.

REFERENCES

Alamgir, A. N. M. 2018. *Biotechnology, in Vitro Production of Natural Bioactive Compounds, Herbal Preparation, and Disease Management (Treatment and Prevention). Progress in Drug Research.* Vol. 74. Springer International Publishing. https://doi.org/10.1007/978-3-319-92387-1_7.

An, Xueli, Shaowei Zhang, Yilin Jiang, Xinze Liu, Chaowei Fang, Jing Wang, Lina Zhao, Quancan Hou, Juan Zhang, and Xiangyuan Wan. 2024. "CRISPR/Cas9-Based Genome Editing of 14 Lipid Metabolic Genes Reveals a Sporopollenin Metabolon ZmPKSB-ZmTKPR1-1/-2 Required for Pollen Exine Formation in Maize." *Plant Biotechnology Journal* 22 (1): 216–232. https://doi.org/10.1111/pbi.14181.

Babu, B. Kishore. 2024. "Personalized Medicine and Advancements in Pharmacology: Shaping the Future of Healthcare." *International Journal of Pharmaceutical Investigation* 14 (2): 338–342. https://doi.org/10.5530/ijpi.14.2.41.

Barkha, S., and S. Bharti. 2016. "Green Biotechnology and Scope of Genetically Modified Crops: Facts and Prejudices." *Indian Journal of Agriculture Business* 2 (March): 1–11.

Bayat, Ardeshir. 2002. "Clinical Review Science, Medicine, and the Future Bioinformatics." *BMJ* 324 (April): 1018–1022. www.ebi.ac.uk.

Bhatia, Saurabh. 2018. *History, Scope and Development of Biotechnology. Introduction to Pharmaceutical Biotechnology.* Vol. 1. https://doi.org/10.1088/978-0-7503-1299-8ch1.

Burlec, Ana Flavia, Andreia Corciova, Monica Boev, Denisa Batir-Marin, Cornelia Mircea, Oana Cioanca, Gabriela Danila, Marius Danila, Anca Florentina Bucur, and Monica Hancianu. 2023. "Current Overview of Metal Nanoparticles' Synthesis, Characterization, and Biomedical Applications, with a Focus on Silver and Gold Nanoparticles." *Pharmaceuticals* 16 (10). https://doi.org/10.3390/ph16101410.

Cecchin, Erika, and Gabriele Stocco. 2020. "Pharmacogenomics and Personalized Medicine." *Genes* 11 (6): 1–5. https://doi.org/10.3390/genes11060679.

Chung, Young Hun, Derek Church, Edward C. Koellhoffer, Elizabeth Osota, Sourabh Shukla, Edward P. Rybicki, Jonathan K. Pokorski, and Nicole F. Steinmetz. 2022. "Integrating Plant Molecular Farming and Materials Research for Next-Generation Vaccines." *Nature Reviews Materials* 7 (5): 372–388. https://doi.org/10.1038/s41578-021-00399-5.

Collard, Bertrand C. Y., and David J. Mackill. 2008. "Marker-Assisted Selection: An Approach for Precision Plant Breeding in the Twenty-First Century." *Philosophical Transactions of the Royal Society B: Biological Sciences* 363 (1491): 557–572. https://doi.org/10.1098/rstb.2007.2170.

Daniotti, Sara, and Ilaria Re. 2021. "Marine Biotechnology: Challenges and Development Market Trends for the Enhancement of Biotic Resources in Industrial Pharmaceutical and Food Applications: A Statistical Analysis of Scientific Literature and Business Models." *Marine Drugs* 19 (2). https://doi.org/10.3390/md19020061.

Deckers, Marie, Dieter Deforce, Marie Alice Fraiture, and Nancy H. C. Roosens. 2020. “Genetically Modified Micro-Organisms for Industrial Food Enzyme Production: An Overview.” *Foods* 9 (3). https://doi.org/10.3390/foods9030326.

Dong, Oliver Xiaoou, and Pamela C. Ronald. 2019. “Genetic Engineering for Disease Resistance in Plants: Recent Progress and Future Perspectives.” *Plant Physiology* 180 (1): 26–38. https://doi.org/10.1104/pp.18.01224.

Efferth, Thomas. 2019. “Biotechnology Applications of Plant Callus Cultures.” *Engineering* 5 (1): 50–59. https://doi.org/10.1016/j.eng.2018.11.006.

Elsayed, Israa G., Osman N. Kanwugu, and Maria N. Ivantsova. 2019. “Red Biotechnology: A Healthy World.” *AIP Conference Proceedings* 2174 (December). https://doi.org/10.1063/1.5134362.

Ezeonu, Chukwuma S., Richard Tagbo, Ephraim N. Anike, Obinna A. Oje, and Ikechukwu N. E. Onwurah. 2012. “Biotechnological Tools for Environmental Sustainability: Prospects and Challenges for Environments in Nigeria—A Standard Review.” *Biotechnology Research International* 2012: 1–26. https://doi.org/10.1155/2012/450802.

Garner, Kathryn L. 2021. “Principles of Synthetic Biology.” *Essays in Biochemistry* 65 (5): 791–811. https://doi.org/10.1042/EBC20200059.

Gavrilescu, Maria. 2010. “Environmental Biotechnology: Achievements, Opportunities and Challenges.” *Dynamic Biochemistry, Process Biotechnology and Molecular Biology* 4 (May): 1–36.

Haas, Thomas, Manfred Kircher, Tim Köhler, Günter Wich, Ulrich Schörken, and Rainer Hagene. 2009. “White Biotechnology.” *RSC Green Chemistry*, 436–478. https://doi.org/10.1201/9781351167444-15.

Halder, Koushik, Abira Chaudhuri, Malik Z. Abdin, Manoj Majee, and Asis Datta. 2022. “RNA Interference for Improving Disease Resistance in Plants and Its Relevance in This Clustered Regularly Interspaced Short Palindromic Repeats-Dominated Era in Terms of DsRNA-Based Biopesticides.” *Frontiers in Plant Science* 13 (May). https://doi.org/10.3389/fpls.2022.885128.

Han, Fengxuan, Jiayuan Wang, Luguang Ding, Yuanbin Hu, Wenquan Li, Zhangqin Yuan, Qianping Guo, et al. 2020. “Tissue Engineering and Regenerative Medicine: Achievements, Future, and Sustainability in Asia.” *Frontiers in Bioengineering and Biotechnology* 8 (March): 1–35. https://doi.org/10.3389/fbioe.2020.00083.

Haque, Md Azizul, Dhirendra Nath Barman, Aminur Rahman, Md Shohorab Hossain, Sibdas Ghosh, Most Aynun Nahar, Mst Nur E. Nazmun Nahar, Joyanta K. Saha, Kye Man Cho, and Han Dae Yun. 2023. “Molecular Cloning, in Silico Analysis, and Characterization of a Novel Cellulose Microfibril Swelling Gene Isolated from Bacillus Sp. Strain AY8.” *Microorganisms* 11 (12): 1–20. https://doi.org/10.3390/microorganisms11122857.

Hughes, Randall A., and Andrew D. Ellington. 2017. “Synthetic DNA Synthesis and Assembly: Putting the Synthetic in Synthetic Biology.” *Cold Spring Harbor Perspectives in Biology* 9 (1). https://doi.org/10.1101/cshperspect.a023812.

Kordi, Masoumeh, Robab Salami, Parisa Bolouri, Nasser Delangiz, Behnam Asgari Lajayer, and Eric D. van Hullebusch. 2022. “White Biotechnology and the Production of Bio-Products.” *Systems Microbiology and Biomanufacturing* 2 (3): 413–429. https://doi.org/10.1007/s43393-022-00078-8.

Kowalczyk, Tomasz. 2022. “Genetic Manipulation and Bioreactor Culture.” *Molecules* 27: 795.

Kushanov, Fakhriddin N., Ozod S. Turaev, Dilrabo K. Ernazarova, Bunyod M. Gapparov, Barno B. Oripova, Mukhlisa K. Kudratova, Feruza U. Rafieva, et al. 2021. “Genetic Diversity, QTL Mapping, and Marker-Assisted Selection Technology in Cotton (Gossypium Spp.).” *Frontiers in Plant Science* 12 (December). https://doi.org/10.3389/fpls.2021.779386.

Latter, Gemma, E. Grice, Yousuf Mohammed, Michael S. Roberts, and Heather A. E. Benson. 2019. “Targeted Topical Delivery of Retinoids in the Management of Acne Vulgaris: Current Formulations and Novel Delivery Systems.” *Pharmaceutics* 11 (490): 1–26. https://doi.org/10.3390/pharmaceutics11100490.

Li, Tianxiang, Yanyan Yang, Hongzhao Qi, Weigang Cui, Lin Zhang, Xiuxiu Fu, Xiangqin He, Meixin Liu, Pei feng Li, and Tao Yu. 2023. “CRISPR/Cas9 Therapeutics: Progress and Prospects.” *Signal Transduction and Targeted Therapy* 8 (1). https://doi.org/10.1038/s41392-023-01309-7.

Malik, Mohammed, Afroz Archana, Pokala Praveen, Kumar Neela, and T. K. Shyamilee Vammi. 2021. “DNA Pro Fi Ling in Forensic Science: A Review.”

Mustafa, Mujahed, Ahmed Mohammed, Awad Alzebair, Abdelrafie Makhawi, and Google Scholar. 2023. “Integrating Bioinformatics in Drug Discovery.” https://doi.org/10.20944/preprints202310.1465.v1.

Najmi, Asim, Sadique A. Javed, Mohammed Al Bratty, and Hassan A. Alhazmi. 2022. “Modern Approaches in the Discovery and Development of Plant-Based Natural Products and Their Analogues as Potential Therapeutic Agents.” *Molecules* 27 (2). https://doi.org/10.3390/molecules27020349.

Nasim, Noohi, Inavolu Sriram Sandeep, and Sujata Mohanty. 2022. "Plant-Derived Natural Products for Drug Discovery: Current Approaches and Prospects." *Nucleus (India)* 65 (3): 399–411. https://doi.org/10.1007/s13237-022-00405-3.

Niewiadomski, Przemysław, and Agnieszka Stachowiak. 2024. "The Sustainable Production and Well-Being of Employees as a Derivative of the Concept of Sustainable Production." *Sustainability (Switzerland)* 16 (1): 1–16. https://doi.org/10.3390/su16010465.

Ormandy, Elisabeth H., Julie Dale, and Gilly Griffin. 2011. "Animal Welfare Bien-Être Des Animaux Genetic Engineering of Animals: Ethical Issues, Including Welfare Concerns." *Canadian Veterinary Journal* 52 (5): 544–550.

Althobaiti, Fayez, Abdulaziz Lafi T. Alanazi, Ala'a Ali Abuharbah, Lojain Abdullah A. Alharbi, Yara Ali S. Alhussaini, and Ghadi Abdulaziz Alhazmi. n.d. "Biotechnological Applications in Forensics: Questioning the Effect of Sun Light Conditions on DNA Integrity Extracted from Blood and Hair Samples." *Journal of Population Therapeutics & Clinical Pharmacology* 31 (2): 2000–2007. https://doi.org/10.53555/jptcp.v31i2.4569.

Rotter, Ana, Michéle Barbier, Francesco Bertoni, Atle M. Bones, M. Leonor Cancela, Jens Carlsson, Maria F. Carvalho, et al. 2021. "The Essentials of Marine Biotechnology." *Frontiers in Marine Science* 8 (March): 1–53. https://doi.org/10.3389/fmars.2021.629629.

Singh, Tejasvi, Athira Jayaprakash, Mohammad Alsuwaidi, and Asha Anish. 2021. "Materials Today: Proceedings Green Synthesized Gold Nanoparticles with Enhanced Photocatalytic Activity." *Materials Today: Proceedings* 42: 1166–1169. https://doi.org/10.1016/j.matpr.2020.12.531.

Tadic, Marijana, Carla Sala, Guido Grassi, Giuseppe Mancia, Stefano Taddei, Wolfgang Rottbauer, and Cesare Cuspidi. 2021. "Omega-3 Fatty Acids and Coronary Artery Disease: More Questions Than Answers." *Journal of Clinical Medicine* 10 (Ldl): 1–11.

Trendafilova, Antoaneta, Moujir M. Laila, Sousa M. C. Pedro, and Seca M. L. Ana. 2021. "Research Advances on Health Effects of Edible Artemisia Species and Some Sesquiterpene Lactones Constituents." *Foods* 10 (65): 1–38.

Tyczewska, Agata, Tomasz Twardowski, and Ewa Woźniak-Gientka. 2023. "Agricultural Biotechnology for Sustainable Food Security." *Trends in Biotechnology* 41 (3): 331–341. https://doi.org/10.1016/j.tibtech.2022.12.013.

Wijerathna-Yapa, Akila, and Jayeni Hiti-Bandaralage. 2023. "Tissue Culture—A Sustainable Approach to Explore Plant Stresses." *Life* 13 (3): 1–16. https://doi.org/10.3390/life13030780.

Zahmanova, Gergana, Alaa A. Aljabali, Katerina Takova, Valentina Toneva, Murtaza M. Tambuwala, Anton P. Andonov, Georgi L. Lukov, and Ivan Minkov. 2023. "The Plant Viruses and Molecular Farming: How Beneficial They Might Be for Human and Animal Health?" *International Journal of Molecular Sciences* 24 (2). https://doi.org/10.3390/ijms24021533.

Zarganes-Tzitzikas, Tryfon, and Constantinos G. Neochoritis. 2019. "Atorvastatin (Lipitor) by MCR." *ACS Medicinal Chemistry Letters* 10: 389–392. https://doi.org/10.1021/acsmedchemlett.8b00579.

2 Requirements in Setting Up a Plant Biotechnology Laboratory and a General Laboratory

Shilpa Malakar, Priya Sutaoney,*
Durgesh Nandini Chauhan, Kamal Shah,
and Nagendra Singh Chauhan
* Corresponding author: priyasut24@gmail.com

2.1 INTRODUCTION

The development of plant biotechnology has had a significant impact on the growth of the human civilization. Domestication of plants has enabled the production of more food, which has been used to feed large populations as well as provide the majority of calories in human diets. Additionally, plants have been used as a source of therapeutic drugs as well as industrial feedstocks. More recently, they have been employed in the production of pharmaceutical proteins, biofuels, and other products (Larson et al. 2014). However, there are still many areas of plant improvement that can be achieved through genetic manipulation, and there are compelling reasons this should be done. Conventional plant breeding can be employed to introduce resistance to insect and microbial attack; however, the identification of loci that result in resistance is difficult and time-consuming, thus necessitating the use of transgenic technology. This is especially the case for insect predation resistance; however, field studies have revealed that insect pests have evolved resistance to Cry1, Cry3 proteins, necessitating the development of improved insecticide proteins to address this threat (Smith 2021). A few transgenic traits that result in microbial disease resistance have already been commercially available, and numerous others have been demonstrated to be successful in model crops. An example of this is the Rainbow papaya, which is a transgenic crop grown in Hawaii and is resistant to the disease papaya ringspot, a disease for which there is no natural resistance. In order to reduce yield losses caused by microbial pathogens, it is necessary for many crops to incorporate genetic variations that result in disease resistance. *Bioprospecting* is the process of identifying and commercializing new pharmaceuticals. Genetic engineering can be employed to augment the presence of these compounds in plants, while synthetic biology techniques can be employed to replicate these pathways in other organisms (Salmeron and Vernooij 1998). Additionally, other protein-based pharmaceuticals are highly effective in plant production and can significantly influence the development of diseases in both humans and animals.

Over the past 50 years, traditional plant breeding has been a huge help in increasing food production and keeping countries ahead of the curve. But now, it has become more important than ever to keep up with population growth and climate change. Crop yields have actually decreased in most of our major crops over the past 20 years. It is difficult to produce crops that meet the demands of modern agriculture and gardening while providing sustainable food, healthy nutrition, eco-friendliness, and resistance to climate change (Steinwand and Ronald 2020). Plant biotechnology involves the manipulation of cells and tissues, the use of markers for crop selection, the use of genomics, and

DOI: 10.1201/9781003326939-2

the use of molecular farming to create novel varieties and traits. Various cell and tissue culture techniques can be applied in the creation, preservation, and utilization of genetic variability in crops. These techniques include *in vitro* pollination, embryo rescue for remote hybridization, haploid and double haploid production, mutagenesis, somatic variations, *in vitro* selection, germplasm conservation (medium- and long-term), exchange, protoplasm fusion, somatic hybridization, gene manipulation, and transgenic or genetically modified plants (Honegger 2001). When performing tissue culture, it is important to ensure that the environment is aseptic and that the tissues, equipment, culture media, and room are free of any microorganisms. Various methods of sterilization are available, such as dry heat, wet heat, ultrasonic, chemical, and autoclave sterilization. These methods are used to sterilize plant materials, glass equipment, other instruments, and culture media at high temperatures using chemicals such as alcohol. It should be noted that culture media can contain microorganisms right from the start. Sugar-based nutrient media can stimulate the growth of rapidly growing microorganisms, which can cause damage to the tissue. To ensure the correct growth of slower-growing microorganisms, the tissue should be aerated within the culture media. The growth of cells in semi-liquid media does not require any specialized aeration techniques; however, cells cultivated in liquid media must remain submerged and require the use of an automated shaker to ensure the appropriate oxygen levels are maintained. Cotton-coated cheesecloth can also be used to enclose culture vials and facilitate aeration, as well as the division of the tissue into distinct cells (Anjanappa and Gruissem 2021).

In recent years, advances in biotechnology have enabled the physical mapping and cloning of traits of interest through high-resolution genetic analysis. Molecular markers are also being used to characterize germplasm and identify duplicates in collections. These advances are becoming increasingly important in breeding programs, as they allow for the early removal of unwanted material from perennial crops, the selection of marker-assisted selection, the creation of saturated linkage maps, and the pyramiding of genes in intragenerative breeding (Jaganathan et al. 2020). Despite stringent laws, genetically modified crops, such as maize, soybeans, rapeseeds, cotton, tomatoes, potatoes, papayas, etc., occupy more than 190 million hectares in 26 countries, which are cultivated by 17 million farmers. These crops bring economic and environmental benefits but can also be associated with social and environmental issues. This special issue will explore all these biotechnology advances in crop improvement, characterization, and conservation, as well as the implications of plant biotechnology for the social, environmental, and global aspects of commercialization (Raman 2017). Ultimately, it is hoped that a balance will be achieved between highly domesticated old crop species and new, underutilized crops, both of which have the potential to meet the future challenges of agricultural production in a fluctuating climate and a shrinking area for cultivation.

2.2 GENETIC ASSAYS OF TISSUE CULTURE–RAISED PLANTS

Tissue culture is a type of aseptic culture in vitro which is used to produce clones of plants that are true to type of the chosen genotype. This culture is maintained in a controlled environment, such as a nutrient supply, pH medium, temperature, and gas and liquid environment, and is increasingly used in large-scale plant multiplication. In addition to their research application, tissue culture techniques have become increasingly important for plant propagation, disease control, plant improvement, and secondary metabolite production (Cardoza 2008). Small pieces of tissue known as explants can be used to create hundreds or thousands of plants over a period of time and space in controlled conditions, regardless of the season or weather, all year round. Tissue culture is a great way to get plants that do not grow naturally. It is a quick and easy process, and you can get thousands of seeds in a short amount of time. Plus, it can help you produce new crops with better traits, faster, since you do not need as much space or time to grow them. You can also use it to grow perennial plants, no matter what the weather or season is like. It can also help you save plants from diseases and germinals, and the plants you get from it are sterile (Vargas et al. 2018).

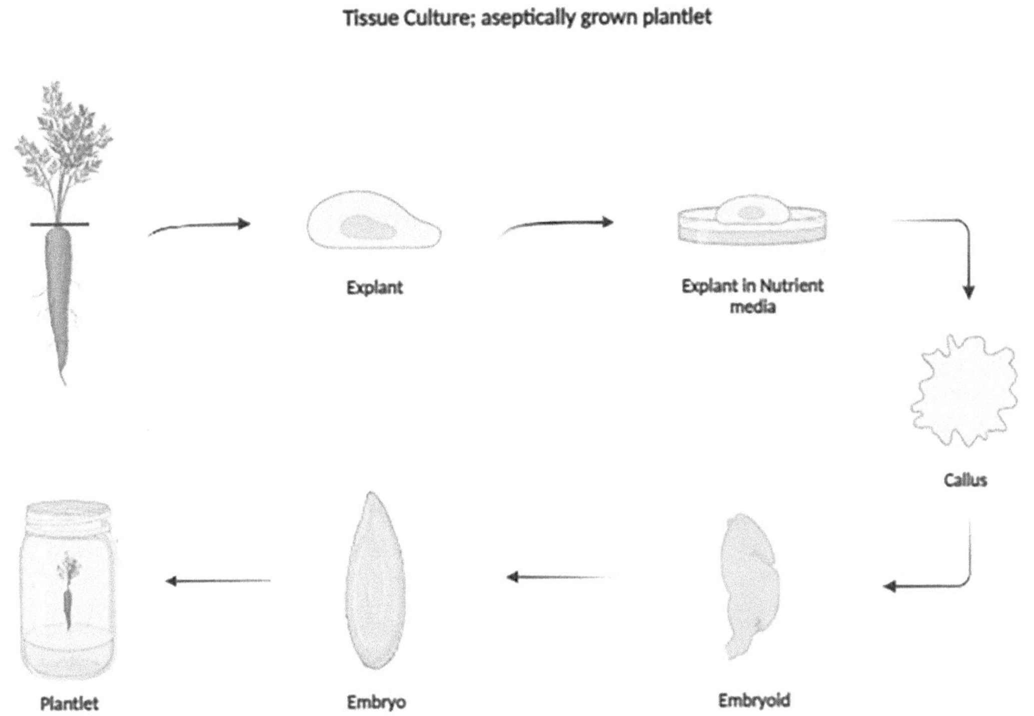

FIGURE 2.1 The process of artificially propagating plants through the use of tissue culture and cell culture techniques.

Source: Created by Biorenderapk.

Some plants have resistant seeds that cannot be kept for long, so you can use tissue culture to preserve them in a vegetative state or to cryopreserve them. And in some cases, you can even get hybrids of different species using embryo rescue, which is not possible with conventional methods. For more than half a century, tissue culture has been a widely used reproductive technique in breeding programs. The resulting hybrids are often sterile through embryo abortion but can be "rescued" through culturation or transplantation of the embryos (De Vitis et al. 2020). It is possible to alter the media conditions of the cell culture to achieve more suitable outcomes for the experiments conducted.

2.3 TECHNIQUES FOR *IN VITRO* GENETIC TRANSFORMATION IN PLANT TISSUE CULTURE

The PTC and GE research facility is a high-throughput plant tissue culture and genetic engineering facility located in a 7,390 ft^2 area. Due to its high multiplication rate and low demands on initial plant numbers and space, it is dedicated to the cultivation and conservation of endangered, threatened, and rare species through micropropagation (Atkinson and Wells 2017). The facility is equipped with Class 10,000 clean air areas, which are equipped with HEPA (high-efficiency particulate air) filters and aseptic epoxy flooring. An air shower has been installed to help remove dirt from the outside environment and maximize cleanliness. All rooms in the facility are well-connected, with pass boxes that minimize extraneous contamination. Outside the clean air area is a well-planned area for glassware washing and autoclaving, as well as areas for general storage and growth chambers. During power outages, the facility is equipped with a UPS and an emergency

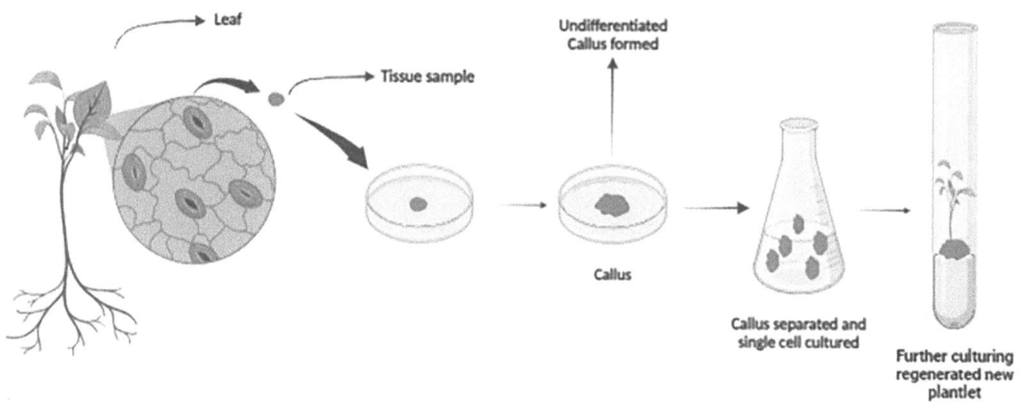

FIGURE 2.2 Basic steps in plant tissue culture to preserve them in a vegetative state using media from time to time.

Source: Created by Biorenderapk.

generator, and fire safety provision has been taken into account. This facility can be used to create technology platforms to provide top-notch services across the country. It can be used to work with industries/research organizations in assessing GMO biosafety risks, validating genomic resources, and creating transgenic crops to improve quality (Wikandari et al. 2021). They have already got the general and specialized lab equipment to do top-notch research into plant tissue culture and the genetic engineering side of agrobiotechnology.

In order to conduct genomic research, it is necessary to have access to a range of tools and equipment. These include a bioinformatics abbreviated identifier (BIO) (3730xl), a DNA analyzer (LICOR), a gel-based sequencing and genotyping system (RT-PCR), a microarray complete fluid station (Affymetrix), a mass array, a sequenom for SNP genotyping and methylation study, and a tissue embedding system and microtome and cryo-microtome for tissue dissection (Mittler and Blumwald 2010). Additionally, there are microscopy facilities, such as stereomicroscopes, fluorescence microscopes, and laser capture microscopes, which can be used to extract DNA, RNA, and protein from tissues. A speed vac system is also available for concentrating fluid, as well as a del doc system, a deep freezer (−80°C), a bioanalyzer, a liquid chromatography system, an HPLC analytical and preparedness system, and a QTRAP, which is high-performance thin layer chromatography (Galluzzi et al. 2021).

2.4 LABORATORY SETUP FOR PLANT BIOTECHNOLOGY

Biotechnology research is essential in the development of agriculture as it provides valuable information on the genetic and physiological properties of crops (Kumar and Loh 2012). As a result of this technology, scientists have realized the importance of transgenes, not only for stress-tolerant crops, but also in the development of new types of medicinal products through molecular agriculture. In the field of biotechnology and crop improvement, tissue culture, and transgenic approaches, important field crops are examined to demonstrate the use of germplasm in the development of resistance to novel diseases and pests, nutrient and water utilization, root traits, and temperature tolerance.

By examining the genetic and biochemical makeup of crops, researchers are able to create novel varieties and bioinformation that can increase crop resilience to environmental conditions, increase yields, and improve nutritional content. Furthermore, biological research plays an important role in the development of sustainable agricultural practices by elucidating the complex interactions between organisms and their environment (Tabei and Muranaka 2020). A biotechnology laboratory requires a variety of components to ensure the successful functioning of the laboratory. These components include storage for chemicals, as well as glassware, cleaning and washing rooms for general cleaning, general laboratories for routine laboratory experimentation and dedicated laboratories for preparing and sterilizing, laminar flow rooms, sterile storage and cultures rooms, general instrumentation rooms for PCR equipment, gel documentation systems, electrophoresis units, centrifuges, pH meters, balances, laminar streams and freezers, and ice machines. Additionally, the laboratory must have a good computer system, with printers, autoclaves, sinks, deionizers, hot air ovens, radiation shields, incubators, dark rooms for photographic film production, transilluminators, water baths, refrigerated centrifuges, microcentrifuges, blotting apparatuses, hot plates, microwave ovens, and various other items. To ensure the successful operation of the laboratory, at least two large and one small room are necessary. Media and cultures must be properly disposed of. The bigger room is used for general lab stuff, like preparing media, autoclaving, and distilling water, while the smaller one is used for asepsis and storing autoclaved stuff (Espinosa et al. 2018).

The lab also has a washing area with a big sink, a bunch of brushes, and some detergent, plus a bucket with some distilled water to rinse off the glass stuff. Prior to commencing work, a laminar airflow is initiated for a period of 10 to 15 minutes. This airflow does not extinguish the flame of the spirit lamp, thus allowing for the use of the lamp while work is in progress. The culture room is equipped with double doors to ensure dust-free entry and a consistent temperature within the room. The temperature should remain 25 ± 2°C within the culture room, with air coolers and specially designed stands for the preservation of culture vessels. Glass or plywood shelves are available for the placement of flasks, bottles, and jars, as well as petri dishes. Each culture rack is equipped with fluorescent lamps, which are regularly photographed and controlled by an automated timer. To monitor the temperature and humidity, both thermometers and hygrometers are mounted on the wall in the safety corner. Humidity should remain above 50% (Ranjha et al. 2022). The global agribusiness biotechnology market is estimated to be worth USD100 billion by 2020 and is projected

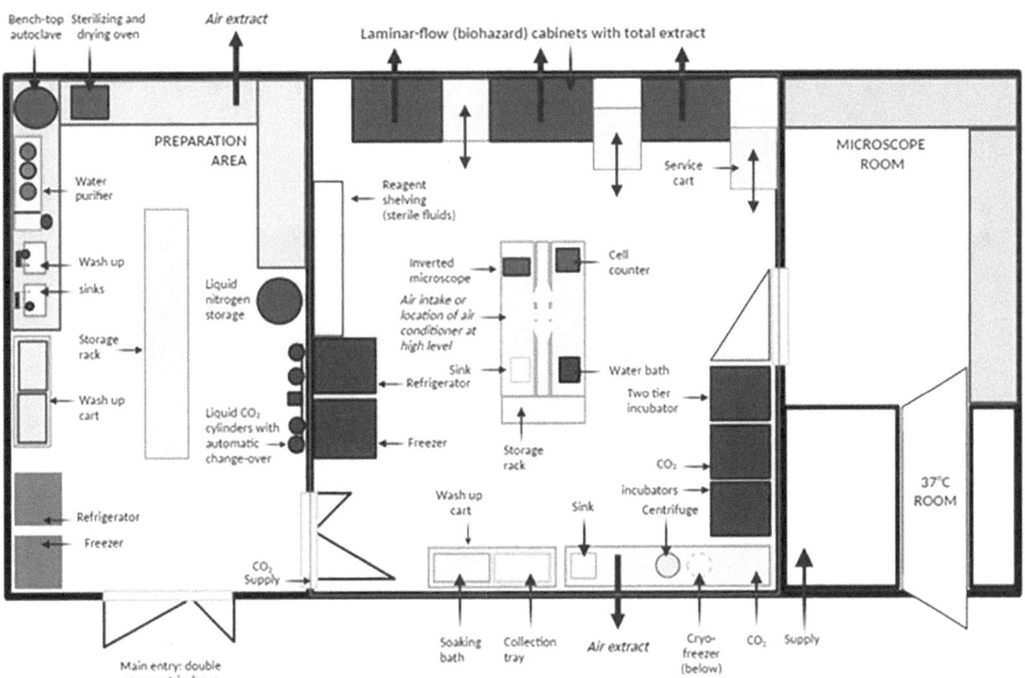

FIGURE 2.3 Cultivation and development of plant tissue culture labs with biosafety cabinets for the proper development and propagation of plant materials.

Source: Created by Biorenderapk.

to grow at a CAGR of 8% in the following five years. The growing adoption and advancement of biotechnology tools will contribute to the growth of the market by allowing for the development of new breeding techniques.

Biotechnology tools that modify or alter the characteristics of organisms, such as plant, animal, or microbial traits, are increasingly becoming popular in agricultural applications. In 2019, biotechnology was cultivated on a large scale across 29 countries, resulting in a significant impact on food security, sustainable development, and climate mitigation. Additionally, biotechnology has had a positive impact on the lives of 17 million farmers and their family members globally (Kamthan et al. 2016). Developing countries such as India, Vietnam, and the Philippines experienced double-digit growth in their biotech crop areas. The United States, Canada, Brazil, and India had the highest adoption rates of major biotech crops in 2019, accounting for 26% of global biotechnology use. Biotech products are a major source of fiber, food, feed, and fuel and are expected to be a major driver of the global agricultural biotechnology industry in the decades to come. The use of genetically modified crops (GMOs) has enabled the advancement of GM technologies, allowing for the improvement of existing crop varieties, rather than the introduction of new ones (Gaj et al. 2016). Recent evidence suggests that farmers and consumers are increasingly using genome editing to improve local crops.

2.5 GENETICALLY INDUCED TISSUE CULTURE TERMINOLOGIES

Heterologous DNA has been introduced into plants through direct transformation methods and indirect methods which were first developed decades ago. Ten years ago, a revolutionary technique came along that allowed scientists to edit almost any gene in any cell type or organism. This

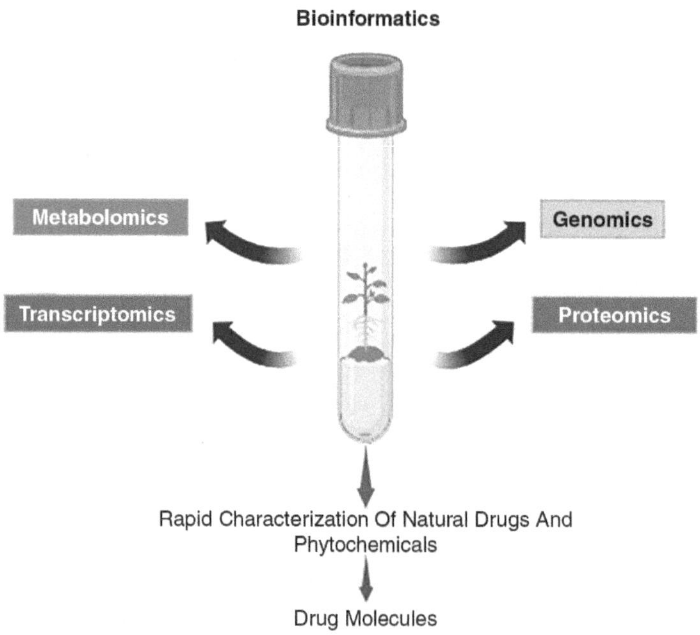

FIGURE 2.4 Bioinformatics, next-generation sequencing (NGS), and various omics technologies provide data integration to uncover genomic, metabolomic, proteomic, and transcriptomic information for better crop production.

Source: Created by Biorenderapk.

technique, known as "genome editing," involves inserting, deleting, or mutating genes of interest. The key to this is the use of engineered nucleases, which are made up of DNA-binding domains that attach to other DNA cleavage modules (Liu et al. 2024). This allows for the precise manipulation of genetic information and the development of better plants, as well as the ability to create new phenotypes and to study biological mechanisms and gene function. The ability to cut certain DNA sequences and to use different DNA repair mechanisms can be used to make a variety of changes to the genome, from single-nucleotide mutations to large-scale deletions, regroups, and integrations. This produces the desired specific alteration in the plant's genome, and the resulting oligonucleotides are then broken down in the cell. In theory, oligonucleotide-directed mutagenesis (ODM) is superior to conventional breeding and conventional mutagenesis because it works in a controlled and precise way, avoiding random mutations and avoiding the need for the introduction of recombinant DNA. The end product is, therefore, often similar to those of conventional breeding or conventional mutagenesis (Liao et al. 2023). The development of agricultural technology has enabled the emergence of new products and solutions to a more sustainable future; however, these developments have also brought with them new anxieties and questions to be addressed. Principles, processes, and policies for the assessment of biosecurity risks have been established for the purpose of protecting the environment and human health from the effects of GMOs. GM crops present a variety of biosafety issues, and those who breed them must demonstrate their products' safety before releasing them to the public for commercial use. Non-tradable biological techniques (NBTs) have been designed to facilitate a more accurate genetic alteration of plants than is conventional and to facilitate some of the earliest genetic modification techniques (Biscotti and Olmo 2015). It is yet to be determined whether the crops obtained through certain NBT techniques should be categorized as GMOs or treated in the same way as those obtained from traditional breeding and mutagenesis, particularly at the EU level, and the global scientific community has suggested that the assessment of plants obtained through NBTs must focus on the alterations to the plant and the finished products obtained. In order to evaluate the products derived from the utilization of these technologies, a simplified approach should be employed, primarily to determine if the induced genomic alterations are within the natural genetic variation of the organism (Turnbull et al. 2021).

When assessing the risk associated with genetically modified plants, one of the primary concerns is the potential for the generation of a novel protein within the modified plant. The cisgenic approach, on the other hand, is designed to avoid this concern, as all components are derived from the same species, or a sexually compatible species. This means that the recipient GM plant does not produce a new protein and thus does not yield any different result than traditional breeding. The full genome sequences available for many crops provide a great opportunity to identify useful genes or promoters from the same species that can be transferred to improve commercial cultivars (Bratlie et al. 2019). The transformation approach can be restricted by the availability of effective selectable markers that can be used to replace commonly used antibiotic and herbicide resistance markers. However, it is important to consider the impact on the RNAi gene target and the potential off-target effects of the GM plant. The potential for off-target mutations to occur in genetically engineered fruit trees is a concern for biosafety. Taking into account the various gene editing approaches, such as CRISPR, ZFN, and ODM, it has been shown that their application can lead to similar mutations to those found in nature or in traditional breeding techniques, but with a much more precise effect (Jones and Sparks 2009).

2.6 GENETIC MODELING USING GENOME ALTERATION AND DNA SEQUENCING FOR GENETICALLY MODIFIED ORGANISMS

A particular product of this novel approach can be classified as a point mutation due to the precision of the double-strand break, which is indistinguishable from natural mutation. Bioinformatics tools, used in the design of sgRNA creation, select the gene regions for off-target sites based on sequence characteristics, such as size (normally 21–23 bases) and nucleotide constitution, thereby reducing the risk of off-target mutation (Yau and Stewart 2013). The scientific community has generally endorsed the notion that genetically modified organisms (GMOs) should not include genetically modified plants that have been edited with the genome-altering gene (CRISPR) or the Cas9 genome-modifying system (CRISPR-Cas9). However, the most common method of introducing this system is through *Agrobacterium*-mediated transformation, in which the first product produced is a genetically modified product (Zaidi et al. 2020).

As a result, some authorities have proposed that the product be regulated as a GMO. Biotechnology laboratories are typically designed in a variety of environments; however, there are certain types of laboratory equipment that are commonly used in biotechnology laboratories regardless of the type of research being conducted. Depending on the purpose of the laboratory, the type of equipment used may vary (Gupta et al. 2016). For example, a laboratory used for vaccine production may require different equipment than a laboratory that is modifying crops to be more resistant to pests or droughts. Regardless of the purpose, all biotechnology laboratories will possess DNA sequencing equipment, standard laboratory equipment, and certain equipment that is essential for the successful functioning of biotechnology (Erickson et al. 2012). Genetic engineering and DNA sequencing are at the core of biotechnology.

The modern biotechnology industry was born in the late 1960s and exploded in the late 1970s and early 1980s with early experiments in cross-species gene inserting and genetic manipulation. This emphasis on DNA technology continues to be a major part of biotechnology today (Nambisan 2017). As a result, there is a lot of specialized equipment involved in dealing with DNA, and you will find DNA lab equipment in biotechnology laboratories across various industries.

Some examples of DNA lab equipment include, but are not limited to, the following: *DNA analyzer.* A DNA analyzer analyzes DNA samples as well as DNA generated by electrophoresis and

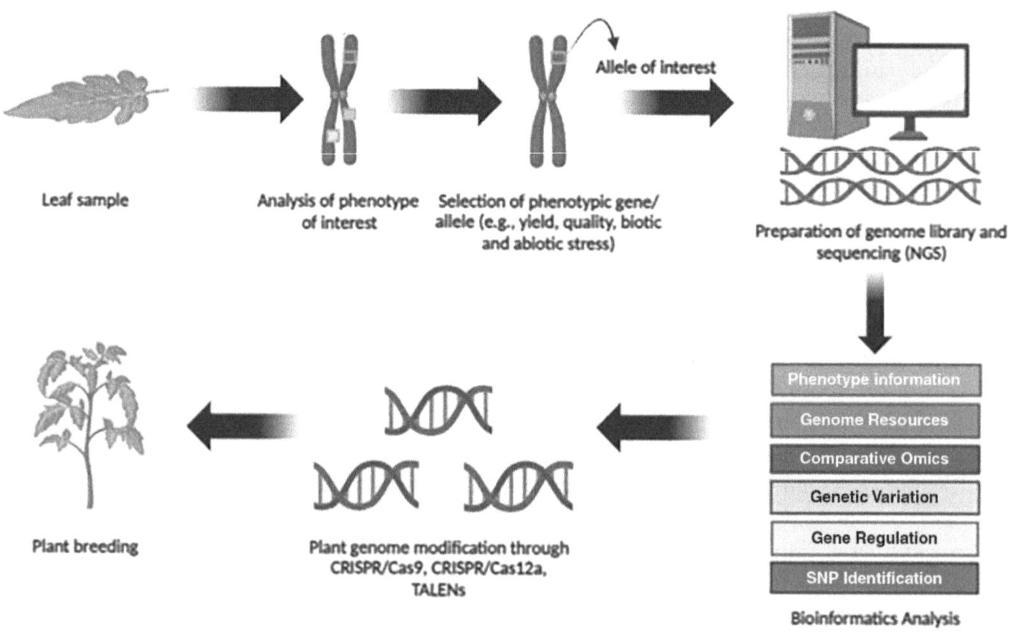

FIGURE 2.5 Bioinformatics analysis; by deciphering the information associated with the relevant genes, bioinformatics can assist in the development of plants with enhanced characteristics, such as disease tolerance, stress tolerance, and increased nutritional crop value.

Source: (Created by Biorenderapk.)

PCR and other techniques. Today's DNA analyzers are capable of high-throughput DNA fragment analysis, sometimes up to hundreds of samples per day. A modern DNA analyzer is also capable of fragment analysis, microsatellite analysis, SNAS analysis, mutation detection, and Sanger sequence material analysis (Wang et al. 2021). The automation of DNA sequencing by DNA sequencers enables scientists to create a comprehensive list of DNA base pairs, including the order in which they are sequenced. This is the initial step in the process of editing a genome and performing any form of genetic analysis and genetic engineering. *Fluorometer.* A fluorometer, on the other hand, is a device that is able to accurately measure the concentration of extremely low nucleic acids in a sample. This allows for the fluorization of the sample with an electromagnetic spectroscope, which allows for the identification of nucleic acid or other chemical groups with fluorescent chemicals. PCR machines are also used to amplify small segmented DNA or RNA molecules, allowing for the creation of multiple copies of the same segment, thus providing scientists with a large amount of DNA to study (Hixson and Ward 2022).

2.7 AN ASEPTIC ENVIRONMENTAL CONDITION FOR THE GROWTH OF GENETICALLY MODIFIED PLANTS

The quality of the laboratory environment is a key factor in the success of any plant tissue culture experiment; hence, the location of a plant tissue culture lab is of paramount importance. It should be situated in an area where there is minimal human activity, dust, or other contaminants (Sharma et al. 2022). Additionally, the laboratory should be away from any electromagnetic interference or vibrations that could impede the growth of the tissue cultures. The size and design of the laboratory should be proportionate to the amount of tissue cultures that will be produced (Kamle et al. 2017).

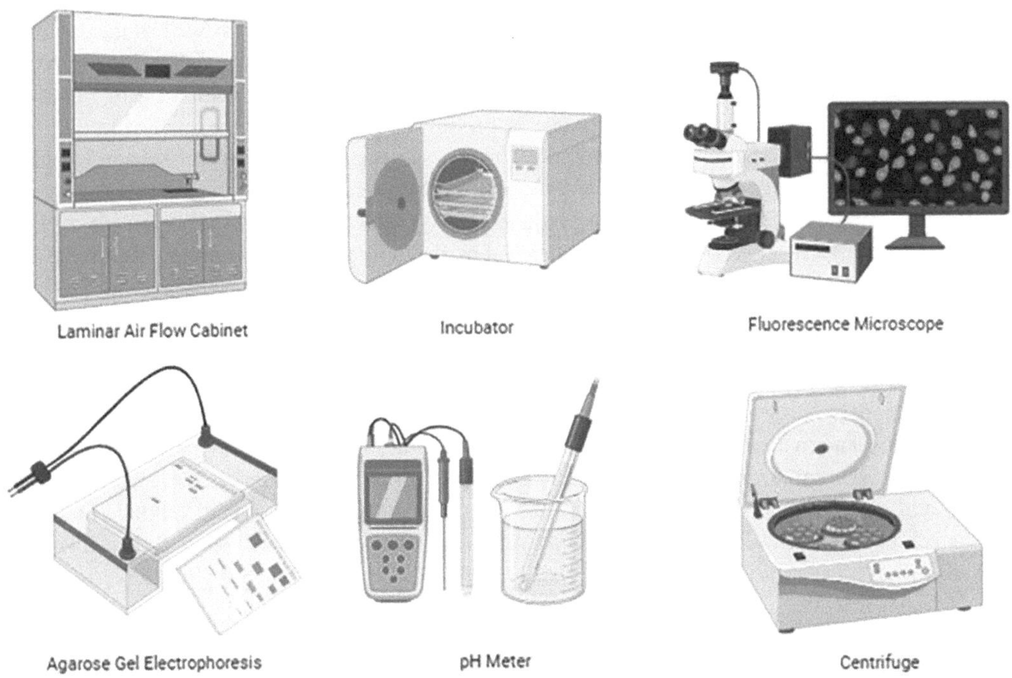

FIGURE 2.6 Common types of equipment used in modern plant biotechnology labs.

Source: Created by Biorenderapk.

This should ensure that the laboratory is designed to facilitate efficient workflow and minimize contamination. Separate rooms should be available for different activities, such as preparation and sterilization, as well as for culture growth (Anklam et al. 2002).

Proper ventilation and lighting should be present, as well as temperature control. Sterilization is essential for the success of the laboratory and should be included in the design of the laboratory. This includes the use of autoclaves for sterilizing the culture media, as well as the use of laminar hoods or biosafety cabinets for handling the cultures. Additionally, a wide selection of culture media should be available (Gupta and Singh 2017). In order to be successful in the field of plant tissue culture, a laboratory must possess a suitable location, size, and design, as well as the necessary sterilization and culture media (Roberts 2019). Additionally, the laboratory must have the necessary equipment, such as a laminar flow hood, microscopes, autoclaves, incubators, water purification systems, plant growth chambers, pH meters, weighing scales, forceps, scalpels, vials, flasks, agarose gel electrophoresis system, spectrophotometers, centrifuges, freeze dryers, and safety equipment. By providing the necessary tools and expertise, a laboratory can play a vital role in the advancement of plant biology and the production of highly sustainable products (Anju et al. 2020).

In order to establish a tissue culture laboratory in a sterile environment, it is essential to carefully select the location. It is important to avoid close proximity to facilities that handle microorganisms, such as bacteria and fungi, as well as other laboratories that handle insects. Additionally, the laboratory should be located away from storage facilities that store seeds and other plant material (Charis and Vallier 2017). If these cannot be avoided, additional precautions must be taken to ensure the lab is free of contamination. To ensure optimal functioning and experimentation, it is necessary to install basic equipment in the following areas: quarantine area, washing area, media preparation area, antiseptic transfer area, culture/grow rooms, and data collection area (Drexler and Uphoff 2002).

This quarantine area is essential for the prevention of contamination from humans in the laboratory, either at the start or at the end of the laboratory. It is primarily used for the purpose of changing clothing and shoes, with separate lab coats for use in the laboratory (Pamies et al. 2022). Additionally, it is necessary for washing explants and other plant materials, as well as for the cleaning of all beakers, cylinders, and flasks which are used for the preparation of media or the measurement of ingredients (Schmidt 1972). The area is equipped with a large sink and draining boards, as well as distilled water for drying, to facilitate the preliminary cleaning of the surface to eliminate microorganisms. Additionally, a media preparation area is necessary for the preparation of culture media recipes, the storage of chemicals, and the use of benches for media preparation (Savvidis et al. 2022). Hot plates, pH meters, weighing balances, and water baths are also present. In order to ensure complete sterile conditions for the transfer of explants into media-containing culture vessels, it is necessary to have a dedicated media dispensing area. This area can be a separate room adjacent to the preparation area, or a separate one (Paterniani et al. 2000). The equipment required for media preparation should include a magnetic stirrer, analytical and high loading balances, a pH meter, a refrigerator with or without a freezer, a water purification system, and the space for storing purified water. Additionally, a sterilization system should be developed to ensure the proper preparation of media, glasses, and instruments. This is especially important for small laboratory setups, as well as for research and commercial tissue culture settings. In order to ensure that bacteria-free, filtration-free air and other transfers are conducted in the area, it is necessary to install laminar flow cabs. This is achieved by introducing air into the laminar unit through a dust filter and then through a high-efficiency pertussis (HEPA) filter. There are two types of laminar flow units: vertical flow units, which direct the air from the HEPA filter downward, and horizontal flow units, which drive the air outward (Dunwell 2000). Additionally, an automated climate and air filtration system is the most effective way to regulate the conditions of the culture/grow room. It is essential to ensure that the explants and shoots are stored in an environment that is conducive to their growth and development, and that the temperature, humidity, air circulation, light quality, and duration of light are controlled (Ahmad et al. 2012.).

FIGURE 2.7 The use of plant biology to combat climate change and the potential for revolutionizing agriculture through the implementation of low-emission crops.

Source: Created by Biorenderapk.

The environment in which plant cells are grown is essential for the growth and differentiation of healthy shoots and roots in later stages. These conditions vary from species to species, so it is important to ensure that the growth environment is conducive to the development of tissue cultures. Additionally, tissue culture protocols for various plant species and varieties necessitate a great deal of trial and error, as they must be adapted to different growth stages and culture/grow room conditions (Altpeter et al. 2016). Therefore, it is essential to have an area for data collection that is conducive to an efficient record-keeping for different types of tissue culture experiments, as well as for product workflow and human workflow (Yu et al. 2022). By setting up these areas, it is possible to efficiently set up a tissue culture laboratory for commercial production. Furthermore, research and development for each plant species is also essential for the success of the plant tissue culture work.

2.8 CONCLUSION/SUMMARY

Recent technological developments in the field of genetic engineering have significantly contributed to the emergence and implementation of the plant biotechnology sector in the agricultural industry. Genetically assisted breeding, next-generation sequencing, and genome editing have enabled the creation of crop varieties that are highly adaptable to climate change. This is considered as a critical factor in meeting global food requirements, as the adaptation of agriculture to climate variability is necessary for the attainment of sustained increases in agricultural efficiency. Consequently, genetic engineering techniques and novel instruments are employed to enhance plants that can cope with the growing number of issues posed by climate variability. Despite the current efforts to design crop varieties that are climate-compatible through contemporary biotechnology, enhanced research efforts and the bridging of technical and knowledge barriers can significantly accelerate the development of crop varieties that can adapt to an ever-changing climate.

REFERENCES

Ahmad, Parvaiz, et al. 2012. Role of transgenic plants in agriculture and biopharming. Biotechnology Advances 30: 524–540. https://doi.org/10.1016/j.biotechadv.2011.09.006.

Altpeter, F., Springer, N. M., Bartley, L. E., et al. 2016. Advancing crop transformation in the era of genome editing. The Plant Cell 28: 1510–1520. https://doi.org/10.1105/tpc.16.00196.

Anjanappa, Ravi B., and Gruissem, W. 2021. Current progress and challenges in crop genetic transformation. Journal of Plant Physiology 261: 1–13. https://doi.org/10.1016/j.jplph.2021.153411.

Anju, V., Verma, M., and Singh, A. 2020. Animal tissue culture principles and applications. In Animal Biotechnology. Models in Discovery and Translation, 2nd ed., ed. A. S. Verma, and A. Singh, 269–293. Academic Press. https://doi.org/10.1016/B978-0-12-811710-1.00012-4.

Anklam, E., Gadani, F., and Heinze, P. 2002. Analytical methods for detection and determination of genetically modified organisms in agricultural crops and plant derived food products. European Food Research and Technology 214: 3–26. https://doi.org/10.1007/s002170100415.

Atkinson, J. A., and Wells, D. M. 2017. An updated protocol for high throughput plant tissue sectioning. Frontiers in Plant Science 8: 1–8. https://doi.org/10.3389/fpls.2017.01721.

Biscotti, M. A., Olmo, E., and Heslop-Harrison, J. S. 2015. Repetitive DNA in eukaryotic genomes. Chromosome Research: An International Journal on the Molecular, Supramolecular and Evolutionary Aspects of Chromosome Biology 23: 415–420. https://doi.org/10.1007/s10577-015-9499-z.

Bratlie, S., Halvorsen, K., Myskja, B. K., et al. 2019. A novel governance framework for GMO. EMBO Reports 20: 1–4. https://doi.org/10.15252/embr.201947812.

Cardoza, V. 2008. Tissue culture: The manipulation of plant development. In Plant Biotechnology and Genetics: Principles, Techniques, and Applications, ed. C. Neal Stewart Jr., 113–134. Wiley Press. https://doi.org/10.1002/9780470282014.ch5.

Charis, P. S., and Vallier, L. 2017. Cell culture: Growing cells as model systems in vitro. In Basic Science Methods for Clinical Researchers, ed. M. Jalali, Francesca Y. L. Saldanha, and J. Mehdi, 151–172. Cambridge University Academic Press.

De Vitis, M., Hay, F. R., Dickie, J. B., et al. 2020. Seed storage: Maintaining seed viability and vigor for restoration use. The Journal of the Society for Ecological Restoration 28: S249–S255. https://doi.org/10.1111/rec.13174open_in_new.

Drexler, H. G., and Uphoff, C. 2002. Mycoplasma contamination of cell cultures: Incidence, sources, effects, detection, elimination, prevention. Cytotechnology 39: 75–90. https://doi.org/10.1023/A:1022913015916.

Dunwell, J. M. 2000. Transgenic approaches to crop improvement. Journal of Experimental Botany 51: 487–496. https://doi.org/10.1093/jexbot/51.suppl_1.487.

Erickson, B., Nelson, J. E., and Winters, P. 2012. Perspective on opportunities in industrial biotechnology in renewable chemicals. Biotechnology Journal 7: 176–185. https://doi.org/10.1002/biot.201100069.

Espinosa-Leal, C. A., Puente-Garza, C. A., and García-Lara, S. 2018. In vitro plant tissue culture: Means for production of biological active compounds. Planta 248: 1–18. https://doi.org/10.1007/s00425-018-2910-1.

Gaj, T., Sirk, S. J., Shui, S. L., et al. 2016. Genome-editing technologies: Principles and applications. Cold Spring Harbor Perspectives in Biology 8: 1–20. https://doi.org/10.1101/cshperspect.a023754.

Galluzzi, M., Zhang, B., Zhang, H., et al. 2021. Unveiling a hidden event in fluorescence correlative microscopy by AFM nanomechanical analysis. Frontiers in Molecular Biosciences 8: 1–13. https://doi.org/10.3389/fmolb.2021.669361.

Gupta, R., and Singh, R. L. 2017. Genetically Modified Organisms (GMOs) and environment. In Principles and Applications of Environmental Biotechnology for a Sustainable Future, ed. R. L. Singh, 425–465. Applied Environmental Science and Engineering for a Sustainable Future. Springer. https://doi.org/10.1007/978-981-10-1866-4_14.

Gupta, V., Sengupta, M., Prakash, J., et al. 2016. An introduction to biotechnology. Basic and Applied Aspects of Biotechnology: 1–21. https://doi.org/10.1007/978-981-10-0875-7_1.

Hixson, J. L., and Ward, A. S. 2022. Hardware selection and performance of low-cost fluorometers. Sensors (Basel, Switzerland) 22: 1–13. https://doi.org/10.3390/s22062319.

Honegger, P. 2001. Overview of cell and tissue culture techniques. In Current Protocols in Pharmacology. Chapter: 12; Unit:12.1. https://doi.org/10.1002/0471141755.ph1201s04.

Jaganathan, D., Bohra, A., Thudi, M., et al. 2020. Fine mapping and gene cloning in the post-NGS era: Advances and prospects. Theoretical and Applied Genetics. Theoretische und angewandte Genetik 133: 1791–1810. https://doi.org/10.1007/s00122-020-03560-w.

Jones, H. D., and Sparks, C. A. 2009. Selection of transformed plants. Methods in Molecular Biology (Clifton, N.J.) 478: 23–37. https://doi.org/10.1007/978-1-59745-379-0_2.

Kamle, M., Kumar, P., Patra, J. K., et al. 2017. Current perspectives on genetically modified crops and detection methods. 3 Biotech 7: 1–15. https://doi.org/10.1007/s13205-017-0809-3.

Kamthan, A., Chaudhuri, A., Kamthan, M., et al. 2016. Genetically modified (GM) crops: Milestones and new advances in crop improvement. TAG: Theoretical and Applied Genetics. Theoretische und angewandte Genetik 129: 1639–1655. https://doi.org/10.1007/s00122-016-2747-6.

Kumar, P. P., and Loh, C. S. 2012. Plant tissue culture for biotechnology. In Plant Biotechnology and Agriculture, ed. A. Altman, and Paul M. Hasegawa, 131–138. Academic Press. https://doi.org/10.1016/B978-0-12-381466-1.00009-2.

Larson, G., Piperno, D. R., Allaby, R. G., et al. 2014. Current perspectives and the future of domestication studies. Proceedings of the National Academy of Sciences 111: 6139–6146. https://doi.org/10.1073/pnas.1323964111.

Liao, X., Zhu, W., Zhou, J., et al. 2023. Repetitive DNA sequence detection and its role in the human genome. Communications Biology 6: 1–21. https://doi.org/10.1038/s42003-023-05322-y.

Liu, B., Dong, X., Zheng, C., Keener, D., et al. 2024. Targeted genome editing with a DNA-dependent DNA polymerase and exogenous DNA-containing templates. Nature Biotechnology 42: 1039–1045. https://doi.org/10.1038/s41587-023-01947-w.

Mittler, R., and Blumwald, E. 2010. Genetic engineering for modern agriculture: Challenges and perspectives. Annual Review of Plant Biology 61: 443–462. https://doi.org/10.1146/annurev-arplant-042809-112116.

Nambisan, P. 2017. Laboratory biosafety and good laboratory practices. In An Introduction to Ethical, Safety and Intellectual Property Rights Issues in Biotechnology, ed. K. Jones, and E. H. Parks, 253–271. Academic Press. https://doi.org/10.1016/b978-0-12-809231-6.00011-9.

Pamies, D., Leist, M., Coecke, S, et al. 2022. Guidance document on Good Cell and Tissue Culture Practice 2.0 (GCCP 2.0). ALTEX 39: 30–70. https://doi.org/10.14573/altex.2111011.

Paterniani, E., Perez, F., Reinach, F., et al. 2000. Examples of GM technology that would benefit world agriculture. In Transgenic Plants and World Agriculture, ed. The Royal Society, 7–14. National Academic Press (US).

Raman, R. 2017. The impact of Genetically Modified (GM) crops in modern agriculture: A review. GM Crops & Food 8: 195–208. https://doi.org/10.1080/21645698.2017.1413522.

Ranjha, M. M. A. N., Shafique, B., Khalid, W., et al. 2022. Applications of biotechnology in food and agriculture: A mini-review. Proceedings of the National Academy of Sciences, India 92: 11–15. https://doi.org/10.1007/s40011-021-01320-4.

Roberts, M. A. J. 2019. Recombinant DNA technology and DNA sequencing. Essays in Biochemistry 63: 457–468. https://doi.org/10.1042/EBC20180039.

Salmeron, John M., and Vernooij, B. 1998. Transgenic approaches to microbial disease resistance in crop plants. In Current Opinion in Plant Biology, ed. C. Kohle, and D. Wagner, 347–352. Elsevier.

Savvidis, S., Gerli, M. F. M., Pellegrini, M., et al. 2022. Monitoring tissue engineered constructs and protocols with laboratory-based x-ray phase contrast tomography. Acta Biomaterialia 141: 290–299. https://doi.org/10.1016/j.actbio.2022.01.022.

Schmidt, N. J. 1972. Tissue culture in the laboratory diagnosis of viral infections. American Journal of Clinical Pathology 57: 820–828. https://doi.org/10.1093/ajcp/57.6.820.

Sharma, P., Singh, S. P., Iqbal, H. M. N., et al. 2022. Genetic modifications associated with sustainability aspects for sustainable developments. Bioengineered 13: 9509–9521. https://doi.org/10.1080/21655979.2022.2061146.

Smith, C. M. 2021. Conventional breeding of insect-resistant crop plants: Still the best way to feed the world population. Current Opinion in Insect Science 45: 7–13. https://doi.org/10.1016/j.cois.2020.11.008.

Steinwand, M., and Ronald, P. 2020. Crop biotechnology and the future of food. Nature Food 1: 273–283. https://doi.org/10.1038/s43016-020-0072-3.

Tabei, Y., and Muranaka, T. 2020. Preface to the special issue "technology in tissue culture toward horizon of plant biotechnology". Plant Biotechnology (Tokyo, Japan) 37: 117–120. https://doi.org/10.5511/plantbiotechnology.20.0000p.

Turnbull, C., Lillemo, M., and Hvoslef-Eide, T. A. K. 2021. Global regulation of genetically modified crops Amid the Gene edited crop boom—a review. Frontiers in Plant Science 12: 1–19. https://doi.org/10.3389/fpls.2021.630396.

Vargas, L., Victor, M., and Neftalí, O. A. 2018. An introduction to plant tissue culture: Advances and perspectives. Methods in Molecular Biology (Clifton, N.J.) 1815: 3–13. https://doi.org/10.1007/978-1-4939-8594-4_1.

Wang, Q., Shen, X., Qiu, T., et al. 2021. Evaluation and application of an efficient plant DNA extraction protocol for laboratory and field testing. Journal of Zhejiang University Science. B 22: 99–111. https://doi.org/10.1631/jzus.B2000465.

Wikandari, R., Manikharda, Baldermann, S., et al. 2021. Application of cell culture technology and genetic engineering for production of future foods and crop improvement to strengthen food security. Bioengineered 12: 11305–11330. https://doi.org/10.1080/21655979.2021.2003665.
Yau, Y. Y., and Stewart, C. N., Jr. 2013. Less is more: Strategies to remove marker genes from transgenic plants. BMC Biotechnology 13: 1–23. https://doi.org/10.1186/1472-6750-13-36.
Yu, H., Yang, Q., Fu, F., et al. 2022. Three strategies of transgenic manipulation for crop improvement. Frontiers in Plant Science 13: 1–15. https://doi.org/10.3389/fpls.2022.948518.
Zaidi, S. S., Mahas, A., Vanderschuren, H., et al. 2020. Engineering crops of the future: CRISPR approaches to develop climate-resilient and disease-resistant plants. Genome Biology 21: 1–19. https://doi.org/10.1186/s13059-020-02204-y.

3 Recombinant DNA Technology and Its Application in Plant Biotechnology

Jyoti, Preeti Garg, Vandana Garg, Rohit Dutt, Manju Rawat Singh, Deependra Singh, and Rahul Pratap Singh*
* Corresponding author: drvandana.pharma@mdurohtak.ac.in

3.1 INTRODUCTION

Biology underwent a revolutionary change with the introduction of recombinant DNA technology, resulting in significant paradigm shifts and other consequential transformations. The advancements in genetic engineering have facilitated the expeditious use of many medicinal compounds in the domains of medical genetics and biomedicine. This has been achieved by the manipulation of microbes, animals, and plants to produce pharmacologically beneficial molecules. Most bioengineering medications employed in the field of medicine are of recombinant origin, playing a pivotal role in combating life-threatening disorders.

The advent of recombinant DNA technology has had a profound impact on the quality of human existence. The number of recombinant pharmaceuticals approved by the US Food and Drug Administration (FDA) in 1997 exceeded the cumulative total of approvals in the preceding years. The medications were employed for the treatment of many medical conditions, such as anemia, cancers (including Kaposi's sarcoma, leukemia, kidney cancer, colorectal cancer, and ovarian cancer), AIDS, as well as genetic disorders (including cystic fibrosis, familial hypercholesterolemia, Gaucher disease, and hemophilia A). The utilization of multigene transfer, site-specific integration, and precisely regulated gene expression represents significant advancements in the study of plant growth mechanisms. The transcriptional regulation of endogenous genes, ensuring their functional adaptability in novel environments, and the precise management of transgenic expression pose significant challenges in the field of plant biotechnology, necessitating resolution before their efficient utilization (Cardi & Neal Stewart, 2016).

Recombinant DNA technology significantly enhances health conditions through the development of novel vaccines and medications. In addition to enhancing treatment plans, there is a development of diagnostic kits, monitoring equipment, and novel therapy procedures. An illustrative instance of genetic engineering in the field of health pertains to the utilization of genetically modified bacteria to produce synthetic human insulin and erythropoietin. Another approach involves the utilization of genetically engineered microorganisms to generate novel strains of experimental mutant mice for research purposes. Similarly, genetic engineering has been employed to address environmental challenges, such as the conversion of waste materials into biofuels and bioethanol, remediation of oil spills, carbon sequestration, and the detection of hazardous substances like arsenic in drinking water. Genetically modified microorganisms also play a significant role in the fields of biomining and bioremediation (Rehman et al., 2020).

The projected value of the subject is anticipated to experience further growth in the 21st century. The following enumeration presents various domains in which the application of recombinant DNA technology is anticipated to yield significant effects. Certain plant species can produce and release their own pesticide compounds. The global population is on a continuous growth trajectory, necessitating

DOI: 10.1201/9781003326939-3

a substantial increase in food production that surpasses current levels of attainability (Mendez et al., 2005). Moreover, a multitude of health risks is accountable for a significant proportion of global mortality. The escalating pollution stemming from the rapid process of industrialization is an additional matter of anxiety (Mani & Kumar, 2014). Despite extensive efforts to tackle these difficulties, the outcomes are much below the necessary threshold. To address these issues, it is imperative to accurately ascertain, delineate, and ultimately manipulate the genetic components implicated in biological pathways (Cronin & Yu, 2023; Misra, 2011). This chapter provides an overview of the fundamental processes employed in recombinant DNA technology and focuses on its current applications.

3.2 FORMATION OF RECOMBINANT DNA TECHNOLOGY

Recombinant DNA technology encompasses a collection of approaches utilized to proficiently merge segments of DNA. The formation of a recombinant DNA molecule occurs when segments from two or more distinct DNA molecules are combined. Under certain circumstances, a recombinant DNA molecule can enter a cell and replicate within the cell, either independently or via integrating into a chromosome. In laboratory settings, the synthesis of recombinant DNA (rDNA) molecules is achieved by the utilization of genetic recombination techniques, such as molecular cloning. These technologies facilitate the combination of genetic material from different origins, leading to the generation of sequences that do not naturally occur in the genome. The successful creation of recombinant DNA was initially accomplished in 1973 by Herbert Boyer from the University of California in San Francisco and Stanley Cohen from Stanford University. Their method involved the utilization of restriction enzymes from *E. coli* to facilitate the integration of exogenous DNA into plasmids (Barman & Kundu, 2019).

Recombinant DNA pertains to a DNA molecule that has been formed by the fusion of two or more distinct DNA strands. The potential for utilizing recombinant DNA is derived from the observation that DNA molecules in all organisms have a consistent chemical makeup, with changes only present in the nucleotide sequence within this common structural framework. Recombinant DNA molecules are sometimes denoted as chimeric DNA owing to their composition, which is formed from genetic material originating from two distinct species, like the mythical chimera. The utilization of palindromic sequences in recombinant DNA (rDNA) technology leads to the production of both sticky and blunt ends (Alfermann & Petersen, 1995).

3.3 PRODUCTION OF RECOMBINANT DNA

The DNA sequences used in the creation of recombinant DNA molecules might come from several animal sources. For example, the genetic material of plants can be amalgamated with that of bacteria, or the genetic material of people can be merged with that of fungi. Moreover, it is feasible to produce DNA sequences, via chemical synthesis, that do not occur naturally in any organism. Subsequently, these synthetically produced DNA sequences can be incorporated into recombinant molecules. The utilization of synthetic DNA technology and recombinant DNA allows for the generation and subsequent incorporation of nearly any DNA sequence into a wide range of living organisms (Abdul Aziz et al., 2022; Twyman et al., 2002).

The production of recombinant DNA can be achieved through three distinct methodologies. The process of transformation, the introduction of phages, and non-bacterial transformation are three key concepts in the field of molecular biology (Shah, 2019).

The initial stage of the transformation process involves the careful selection of a specific segment of DNA that will subsequently be introduced into a vector. The subsequent procedure involves cleaving the DNA fragment using a restriction enzyme, followed by the integration of the DNA insert into the vector using DNA ligase. The insert is equipped with a selectable marker that facilitates the identification of recombinant molecules. The utilization of an antibiotic marker is a common practice in which a host cell lacking a vector undergoes cell death upon exposure to a certain antibiotic, while the host cell containing the vector survives due to its inherent resistance (Twyman et al., 2002).

The vector is introduced into a recipient cell by a biological process known as transformation. The host cells, such as *E. coli*, necessitate certain preparations to facilitate the uptake of exogenous DNA. Selectable markers can encompass antibiotic resistance, alterations in color, or any other discernible trait that enables differentiation between transformed and untransformed hosts. Various vectors possess distinct characteristics that render them appropriate for diverse purposes. Several characteristics that can be observed in certain qualities are symmetrical cloning sites, size, and high copy number (Koçoglu & Yalçinkaya, 1992).

Non-bacterial transformation refers to a process in which genetic material is introduced into a host organism without the involvement of bacterial cells. This procedure bears a striking resemblance to the transformation process. The sole distinction between the two is that non-bacterial variants do not rely on bacterial hosts, such as *E. coli*. Microinjection involves the direct injection of DNA into the nucleus of the target cell during the process of cellular transformation. In the biolistic technique, host cells are subjected to bombardment by microprojectiles with high velocity. These microprojectiles, typically composed of gold or tungsten, are coated with DNA (Bose & Bose, 2022).

Phage introduction refers to the method of transfection wherein a phage is employed instead of bacteria, hence serving as an equivalent to the process of transformation. The technique of in vitro packaging is employed for vector encapsulation. Lambda or MI3 phages are utilized in this process to generate phage plaques that harbor recombinant entities. Differences between recombinants and non-recombinants can be discerned through the application of diverse selection procedures, hence facilitating the identification of the generated recombinants (Bose & Bose, 2022; Koçoglu & Yalçinkaya, 1992).

3.4 THE OPERATIONAL MECHANISMS OF RECOMBINANT DNA (RDNA) INTO PLANTS

The efficacy of recombinant DNA technology is demonstrated by the host cell's proficient synthesis and expression of proteins obtained from the injected recombinant genes. The addition of expression factors is a prerequisite for the host organism to produce a substantial amount of recombinant protein (Kumar & Kumar, 2015). The production of proteins relies on the existence of a certain collection of signals that envelop the gene. These signals function as instructions for the cell's transcription and translation processes. The signals included within this category comprise the promoter, the ribosome binding site, and the terminator (Ma et al., 2003; Mani & Kumar, 2014). Expression vectors commonly include these signals, which aid in the incorporation of exogenous DNA. Signals demonstrate species specificity. In the context of *Escherichia coli*, it is crucial to employ signals that are peculiar to this bacterium, as *E. coli* is unlikely to interpret the signals commonly associated with human promoters and terminators. Difficulties may occur in cases when the gene possesses introns or encompasses signals that function as terminators for a bacterial host (Ferrer-Miralles et al., 2009). As a result, the early end of protein synthesis may occur, which has the potential to result in inadequate processing, erroneous folding, or destruction of the recombinant protein. Recombinant protein production in eukaryotic systems is frequently observed in yeast and filamentous fungus. The obstacles associated with the exploitation of animal cells mostly stem from their need for a solid support surface, which distinguishes them from bacteria, as well as their complex growth needs (Kumar & Kumar, 2015). However, there are certain proteins that possess a degree of complexity that surpasses the synthesis capabilities of bacteria, thereby requiring the use of eukaryotic cells (Ma et al., 2003).

3.5 TRANSGENIC PLANTS

Recombinant DNA technology can create genetically modified plants that possess specific traits, such as higher production and nutritional value, enhanced resistance to both living organisms and environmental factors, and consequently demonstrate improved adaptation for improved longevity (Parmar et al., 2017). Recombinant DNA technology enables the rapid, cost-effective, and precise incorporation of desired features from many sources into the genetic makeup of plants, surpassing the efficiency of natural recombination or traditional breeding methods.

Recombinant DNA technology involves the fusion of DNA molecules from selected organisms and their introduction into a host organism, resulting in the production of advantageous recombinant products with applications in agriculture, medicine, and industry. Efficient transformation technologies are necessary prerequisites for the generation of sustainable and commercially viable transgenic plants. Various techniques have been devised thus far to introduce the required genes into the host plants.

Gene transfer can be achieved through various methods. One approach involves the use of vectors, such as *Agrobacterium tumefaciens*. However, there are also vector-less methods available, such as particle gun bombardment (Furth, 1997; Sanford, 1988), microinjection (De La Peña et al., 1987), polyethylene glycol–mediated (PEG) transfer (Uchimiya et al., 1986), and electroporation (Arencibia et al., 1995). Among the different transformation techniques available, the most frequently utilized for creating transgenic plants are *Agrobacterium tumefaciens*–mediated gene transfer, which enables the direct transmission of DNA into cells or protoplasts, and the particle bombardment approach, as well as osmotic or electric shock. After a particular cell or protoplast is modified, standardized tissue culture techniques are used to regenerate transgenic plants (Nalluri & Karri, 2020). Figure 3.1 illustrates the application of recombinant DNA technology to enhance plant growth and productivity in a sustainable manner.

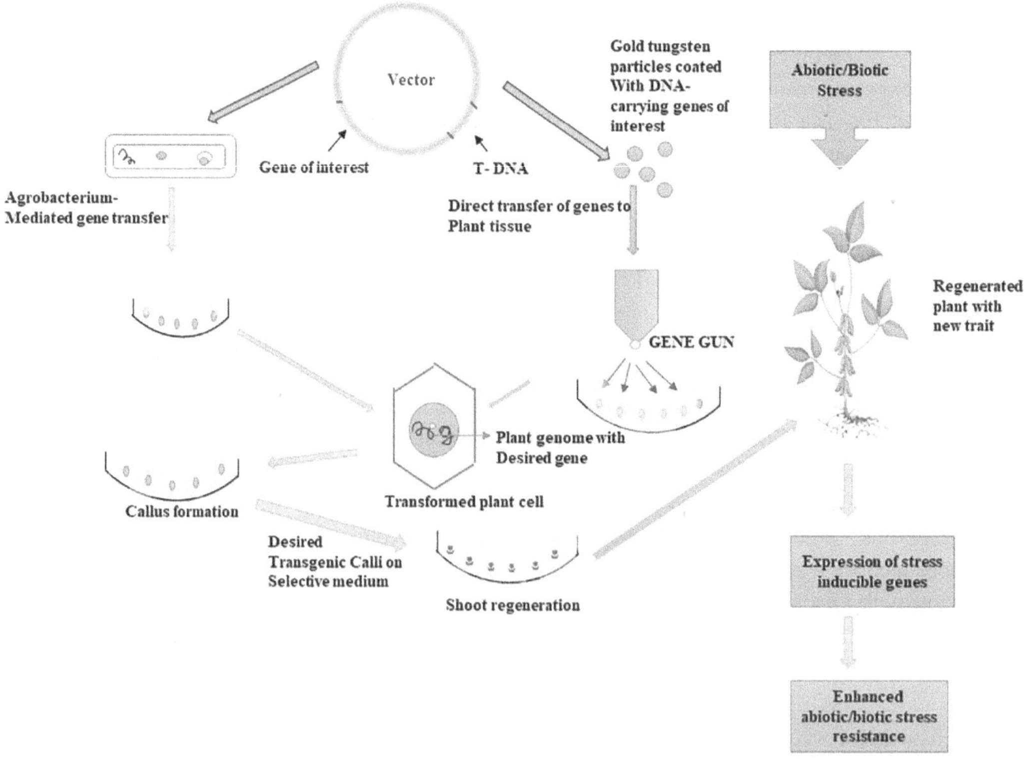

FIGURE 3.1 Recombinant DNA technology for the genetic modification of plants.

3.6 THE SUSTAINABLE DEVELOPMENT OF PLANTS USING RDNA TECHNOLOGY

In 2015, the United Nations approved a new Sustainable Development agenda that includes 17 Sustainable Development Goals (SDGs). The goals are to end poverty, guarantee food security, and protect natural resources in order to bring about world peace and prosperity by 2030. Due to the interdependence of the goals, the course of action in one field will affect the result in another. The primary focus of development should be on ensuring social, economic, and environmental sustainability (Hák et al., 2016).

Biotechnology contributes to achieving the objectives for sustainable development by enhancing the health and yield of plant, offering enriched food alternatives, and decreasing greenhouse gas emissions. Biotechnology is significant in achieving several sustainable development objectives, as shown in Figure 3.2.

Recombinant DNA technology is a promising tool in biotechnology that can help balance environmental protection and the developmental demands of the world's increasing population. This new technology can trigger another green revolution and ultimately eradicate world hunger. Recombinant DNA technology has significantly boosted crop output, yield, and disease resistance in recent years (Babiye et al., 2020).

Genetic engineering (recombinant DNA technology) is the most widely used technology globally for enhancing crops, with a 113-fold expansion in the cultivated land area since its commercial debut in 1996. Genetically altered crop cultivation has expanded to 26 countries, comprising 21 developing nations and 5 industrialized countries. Additionally, 43 non-planting countries have utilized an imported genetically altered crop. Genetically modified crops have a significant role

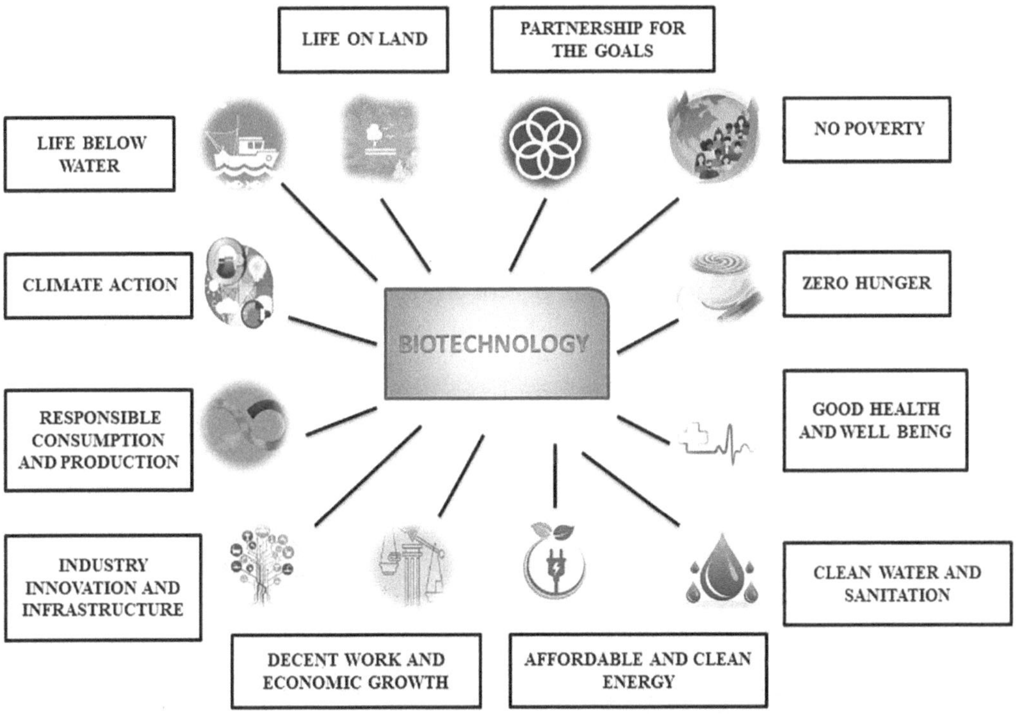

FIGURE 3.2 Graphical depiction of the correlation between biotechnology and sustainable development objectives.

in enhancing global food security and sustainability and in addressing climate change. In 2017, GM crops, including cotton, soybean, maize, and canola, covered almost 80%, 77%, 32%, and 30% of the worldwide agricultural area, respectively. The United States of America, Canada, and Argentina are the leading nations in terms of production and exportation of genetically modified crops and their derivatives. India, China, Brazil, and Argentina are major contributors to the production of genetically modified crops in developing countries. Genetically modified crops have resulted in a decrease of 670 million kilograms in the use of pesticides (Harsh Environment and Plant Resilience, 2021).

Developing crops that can withstand both biotic and abiotic stressors is essential for the progress of sustainable agriculture. The application of recombinant DNA technology has extensively been employed to augment plants' resilience against both living organisms and environmental factors, as well as to amplify the overall quality and productivity of plants (Nam et al., 2020; Parmar et al., 2017).

Numerous studies in genetic engineering have resulted in the insertion of genes that enhance resistance to various stresses, improve nutritional value, extend shelf life, and modify traits like yield, color, and texture in plants. The researcher has focused on various genes, including chitinase, osmotin, protease inhibitors, cystatin, defensin, and glucanase, to create plant varieties resistant to biotic stress. Furthermore, genes involved in the production of stress-protecting compounds, such as mannitol, heat shock proteins, glycine, and betaine, are being focused on to boost plants' ability to withstand abiotic stress (Parmar et al., 2017).

3.7 THE DEVELOPMENT OF ABIOTIC STRESS–TOLERANT PLANTS USING RDNA TECHNOLOGY

Plants' ability to adapt is significantly impacted by the swiftly changing environmental conditions primarily caused by human activities (Amedie, 2013; Anderson & Song, 2020; Parmesan & Hanley, 2015). Plants are often subjected to various uncommon environmental conditions, such as waterlogging, UV radiation, drought, nutrient imbalances, heat, salinity, freezing, metal toxicities, and high light intensity (He et al., 2018; Wani et al., 2016). These conditions are collectively known as abiotic stresses. Multiple reports have indicated that abiotic stresses are significant factors that limit the yield of food crops, resulting in a decline of up to 70% in the yield of important food crops (Mantri et al., 2012; Zörb et al., 2019). The primary abiotic factor is drought, which impacts 25% of arable land, followed by salinity, affecting 10% (Roy et al., 2011). Wang *et al.* found that elevated salinity levels in arable land could result in a loss of approximately 50% of the land by 2050 (Chang et al., 2020; Wang et al., 2003). Genetic engineering techniques are efficient methods that allow for the direct modification of higher plant genomes, resulting in altered plant metabolism and improved growth and yield under stressful conditions (Ahanger et al., 2017; Gilliham et al., 2017).

Plants can adapt to a variety of changes in their environment at physiological, morphological, and genetic levels to overcome unfavorable conditions. The most effective method to improve stress resistance in plants is to identify and transfer stress-related genes in higher plants (Ahmad & Wani, 2014; Lamalakshmi Devi et al., 2017; Tzfira et al., 2012).

3.8 THE DEVELOPMENT OF BIOTIC STRESS–TOLERANT PLANTS USING RDNA TECHNOLOGY

Biotic stresses, such as pests, bacteria, fungi, viruses, and insects, cause plant damage. Weeds cause the most damage among biotic stresses at 34%, trailed by insects at 18%, and pathogens at 16% (Oerke, 2006). Various plants like cotton, alfalfa, rice, canola, maize, fodder, beetroot, soybean, and coffee have undergone genetic modification through recombinant DNA technology with genes that aid in breaking down a herbicide's active component, rendering it harmless (Jin et al., 2022; Kim et al., 2009; Mazur & Falco, 1989; Nakka et al., 2019; Vencill et al., 2012). The most used methods for efficient genetic transformation in developing herbicide-tolerant plants are *Agrobacterium*-mediated and particle bombardment techniques (Gosavi et al., 2022; Tsaftaris, 1996). Transgenic plants offer sustainable solutions for weed control by reducing herbicide use, improving the environment, promoting low-tillage practices, allowing flexibility in spraying time, reducing labor and fuel inputs, and benefiting carbon emissions and soil structure (Brookes & Barfoot, 2015). Similarly, cold-resistant and insect pest–resistant strains of plants are produced using rDNA technology.

3.9 CONCLUSION

Employing recombinant DNA technology to enhance plant productivity, development, and resilience against both living organisms and non-living environmental factors is a viable and enduring resolution to the challenges arising from the increasing global population. It maintains equilibrium by both safeguarding the environment and fulfilling the developmental requirements of the world's expanding population. Also, new technologies for sequencing and changing genes have made it possible to quickly and accurately sequence the genome, find the genes that control the traits that are wanted, and change into host plants.

REFERENCES

Abdul Aziz, M., Brini, F., Rouached, H., & Masmoudi, K. (2022). Genetically engineered crops for sustainably enhanced food production systems. *Frontiers in Plant Science, 13*. https://doi.org/10.3389/fpls.2022.1027828

Ahanger, M. A., Akram, N. A., Ashraf, M., Alyemeni, M. N., Wijaya, L., & Ahmad, P. (2017). Plant responses to environmental stresses-from gene to biotechnology. *AoB Plants, 9*(4). https://doi.org/10.1093/aobpla/plx025

Ahmad, P., & Wani, M. R. (2014). Physiological mechanisms and adaptation strategies in plants under changing environment: Volume 1. In *Physiological Mechanisms and Adaptation Strategies in Plants under Changing Environment: Volume 1*. https://doi.org/10.1007/978-1-4614-8591-9

Alfermann, A. W., & Petersen, M. (1995). Natural product formation by plant cell biotechnology. *Plant Cell, Tissue and Organ Culture, 43*(2). https://doi.org/10.1007/bf00052176

Amedie, F. A. (2013). Impacts of climate change on plant growth, ecosystem services, biodiversity, and potential adaptation measure. *Master Thesis in Atmospheric Science with Orientation towards Environmental Science (60 HEC).*

Anderson, J. T., & Song, B. H. (2020). Plant adaptation to climate change—where are we? In *Journal of Systematics and Evolution, 58*(5). https://doi.org/10.1111/jse.12649

Arencibia, A., Molina, P. R., de la Riva, G., & Selman-Housein, G. (1995). Production of transgenic sugarcane (Saccharumofficinarum L.) plants by intact cell electroporation. *Plant Cell Reports, 14*(5). https://doi.org/10.1007/BF00232033

Babiye, B., Haile, G., & Adamu, M. (2020). Major achievements of plant biotechnology in crop improvements. *American Journal of Life Sciences, 8*(5). https://doi.org/10.11648/j.ajls.20200805.13

Barman, M., & Kundu, S. (2019). Molecular markers and a new vista in plant breeding: A review. *International Journal of Current Microbiology and Applied Sciences, 8*(12). https://doi.org/10.20546/ijcmas.2019.812.230

Bose, R., & Bose, K. (2022). A brief introduction to recombinant DNA technology. In *Textbook on Cloning, Expression and Purification of Recombinant Proteins*. https://doi.org/10.1007/978-981-16-4987-5_1

Brookes, G., & Barfoot, P. (2015). Environmental impacts of genetically modified (GM) crop use 1996–2013: Impacts on pesticide use and carbon emissions. *GM Crops & Food, 6*(2), 103. https://doi.org/10.1080/21645698.2015.1025193

Cardi, T., & Neal Stewart, C. (2016). Progress of targeted genome modification approaches in higher plants. *Plant Cell Reports, 35*(7). https://doi.org/10.1007/s00299-016-1975-1

Chang, Y. N., Zhu, C., Jiang, J., Zhang, H., Zhu, J. K., & Duan, C. G. (2020). Epigenetic regulation in plant abiotic stress responses. *Journal of Integrative Plant Biology, 62*(5). https://doi.org/10.1111/jipb.12901

Cronin, J. M., & Yu, A. M. (2023). Recombinant technologies facilitate drug metabolism, pharmacokinetics, and general biomedical research. *Drug Metabolism and Disposition*, *51*(6). https://doi.org/10.1124/dmd.122.001008

De La Peña, A., Lörz, H., & Schell, J. (1987). Transgenic rye plants obtained by injecting DNA into young floral tillers. *Nature*, *325*(6101). https://doi.org/10.1038/325274a0

Ferrer-Miralles, N., Domingo-Espín, J., Corchero, J., Vázquez, E., & Villaverde, A. (2009). Microbial factories for recombinant pharmaceuticals. *Microbial Cell Factories*, *8*. https://doi.org/10.1186/1475-2859-8-17

Furth, P. A. (1997). Gene transfer by biolistic process. *Applied Biochemistry and Biotechnology—Part B Molecular Biotechnology*, *7*(2). https://doi.org/10.1007/BF02761749

Gilliham, M., Able, J. A., & Roy, S. J. (2017). Translating knowledge about abiotic stress tolerance to breeding programmes. *Plant Journal*, *90*(5). https://doi.org/10.1111/tpj.13456

Gosavi, G., Ren, B., Li, X., Zhou, X., Spetz, C., & Zhou, H. (2022). A new era in herbicide-tolerant crops development by targeted genome editing. *ACS Agricultural Science and Technology*, *2*(2). https://doi.org/10.1021/acsagscitech.1c00254

Hák, T., Janoušková, S., & Moldan, B. (2016). Sustainable development goals: A need for relevant indicators. *Ecological Indicators*, *60*. https://doi.org/10.1016/j.ecolind.2015.08.003

Harsh Environment and Plant Resilience. (2021). *Harsh Environment and Plant Resilience*. https://doi.org/10.1007/978-3-030-65912-7

He, M., He, C. Q., & Ding, N. Z. (2018). Abiotic stresses: General defenses of land plants and chances for engineering multistress tolerance. *Frontiers in Plant Science*, *871*. https://doi.org/10.3389/fpls.2018.01771

Jin, M., Chen, L., Deng, X. W., & Tang, X. (2022). Development of herbicide resistance genes and their application in rice. *Crop Journal*, *10*(1). https://doi.org/10.1016/j.cj.2021.05.007

Kim, H. A., Utomo, S. D., Kwon, S. Y., Min, S. R., Kim, J. S., Yoo, H. S., & Choi, P. S. (2009). The development of herbicide-resistant maize: Stable Agrobacterium-mediated transformation of maize using explants of type II embryogeniccalli. *Plant Biotechnology Reports*, *3*(4). https://doi.org/10.1007/s11816-009-0099-2

Koçoglu, T., & Yalçinkaya, T. (1992). Recombinant DNA technology. *Mikrobiyolojibülteni*, *26*(2). https://doi.org/10.1038/nbt0184-86b

Kumar, S., & Kumar, A. (2015). Role of genetic engineering in agriculture. *Plant Archives*, *15*(1).

Lamalakshmi Devi, E., Kumar, S., Basanta Singh, T., Sharma, S. K., Beemrote, A., Devi, C. P., Chongtham, S. K., Singh, C. H., Yumlembam, R. A., Haribhushan, A., Prakash, N., & Wani, S. H. (2017). Adaptation strategies and defence mechanisms of plants during environmental stress. *Medicinal Plants and Environmental Challenges*. https://doi.org/10.1007/978-3-319-68717-9_20

Ma, J. K. C., Drake, P. M. W., & Christou, P. (2003). The production of recombinant pharmaceutical proteins in plants. *Nature Reviews Genetics*, *4*(10). https://doi.org/10.1038/nrg1177

Mani, D., & Kumar, C. (2014). Biotechnological advances in bioremediation of heavy metals contaminated ecosystems: An overview with special reference to phytoremediation. *International Journal of Environmental Science and Technology*, *11*(3). https://doi.org/10.1007/s13762-013-0299-8

Mantri, N., Patade, V., Penna, S., Ford, R., & Pang, E. (2012). Abiotic stress responses in plants: Present and future. In *Abiotic Stress Responses in Plants*. https://doi.org/10.1007/978-1-4614-0634-1_1

Mazur, B. J., & Falco, S. C. (1989). The development of herbicide resistant crops. *Annual Review of Plant Physiology and Plant Molecular Biology*, *40*(1). https://doi.org/10.1146/annurev.pp.40.060189.002301

Mendez, C., Weitnauer, G., Bechthold, A., & Salas, J. (2005). Structure alteration of polyketides by recombinant DNA technology in producer organisms prospects for the generation of novel pharmaceutical drugs. *Current Pharmaceutical Biotechnology*, *1*(4). https://doi.org/10.2174/1389201003378861

Misra, A. (2011). Challenges in delivery of therapeutic genomics and proteomics. In *Challenges in Delivery of Therapeutic Genomics and Proteomics*. https://doi.org/10.1016/C2010-0-65663-X

Nakka, S., Jugulam, M., Peterson, D., & Asif, M. (2019). Herbicide resistance: Development of wheat production systems and current status of resistant weeds in wheat cropping systems. *Crop Journal*, *7*(6). https://doi.org/10.1016/j.cj.2019.09.004

Nalluri, N., & Karri, V. R. (2020). Recent advances in genetic manipulation of crops: A promising approach to address the global food and industrial applications. *Plant Science Today*, 7(1). https://doi.org/10.14719/PST.2020.7.1.659

Nam, K. H., Kim, D. Y., Moon, Y. S., Pack, I. S., Jeong, S. C., Kim, H. B., & Kim, C. G. (2020). Performance of hybrids between abiotic stress-tolerant transgenic rice and its weedy relatives under water-stressed conditions. *Scientific Reports*, *10*(1). https://doi.org/10.1038/s41598-020-66206-3

Oerke, E. C. (2006). Crop losses to pests. *Journal of Agricultural Science*, *144*(1). https://doi.org/10.1017/S0021859605005708

Parmar, N., Singh, K. H., Sharma, D., Singh, L., Kumar, P., Nanjundan, J., Khan, Y. J., Chauhan, D. K., & Thakur, A. K. (2017). Genetic engineering strategies for biotic and abiotic stress tolerance and quality enhancement in horticultural crops: A comprehensive review. *3 Biotech*, *7*(4). https://doi.org/10.1007/s13205-017-0870-y

Parmesan, C., & Hanley, M. E. (2015). Plants and climate change: Complexities and surprises. *Annals of Botany*, *116*(6). https://doi.org/10.1093/aob/mcv169

Rehman, S., Ul Rehman, I., Jan, B., Rashid, I., Ah Reshi, Z., & Ganie, A. H. (2020). Genome editing: Applications for medicinal and aromatic plants. In *Medicinal and Aromatic Plants: Expanding Their Horizons through Omics*. https://doi.org/10.1016/B978-0-12-819590-1.00006-9

Roy, B., Noren, S. K., Mandal, A. B., & Basu, A. K. (2011). Genetic engineering for abiotic stress tolerance in agricultural crops. *Biotechnology*, *10*(1). https://doi.org/10.3923/biotech.2011.1.22

Sanford, J. C. (1988). The biolistic process. *Trends in Biotechnology*, *6*(12). https://doi.org/10.1016/0167-7799(88)90023-6

Shah, N. J. (2019). Recombinant DNA technology. In *Introduction to Basics of Pharmacology and Toxicology: Volume 1: General and Molecular Pharmacology: Principles of Drug Action*. https://doi.org/10.1007/978-981-32-9779-1_34

Tsaftaris, A. (1996). The development of herbicide-tolerant transgenic crops. *Field Crops Research*, *45*(1–3). https://doi.org/10.1016/0378-4290(95)00064-x

Twyman, R., Christou, P., & Stoger, E. (2002). Genetic transformation of plants and their cells. In *Plant Biotechnology and Transgenic Plants*. https://doi.org/10.1201/9780203910849.ch6

Tzfira, T., Weinthal, D., Marton, I., Zeevi, V., Zuker, A., & Vainstein, A. (2012). Genome modifications in plant cells by custom-made restriction enzymes. *Plant Biotechnology Journal*, *10*(4). https://doi.org/10.1111/j.1467-7652.2011.00672.x

Uchimiya, H., Fushimi, T., Hashimoto, H., Harada, H., Syono, K., & Sugawara, Y. (1986). Expression of a foreign gene in callus derived from DNA-treated protoplasts of rice (Oryzasativa L.). *MGG Molecular & General Genetics*, *204*(2). https://doi.org/10.1007/BF00425499

Vencill, W. K., Nichols, R. L., Webster, T. M., Soteres, J. K., Mallory-Smith, C., Burgos, N. R., Johnson, W. G., & McClelland, M. R. (2012). Herbicide resistance: Toward an understanding of resistance development and the impact of herbicide-resistant crops. *Weed Science*, *60*(SP1). https://doi.org/10.1614/ws-d-11-00206.1

Wang, W., Vinocur, B., & Altman, A. (2003). Plant responses to drought, salinity and extreme temperatures: Towards genetic engineering for stress tolerance. *Planta*, *218*(1). https://doi.org/10.1007/s00425-003-1105-5

Wani, S. H., Kumar, V., Shriram, V., & Sah, S. K. (2016). Phytohormones and their metabolic engineering for abiotic stress tolerance in crop plants. *Crop Journal*, *4*(3). https://doi.org/10.1016/j.cj.2016.01.010

Zörb, C., Geilfus, C. M., & Dietz, K. J. (2019). Salinity and crop yield. *Plant Biology*, *21*. https://doi.org/10.1111/plb.12884

Tissue Culture Techniques

4 Protoplast Isolation and Fusion

*Shalini Mathpal, Tushar Joshi, Priyanka Sharma, Priyanka Maiti, Mahesha Nand, and Subhash Chandra**
* Corresponding author: scjnu@yahoo.co.in

4.1 INTRODUCTION

Advances in biotechnological research over the past decades have opened up several opportunities in many areas. At present, the protoplast fusion technique is widely used to improve the propagation and genetic qualities of plant species. Plant protoplasts provide a unique single-cell system to support several aspects of modern biotechnology. Basically, "protoplasts" are the cells from which the cell wall has been detached either mechanically or enzymatically (Cocking, 1960; Kaladharan, 1998; Verma et al., 2000a). They have the capability to dedifferentiate, re-enter the cell cycle, and undergo multiple rounds of mitotic divisions. Protoplasts can proliferate or regenerate into various organs—that is, protoplasts are potentially totipotent. Nagata and Takebe (1971) developed the first complete fertile plant derived by protoplast from mesophyll protoplasts of tobacco.

The protoplast technique involves the fusion of two genetically distinct species, resulting in the formation of hybrid protoplast with two fused parent nuclei. This successful fusion of two different species can result in progeny that expresses both parental cell traits and a new hybrid. By the fusion of protoplasts, it is feasible to transfer some beneficial genes, such as nitrogen fixation, disease resistance, rapid growth rate, drought resistance, herbicide resistance, protein quality, cold and heat resistance genes, from one species to another. This technique allows for the genetic recombination of both inter- and intraspecific genes, leading to organisms possessing improved traits of interest. The transformation of protoplasts with foreign genes is a relatively new versatile method that has the potential to significantly improve crops (Mazarei et al., 2008; Rhodes et al., 1988).

The process of protoplast fusion requires cell wall removal, protoplast isolation and purification, electrofusion or chemical fusion, culture of hybrid cell, and characterization of hybrids (Tomar & Dantu, 2010).

There are several strategies for the isolation of protoplasts from various microorganisms, including both prokaryotes and eukaryotes. Isolated protoplasts may be utilized for a broad range of experimental procedures, such as the study of membrane dynamics and chemical as well as physical properties of cell (Aoyagi, 2011; Davey et al., 2005; Yang et al., 2015; Yu et al., 2014). By isolating organelles from the protoplast, the metabolite pathway between different intracellular compartments may be investigated. Cells used for the isolation of protoplast may be obtained from several sources, of which plant tissue, suspension culture, and callus are the most commonly used.

Protoplast fusion will continue to be a current area of research in modern biotechnology. This technology will become one of the most regularly used research tools for tissue culture, molecular biologists, and biotechnologists in the future. This chapter includes the methods of isolation, the culture of protoplasts, and their application in modern biotechnology.

4.2 SOURCES OF PROTOPLASTS

Protoplasts can be isolated from various tissues and organs, such as leaves, shoot, roots, fruits, apices, cotyledons, flower petals, xylem embryos, and microspores. Among these, the mesophyll tissues of fully expanded leaves from young plants or new shoots are the most convenient and popular

DOI: 10.1201/9781003326939-5

source. The isolation of protoplast is affected by a variety of factors, like temperature, nutrition, and the age of plants. All these parameters vary from species to species.

The isolation of protoplasts from leaves usually involves the following steps:

1. Leaf sterilization
2. Removal of the epidermal cell layer
3. Treatment with enzymes
4. Isolation by filtration and centrifugation techniques

The precise procedure for isolating protoplasts can vary depending on the genotype, but there are some standard features. The general protocol for isolating protoplasts from leaves is mentioned here. Before the isolation of protoplast, the source material should be adequately surface-sterilized. Leaves are typically sterilized by immersing them in 70% ethanol for 1–5 minutes. Following this, they are surface-sterilized in a 1–2% solution of sodium hypochlorite for 10–30 minutes.

The cell walls of plant leaves are usually removed through enzymatic digestion. The cellulase enzyme allows the digestion of the cellulose in plant cell walls, while pectinase breaks down the pectin that holds cells together. To prevent protoplasts from rupturing or shrinking during enzymatic digestion, optimal parameters for the osmotic potential of the digestion medium must be maintained. This is accomplished by adding enzymes to the vitamin and salt solution and adding osmotic potential using sorbitol, mannitol, or a mixture of the two. It is also necessary to adjust the pH of the digestion medium to optimize cell growth and enzyme activity.

Once the cell wall of plant leaves has been removed, the resulting protoplasts have a distinctive spherical appearance. Following liberation from the enzyme solutions, protoplasts undergo purification to eliminate contaminating materials, such as cell wall debris (Eriksson, 1985). Besides leaves, young callus culture and cell suspension cultures also represent valuable sources for obtaining large quantities of protoplasts.

4.3 METHODS FOR THE ISOLATION AND CULTURE OF PROTOPLASTS

Protoplast isolation may be accomplished by two different methods: mechanical method or enzymatic method. Of these two methods, an enzymatic method is the safer method because it provides higher protoplast yield with less tissue damage and, at the same time, mechanical isolation, often leading to lower yields and maximum tissue damage. Both methods are described in the next sections.

4.3.1 Mechanical Method

The first mechanical method for the isolation of protoplast was performed by Klercker. In this method, the cells or tissues are placed in a suitable plasmolytic medium. Once plasmolysis is complete, the tissue is finely cut or chopped into pieces. The plasmolyzed are released into the medium from the cut surface. The suspension is then allowed to undergo deplasmolysis. The protoplasts swell and come out of the cell wall. In this process, the yield and viability of protoplasts are meager.

4.3.2 Enzymatic Method

The development of enzymatic approaches to protoplast isolation has largely overcome the limitations of mechanical methods. Plant cell walls comprise intricate combinations of celluloses, hemicelluloses, pectins, proteins, and various polymers. Commercially available enzymes for protoplast isolation are typically blends of cellulases, hemicellulases, and pectinases. The enzymes can be used either sequentially in a two-step method or in a single step using a mixed enzymatic method.

TABLE 4.1
List of Some Widely Used Commercial Enzymes for Protoplast Isolation

S. No.	Enzyme	Source Organism
1	Macerase	*Rhizopus arrhizus*
2	Cellulose onozuka	*Trichoderma viride*
3	Zymolyase	*Arthrobacter luteus*
4	Helicase	*Helix pomatia*
5	Hemicellulase	*Aspergillus niger*
6	Cellulysin	*Trichoderma viride*
7	PATE	*Bacillus polymyxa*
8	Pectolyase	*Aspergillus japonica*
9	Macerozyme R-10	*Rhizopus arrhizus*
10	Rhozyme HP150	*Aspergillus niger*
11	Drisclease	*Irpexlacteus*

4.3.2.1 Sequential Enzymatic Method

Takebe and other scientists used two commercially available enzymes to employ a sequential enzymatic approach for the first time in 1968 (Nagata & Takebe, 1971). The tissue was first incubated with pectinase. This breaks up the cell aggregates into individual cells, and these free cells are later treated with cellulose to release protoplasts.

4.3.2.2 Mixed Enzymatic Method

This method was used by Power and Cocking for the isolation of protoplasts. This is a one-step procedure that gives high yields, around 2.5×10^6 protoplasts/gram of leaf tissue (Power & Cocking, 1969). In the mixed enzymatic method, plant tissues are plasmolyzed in a mixture of both the enzymes pectinase and cellulase. It reduces time and the chances of contamination.

4.3.3 Purification of Protoplasts

For successful cultivation and fusion, the protoplasts must be isolated from debris (Tomar & Dantu, 2010). So, the protoplasts suspended in the osmoticum are separated by centrifugation using a sucrose (20%) solution. The viable protoplasts float on the top surface of the sucrose solution, forming a band. These protoplasts are then collected and re-suspended in an osmoregulatory solution.

Finally, after washing, the yield of protoplast is estimated with the help of a hemocytometer. The competence and viability of protoplast are determined by fluorescin diacetate staining, phenosafranine, or calcofluore white.

4.4 FUSION TECHNIQUE USED IN PLANT PROTOPLASTS

According to the fluid mosaic model, the plant membrane is dynamic and asymmetric with integrated protein into the plasma membrane. Protoplast fusion requires direct contact between two cells. Several chemicals and physical agents are known as fusogen, or fusion agents, which increase fusion between protoplasts, for example, sodium nitrate, PEG, glycol, and dextran. The concentration of calcium ions and the pH of the mixture have a profound effect on fusion.

Protoplast fusion is carried out in three sequential steps: (1) placing the protoplasts in close contact, (2) disrupting the cell wall and fusing the protoplasts, and (3) forming bridges between protoplasts to allow cytoplasmic continuity between the cells (Navrátilová, 2004).

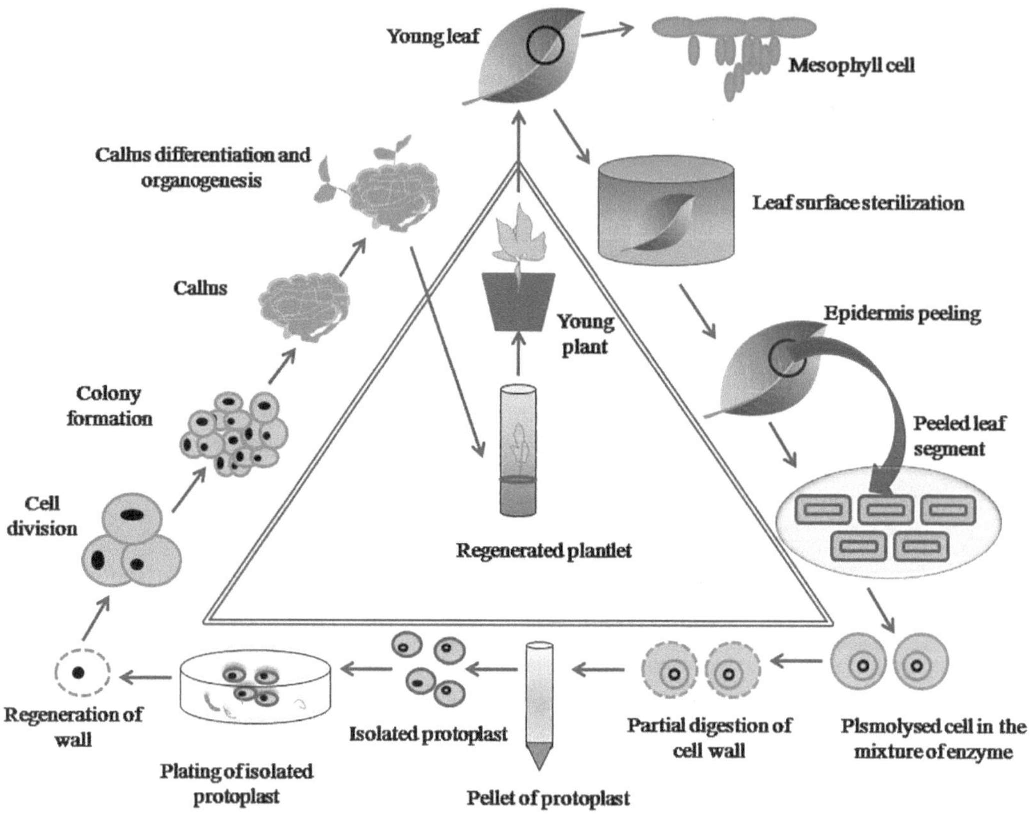

FIGURE 4.1 An overview of protoplast isolation.

Isolated protoplasts usually do not appear to fuse due to the negative charge present on their surface outside the plasma membrane, which ranges between −10 mV and −30 mV, leading to a strong tendency for repulsion. Therefore, fusion requires the presence of a chemical inducer agent that reduces the electronegativity of isolated protoplasts and allows them to interact closely (Verma et al., 2000b). The two primary techniques for protoplast fusion are chemofusion and electrofusion. In addition, the microinjection technique for nuclear fusion can also be explored. This technique consists of injecting foreign material into the cell by inserting a glass capillary in the intracellular environment (Zhou et al., 2017).

4.4.1 Chemofusion

There are several fusogenic agents that have been used to achieve the fusion of protoplasts, which can act as divalent cations and osmotic regulators (Muralidhar & Panda, 2000; Navrátilová, 2004). Polyethylene glycol (PEG) provides the awaited tool because of its brilliant stability and binding properties. For PEG, the most widely utilized divalent ions are calcium ions (Ca^{2+}). Calcium ions play an essential role in protoplast fusion. The addition of high pH, alone or in combination with other chemicals, makes fusion more effective. The fundamental role of PEG in protoplast fusion is still to be clearly defined. Additionally, DMSO combination with PEG has a stimulatory impact on protoplast fusion. It has been proposed that the main cause of protoplastic fusion by PEG is the formation of a bridge, which is mediated by Ca^{2+}.

4.4.2 Electrofusion

Electrofusion is a biophysical mechanism involving the generation of AC (alternative current) and DC (direct current), resulting in the disruption of cell membranes. This disruption occurs when a membrane is polarized in an electric field at voltages typically ranging between 0.5 and 1.5 V over an extremely short time interval, spanning from nanoseconds to microseconds (Zimmermann & Scheurich, 1981). The exposure of the protoplast to the electric field enhances the permeability of membranes.

Electrofusion involves two steps.

During the initial phase, the protoplasts are subjected to an alternating electric field typically ranging from 0.5 to 1.5 MHz, resulting in the generation of dipoles through a condition known as dielectrophoresis. Because of dielectrophoresis, the protoplast moves in the direction of increasing field strength. The mobilization of charge takes place within protoplasts. This process ultimately facilitates the fusion between the nearest protoplasts.

In the second step, a short pulse (10–100 μs) of direct current is applied that causes the reversible breakdown of the membrane and the formation of pores. This process allows for the fusion of contacting membranes, paving the way for the formation of hybrid cells.

To increase fusion frequency, several cycles of AC and DC current are typically given. In achieving protoplast fusion, the length and nature of the DC pulse are essential. Long pulse duration and high voltage can harm the protoplast and may cause cell lysis. The electrofusion technique offers easy control, with fusion frequencies of up to 100%. It provides high reproducibility and less-cytotoxic culture environment.

4.5 PROTOPLAST CULTURE

Just after isolation, protoplasts may be cultured in liquid or on semisolid media. Protoplastic cultivation on solid media offers the physical isolation required to grow genetically identical colonies.

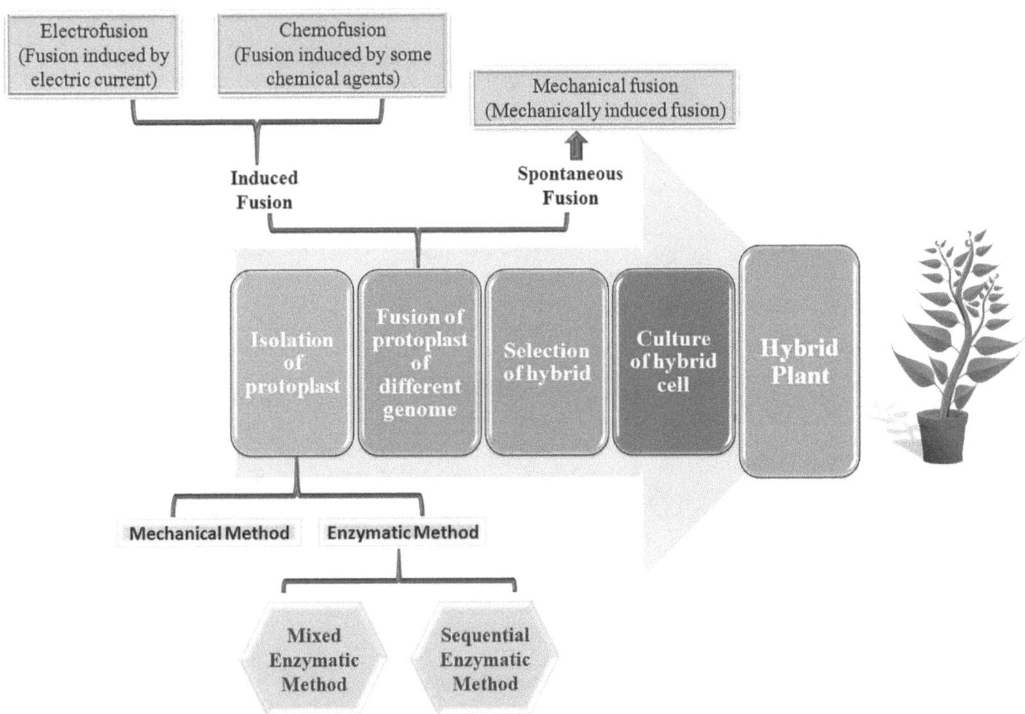

FIGURE 4.2 A schematic representation of hybrid plant production via the protoplast isolation method.

However, the liquid medium has also been used for the protoplasts of some species. For protoplast cultivation, low light levels are preferred, possibly because chloroplasts are bleached by high light. The osmotic protection of isolated protoplasts in the culture medium should be adjusted until they regenerate a strong wall. The addition of auxin and cytokinin in the Murashige and Skoog medium helps the growth of protoplasts. The protoplast culture medium requires the same components as the callus or suspension culture. Synthesis of the cell wall begins within hours, and some regenerated cells may start to divide within 1 to 2 weeks (David et al., 1984). Sequential cell division contributes to the development of calluses. A clear connection exists between cell wall formation and the division of cells. Protoplasts that do not regenerate a proper cell wall fail to undergo mitosis (Bala Bawa, 1971; Kohlenbach & Bohnke, 1975; Meyer & Abel, 1975). Cell divisions are asynchronous in protoplast cultures.

Some modifications, like higher osmotic potential, alteration of nitrogen nutrition, growth hormones auxin and cytokinin, and calcium concentration, are required for continued cell division and growth. When subcultured on a medium lacking auxin and mannitol, the callus can undergo organogenesis to finally produce the whole plant. Nagata and Takebe (1971) achieved the first success of plant regeneration from *Nicotiana tabacum* protoplast cultures.

4.6 BIOTECHNOLOGICAL APPLICATIONS OF PROTOPLAST FUSION

4.6.1 Useful Metabolite Production

Several metabolites are present in the cell walls of plants that can be used as anticancer agents, functional proteins, enzymes, and antiviral proteins (Bonness et al., 1994; Benjamin et al., 2019; Millar

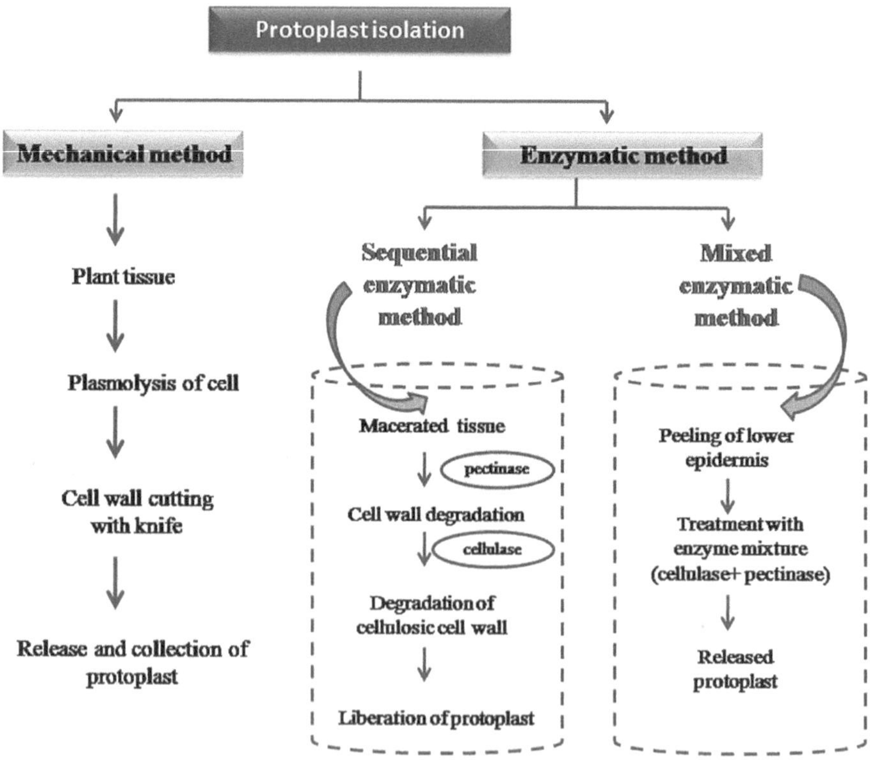

FIGURE 4.3 Mechanical and enzymatic isolation of the protoplast.

et al., 2009). But the problem is that these metabolites typically have low accumulation. The protoplast fusion technique can be used for the production of metabolites in high amounts (Aoyagi, 2011).

The advantage of using protoplasts to produce essential metabolites is that the product is easily released into the culture medium, with the advantage of increasing overall productivity and facilitating the process of downstream in situations where the cell wall's useful products limit secretions. The major challenge is that protoplasts are very fragile and difficult to produce over a long period of time, and the cell walls of active protoplasts are easily dispersed.

4.6.2 Improvement of the Nutrition Content of Food

Chlorella vulgaris is one example of microalgae that denotes the major natural source of nutrition. Microalgae are considered to be one of the organisms that generate various bioactive chemical compounds, mainly vitamins, pigments, proteins, minerals, lipids, and polysaccharides.

The protoplast fusion technique has been found to be an effective method for enhancing the production and diversification of its nutrition. Lee and Tan (1988) investigated the genetically improved algae strain by somatic fusion and hybridization. Algae–algae protoplast fusion has been proven to be a useful technique for the improvement of their nutrition. The protoplast fusion research was carried out by Kusumaningrum & Zainuri (2017) on interspecific microalgae of *C. vulgaris* to improve their carotenoid. The fusant was subjected to nutrition content analysis by GCMS on *C. vulgaris* powder obtained from 100 L liquid cultivation of the fusant. Nutritional analysis of the fusant reveals the presence of 17 amino acids, with a high concentration of glutamic, leucine, and aspartic acid, respectively. The study outcome showed the potential acquisition of nutrition improvement through the application of protoplast fusion on microalgae.

4.6.3 Nitrogen Fixation

Nitrogen is very essential for the growth and production of plants. Most of the nitrogen plants use is present in the atmosphere. However, certain *nitrogen-fixing bacteria* can transform atmospheric *nitrogen* into fixed *nitrogen*. Scientific research has been performed over the last few decades on bacteria that have nitrogen-fixing capacity, most of which are made of the following genera: *Sinorhizobium*, *Mesorhizobium*, *Bradyrhizobium*, *Rhizobium*, and *Azorhizobium* (Hakoyama et al., 2012; MacLean et al., 2007). Many leguminous plants can produce nodules that fix nitrogen through contact with symbiotic bacteria. *Frankia* from actinomycete forms nitrogen-fixing root nodules on non-legume plants. Still, the genetic basis of the symbiotic relationships between *Frankia* strains and host plants is poorly understood (Normand et al., 2007). The successful use of protoplast fusion to generate novel actinomycetes capable of fixing atmospheric nitrogen was reported by Prakash and Cummings in 1988 (Prakash & Cummings, 1988). Protoplasts from *Streptomyces griseofuscus*, a fast-growing actinomycete, and *Frankia*, a slow-growing actinomycete, were both allowed to fuse and regenerate on non-nitrogen-supplied media. The regenerated colonies were able to obtain the rapid growth features of streptomyces and the ability of *Frankia* to expand on nitrogen-deficient media. Sabir and El-Bestawy (2009) reported that protoplast fusion has a powerful role in enhancing the nodulation of Rhizobial species. The nodulation capacity of bacteria was observed on two weak strains (RtI1 and RtI2) and one successful strain (RtA1) which were selected for protoplast fusion, and the number of nodules produced by the intraspecific strains of protoplast fusion was observed. When compared with that of their parent strains, the protoplast fusion of indigenous *R. leguminosarum biovar trifolii* strains resulted in nodulation rise of 1.93- to 5.67-fold. This is an outstanding outcome for the formation of nitrogen-fixing root nodules on leguminous crops.

4.6.4 Production of Herbicide and Disease-Resistant Plants

Protoplast fusion can be used to grow herbicide-resistant plants. Many herbicides work by inactivating certain plant enzymes (target proteins) that are very important for plant-specific functions, such as photosynthetic or other biosynthetic pathways (Botterman & Leemans, 1988). For the production of herbicide-resistant rice plants, Rathore et al. (1993) (Rathore et al., 1993) reported the use of bar gene in combination with the herbicide Basta.

Resistance to certain diseases, such as potato leaf roll virus (PLRV), potato virus X (PVX), and potato virus Y (PVY), has been introduced into *Solanum tuberosum* from *Solanum brevidens* and *Solanum phurejavia* protoplast fusion techniques (Nouri-Ellouz et al., 2006; Rokka et al., 1994). Similarly, protoplast fusion was also used in the development of resistance against *Phoma lingam* disease in *Brassica* species (Sjödin & Glimelius, 1989) and tristeza virus (CTV) in citrus (Mendes et al., 2001).

Protoplast fusion is also useful in the transfer of disease-resistant genes from different plants. In breeding, disease resistance may come from either more distantly related species or closely related species. The application of protoplast fusion to overcome sexual incompatibility between cultivated potatoes and diploid solanum was reported by Chen et al. (2008). They developed a protocol for the isolation of a large number of protoplasts from different Mexican wild species of late blight. Using this protocol, successful somatic hybrids were produced from the cell fusion of *Solanum tuberosum* and the diploid species, which show a higher level of resistance to late blight than was found in *S. tuberosum*. Rootstocks produced by cybridizations and somatic hybridizations in citrus have been found to be resistant against abiotic and biotic factors, thereby increasing fruit yield and quality (Dambier et al., 2011). Xiao et al. (2004) also reported the production of the first resistant raphanobrassica asymmetric and symmetric hybrids.

4.6.5 Cytoplasmic Male Sterility

Cytoplasmic male sterility is an important mechanism in a large number of plant breeding programs. The location of genes causing cytoplasmic male sterility is not known. However, mutations or modifications in the mitochondria or the chloroplast have been involved in cytoplasmically inherited phenotype alteration (Sager, 1972). In this regard, plant protoplasts have been used **as** a model system to test this hypothesis (Vasil & Vasil 1979). Chloroplasts or mitochondria from cytoplasmic male sterile plants can be introduced into the normal male fertile protoplasts.

Regenerated tissues or cells were then investigated for structural or biochemical changes. Plants produced from this technique could be used to determine the role of mitochondrial or chloroplast DNA in cytoplasmic male sterility. Vasil and Vasil (1974) have regenerated the first plants from isolated protoplast from normal and cytoplasmic male sterile *Petunia hybrid.*

4.6.6 Somaclonal Variation

The somatic hybridization method provides valid approaches for the modification of targeted genomes by combining interspecific and intergeneric traits. In the case of protoplasts, these procedures are known as protoclonal variation. Mizuhiro et al. (2001) reported that, in comparison with in vitro plants, protoplast-derived plants of *Primula malacoides* showed variation in morphological characteristics. These changes included significantly reduced petioles and alterations in flower morphology, pedicle and peduncle lengths, and the number of inflorescences.

Generally, plants derived from protoplast develop compact inflorescences with multiple small flowers. Likewise, Jelenic et al. (2001) reported that potato plants regenerated from protoplast-derived callus have morphological variation in leaf morphology and tuber characteristics. The main advantage of protoclonal variation is that it requires no prior knowledge of the genetics of specific traits, and it does not require recombinant DNA techniques, mutagenic agents, specialized apparatus, or containment measures.

4.6.7 Other Applications

Genomic variation has significant importance in agriculture to improve the quality and yield of plants. The most common goal of protoplast fusion is the development of salt tolerance and disease resistance in plants. Applying protoplast fusion on somatic cells can enable genetic analysis, which is impossible to achieve through conventional sexual reproduction methods. Protoplast fusion also allows the investigation of combined genomes. This technique involves the transfer of agronomically beneficial traits without the need for detailed molecular knowledge. Isolated protoplasts are highly appropriate for cytoplasmic and genetic alterations. According to reports, plant protoplasts have the ability to uptake isolated nuclei, DNA, chromosomes, chloroplasts, nitrogen-fixing bacteria, cyanobacteria, and virus particles.

The technique of protoplast fusion has a significant potential in the improvement of strain and genetic analysis. It is particularly helpful for microorganisms that are useful in industry (Muralidhar & Panda, 2000). These methods have been commonly employed to increase ethanol yield (Knowles et al., 1987), improve strains for alcohol fermentation (Lima et al., 1995), and enhance citric acid production in strains of *Aspergillus niger*.

In addition, this approach is a useful tool in the genetic modification of industrially important fungi. Ogawa et al. (1989) described an increased production of cellulase in *Trichoderma reesei* with interspecific protoplast fusion, while Prabavathy et al. (2006) reported enhanced chitinase production and biocontrol activity in *Trichoderma harzianum* via self-fusion of protoplasts.

In biotechnology, the protoplast is typically used to evaluate the efficiency of target site mutagenesis. Protoplast regeneration is a big challenge when CRISPR mutagenesis is used because it requires high-regeneration-efficiency protocols. Currently, the CRISPR/Cas9 technology is a beneficial method in research and plant breeding. Lin et al. (2018) described the protocol for analyzing DNA from individual mutagenized protoplasts and also demonstrated that CRISPR-based mutagenesis of protoplasts is a beneficial tool for polyploid crops.

At present, the protoplast-based transient expression system (PTES) has emerged as a great experimental tool in identifying gene function, regulation, protein localization, and protein–protein interactions. Protoplast-based transformation systems have been reported in various plant species, including rice (He et al., 2016), wheat (Jia et al., 2016), perennial ryegrass (Yu et al., 2017), pineapple (Priyadarshani et al., 2018), soybean (Xiong et al., 2019), and ma bamboo (Ye et al., 2020).

Plant protoplasts have also generated interest in various fields of plant biotechnology, including genetic manipulation, transient gene expression in protoplasts, functional characterization of plant genes, and genome editing (Davey et al., 2005; Lin et al., 2018; Subburaj et al., 2016; Xu et al., 2020). Protoplast application in biotechnology demonstrates the rapid and remarkable progress in isolation, culture, and fusion techniques of plant protoplasts. To understand the great potential of this technique, there is an urgent need for comprehensive research on protoplasts.

4.7 CONCLUSION

Over the last few years, plentiful research has been made on protoplast culture and isolation. The progression is particularly noteworthy from isolated protoplasts with regard to the number of species for which plant regeneration has been achieved. This has been possible because of the collection of the correct source tissue and plant genotype. At present, the protoplast fusion technique contributes to exploring most aspects of plant cell physiology, including proteomic and genomic research. Protoplast fusion technology allows the insertion of new genes into plants without modification of the plant genome, which is the fear of the common man. Therefore, protoplast fusion provides an alternative to crop improvement. Although significant progress has been made in this field, several important challenges still remain. It is clear that in the future, protoplast research will observe some degree of growth and development. Such advances are not likely to be limited to higher plant protoplasts but will also include fungal and algal protoplasts and other products that are useful to humans.

REFERENCES

Aoyagi, H. (2011). Application of plant protoplasts for the production of useful metabolites. *Biochemical Engineering Journal*, *56*(1–2), 1–8. https://doi.org/10.1016/j.bej.2010.05.004

Bala Bawa, S., & Torrey, J. G. (1971). Budding and nuclear division in cultured protoplasts of corn, convolvulus and onion. *Botanical Gazette*, *132*, 240–245.

Benjamin, E. D., Ishaku, G. A., Peingurta, F. A., & Afolabi, A. S. (2019). Callus culture for the production of therapeutic compounds. *American Journal of Plant Biology*, *4*(4), 76–84.

Bonness, M. S., Ready, M. P., Irvin, J. D., & Mabry, T. J. (1994). Pokeweed antiviral protein inactivates pokeweed ribosomes: Implications for the antiviral mechanism. *The Plant Journal: For Cell and Molecular Biology*, *5*(2), 173–183. https://doi.org/10.1046/j.1365-313x.1994.05020173.x

Botterman, J., & Leemans, J. (1988). Engineering herbicide resistance in plants. *Trends in Genetics: TIG*, *4*(8), 219–222. https://doi.org/10.1016/s0168-9525(88)80004-0

Chen, Q., Li, H. Y., Shi, Y. Z. B. D., & Bizimungu, B. G. M. (2008). Development of an effective protoplast fusion system for production of new potatoes with disease and insect resistance using Mexican wild potato species as gene pools. *Canadian Journal of Plant Science*, *88*, 611–619.

Cocking, E. C. (1960). A method for the isolation of plant protoplasts and vacuoles. *Nature*, *187*(4741), 962–963. https://doi.org/10.1038/187962a0

Dambier, D., Benyahia, H., Pensabene-Bellavia, G., Aka Kaçar, Y., Froelicher, Y., Belfalah, Z., Lhou, B., Handaji, N., Printz, B., Morillon, R., Yesiloglu, T., Navarro, L., & Ollitrault, P. (2011). Somatic hybridization for citrus rootstock breeding: An effective tool to solve some important issues of the Mediterranean citrus industry. *Plant Cell Reports*, *30*(5), 883–900. https://doi.org/10.1007/s00299-010-1000-z

Davey, M. R., Anthony, P., Power, J. B., & Lowe, K. C. (2005). Plant protoplasts: Status and biotechnological perspectives. *Biotechnology Advances*, *23*(2), 131–171. https://doi.org/10.1016/j.biotechadv.2004.09.008

David, H., Jarlet, E., & David, A. (1984). Effects of nitrogen source, calcium concentration and osmotic stress on protoplasts and protoplast-derived cell cultures of Pinus pinaster cotyledons. *Physiology Plant*, *61*, 477–481.

Eriksson, T. (1985). Plant transformation and protoplasts as recipients of foreign genes. *Hereditas Siippl*, *3*, 93–95.

Hakoyama, T., Niimi, K., Yamamoto, T., Isobe, S., Sato, S., Nakamura, Y., Tabata, S., Kumagai, H., Umehara, Y., Brossuleit, K., Petersen, T. R., Sandal, N., Stougaard, J., Udvardi, M. K., Tamaoki, M., Kawaguchi, M., Kouchi, H., & Suganuma, N. (2012). The integral membrane protein SEN1 is required for symbiotic nitrogen fixation in Lotus japonicus nodules. *Plant & Cell Physiology*, *53*(1), 225–236. https://doi.org/10.1093/pcp/pcr167

He, F., Chen, S., Ning, Y., & Wang, G.-L. (2016). Rice (Oryza sativa) protoplast isolation and its application for transient expression analysis. *Current Protocols in Plant Biology*, *1*(2), 373–383. https://doi.org/10.1002/cppb.20026

Kusumaningrum, H. P., & Zainuri, M. (2017). Improvement of nutrition production by protoplast fusion techniques in chlorella vulgaris. *Journal of Food Processing & Technology*, *9*(1), 711. https://doi.org/10.4172/2157-7110.1000711

Jelenic, S., Berljak, J., & Papes, D. J. S. (2001). Mixoploidy and chimeric structures in somaclones of potato (Solanum tuberosum L) cv. Bintje. *Food Technol Biotechnol*, *39*.

Jia, X. Y., Zhang, X. H., & Qu, J. M. H. R. (2016). Optimization conditions of wheat mesophyll protoplast isolation. *Agricultural Science*, *7*, 850–858.

Kaladharan, P. (1998). Protoplast—a powerful tool in ge-netic manipulation of commercial seaweeds. *Proc. First Natl. Semi. Mar. Biotech*, 83–88.

Knowles, J., Lehtovaara, P., Penttila, M., Salovuori, I., & Andre, L. (1987). The application of recombinant-DNA technology to cellulases and lignocellulosic wastes. *Philosophical Transactions of the Royal Society Lond A*, *321*, 449–454.

Kohlenbach, H. W., Bohnke, E. (1975). Isolation and culture of mesophyll protoplasts of Hyoscyamus Niger L. *Var Annuus Sims. Experientia*, *31*, 1281–1283.

Lee, Y.-K., & Tan, H.-M. (1988). Interphylum protoplast fusion and genetic recombination of the algae Porphyridium cruentum and Dunaliella spp. *Microbiology*, *134*(3), 635–641.

Lima, N., Moreira, C., Teixeira, J. A., & Mota, M. (1995). Introduction of flocculation into industrial yeast, Saccharomyces cerevisiae saké, by protoplast fusion. *Microbios*, *81*(328), 187–197.

Lin, H.-Y., Chen, J.-C., & Fang, S.-C. (2018). A protoplast transient expression system to enable molecular, cellular, and functional studies in phalaenopsis orchids. *Frontiers in Plant Science*, *9*, 843. https://doi.org/10.3389/fpls.2018.00843

MacLean, A. M., Finan, T. M., & Sadowsky, M. J. (2007). Genomes of the symbiotic nitrogen-fixing bacteria of legumes. *Plant Physiology*, *144*(2), 615–622. https://doi.org/10.1104/pp.107.101634

Mazarei, M., Al-Ahmad, H., Rudis, M. R., & Stewart, C. N. (2008). Protoplast isolation and transient gene expression in switchgrass, Panicum virgatum L. *Biotechnology Journal*, *3*(3), 354–359. https://doi.org/10.1002/biot.200700189

Mendes, B. M. J., Mourão Filho, F. A., Farias, P. M., & Benedito, V. A. (2001). Citrus somatic hybridization with potential for improved blight and CTV resistance. *In Vitro Cellular & Development Biology Plant*, *37*, 490–495.

Meyer, Y., & Abel, W. O. (1975). Budding and cleavage division of tobacco mesophyll protoplasts in relation to pseudo-wall and wall formation. *Planta*, *125*(1), 1–13. https://doi.org/10.1007/BF00388868

Millar, D. J., Whitelegge, J. P., Bindschedler, L. V., Rayon, C., Boudet, A.-M., Rossignol, M., Borderies, G., & Bolwell, G. P. (2009). The cell wall and secretory proteome of a tobacco cell line synthesising secondary wall. *Proteomics*, *9*(9), 2355–2372. https://doi.org/10.1002/pmic.200800721

Mizuhiro, M., Kenichi, Y., Ito, K., Kadowaki, S., Ohashi, H., & Mii, M. (2001). Plant regeneration from cell suspension-derived protoplasts of Primula malacoides and Primula obconica. *Plant Science: An International Journal of Experimental Plant Biology*, *160*(6), 1221–1228. https://doi.org/10.1016/s0168-9452(01)00390-9

Muralidhar, R. V., & Panda, T. (2000). Fungal protoplast fusion: A revisit. *Journal of Bioprocess and Biosys-Tems Engineering*, *22*, 429–431.

Nagata, T., & Takebe, I. (1971). Plating of isolated tobacco mesophyll protoplasts on agar medium. *Planta*, *99*(1), 12–20. https://doi.org/10.1007/BF00392116

Navrátilová, B. (2004). Protoplast cultures and protoplast fusion focused on Brassicaceae—a review. *Horticultural Science (Prague)*, *4*, 140–157.

Normand, P., Lapierre, P., Tisa, L. S., Gogarten, J. P., Alloisio, N., Bagnarol, E., Bassi, C. A., Berry, A. M., Bickhart, D. M., Choisne, N., Couloux, A., Cournoyer, B., Cruveiller, S., Daubin, V., Demange, N., Francino, M. P., Goltsman, E., Huang, Y., Kopp, O. R., . . . Benson, D. R. (2007). Genome characteristics of facultatively symbiotic Frankia sp. strains reflect host range and host plant biogeography. *Genome Research*, *17*(1), 7–15. https://doi.org/10.1101/gr.5798407

Nouri-Ellouz, O., Gargouri-Bouzid, R., Sihachakr, D., Triki, M. A., Ducreux, G., Drira, N., & Lakhoua, L. (2006). Production of potato intraspecific somatic hybrids with improved tolerance to PVY and Pythium aphanidermatum. *Journal of Plant Physiology*, *163*(12), 1321–1332. https://doi.org/10.1016/j.jplph.2006.06.009

Ogawa, K., Ohara, H., Koide, T., & Toyama, N. (1989). Interspecific hybridization of T. reesei by protoplast fusion. *Journal of Fermentation and Bioengineering*, *67*, 207–209.

Power, J. B., & Cocking, E. C. (1969). A simple method for the isolation of very large numbers of leaf protoplasts by using mixtures of cellulase and pectinase. *The Biochemical Journal*, *111*(5), 33P. https://doi.org/10.1042/bj1110033pa

Prabavathy, V. R., Mathivanan, N., Sagadevan, E., Murugesan, K., & Lalithakumari, D. (2006). Self-fusion of protoplasts enhances chitinase production and biocontrol activity in Trichoderma harzianum. *Bioresource Technology*, *97*(18), 2330–2334. https://doi.org/10.1016/j.biortech.2005.10.031

Prakash, R. K., & Cummings, B. (1988). Creation of novel nitrogen-fixing actinomycetes by protoplast fusion of Frankia with streptomyces. *Plant Molecular Biology*, *10*(3), 281–289. https://doi.org/10.1007/BF00027405

Priyadarshani, S. V. G. N., Hu, B., Li, W., Ali, H., Jia, H., Zhao, L., Ojolo, S. P., Azam, S. M., Xiong, J., Yan, M., Ur Rahman, Z., Wu, Q., & Qin, Y. (2018). Simple protoplast isolation system for gene expression and protein interaction studies in pineapple (Ananas comosus L.). *Plant Methods*, *14*, 95. https://doi.org/10.1186/s13007-018-0365-9

Rathore, K. S., Chowdhury, V. K., & Hodges, T. K. (1993). Use of bar as a selectable marker gene and for the production of herbicide-resistant rice plants from protoplasts. *Plant Molecular Biology*, *21*(5), 871–884. https://doi.org/10.1007/BF00027118

Rhodes, C. A., Pierce, D. A., Mettler, I. J., Mascarenhas, D., & Detmer, J. J. (1988). Genetically transformed maize plants from protoplasts. *Science (New York, N.Y.)*, *240*(4849), 204–207. https://doi.org/10.1126/science.2832947

Rokka, V. M., Xu, Y. S., Kankila, J., Kuusela, A., Pulli, S., & E. P. (1994). Identification of somatic hybrids of dihaploid Solanum tuberosum lines and S. brevidens by species specific RAPD patterns and assessment of disease resistance of the hybrids. *Euphytica*, *80*, 207–217.

Sabir, J. M. S., & El-Bestawy, E. (2009). Enhancement of nodulation by some arid climate strains of Rhizobium leguminosarum biovar trifolii using protoplast fusion. *World Journal of Microbiology and Biotechnology*, *25*, 545–552.

Sager, R. (1972). *Cytoplasmic Genes and Organelles*. Academic Press, New York.

Sjödin, C., & Glimelius, K. (1989). Transfer of resistance against Phoma lingam to Brassica napus by asymmetric somatic hybridization combined with toxin selection. *TAG: Theoretical and Applied Genetics. Theoretische Und Angewandte Genetik*, *78*(4), 513–520. https://doi.org/10.1007/BF00290835

Subburaj, S., Chung, S. J., Lee, C., Ryu, S.-M., Kim, D. H., Kim, J.-S., Bae, S., & Lee, G.-J. (2016). Site-directed mutagenesis in Petunia × hybrida protoplast system using direct delivery of purified recombinant Cas9 ribonucleoproteins. *Plant Cell Reports*, *35*(7), 1535–1544. https://doi.org/10.1007/s00299-016-1937-7

Tomar, U. K., & Dantu, P. K. (2010). Protoplast culture and somatic hybridization. *Cellular and Biochemical Science*, 876–891.

Vasil, V. V., & Vasil, I. K. (1974). Regeneration of tobacco and petunia plants from protoplasts an culture of corn protoplasts. *In Vitro*, *10*(1–2), 83–96. https://doi.org/10.1007/BF02615342

Vasil, V. V., & Vasil, I. K. (1979). Isolation and culture of cereal pro top lasts I. Callus formation from Pearl Millet (*Pennisetum americanum*) protoplasts. *Zeitschrift für Pflanzenphysiologie*, *92*(5), 379–384. https://doi.org/10.1016/s0044-328x(79)80183-x

Verma, N., Kumar, V., & Bansal, M. (2000a). Protoplast fusion technology and its biotechnological. *Industrial Biotechnology International Conference*.

Verma, N., Kumar, V., & Bansal, M. (2000b). Protoplast fusion technology and its biotechnological. *International, Industrial Biotechnology Conference*.

Xiao, Z., Wan, L., & Han, B. (2004). An interspecific somatic hybrid between Actinidia chinensis and Actinidia kolomikta and its chilling tolerance. *Plant Cell, Tissue and Organ Culture*, *79*(3), 299–306. https://doi.org/10.1007/s11240-004-4611-7

Xiong, L., Li, C., Li, H., Lyu, X., Zhao, T., Liu, J., Zuo, Z., & Liu, B. (2019). A transient expression system in soybean mesophyll protoplasts reveals the formation of cytoplasmic GmCRY1 photobody-like structures. *Science China Life Sciences*, *62*(8), 1070–1077. https://doi.org/10.1007/s11427-018-9496-5

Xu, J., Kang, B.-C., Naing, A. H., Bae, S.-J., Kim, J.-S., Kim, H., & Kim, C. K. (2020). CRISPR/Cas9-mediated editing of 1-aminocyclopropane-1-carboxylate oxidase1 enhances Petunia flower longevity. *Plant Biotechnology Journal*, *18*(1), 287–297. https://doi.org/10.1111/pbi.13197

Yang, B., Liu, J., Liu, B., Sun, P., Ma, X., Jiang, Y., Wei, D., & Chen, F. (2015). Development of a stable genetic system for Chlorella vulgaris—a promising green alga for CO2 biomitigation. *Algal Research*, *12*, 134–141. https://doi.org/10.1016/j.algal.2015.08.012

Ye, S., Chen, G., Kohnen, M. V., Wang, W., Cai, C., Ding, W., Wu, C., Gu, L., Zheng, Y., Ma, X., Lin, C., & Zhu, Q. (2020). Robust CRISPR/Cas9 mediated genome editing and its application in manipulating plant height in the first generation of hexaploid Ma bamboo (Dendrocalamus latiflorus Munro). *Plant Biotechnology Journal*, *18*(7), 1501–1503. https://doi.org/10.1111/pbi.13320

Yu, C., Wang, L., Chen, C., He, C., Hu, J., Zhu, Y., & Huang, W. (2014). Protoplast: A more efficient system to study nucleo-cytoplasmic interactions. *Biochemical and Biophysical Research Communications*, *450*(4), 1575–1580. https://doi.org/10.1016/j.bbrc.2014.07.043

Yu, G., Cheng, Q., Xie, Z., Xu, B., Huang, B., & Zhao, B. (2017). An efficient protocol for perennial ryegrass mesophyll protoplast isolation and transformation, and its application on interaction study between LpNOL and LpNYC1. *Plant Methods*, *13*, 46. https://doi.org/10.1186/s13007-017-0196-0

Zhou, X., Zhang, X., Boualavong, J., Durney, A. R., Wang, T., Kirschner, S., Wentz, M., & Mukaibo, H. (2017). Electrokinetically controlled fluid injection into unicellular microalgae. *Electrophoresis*, *38*(20), 2587–2591. https://doi.org/10.1002/elps.201600548

Zimmermann, U., & Scheurich, P. (1981). High frequency fusion of plant protoplasts by electric fields. *Planta*, *151*(1), 26–32. https://doi.org/10.1007/BF00384233

5 *In vitro* Propagation of Medicinal and Aromatic Plants

The Current Status on Concepts and Methods

*Hemavathi Brijesh, Renuka Madhu, Brijesh Ajjappala, Ramya Manjunath, and Santosh Anand**
* Corresponding author: sanand.biotech@gmail.com

5.1 INTRODUCTION

Plants with medicinal and aromatic properties are widely cultivated owing to their application in the pharmacological industry. These therapeutic properties of MAPs are attributed to the secondary metabolites present in different plants or closely linked species. Low yield, a lack of elite genotypes, and the generation of disease-free vegetative planting materials are major constraints which limit the production of MAPs (Gantait et al., 2011). As a result, the demand for species-specific MAPs in the market and the need to preserve native populations make their proliferation crucial. The development of genetically similar genotypes with economic value could be accomplished through asexual reproduction of MAPs. The rapid multiplication of identified rare and important plants with commercial value using tissue culture or micropropagation techniques would be made possible without the need for seasonal restrictions, increasing the chances of success if the species is at risk for extinction (Altpeter et al., 2016). *Micropropagation* is defined as an aseptic process of cultivating isolated plant tissues, cells, organs, or even whole plants that takes place in culture vessels with a specified culture media under controlled light and temperature conditions. The shoots or plantlets under in vitro condition can be produced directly or indirectly while maintaining low genetic diversity between the source material and the micropropagated plantlets. The following are a few advantageous features of using tissue culture techniques for the micropropagation of MAPs:

1. Rapid clonal propagation of MAPs, particularly those incapable of producing seeds
2. Increase in the clonal propagation rate within a short period of time
3. Development of pest-resistant plants and enablement of the availability of planting materials throughout the year despite seasonal variation
4. Production of plantlets with genetically and physiologically similar traits and with a heterogeneous genetic makeup (tetraploids, haploids, hybrids)
5. Preservation of rare and endangered plants
6. Generation of plants that are difficult to cultivate, with valuable secondary metabolite production

The strategies used during the in vitro propagation of MAPs encompass axillary branching and adventitious shoot generation by organogenesis, and shoot and root induction through somatic embryogenesis (Jwala & Shekhar, 2013). This chapter attempts to present a current overview of the methodologies employed in the generation of in vitro–grown MAPs and their major economically relevant metabolites.

DOI: 10.1201/9781003326939-6

5.2 MICROPROPAGATION OF MAPS *IN VITRO* BY ORGANOGENESIS AND SOMATIC EMBRYOGENESIS

Many medicinal plants are difficult to propagate through conventional propagation methods. The various obstacles are due to dormancy and the viability of the seed, long breeding cycles, the regeneration of roots from stem cuttings, etc. One such plant is *Chonemorpha fragrans*, extensively used in Indian medicinal systems. The plant is known for its various medicinal properties and is utilized in treating skeletal issues, gynecological disorders, diabetes, rheumatism, skin-related diseases, etc. (Yadav & Agarwal, 2011). The anticancer property of this plant has also been reported (Pradnya & Malpathak, 2013). The plant is shortlisted by the IUCN, or the International Union for Conservation of Nature (Kolammal, 1978), and is considered as an endangered species in Karnataka and as vulnerable in Kerala, India. The plant is conventionally propagated by means of seeds and stem cuttings. Limitations include its poor seed viability, poor germination rate, difficulty in rooting of the cuttings, and the limited survival rate. Madhu et al. (2017) worked on the micropropagation of *Chonemorpha fragrans*, where the ideal explants for direct regeneration were found to be a single node and shoot tip. Somatic embryogenesis was initiated from callus culture.

The somatic embryogenesis route of micropropagation is widely used in the clonal proliferation of medicinal and aromatic plants. This method involves the differentiation of somatic embryos from a single somatic cell either in a direct manner from the explant or from the callus initiated from the explant. In case of *Thymus hyemalis*, an important aromatic and medicinal plant grown in the Mediterranean region, MS media supplemented with a higher concentration of cytokinin with

FIGURE 5.1 (a) *Chonemorpha fragrans* plant in natural habitat. (b) *Chonemorpha fragrans* flowers. (c) Fruit of plant. (d) Plantlet.

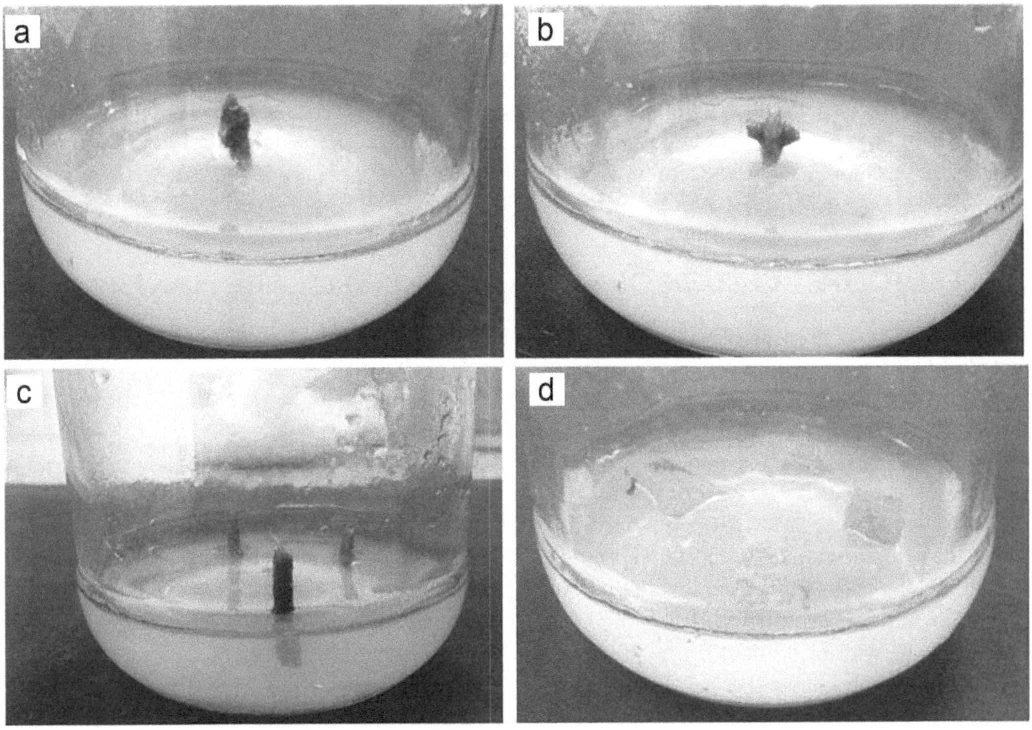

FIGURE 5.2 (a) Explants inoculated to medium. (b) Shoot tip. (c) Shoot node and stem. (d) Leaves.

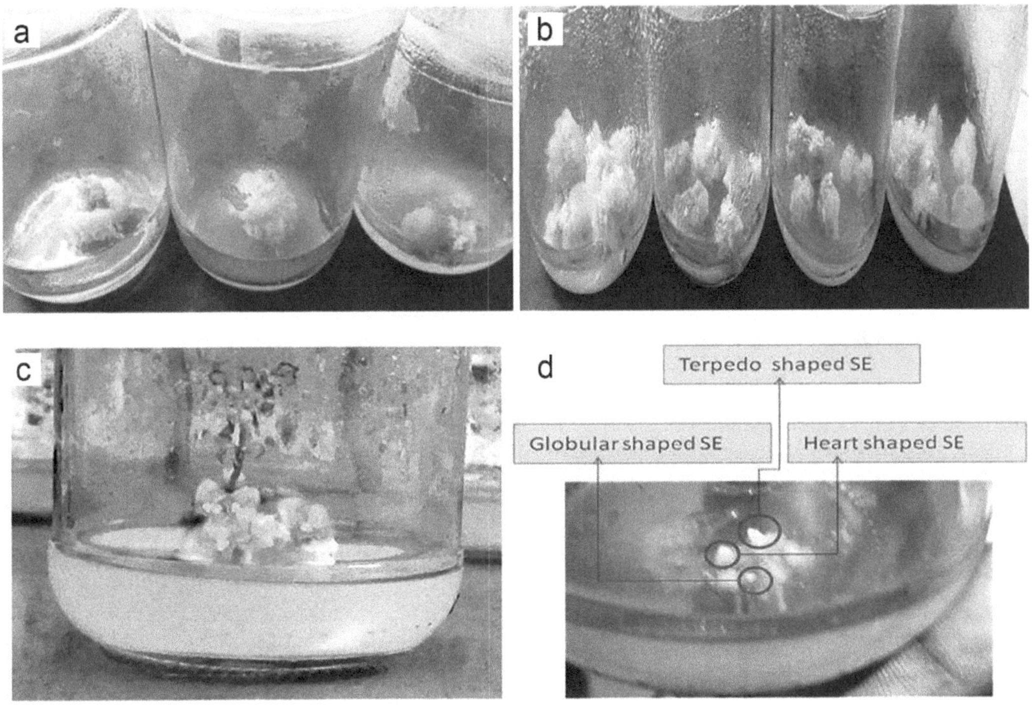

FIGURE 5.3 Indirect regeneration of plant: (a) Callus from stem explants on various treatments. (b) Callus from leaf explants in various treatments. (c) Shoot regeneration on various treatment. (d) Developmental stages in somatic embryogenesis from callus.

6-benzylaminopurine (BAP), 4.65 mu M KIN, and low concentration of auxin, 0.54 mu M NAA, gave the maximum number of somatic embryos with the output of 27 embryos/45 mg of callus induced. The ideal production of somatic embryos happened when incubated in the dark. Nodal culture gave the best results with respect to callus production and somatic embryogenesis (Nordine et al., 2014). Somatic embryogenesis as an efficient tool of micropropagation was also validated in the Gentianaceae family. These embryos were also used for *Agrobacterium*-mediated gene transformation in *G. cruciate* and *G. dahurica* Fisch. (Tomiczak et al., 2019).

5.3 PROTOPLAST CULTURE OF MEDICINAL AND AROMATIC PLANTS

Complete hybrids between targeted individuals have never been established when extremely distant species are crossed, because of difficulties caused by sexual incompatibility. This problem was solved by isolating protoplasts from tobacco plant leaf cells employing enzymes, and the regeneration of entire plants demonstrated the potential of the protoplast culture technology (Cocking, 1960; Takebe et al., 1971). Plant protoplasts are referred to as cellulose cells, where the tough cell wall is taken off without disrupting the outer cell membrane that encloses the cytoplasm and the nucleus. Cell wall–free protoplasts are obtained by dissolving the plant material with the required enzymes. The induction of the fusion of isolated protoplasts from various species is then carried out by combining the protoplasts in a liquid medium together with polyethylene glycol (fusogenic agent) or by physically administering an electric shock. The dividing cell aggregates are then transferred to solid media, either with or without hormones, to allow the regeneration of somatic hybrid plants. The sprouting green plantlets that result from this process are then planted in the soil. Protoplast culture technology can enhance the synthesis of alkaloids and other secondary metabolites in plants. Somatic hybrid plants from the fusion of the protoplasts of *Hyoscyamus muticus* and *Nicotiana tabacum* have resulted in increased growth rate (Jia et al., 1983). Since then, a number of therapeutic plants and somatic hybrids have been successfully established from isolated protoplasts (Rout et al., 2000; Aziz et al., 2006). Cell lines derived from callus induced by protoplast fusion have resulted in enhanced shikonin content in fennel (Miura & Tabata, 1986), asparagus (Elmer et al., 1989), and *Lithospermum erythrorhizon* (Kumar, 1992) plants, respectively. Successful protoplast isolation and plant regeneration under in vitro condition have been achieved in medicinal plants such as *Artemisia judaica*, *Echinops spinosissimus*, and *Rhodiola sachalinensis* (Jian-Feng et al., 2009). Ginsenoside, a secondary metabolite produced in the hybrid calli, aided the protoplast fusion between carrot and American ginseng (*Panax quinquefolius*) (Han et al., 2009). Protoplasts were successfully isolated from *Maesa lanceolata*, a medicinal plant traditionally a member of the family Maesaceae that has historically been used as a remedy for leishmania in Africa by Lambert and Geelen (2010).

5.4 PRODUCTION OF MAPS USING ANTHER CULTURE METHOD

The process of "anther culture" entails aseptically removing developing anthers from an unopened flower bud and culturing them on a nutritional medium. The microspores in the cultured anther undergo organogenesis or embryogenesis to become callus tissue or embryoids that give rise to haploid plantlets. Anther and pollen culture is the technique employed most frequently to develop haploid plants, by utilizing the totipotency property of microspores, which also carry only one set of chromosomes (n) present in microspores. The gametophytic developmental pathway of microspores is usually diverted into the sporophytic pathway under predetermined ideal and induced culture conditions in anther culture technology. The type of genotype, the culture condition, and the culture medium govern the efficacy of the anther culture method in producing haploids. The first haploid plant regeneration was achieved in purple coneflower when the callus induction medium was supplemented with the combination of BA and NAA, which resulted in the production of embryogenic calluses (Zhao et al., 2006). Maximum regeneration (100%) frequency was achieved, resulting in large amounts of plantlets with successful callus induction when anthers were cultured from *Datura innoxia* (Trigiano & Gray, 2011). Doi et al. (2013) investigated the response of the medicinal plant *Gentiana triflora* to haploid production using anther culture technique. The induction of haploid and double haploid lines has been done in many species of major medicinal and aromatic plants belonging to the family Umbelliferae (Ferrie & Caswell, 2011). Haploid plants were developed from a succulent herb, *Portulaca oleracea*, commonly known as purslane, belonging to the family Portulacaceae and the richest source of vital omega-3 and omega-6 fatty acids (Alam et al., 2014). Chen et al. (2016) produced double haploid plants by generating tetraploid plants. Recently, the anther culture technique was applied in investigating callus induction using different culture media, such as AT3 and B5-NLN, which produced the greatest number of embryo-like structures (ELSs) (Hoveida et al., 2017).

5.5 SECONDARY METABOLITE PRODUCTION USING CELL SUSPENSION CULTURES OF MAPS

A plethora of major secondary metabolites can now be produced using plant cell culture obtained from MAPs (Fischer et al., 2015). *Cell suspension culture* is the process of growing single cells or tiny clumps of cells through multiplication when suspended in a liquid medium under agitated conditions. Sterile cells and tissue cultures could be established using various explants, such as leaves, stems, roots, and meristems, for callus induction and secondary metabolite accumulation in the liquid medium supplemented with a high concentration of auxin, either alone or in combination with cytokinin. The most prolific callus lines are then chosen after being evaluated for productivity, and they are employed in establishing batch or continuous fermentation cell suspension cultures to produce the appropriate secondary metabolites (Xu et al., 2011). Secondary metabolites, such as α-tocopherol, flavonoids, serpentine, ajmaline, scopolamine, reserpine, stilbene, anthocyanins, resveratrol, and the anticancer agent paclitaxel, have been produced using suspension cultures (Efferth, 2019). The most effective method in enhancing the synthesis of secondary metabolites in cell suspension cultures is often due to the elicitation process. Several elicitors each with distinct function, including plant cell wall constituents, such as pectin and cellulose; microbial-origin elicitors, like chitin and glucan; and signaling molecules, such as salicylic acid and methyl jasmonate, are involved in the enhanced production of secondary metabolites (Sharma et al., 2011; Ochoa-Villarreal et al., 2016). The proliferation of the explants and the accumulation of secondary metabolites are both greatly influenced by the optimum nutrient concentration (Murthy et al., 2014; Fargoso et al., 2018). The possible reason for the synthesis of these bioactive compounds is the process of totipotency of the plant cells, which ensures the maintenance of intact genetic characteristics. However, the usage of cell culture techniques is restricted to a small number of applications, including the production of ginsenosides from *Panax ginseng*, shikonin from *Lithospermum erythrorhizon*, and berberine from *Coptis japonica*. The best approach in producing active metabolites is still the use of cell suspension culture, especially natural compounds, such as paclitaxel from *Taxus* spp. (Atanasov et al., 2015).

5.6 SECONDARY METABOLITE PRODUCTION FROM ORGAN CULTURES OF MAPS

Plant propagation and secondary metabolite production are both achievable using organ cultures. In producing a target metabolite that can accumulate in a particular type of plant tissue or organ, well-differentiated organ cultures are used.

5.6.1 SHOOT CULTURES

Shoot cultures developed in vitro are analogous to cell cultures and can be employed in the production of target metabolites that are of commercial value. The tissue differentiation ability retained by the parent plant in shoot cultures has rendered them capable of producing secondary metabolites that are not present in cell suspension cultures. Various approaches have been employed in producing secondary metabolites using root, embryo, and shoot cultures for different MAPs, because cell suspension culture is not always assured in bioactive compound production (Baque et al., 2012; Murthy et al., 2008a). Shoot cultures established from *Bacopa monnieri* have resulted in increased (threefold) bacoside A, as opposed to those of conventionally field-grown plants (Praveen et al., 2009). Different bioreactor systems were analyzed for the production of the artemisinin compound derived from *Artemisia annua* shoots, highlighting the effectiveness of bioreactors in the generation of organ culture (Grech-Baran & Pietrosiuk, 2012). For the first time, dimeric indole alkaloids were produced using in vitro shoot cultures derived from *C. roseus* (Pietrosiuk et al., 2007) in bioreactor conditions; however, scale-up studies were hampered due to reduced yield. Similarly, using bioreactor cultivation, in vitro shoot cultures were established to produce another alkaloid, securinine, from the medicinal plant *Securinega suffruticosa*, utilized in treating neurological illnesses (Raj et al., 2014). However, a noteworthy exception to this trend was observed when in vitro shoot cultures of *Genista tinctoria* accumulated the genistin metabolite four times higher than that of intact plant material (Łuczkiewicz & Kokotkiewicz, 2005).

5.6.2 Root Cultures

Adventitious roots are those that grow from regions of the plant other than the radicles or the root axis. They are usually formed in response to wounds and exogeneous phytohormones, especially auxins, under *in vitro* conditions. Due to their rapid development and consistency in producing secondary metabolites, they are regarded as the most promising approach in producing biomass, especially for medicinal plants (Carvalho & Curtis, 1998). Another study indicated that medicinal plant *Glycyrrhiza uralensis*–induced adventitious roots under the influence of elicitor salicylic acid produced significant amounts of secondary metabolites, such as glycyrrhizic acid (0.31 mg/g) and glycyrrhetinic acid (0.14 mg/g), along with polysaccharide (159.29 mg/g). The laboratory-scale to bioreactor-scale cultivation of secondary metabolite was achieved in various medicinal plants induced from adventitious roots (Murthy et al., 2008a). When compared to plants grown in the field or in a greenhouse, adventitious root cultures of *Morinda citrifolia* produced in bioreactors, for instance, demonstrated a severalfold increase of anthraquinone content (Baque et al., 2012). Leaf segments of *Andrographis paniculata* were used to induce adventitious roots, which accumulated andrographolide metabolites within four weeks when roots were cultured in a liquid MS medium supplemented with 2.7 μM NAA (1-naphthaleneacetic acid).

5.7 PRODUCTION OF HAIRY ROOT CULTURES FROM MAPS

The roots that are normally induced by inoculating the wounded parts of plants with *Agrobacterium rhizogenes* aseptically under in vitro conditions are known as hairy roots. This property was adopted to induce hairy roots from different MAPs, as they show potential to produce and secrete complex glycoproteins and other secondary metabolites into the medium they are cultured in. The secondary metabolites accumulated by hairy roots have similar or higher yields than those typically produced in intact parent roots. Due to their tendency to grow rapidly in culture media devoid of phytohormones and their ability to maintain the genetic stability of an intact plant, hairy roots are thought to be favorable for the production of secondary metabolites. Efficient hairy root culture was induced on wounded shoot explants of the medicinal plant *Salvia viridis* and infected with *A. rhizogenes* strain A4 culture (Grzegorczyk-Karolak et al., 2018). The bioactive compound scopolamine induced from *Atropa belladonna* showed an elevated, fivefold higher concentration in hairy roots expressing the 6-hydroxylase gene isolated from *Hyoscyamus muticus* (Hashimoto et al., 1993). Similarly, the therapeutically valuable metabolite camptothecin (0.1% DW) was produced in the hairy roots induced from *Ophiorrhiza pumila* Champ. ex Benth. plant (Saito et al., 2001). Additionally,

Withania somnifera–derived terpenoid molecule withanolide A was efficiently produced in hairy root cultures (Murthy et al., 2008b). Secondary metabolites, such as catharanthine, tabersonine, lochnericine, and horhammericine, belonging to indole alkaloids were successfully produced in hairy root cultures obtained from *C. roseus* which were transformed by *A. rhizogenes* (Li et al., 2011).

5.8 *IN VITRO* ROOTING OF MEDICINAL AND AROMATIC PLANTS

The successful establishment of micropropagated shoots involves the prolific rooting of in vitro cultured microshoots before being established in the field. Depending on the plant genotype, the in vitro induction of roots from developing shoots has been accomplished in different formulations of the media with or without auxin in various medicinal plants (Sharafi et al., 2013). Some authors claim that plantlets produced using ex vitro rooting could be superior to those produced through in vitro rooting in terms of their root systems, root numbers, ease of acclimation, and survival rate (Yan et al., 2010). The rooting process is genetically predetermined and regulated by elements in the culture environment, such as the concentration of minerals, the kind of growth regulators, and the amount of sucrose in the culture media. Because the plant growth regulators and sucrose requirements of the morphogenetic responses of in vitro rooting differ from genotype to genotype in sugarcane, genotype-specific protocols are therefore required (Sood et al., 2006). The IAA and IBA in the rooting medium specifically induce a good-rooting clone during in vitro rooting. Surprisingly, the supply of 2,3,5-triiodobenzoic acid, kinetin, and β-chloro-phenoxy-isobutyric acid reduced the percentage of in vitro rooting significantly. In vitro root elongation is predominantly influenced by the substrate and the medium (Díaz et al., 2009). Agar, being a universally used substrate in the plant tissue culture to provide semisolid support to the growing callus, has been reported to pose specific problems, like mineral availability, mineral composition, gel strength, and inhibitory compounds that affect in vitro growth and morphogenesis (Aggarwal & Nirmala, 2012). In the gelled medium, the radial expansion of cells may lead to thicker roots and may constrain the root elongation. The plantlets grown on agar, when transferred to soil, are withered and turn black. Therefore, the liquid medium becomes a savior for the ill-effects of agar substratum. The liquid medium has resulted to the highest rooting percentage and shoot length for *Boswellia serrata* Roxb. (Suthar et al., 2011). It has much more advantages over the substrate medium, having more concentration and availability of oxygen. Improvisation of the medium can be achieved by agitation or bubbling. Porous materials like foam, vermiculite, beads, or paper can be used for solid support (Cuenca et al., 2012).

5.9 ACCLIMATIZATION AND FIELD ESTABLISHMENT OF *IN VITRO* MICROPROPAGATED MEDICINAL AND AROMATIC PLANTS

Acclimatization is the next step in rooting. Since the plantlets should be hardened to get adjusted to the external environment, a crucial stage in micropropagation is acclimatization (Jwala & Shekhar, 2013). All the rooted plantlets of *Bowiea volubilis* were maintained for 4–6 weeks in MS salts with 0.5% sucrose and incubated at 24–30°C. The best survival rates during acclimatization (94%) in N52 and N53 (100%) were attained on substrates combining sand and soil in a 1:1 ratio. Therefore, it may be concluded that this procedure can be utilized to successfully root and acclimatize these genotypes *in vitro* (Tesfa et al., 2016). Transferring tissue-cultured plants to *ex vivo* conditions requires an extensive hardening process to avoid mortality (Ziv & Chen, 2008). A few studies have reported that the survival of orchid seedlings directly in the natural habitat will fail, unless those seedlings do not develop mycorrhizal associations in the environment (Teixeira da Silva et al., 2015b). Hence, acclimatization plays an important step in the mass propagation of tissue-cultured plantlets. This can be achieved by lowering transpiration with antitranspirants or providing increasing concentrations of CO_2 to enhance photosynthetic rate.

5.10 CONCLUSIONS

The usage of medicinal and aromatic plants in the food, cosmetic, and pharmaceutical industries has increased in the past several years due to customer interest in herbal-based products. Plant tissue culture offers an alternative to the conventional methods of plant production and may be able to help with the challenges associated with domesticating rare and endangered species that are difficult to multiply, especially when the initial plants are limited. For the mass production of MAPs, *in vitro* propagation can be effective, raising the possibility of their sustainable utilization and commercialization. The applications of plant tissue culture include the generation of haploid and diploid plants, the screening and selection of variant genotypes through somaclonal variation studies, and the synthesis of valuable therapeutic phytochemicals. In addition to plant cell cultures, adventitious and genetically modified hairy roots are promising sources for the production of valuable secondary metabolites derived from plants that are employed in the food and pharmaceutical industries. Although different methods have been employed to produce organ cultures from MAPs in vitro, optimization is still required to produce a successful strategy that can be applied to both basic and applied research. New possibilities for the large-scale production of therapeutics, nutraceuticals, and other health-promoting compounds have emerged as a result of the advancements in tissue culture and genetic engineering, particularly transformation technologies. In this chapter, various methods employed for the in vitro propagation of MAPs and their advantages over conventional methods of plant propagation techniques have been highlighted.

ACKNOWLEDGMENTS

The authors acknowledge the Department of Biotechnology, School of Applied Sciences, REVA University for providing the necessary facilities.

REFERENCES

Aggarwal, S., & Nirmala, C. (2012). Utilization of coir fibers as an eco-friendly substitute for costly gelling agents for in vitro orchid seed germination. *Scientia Horticulturae* (Amsterdam), 133, 89–92.

Alam, M. A., Juraimi A. S., Rafii M. Y., Hamid A. A., Zainudin M. A. M., Hasan M., & Uddin M. K. (2014). Evaluation of antioxidant compounds, antioxidant activities and mineral composition of 13 collected purslane (*Portulaca oleracea* L.) accessions. *BioMed Research International*, 10, 296063.

Altpeter, F., Springer, N. M., Bartley, L. E., Blechl, A. E., Brutnell, T. P., Citovsky, V., Conrad, L., Gelvin, S. B., Jackson, D., Kausch, A. P., Lemaux, P. G., Medford, J. I., Orozco-Cardenas, M., Tricoli, D., Van Eck, J., Voytas, D. F., Walbot, V., Wang, K., Zhang, Z. J., & Stewart, C. N. (2016). Advancing crop transformation in the era of genome editing. *Plant Cell*, 28, 1510–1520.

Atanasov, A. G., Waltenberger, B., Pferschy-Wenzig, A. M., Linder, T., Wawrosch, C., Uhrin, P., Temml, V., Wang, L., Schwaiger, S., Heiss, E. H., Rollinger, J. M., Schuster, D., Breuss, J. M., Bochkov, V., Mihovilovic, M. D., Kopp, B., Bauer, R., Dirsch, V. M., & Stuppner, H. (2015). Discovery and resupply of pharmacologically active plant-derived natural products: A review. *Biotechnology Advances*, 33, 582–1614.

Aziz, Z. A., Davey, M. R., Lowe, K. C., & Power, J. B. (2006). Isolation and culture of protoplasts from the medicinal plant *Centella asiatica*. *Revista Brasileira de Plantas Medmedicinais Botucatu*, 8, 105–109.

Baque, M. A., Moh, S. H., Lee, E. J., Zhong, J. J., & Paek, K. Y. (2012). Production of biomass and useful compounds from adventitious roots of high-value added medicinal plants in bioreactor. *Biotechnology Advances*, 30, 1255–1267.

Carvalho, E. B., & Curtis, W. R. (1998). Characterization of fluid-flow resistance in root cultures with a connective flow tubular bioreactor. *Biotechnology Bioengineering*, 60, 375–384.

Chen, X., Nilanthi, D., Yang, Y., & Wu, H. (2016). Anther culture and plant regeneration of tetraploid purple coneflower (*Echinacea purpurea* L.). *Journal of Biosciences and Medicines*, 4, 89–96.

Cocking, E. C. (1960). A method for the isolation of plant protoplasts and vacuoles. *Nature*, 187, 962–963.

Cuenca, B., Sánchez, C., Aldrey, A., & Bogo, B. (2012). Micropropagation of axillary shoots of hybrid chestnut (*Castaneasativa* × *C. crenata*) in liquid medium in a continuous immersion system. *Plant Cell Tissue Organ Culture*, 131, 307–320.

Díaz, K., Valiente, C., Martínez, M., Castillo, M., & Sanfuentes, E. (2009). Root-promoting rhizobacteria in *Eucalyptus globulus* cuttings. *World Journal of Microbiology and Biotechnology*, 25, 867–873.

Doi, H., Hosh, N., Yamada, E., Yokoi, S., Nishihara, M., Hikage, T., & Takahata, Y. (2013). Efficient haploid and doubled haploid production from unfertilized ovule culture of gentians (*Gentiana* spp.). *Breeding Science*, 63, 400–406.

Efferth, T. (2019). Biotechnology applications of plant callus cultures. *Engineering*, 5(1), 50–59.

Elmer, W. H., Ball, T., Volokita, M., Stephens, C. T., & Sink, K. C. (1989). Plant regeneration from callus derived protoplasts of Asparagus. *Journal of the American Society for Horticultural Sciences*, 114, 1019–1024.

Fargoso, M. L. E., Bertolucci, S. K. V., Lima, A. F., de Carvalho, A. A., Mohammed, A., Blank, A. F., & Pinto, J. E. B. P. (2018). Effects of plant growth regulators, different culture media and strength MS on production of volatile fraction composition in shoot cultures of Ocimum basilicum. *Industrial Crops and Products*, 116, 231–239.

Ferrie, A. M. R., & Caswell, K. L. (2011). Isolated microspore culture techniques and recent progress for haploid and doubled haploid plant production. *Plant Cell, Tissue and Organ Culture*, 104, 301–309.

Fischer, R., Vasilev, N., Twyman, R. M., & Schillberg, S. (2015). High-value products from plants: The challenges of process optimization. *Current Opinion in Biotechnology*, 32, 156–162.

Gantait, S., Mandal, N., & Nandy, S. (2011). Advances in micropropagation of selected aromatic plants: A review on vanilla and strawberry. *American Journal of Biochemistry and Molecular Biology*, 1(1), 1–19.

Grech-Baran, M., & Pietrosiuk, A. (2012). Artemisia species in vitro cultures for production of biologically active secondary metabolites. *Biotechnologia*, 4, 371–380.

Grzegorczyk-Karolak, I., Kuźma, Ł., Skała, E., & Kiss, A. (2018). Hairy root cultures of *Salvia viridis L.* for production of polyphenolic compounds. *Indian Crops Production*, 117, 235–244.

Han, L., Zhou, C., Shi, J., Zhi, D., & Xia, G (2009). Ginsenoside Rb1 in asymmetric somatic hybrid calli of *Daucuscarota* with *Panaxquinquefolius*. *Plant Cell Reports*, 28, 627–638.

Hashimoto, T., Yun, D. J., Yamada, Y. (1993). Production of tropane alkaloids in genetically engineered root cultures. *Phytochemistry*, 32, 713–718.

Hoveida, Z. S., Abdollahi, M. R., Mirzaie, A., Moosavi, S. S., & Seguí-Simarro, J. M. (2017). Production of doubled haploid plants from anther cultures of borage (*Borago officinalis* L.) by the application of chemical and physical stress. *Plant Cell, Tissue and Organ Culture*, 130, 369–378.

Jia, J., Potrykus, I., Lazal, G. B., & Saul, M. (1983). Hyoscyamus-Nicotiana fusion hybrid selected via auxotroph concentration and verified by species specific DNA hybridization. In: Protoplasts, Poster Proceeding. Biekhauser, Basel/Boston/Stuttgart, pp. 110–111.

Jian-Feng, L., Yun-Qing, C., & Zhi-Wen, C. (2009). Protoplast isolation and plant regeneration from leaves of *Rhodiolasachalinesis*. *Chinese Traditional and Herbal Drugs*, 7, 2010–2014. (Chinese)

Jwala, A., & Shekhar, K. A. (2013). Plant tissue culture: Modern techniques and practices. In: Encyclopedia of Plant Tissue Culture. vol. 1–5, 1st edn, Campus Books International, New Delhi.

Kolammal, K. (1978). Pharmacognosy of ayurvedic drugs series-1. *Trivandrum, Ayurveda Research Institute*, 1, 333–356.

Kumar, A. (1992). Somatic embryogenesis and high frequency plantlet regeneration in callus cultures of *Thevetiaperuviana*. *Plant Cell Tissue and Organ Culture*, 51, 47–50.

Lambert, E., & Geelen, D. (2010). High efficiency protoplast isolation from *in vitro* cultures and hairy roots of *Maesalanceolata*. *African Journal of Biotechnology*, 9, 7071–7078.

Li, M., Peebles, C. A., Shanks, J. V., & San, K. Y. (2011). Effect of sodium nitroprusside on growth and terpenoidindole alkaloid production in *Catharanthusroseus* hairy root cultures. *Biotechnology Progress*, 27, 625–630.

Łuczkiewicz, M., & Kokotkiewicz, A. (2005). Co-cultures of shoots and hairy roots of *Genistatinctoria* L. for synthesis and biotransformation of large amounts of phytoestrogens. *Plant Science*, 169, 862–871.

Madhu, R., Das, A. J., Lakshmikant, G., & Veena, V. (2017). Established *Chonemorphafragrans* callus exhibited biological activities as that of plant extract. *International Journal of Current Research*, 9(2), 46545–46552.

Miura, Y., & Tabata, M. (1986). Direct somatic embryogenesis from protoplasts of Foeniculumvulgare Miller. *Plant Cell Reports*, 5, 310–313.

Murthy, H. N., Dijkstra, C., Anthony, P., White, D. A., Davey, M. R., Power, J. B., Hahn, E. J., & Paek, K. Y. (2008b). Establishment of *Withaniasomnifera* hairy root cultures for the production of withanolide. *American Journal of Integrated Plant Biology*, 50, 975–981.

Murthy, H. N., Hahn, E. J., & Paek, K. Y. (2008a). Adventitious root and secondary metabolism. *Chinese Journal of Biotechnology*, 24, 711–716.

Murthy, H. N., Lee, E. J., & Paek, K. Y. (2014). Production of secondary metabolites from cell and organ cultures: Strategies and approaches for biomass improvement and metabolite accumulation. *Plant Cell Tissue Organ Culture*, 118, 1–16.

Nordine, A., Tlemcani, C. R., & El Meskaoui, A. (2014). Regeneration of plants through somatic embryogenesis in Thymus hyemalis Lange, a potential medicinal and aromatic plant. *In Vitro Cellular & Developmental Biology-Plant*, *50*(1), 19–25.

Ochoa-Villarreal, M., Howat, S., Hong, S., Jang, M. O., Jin, Y. W., Lee, E. K., & Loake, G. J. (2016). Plant cell culture strategies for the production of natural products. *BMB Reports*, 49, 149–158.

Pietrosiuk, A., Furmanowa, M., & Łata, B. (2007). *Catharanthusroseus*: Micropropagation and in vitro techniques. *Phytochemical Reviews*, 6, 459–473.

Pradnya, K., & Malpathak, N. (2013). Subcellular localization and quantification of camptothecin in different plant parts of *Chonemorphafragrans*. *Advances in Zoology and Botany*, 1(2), 34–38.

Praveen, N., Naik, P. M., Manohar, S. H., Nayeem, A., & Murthy, H. N. (2009). In vitro regeneration of Brahmi shoots using semisolid and liquid cultures and quantitative analysis of bacoside A. *Acta Physiology Plant*, 31, 723–728.

Raj, D., Kokotkiewicz, A., & Luczkiewicz, M. (2014). Production of therapeutically relevant indolizidine alkaloids in Securinega suffruticosa *in vitro* shoots maintained in liquid culture systems. *Applied Biochemistry and Biotechnology*, 175, 1576–1587.

Rout, G. R., Samantary, S., & Das, P. (2000). *In vitro* manipulation and propagation of medicinal plants. *Biotechnology Advances*, 18(2), 91–120.

Saito, K., Sudo, H., Yamazaki, M., Koseki-Nakamura, M., Kitajima, M., Takayama, H., & Aimi, N. (2001). Feasible production of camptothecin by hairy root cultures of *Ophiorrhizapumila*. *Plant Cell Reports*, 20, 267–271.

Sharafi, A., Haleh, H. S., Amir, M., Pejman, A., Bahman, H. K., & Khadijeh, R. (2013). Metabolic engineering of morphinan alkaloids by over-expression of code in one reductase in transgenic hairy roots of Papaverbracteatum, the Iranian poppy. *Biotechnology Letters*, 35(3), 445–453.

Sharma, M., Sharma, A., Kumar, A., & Basu, S. K. (2011). Enhancement of secondary metabolites in cultured plant cells through stress stimulus. *American Journal of Plant Physiology*, 1(6), 50–71.

Sood, N., Gupta, P. K., Srivastava, R. K., & Gosa, S. S. (2006). Comparative studies on field performance of micro propagated conventionally propagated sugarcane plants. *Plant Tissue Culture & Biotechnology*, 16, 25–29.

Suthar, R., Habibi, N., & Purohit, S. (2011). Influence of agar concentration and liquid medium on in vitro propagation of *Boswellia serrata* Roxb. *Indian Journal of Biotechnology*, 10, 224–227.

Takebe, I., Labib, G., & Melchers, G. (1971). Regeneration of whole plants from isolated mesophyll protoplasts of tobacco. *Naturwissenschaften*, 58, 318–320.

Teixeira da Silva, J. A., Tsavkelova, E., Zeng, S. J., Ng, T. B., Dobránszki, J., Parthibhan, S., Cardoso, J. C., & Rao, M. V. (2015b). Symbiotic *in vitro* seed propagation of *Dendrobium*: Fungal and bacterial partners and their influence on plant growth and development. *Planta*, 242, 1–22.

Tesfa, M., Admassu, B., & Bantte, K. (2016). *In vitro* rooting and acclimatization of micropropagated elite sugarcane (Saccharumofficinarum L.) genotypes—N52 and N53. *Journal of Tissue Science and Engineering*, 7, 164.

Tomiczak, K., Mikuła, A., Niedziela, A., Wójcik-Lewandowska, A., Domżalska, L., & Rybczyżski, J. J. (2019). Somatic embryogenesis in the family Gentianaceae and its biotechnological application. *Frontiers in Plant Science*, 10, 762.

Trigiano, R. N., & Gray, D. J. (eds.). (2011). Plant Tissue culture, Development and Biotechnology. CRC Press, Boca Raton, FL.

Xu, J., Ge, X., & Dolan, M. C. (2011). Towards high-yield production of pharmaceutical proteins with plant cell suspension cultures. *Biotechnology Advances*, 29, 278–299.

Yadav, R. N. S., & Agarwal, M. (2011). Phytochemical analysis of some medicinal plants. *Journal of Phytopharmacology*, 3(12), 10–14.

Yan, H., Liang, C., Litao, Y., & Li, Y. (2010). In vitro and ex vitro rooting of *Siratiagrosvenorii*, a traditional medicinal plant. *Acta Physiologiae Plantarum*, 32, 115–120.

Zhao, F.-C., Nilanthi, D., Yang, Y.-H., & Wu, H. (2006). Anther culture and haploid plant regeneration in purple coneflower (*Echinacea purpurea* L.). *Plant Cell Reports*, 86, 55–62.

Ziv, M., & Chen, J. (2008). The anatomy and morphology of tissue cultured plants. In: George, E. F., Hall, M. A., De Klerk, G. J. (Eds.), *Plant Propagation by Tissue Culture*, vol. 1, 3rd ed. The Background. Springer, Dordrecht, pp. 465–447.

6 Sterilization Methods and Techniques Involved in Tissue Culture, with Protocols to Sterilize Different Plant Parts

Kiran, Vandana Garg, Rohit Dutt, Manju Rawat Singh, Deependra Singh, and Rahul Pratap Singh*
* Corresponding author: drvandana.pharma@mdurohtak.ac.in

6.1 INTRODUCTION

6.1.1 Plant Tissue Culture Definition

Plant tissue culture is a technique used to propagate and grow plants under controlled conditions in aseptic (sterile) laboratory conditions. This method involves the culturing of plant cells, tissues, or organs in an artificial nutrient medium under controlled environmental conditions. Plant tissue culture has various applications in plant propagation, genetic improvement, conservation, and the production of secondary metabolites.

Explants, on the other hand, are plant cells, tissues, or organs that are used to begin *in vitro* cultures. Explant selection is based on anticipated research. Explants can be extracted from the different parts of plants, like the apical meristem, seed, embryo, anther, pollen, or protoplast, as shown in Figure 6.1. Explants are always contaminated with microorganisms (Deore, 2018), impurities, and toxins. Thus, due to contamination, the cultured tissues are unsuited for healthy plantlets to continue growing from them. The fact that plants are extremely vulnerable to pathogen infestations makes it difficult to totally eradicate contamination. The contamination of tissue culture might take many distinct forms. A beaker, a pipette, or one of the culture pots may contain microorganisms. The materials and equipment used in plant tissue culture are also quickly contaminated by a single fungus or bacterial cell that comes into contact with the culture media or explants (Stanisavljević, et al., 2017).

Tissue culture media also include sugar and other substances that are more likely to draw bacteria, fungus, and other microorganisms. Plant tissue cultures are therefore difficult to prevent because they are naturally vulnerable to contamination. Scientists have developed a variety of strategies over time to lessen the amount of current pollutants. However, contaminations continue to be one bottleneck in plant tissue culture, resulting in decreased productivity and yield in cultured plants. To overcome this type of problem, explants, media, and equipment should be sterilized to get good productivity with better yield (Bhojwani, et al., 2013).

Sterilization is the process of eliminating, removing, killing, or deactivating all forms of life and other biological agents, such as prions, present in or on a specific surface, object, or fluid. Sterilization can be achieved through various means, including heat, chemicals, irradiation, high pressure, and filtration. There are different types of sterilization methods, namely, the physical, chemical, and gaseous methods, as shown in Figure 6.2. The physical method of sterilization is used for equipment, glassware, and media, while the chemical and gaseous methods are used for the sterilization of different plant parts.

DOI: 10.1201/9781003326939-7

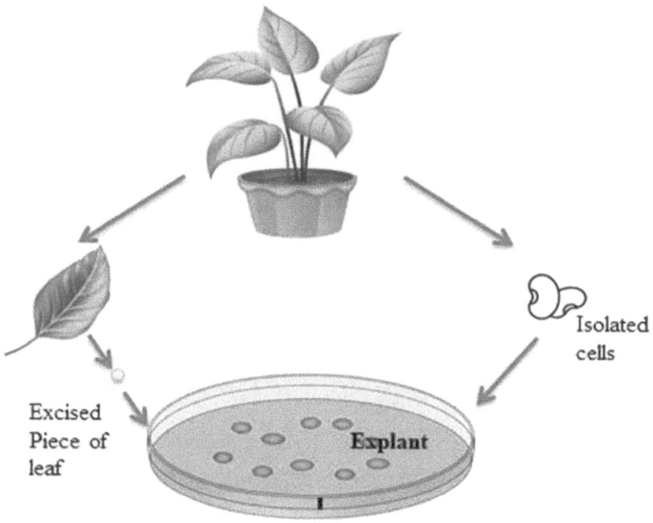

FIGURE 6.1 Collection of explant.

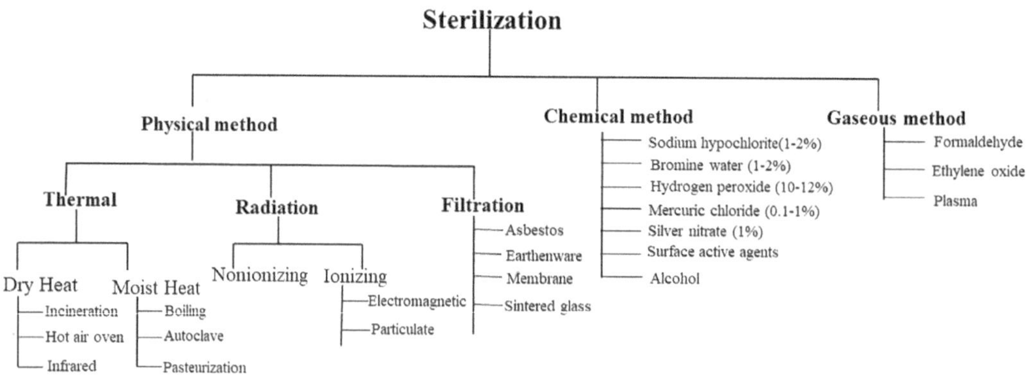

FIGURE 6.2 Different methods of sterilization.

6.1.2 Physical Methods

Physical methods of sterilization involve the use of various physical agents or energy to eliminate or destroy all forms of microbial life, including bacteria, viruses, fungi, and spores. These methods are employed to ensure the complete eradication of pathogens and to achieve sterility in a given object or environment. The choice of a particular physical method often depends on the nature of the items being sterilized and their heat sensitivity. Here are some common physical methods of sterilization and the reasons for their use (Felek, et al., 2015).

6.1.2.1 Thermal/Heat Sterilization

The most popular and effective form of sterilization is thermal/heat sterilization, which involves the destruction of enzymes and other vital cell components with heat. Thermal sterilization is done by two methods: One is dry heat sterilization, and the other is moist heat sterilization. Dry conditions lead to oxidative changes and a higher heat intake. In moist heat, there is greater hydrolysis and denaturation due to high humidity levels; hence, the process is more efficient and requires less heat.

6.1.2.1.1 Dry Heat Sterilization

Dry heat sterilization is a method used to eliminate microorganisms from inanimate objects using hot air. Dry heat sterilization relies on high temperatures without the presence of moisture. This method is particularly suitable for items that are sensitive to moisture or that might be damaged by exposure to steam. Dry heat is suitable for items that can withstand high temperatures without being damaged or undergoing significant changes. This includes glassware, metal instruments, and certain powders or oils. Dry heat sterilization typically involves higher temperatures and longer exposure times compared to moist heat methods. Commonly, temperatures range from 160°C to 180°C (320°F to 356°F), and exposure times can vary from 1 to 2 hours or more. Methods of dry heat sterilization are discussed in what follows (Wendling, et al., 2014).

6.1.2.1.1.1 Hot Air Oven Sterilization by hot air oven, also known as dry heat sterilization, is a method used to eliminate microorganisms from items that are sensitive to moisture, such as glassware, metal instruments, powders, and oils. The process involves heating the items at a high temperature for a specified period, typically between 160°C and 180°C (320°F to 356°F) for 1 to 2 hours. This is the most common method of dry heat sterilization. It involves placing items in an oven where hot air circulates, providing uniform heat distribution. The high temperature disrupts the proteins and cellular structures of microorganisms, rendering them nonviable. Before sterilization, the items to be sterilized must be thoroughly cleaned and dried to remove any debris, contaminants, or moisture. Proper cleaning ensures that the sterilization process is effective. Arrange the cleaned and dried items inside the hot air oven in a manner that allows for adequate airflow around each item. Overcrowding should be avoided to ensure uniform heat distribution and effective sterilization.

6.1.2.1.1.2 Incineration Dry heat sterilization by incineration is a method of sterilization that involves exposing objects or materials to high temperatures in a controlled environment until all microorganisms are destroyed. Incineration utilizes extremely high temperatures to burn organic materials, including microorganisms, to ash. The intense heat denatures proteins, disrupts cellular structures, and effectively kills all microorganisms. Incinerators used for sterilization are designed to reach and maintain high temperatures, typically ranging from 800°C to 1,200°C (1,472°F to 2,192°F). These incinerators may feature insulated chambers, burners, controls for temperature regulation, and systems for managing emissions. The items to be sterilized are placed inside the incinerator chamber. The chamber is then heated to the required temperature and maintained at that temperature for a sufficient duration to ensure the complete destruction of all microorganisms. The organic materials are subjected to combustion, leaving behind sterile ash and gases (Sharuti, et al., 2011).

6.1.2.1.1.3 Infrared Rays Dry heat sterilization by infrared rays is a method used to kill microorganisms on surfaces or within materials through exposure to high temperatures generated by infrared radiation. Infrared radiation is a form of electromagnetic radiation with wavelengths longer than visible light but shorter than radio waves. When objects absorb infrared radiation, it causes their molecules to vibrate, generating heat. Microorganisms are sensitive to heat, and exposure to high temperatures can denature their proteins, disrupt their cellular structures, and ultimately kill them. Dry heat sterilization equipment utilizing infrared radiation usually consists of an infrared emitter or a lamp that emits focused infrared rays onto the target surface or material to be sterilized. The process involves placing the items to be sterilized in the path of the infrared rays. The infrared radiation penetrates the surface of the material, generating heat throughout, effectively killing microorganisms present. The temperature and exposure time are carefully controlled to ensure thorough sterilization without damaging the material being treated.

6.1.2.1.1.4 Advantages Dry heat sterilization has several advantages, including its ability to penetrate materials thoroughly, its lack of dependence on moisture, and its applicability to a wide range of materials that might be damaged by steam.

6.1.2.1.1.5 Disadvantages The main disadvantage of dry heat sterilization is the longer exposure time required compared to moist heat methods. Additionally, certain materials, such as rubber and plastics, may not be suitable for dry heat sterilization due to the risk of damage (Teixeira, et al., 2016).

6.1.2.1.2 Moist Heat Sterilization

Moist heat sterilization is highly effective in destroying a wide range of microorganisms, including bacteria, viruses, and fungi. It involves the use of steam under pressure to kill microorganisms and their spores. It is particularly useful for heat-resistant and moisture-stable materials. It is suitable for a wide range of materials, including glassware, metal instruments, textiles, plastics, and liquids. There are three primary methods of moist heat sterilization, as discussed in the following (Ahmadpoor, et al., 2022)

6.1.2.1.8 Boiling

Sterilization by boiling is a common method used to kill microorganisms and pathogens present on the surface of the equipment and objects. Boiling is a simple and effective way to achieve sterilization, particularly for objects that can withstand high temperatures. Before sterilization, one should ensure that the equipment is clean and free from any visible debris. Place the equipment/objects in a container filled with water, and make sure there is enough water to completely submerge the object. Bring the water to a rolling boil. Once the water reaches a rolling boil, maintain it at that temperature for a specific period of time. The exact duration depends on the object being sterilized and the specific pathogens one is trying to eliminate. Typically, boiling for at least 10–15 minutes is recommended to ensure effective sterilization. After the boiling process is complete, remove the object from the water using sanitized utensils, to avoid recontamination. Allow it to cool and air-dry completely before use or storage.

6.1.2.1.9 Autoclave

Autoclave sterilization is suitable for a wide range of materials, including metal equipment, glassware, and certain types of plastics. Autoclaves consist of a chamber where the items to be sterilized are placed. Autoclaves operate at high temperatures and pressures to achieve sterilization. The standard conditions for sterilization in an autoclave are typically 121°C (250°F) at 15 psi (pounds per square inch) for 15–20 minutes. Autoclaves use steam generated from water to create the necessary conditions for sterilization. Water is heated to produce steam, which is then introduced into the chamber where the objects to be sterilized are placed. The steam penetrates the objects and effectively kills the microorganisms present on their surfaces. After the sterilization cycle is complete, the autoclave is depressurized gradually to prevent rapid cooling, which could cause thermal shock to the sterilized items. Once the chamber has reached a safe temperature, the items can be removed (Sen, et al., 2013).

6.1.2.1.10 Pasteurization

Pasteurization is a process commonly used to sterilize liquids. Pasteurization effectively reduces the microbial load in the liquid and extends shelf life without significantly altering their nutritional quality. It was developed by French scientist Louis Pasteur in the 19th century and has since become a widely used method. Liquid is heated to a specific temperature below its boiling point. This temperature is typically around 145°F to 161°F (63°C to 72°C) for pasteurization. The exact temperature and duration of heating depend on the specific product and regulatory requirements. After reaching the target temperature, the liquid is held at that temperature for a certain period of time. This allows sufficient time for the heat to penetrate the liquid and effectively kill harmful microorganisms, including bacteria, yeasts, and molds. After that, the liquid is rapidly cooled to stop the heating process and prevent overheating or other undesirable changes. However, it is important to note that pasteurization may not completely sterilize the liquid, especially if it is later exposed to contaminants. Therefore, pasteurized products should still be handled and stored properly to minimize the risk of microbial contamination (Naika, et al., 2008).

6.1.2.2 Radiation Sterilization

Sterilization by radiation is a process where ionizing radiation, such as gamma rays, X-rays, or electron beams, is used to kill or inactivate microorganisms and pathogens on various surfaces or objects. Objects to be sterilized are placed in a designated area or container suitable for radiation exposure. The objects are exposed to a controlled dose of ionizing radiation. Gamma rays from a radioactive isotope, X-rays from a machine, or electron beams from an accelerator are common sources of radiation used for sterilization (Phillips, et al., 1981).

6.1.2.2.1 Ionizing Radiation

Sterilization by ionizing radiation is a commonly used method to kill or inactivate microorganisms, pathogens, and other contaminants on various surfaces of equipment and products. Ionizing radiation, such as gamma rays, X-rays, and electron beams, are employed to achieve sterilization without heat or chemicals.

6.1.2.2.2 Electromagnetic Ionizing Radiation

Sterilization by electromagnetic ionizing radiation involves the use of high-energy electromagnetic waves, such as gamma rays and X-rays, to achieve sterilization. Equipment to be sterilized is packed properly and then placed in a chamber equipped with a source of ionizing radiation. Gamma rays emitted from a radioactive isotope (e.g., cobalt-60 or cesium-137), X-rays generated by a machine (e.g., linear accelerator or X-ray irradiator), and electron beams produced by an accelerator are commonly used sources of ionizing radiation for sterilization. The packaged objects are exposed to a controlled dose of ionizing radiation for a specified period of time. The energy from the radiation penetrates the packaging, effectively killing or inactivating microorganisms by damaging their DNA, RNA, or other vital cellular components. A dosimeter device is used to measure the absorbed dose of radiation to ensure that the objects receive the appropriate level of radiation for sterilization. Regular monitoring and calibration of radiation equipment are essential to maintain accuracy and consistency in the sterilization process (Khafagi, et al., 2001).

6.1.2.2.3 Particulate Ionizing Radiation

Sterilization by particulate ionizing radiation typically involves the use of electron beams (e-beams) to achieve sterilization. Electron beams are streams of high-energy electrons generated by specialized equipment, called an accelerator. The linear accelerator and the Van de Graaff generator are the commonly used accelerators in particulate sterilization. The packaged objects are placed in a chamber equipped with the electron beam generator. They are then exposed to the electron beam for a specific period of time. The high-energy electrons achieve sterilization by penetrating the packaging materials and damaging the DNA, RNA, or other cellular components of microorganisms. A dosimeter device is used to measure the absorbed dose of radiation to ensure that the object receives the appropriate level of electron beam radiation for sterilization. Regular monitoring and calibration of the equipment are essential to maintaining accuracy and consistency in the sterilization process (Espinosa-Leal, et al., 2015).

6.1.2.2.4 Non-Ionizing Radiation

Sterilization by non-ionizing radiation refers to the process of eliminating microorganisms, including bacteria, viruses, fungi, and spores, using forms of radiation that do not have sufficient energy to ionize atoms or molecules. This method of sterilization is primarily achieved through ultraviolet (UV) radiation and visible light.

6.1.2.2.5 Ultraviolet (UV) Radiation

UV radiation is commonly used for sterilization purposes. It operates within the wavelength range of approximately 100 to 400 nm. The most germicidal wavelengths fall within the UV-C range (200 to 280 nm). UV-C radiations disrupt the DNA and RNA of microorganisms and prevent reproduction. UV sterilization is widely employed in water treatment, air purification, and surface disinfection processes.

6.1.2.2.6 *Visible Light*

Certain wavelengths within the visible light spectrum can also possess germicidal properties, although to a lesser extent compared to UV radiation. Blue light in particular has shown effectiveness against some bacteria. However, its application in sterilization is not as common as for UV radiation.

6.1.2.3 Filtration

Sterilization by filters is a method used to remove or destroy microorganisms from a fluid, typically air or liquids, by passing it through a filter with fine pores. The process works by physically trapping or inhibiting the growth and reproduction of microorganisms, such as bacteria, viruses, and fungi, present in the fluid. Following are some common types of filters used for sterilization (Garg, et al., 2014).

6.1.2.3.1 *High-Efficiency Particulate Air (HEPA) Filters*

HEPA filters are commonly used in air purification systems to remove airborne particles, including bacteria and viruses. These filters are highly efficient, capable of removing particles as small as 0.3 μm with an efficiency of 99.97%.

6.1.2.3.2 *Membrane Filters*

Membrane filters are thin films made of materials such as cellulose acetate, nitrocellulose, or polyvinylidene difluoride (PVDF). They work by physically trapping microorganisms larger than the pore size of the membrane, which can range from nanometers to micrometers in diameter.

6.1.2.3.3 *Depth Filters*

Depth filters are composed of porous materials, such as cellulose, diatomaceous earth, or activated carbon. They operate by trapping microorganisms within the depth of the filter medium, rather than just on the surface.

6.1.2.3.4 *UV Filters*

UV filters utilize ultraviolet (UV) light to sterilize air or liquids by damaging the DNA of microorganisms, rendering them unable to replicate. While not technically a filter in the traditional sense, UV sterilization is often used in conjunction with physical filtration methods for enhanced sterilization.

6.1.2.3.5 *Carbon Filters*

Carbon filters, also known as activated carbon filters, are primarily used in removing organic compounds, odors, and chlorine from water or air. While they are not designed specifically for microbial sterilization, they can indirectly contribute to improving water quality by removing organic matter that may support microbial growth.

6.1.3 Chemical Methods

Sterilization by chemicals involves the use of specific chemical agents to eliminate all forms of microbial life, including bacteria, viruses, fungi, and spores, from the surface of plant material. These chemicals are known as sterilant liquids when used in appropriate concentration for appropriate duration. Commonly used liquid sterilants are sodium hypochlorite, hydrogen peroxide, mercuric chloride, and bromine water. The chemicals used for the sterilization of different plant parts are listed in Table 6.1, along with exposure time and concentration (Jan, et al., 2013).

TABLE 6.1
Surface Sterilization Solutions Used to Sterilize, along with Exposure Time and Concentration

Name of Chemical	Concentration (%W/V)	Exposure (Min.)	Pant Part
Bromine water	1–2	2–10	Seeds
Benzalkonium chloride	0.01–0.1	5–20	Flower
Sodium hypochlorite	0.5–51	5–30	Stem, fruits
Calcium hypochlorite	9–10	5–30	Seeds
Mercuric chloride	1–2	2–10	Leaves, flowers
Hydrogen peroxide	3–10	5–15	-
Silver nitrate	1–2	5–20	Roots

Bromine water is known for its strong oxidizing properties, which make it effective in killing microorganisms, including bacteria, fungi, and viruses. In plant biology and related fields, bromine water can be used to sterilize plant materials or surfaces to prevent contamination. The plant material to be sterilized is typically immersed in the bromine water solution for a certain period of time. The duration of immersion will vary depending on factors such as the type of plant material, its size, and the level of contamination. Bromine water is prepared by dissolving bromine gas in water. This should be done in a well-ventilated area, preferably under a fume hood, as bromine gas is toxic and can be harmful if inhaled. The concentration of bromine in the water will depend on the specific application and the desired sterilization level. After the sterilization period, the plant material may need to be rinsed with sterile water, or another appropriate solution, to remove any excess bromine residue. Once sterilized, the plant material can be transferred to a sterile culture.

Sodium hypochlorite (NaClO), commonly known as bleach, is another chemical that is frequently used for the sterilization of plants. Plant explants, such as small sections of plant tissue or seeds, are usually surface-sterilized to remove any microorganisms that may be present and to prevent contamination in the subsequent culture. The explants are typically immersed in the sodium hypochlorite solution for a short period of time, usually ranging from a few seconds to a few minutes, depending on the plant species and the size and type of explant. A dilute solution of sodium hypochlorite is prepared by mixing sodium hypochlorite with water. The concentration range of the solution is 1% to 10%, and specific application will depend on the level of sterilization required. After surface sterilization treatment, the explants are rinsed several times with sterile water or a sterile saline solution to remove any residual bleach and to prevent any potential damage to the explants due to the bleach. Once sterilized, the explants can be transferred to a sterile culture medium.

Hydrogen peroxide (H_2O_2) is another chemical commonly used in the sterilization of plant explants. Plant explants are usually surface-sterilized to remove any microorganisms that may be present on the surface of the explant. The explants are typically immersed in the hydrogen peroxide solution from a few seconds to a few minutes, depending on the type of the explant. A dilute solution of hydrogen peroxide is prepared by mixing commercially available hydrogen peroxide solutions (typically 3% concentration) with sterile water. After surface sterilization, the explants are rinsed several times with sterile water or a sterile saline solution to remove any residual hydrogen peroxide and to prevent any potential damage to the explants due to the chemical. Once sterilized, the explants can be transferred to a sterile culture medium.

6.1.3.1 Alcohol

Ethanol or isopropyl alcohol is commonly used for the sterilization of plant explants. Plant explants are usually surface-sterilized to remove any microorganisms. The explants are typically immersed in the alcohol (70%) solution for a short period. Ethanol or isopropyl alcohol is diluted with water to achieve the desired concentration for sterilization. The concentration of alcohol used for sterilization can vary depending on the specific application and the level of sterilization required. After surface sterilization treatment, the explants are rinsed several times with sterile water or a sterile saline solution to remove any residual alcohol and to prevent potential damage to the explants. Once sterilized, the explants can be transferred to a sterile culture medium.

Silver nitrate ($AgNO_3$) has been used in the past for the sterilization of plant explants. However, its use has diminished over time due to a few limitations and safety concerns. The explants are typically immersed in the silver nitrate solution for few seconds. A dilute solution of silver nitrate is prepared by mixing commercially available silver nitrate crystals with sterile water. After surface sterilization, the explants are rinsed several times with sterile water or a sterile saline solution to remove any residual silver nitrate and to prevent any potential damage to the explants due to the chemical. Once sterilized, the explants can be transferred to a sterile culture medium.

Mercuric chloride ($HgCl_2$) was once commonly used for the sterilization of plant explants. However, its use has significantly declined due to safety and environmental concerns associated with mercury compounds. While mercuric chloride is an effective sterilizing agent, it poses serious health risks to humans and the environment. Mercury is highly toxic and can cause severe health problems. Additionally, mercury compounds are persistent environmental pollutants that can accumulate in ecosystems and pose risks to wildlife and human health through bioaccumulation. Due to these risks, the use of mercuric chloride for sterilization is strongly discouraged, and alternative methods and sterilizing agents should be used instead. Common alternatives include bleach (sodium hypochlorite), hydrogen peroxide, and ethanol.

6.1.4 Sterilization of Different Parts of Plants

6.1.4.1 Leaves and Buds

The sterilization of leaf and bud explants is a crucial step in plant tissue culture to eliminate microbial contamination and ensure the successful propagation of sterile plant materials. The collection of leaves as explant, followed by its sterilization process and its incubation, is shown in Figure 6.3. Rinse the leaves and buds under running tap water to remove any debris or dust particles (Gunawardhana, et al., 2022). Trim off any damaged or discolored parts of the leaves and buds using a sterile scalpel

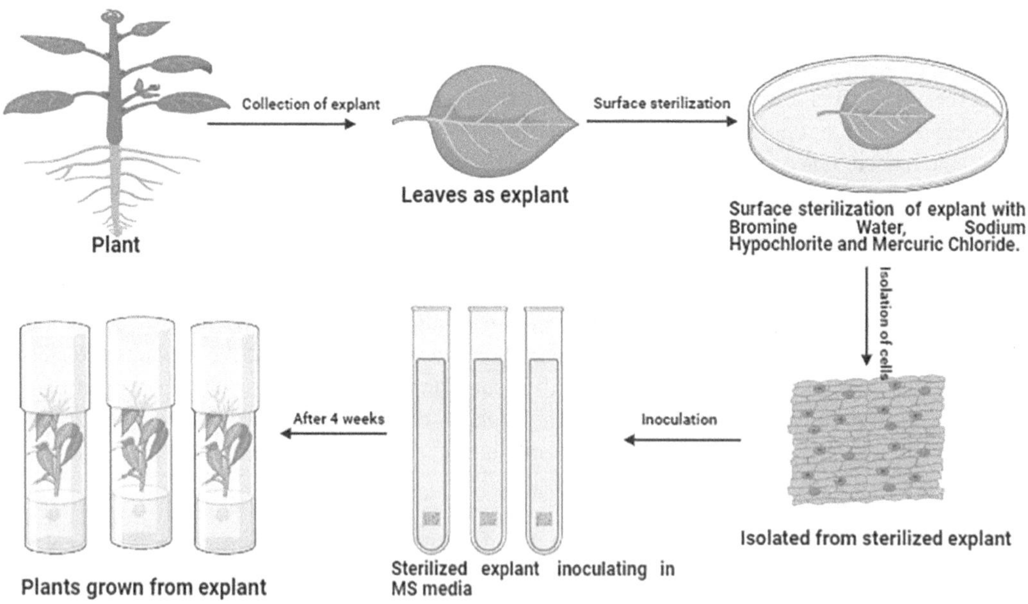

FIGURE 6.3 Collection of leaves as explant, followed by its sterilization process and incubation.

or razor blade. Prepare a sterilizing solution by mixing a suitable sterilizing agent (70% ethanol or a bleach solution diluted in 0.5–1% active chlorine) with sterile distilled water. There are several surface sterilization solutions used to reduce browning and contamination with the highest explant survival rate, as reported in Table 6.2. Add a few drops of Tween 20 (surfactant) to the solution to break surface tension and enhance sterilization. Submerge the leaves and buds in the sterilizing solution for a specific duration, usually 1–3 minutes, depending on the plant species and tissue type. Agitate the solution gently to ensure all surfaces of the plant material are exposed to the sterilizing agent. After the sterilization period, transfer the leaves and buds to the culture medium (Ndakidemi, et al., 2013).

6.1.4.2 Seeds

There are various sterilization solutions used to sterilize seeds, discussed in what follows, and the collection of seeds as explant, followed by its sterilization process and its incubation, is shown in Figure 6.4.

TABLE 6.2
Surface Sterilization Solution Used to Sterilize Leaves and Buds

Sr. No.	Sterilization Solution
1	10 mL of 0.1% $HgCl_2$
2	10 mL of 0.1% $HgCl_2$, and 1 mL of 70% EtOH
3	10 mL of 0.1% $HgCl_2$; 0.5 mL of 70% EtOH and 25 g charcoal
4	15 mL of 10% NaOCl
5	1 mL of 70% EtOH and 15 mL of 10% NaOCl
6	0.5 mL of 70% EtOH and 25 gm Charcoal and 5 mL of 10% NaOCl
7	15 mL of 10% NaOCl
8	15 mL of 20% NaOCl
9	1 mL of 70% EtOH and 20% NaOCl = 15
10	5 mL of 70% EtOH, 25 gm charcoal, and 5 mL of 20% NaOCl
11	1.1 mL of 20% NaOCl

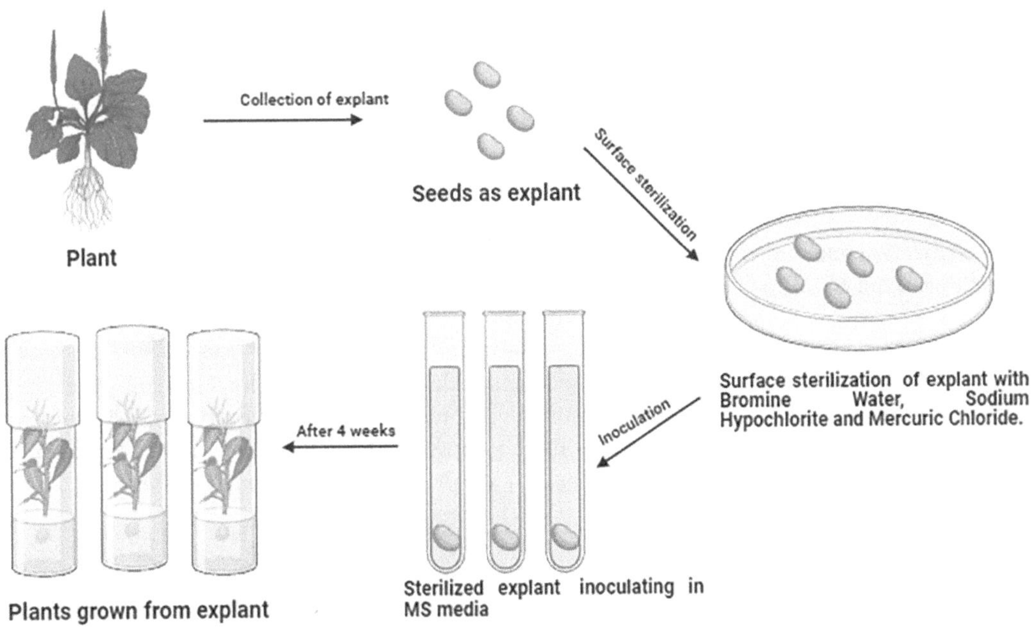

FIGURE 6.4 Collection of seeds as explant, followed by its sterilization process and its incubation.

6.1.4.2.1 3.6% Sodium Hypochlorite or 0.1% Mercuric Chloride

Seeds are first sterilized for 2 minutes in 75% ethanol. After that, seeds are rinsed with distilled water. In 3.6% sodium hypochlorite or 0.1% mercuric chloride, seeds are soaked for 1–2 minutes. MS media will be used to cultivate the seeds. For 20 days, all petri dishes are incubated at 25/10°C, with a photoperiod of 14 hours light and 10 hours dark. When the length of the radicle reaches 5 mm, a seed is deemed to have germinated (*Suaeda aralocaspica*) (Si, et al., 2022).

6.1.4.2.2 Bleach Solution

To sterilize the seeds, a 50% (v/v) bleach solution can also be used. The bleach should be diluted by mixing 100 mL of bleach with 100 mL of distilled water. To the bleach solution, add 50 µl of Tween. After 2 minutes, the bleach solution is removed from the seeds. Rinse the treated seeds with 500 µl of sterile distilled water six times. After that, the seeds are added in the MS media for germination (e.g., *Arabidopsis*) (Lindsey, et al., 2017).

6.1.4.2.3 Mercuric Chloride Solution

The viable seeds are submerged in 70% ethanol for 1 minute; 0.1% (w/v) mercuric chloride solution for 1 minute; and 20%, 40%, 60%, 80%, and 100% sodium hypochlorite for 2 minutes. As a wetting agent, one drop of Tween 20 is used. Following a 20-minute soaking in sterile distilled water to soften the coat, the seeds are rinsed five times in sterile distilled water to completely eliminate all traces of the sterilant. After that, the seeds are added in the MS media for germination (e.g., *Zehneria capillacea*) (Agogbua, et al., 2022).

6.1.4.2.4 Chlorine Gas

chlorine gas is produced by the mixing of 100 mL of bleach and 3 mL of concentrated hydrochloric acid. On the basis of the chlorine gas, seeds are sterilized in four different ways. In the first method, seeds are treated with chlorine for 5 hours before the inoculation. In the second method, seeds are soaked in soapy water for 20 minutes; after that, these seeds are treated with chlorine for 5 hours. These treated seeds are ready for inoculation. In the third method, seeds are exposed to chlorine for 5 hours, washed for 5 minutes with 70% (v/v) ethanol, and then washed twice more with sterile distilled water prior to inoculation on the media. In the fourth method, the seeds are exposed to chlorine for 5 hours, washed for 2 minutes with 70% (v/v) ethanol, and then washed twice more with sterile distilled water; after that, a quick washing with a sterile solution containing 1% (v/v) plant preservative mixture in a 1/2 MS medium (Murashige, et al., 1962) before inoculation on the media. This type of sterilization is given to cotton seeds (Barampuram, et al., 2014).

6.1.4.3 Nodal Explants

The sterilization solution used to sterilize the nodes are discussed in what follows, and the collection of nodes as explant, followed by its sterilization process and its incubation, is shown in Figure 6.5.

6.1.4.3.1 70% Ethanol or 10% Calcium Hypochlorite

Nodes 2–3 cm long are rinsed with liquid detergent under running water for 1 minute. After that, they are submerged in 70% ethanol for 2 minutes, and the surface is sterilized with 10% calcium hypochlorite for 5 minutes and then rinsed three times with sterile distilled water. The bleached ends of sterilized single-node explants are clipped, and they are cultivated on Murashige and Skoog (MS) (e.g., *Solanecio biafrae*) (Bello, et al., 2018).

6.1.4.3.2 0.1% Bavistin

The explants undergo a 5-minute soak in double-distilled water. Afterward, the explants are subjected to a 1-minute treatment with 0.1% Bavistin and a 1-minute rinse in double-distilled water. Additionally, explants are treated with mercuric chloride (0.1%) and sodium hypochlorite (1%). To fully eradicate any remaining chemical residue, all explants are given a 2-minute rinse

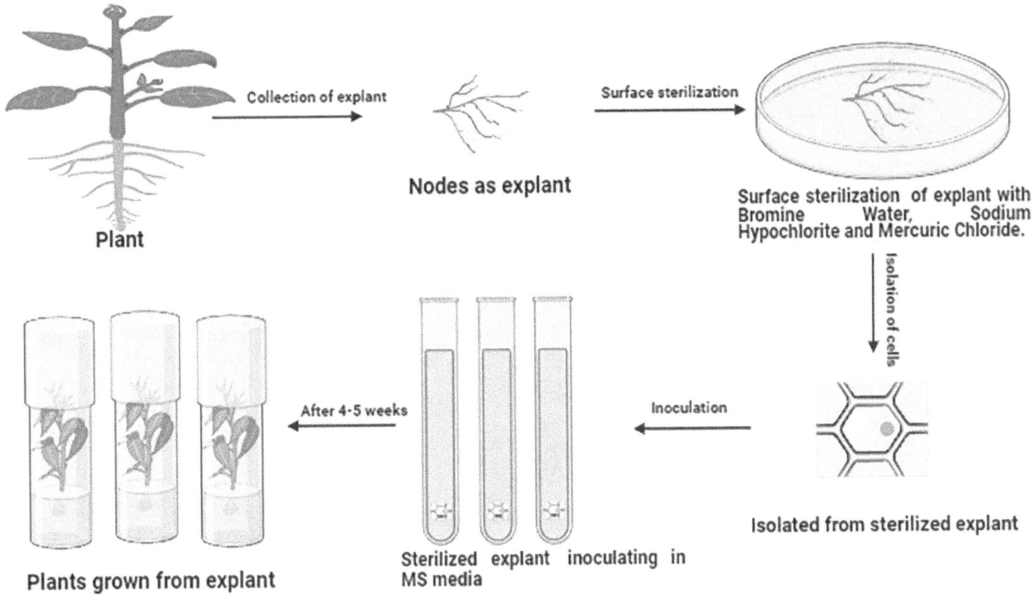

FIGURE 6.5 The collection of nodes as explant, followed by its sterilization process and its incubation.

in double-distilled water. Sterilized single-node explants are clipped, and they are cultivated on Murashige and Skoog (MS) (e.g., *Momordica dioica*) (Dodake, et al., 2021).

6.1.4.3.3 0.1% Mercuric Chloride

Immature stem segments with bud tips are used as explant. After being thoroughly washed with a detergent solution, the explants are washed for 2 minutes under slow-running water. After that, sterilizing stem segments are treated with 75% ethanol for 50 seconds; they are rinsed five times with sterile water and then treated with 0.1% mercuric chloride for 1 minute. In the end, the stem segments are vertically placed into the Murashige and Skoog (MS) medium (e.g., *Cnidoscolus aconitifolius*) (Gu, et al., 2022).

6.1.4.4 Floral Explants

The sterilization solutions used to sterilize floral explants are discussed in what follows, and the collection of the flower as explant, followed by its sterilization process and its incubation, is shown in Figure 6.6.

6.1.4.4.1 The Flower Stalk Segments

Explants of flower stalks are taken from healthy plants and carefully cleaned under running tap water for 2 minutes to remove any remaining dust particles. The explants are then submerged in liquid detergent (Tween 20) for 1 minute. Explants are rinsed under running tap water to remove the detergent. They are submerged in a 0.2% w/v Bavistin solution for five minutes, and then they are again washed three to four times with distilled water. Mercuric chloride in an aqueous solution at a concentration of 0.1% (w/v) is used to sterilize the explants. The explants are then carefully cleaned using sterile distilled water. Around 0.5 mL of 1.0% sodium hypochlorite is applied to these explants. Sterile double-distilled water is used to wash the explants after they have been surface-sterilized. The explants are then inoculated in media (e.g., *Phalaenopsis*) (Madhuri, et al., 2021).

6.1.4.4.2 The Flower Young Capitula

About 3- to 4-day-old capitula of flower are used as explants. Initially, capitula are washed with running water for 2 minutes; after that, the capitula are repeatedly stirred with 10 mL/L of liquid

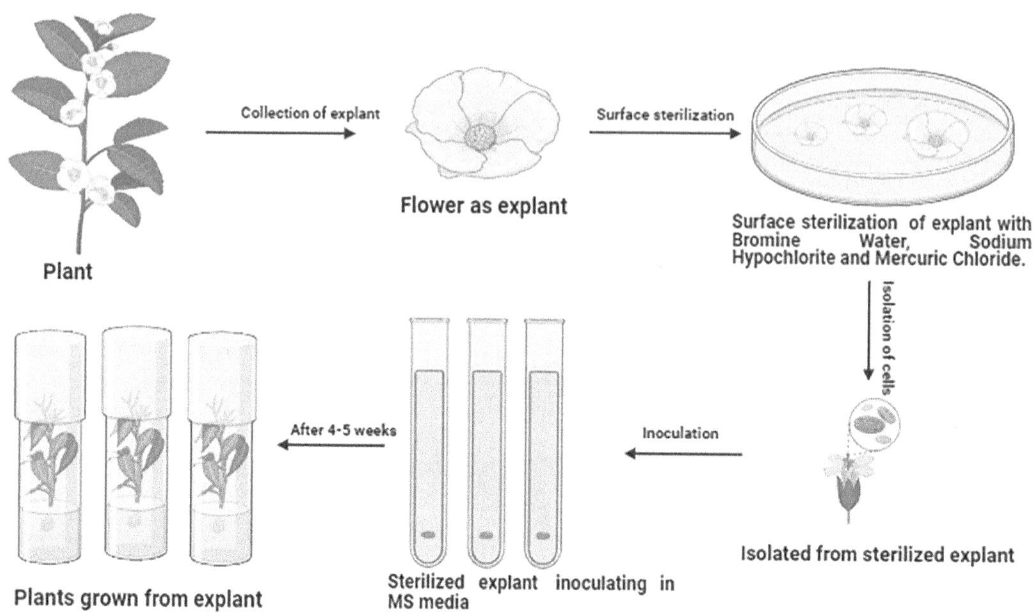

FIGURE 6.6 Collection of flower as explant, followed by its sterilization process and its incubation.

Teepol for the duration of 1 minute. Later, they are again washed for 2 minutes under running water from the faucet. The next step is to treat the explants with 2 g/L of Bavistin for 30 minutes while constantly stirring the solution to lower the risk of fungal infection. Explants are properly cleansed with distilled water to get rid of any remaining chemical after being treated with Bavistin. The sterilized explant inoculates in MS media (e.g., *Gerbera jamesonii*) (Sharma, et al., 2021).

6.2 CONCLUSION

The sterilization of plant material, such as seeds, explants, or equipment used in plant tissue culture, is a common practice to eliminate or reduce microbial contamination. Chemical sterilization involves the use of specific chemicals to kill or inhibit the growth of microorganisms. Always follow proper safety protocols when working with chemicals, and ensure that the chosen method and concentration are suitable for the specific plant material or equipment being sterilized. It is also crucial to maintain aseptic conditions during subsequent handling to prevent recontamination.

REFERENCES

Agogbua, Josephine U., and Bosa E. Okoli. "Procedure for in vitro seed sterilization, germination and aseptic seedling establishment of Zehneriacapillacea (Schumach) C. Jeffrey." *GSC Biological and Pharmaceutical Sciences* 19, no. 1 (2022): 143–148.

Ahmadpoor, Fatemeh, Nasser Zare, Rasool Asghari, and Parisa Sheikhzadeh. "Sterilization protocols and the effect of plant growth regulators on callus induction and secondary metabolites production in in vitro cultures Meliaazedarach L." *AMB Express* 12, no. 1 (2022): 1–12.

Barampuram, Shyam, George Allen, and Sergei Krasnyanski. "Effect of various sterilization procedures on the in vitro germination of cotton seeds." *Plant Cell, Tissue and Organ Culture (PCTOC)* 118 (2014): 179–185.

Bello, Oluwakemi A., Edward B. Esan, and Olawole O. Obembe. "Establishing surface sterilization protocol for nodal culture of Solaneciobiafrae." *IOP Conference Series: Earth and Environmental Science* 210, no. 1 (2018): 012007. IOP Publishing.

Bhojwani, Sant Saran, and Prem Kumar Dantu. *Plant tissue culture: An introductory text*. Vol. 318. India: Springer, 2013.

Deore, S. I. *Plant tissue culture. Pharmacognosy and phytochemistry*. Pharma Med Press, edition second, 2018, 108–120.

Dodake, S. S., A. S. Jadhav, S. V. Pawar, P. Y. Shinde, and N. M. Magar. "Standardizing efficient in vitro surface sterilization protocol for explants of Kankoda (Momordicadioica Roxb.)." *Journal of Pharmacognosy and Phytochemistry* 10, no. 1 (2021): 621–626.

Espinosa-Leal, C., J. F. Treviño-Neávez, R. A. Garza-Padrón, M. J. Verde-Star, C. Rivas-Morales, and M. E. Morales-Rubio. "Total phenols and anti-radical activity of methanolic extracts from wild and in vitro conditions of Leucophyllum frutescens." *Revista Mexicana Ciencias Farmaceuticas* 46 (2015): 52–56.

Felek, Wegayehu, Firew Mekibib, and Belayneh Admassu. "Optimization of explants surface sterilization condition for field grown peach (Prunuspersica L. Batsch. Cv. Garnem) intended for in vitro culture." *African Journal of Biotechnology* 14, no. 8 (2015): 657–660.

Garg, R. K., V. Srivastava, K. Kaur, and S. S. Gosal. "Effect of sterilization treatments on culture establishment in Jatrophacurcas L." *Karnataka Journal of Agricultural Sciences* 27 (2014): 190–192.

Gu, Min, Youli Li, Huier Jiang, Shihu Zhang, Qingmin Que, Xiaoyang Chen, and Wei Zhou. "Efficient in vitro sterilization and propagation from stem segment explants of Cnidoscolusaconitifolius (Mill.) IM Johnst, a multipurpose woody plant." *Plants* 11, no. 15 (2022): 1937.

Gunawardhana, W. P. D. S., P. I. P. Perera, H. M. A. P. Muhandiram, D. B. R. Swarnathilaka, and K. D. N. Priyadarshani. "Surface sterilization protocols of leaf and bud explants for initiating in vitro cultures of *Piper nigrum* L. (Pepper)." *Tropical Agricultural Research and Extension* 25, no. 2 (2022): 132–142.

Jan, Aarifa, K. M. Bhat, S. J. A. Bhat, M. A. Mir, M. A. Bhat, A. Imtiyaz, and J. A. Rather. "Surface sterilization method for reducing microbial contamination of field grown strawberry explants intended for in vitro culture." *African Journal of Biotechnology* 12, no. 39 (2013): 5749–5753.

Khafagi, Ishrak, Ahmed Dewedar, Mimona Kord, and E. Mohammed. "Identification and antibiotic sensitivity of bacteria occasionally isolated from differentiated and undifferentiated cultures of Sinai medicinal plants." *Egyptian Journal of Biology* 3 (2001): 67–78.

Lindsey III, Benson E., Luz Rivero, Chistopher S. Calhoun, Erich Grotewold, and Jelena Brkljacic. "Standardized method for high-throughput sterilization of Arabidopsis seeds." *JoVE (Journal of Visualized Experiments)* 128 (2017): e56587.

Madhuri, Gandamalla, Siddharth Kumar Palai, and Sashikala Beura. "Effect of chemical sterilants on surface sterilization of flower stalk during in vitro propagation of Phalaenopsis hybrids cv. Shagan." *Pharma Innovation* 10, no. 12 (2021): 221–224.

Murashige, T., and F. Skoog. "A revised medium for rapid growth and bio assays with tobacco tissue cultures." *Physiologia Plantarum* 15, no. 3 (1962): 473–497.

Naika, Hanumanaika Raja, and Venkatarangaiah Krishna. "Plant regeneration from callus culture of Clematis gouriana Roxb.—a rare medicinal plant." *Turkish Journal of Biology* 32, no. 2 (2008): 99–103.

Ndakidemi, Cosmas Funguomali, Emerald Mneney, and Patrick Alois Ndakidemi. "Development of sanitation protocol for leaf explants of B. huillensis for in vitro culture." *American Journal of Plant Sciences* 4, no. 12 (2013): 2425–2430.

Phillips, R., S. M. Arnott, and S. E. Kaplan. "Antibiotics in plant tissue culture: Rifampicin effectively controls bacterial contaminants without affecting the growth of short-term explant cultures of Helianthus tuberosus." *Plant Science Letters* 21, no. 3 (1981): 235–240.

Sen, Monokesh Kumer, M. A. H. M. Jamal, and Shamima Nasrin. "Sterilization factors affect seed germination and proliferation of *Achyranthes aspera* cultured *in vitro*." *Environmental and Experimental Biology* 11 (2013): 119–123.

Sharma, R., and S. Beura. "Surface sterilization and in vitro callusing of Gerbera jamesonii Bolus cv. Balance." *Biological Forum—An International Journal* 13, no. 4 (2021): 191–195.

Sharuti, Verma, Yadav Kuldeep, and Singh Narender. "Optimization of the protocols for surface sterilization, regeneration and acclimatization of Stevia rebaudiana Bertoni." *American-Eurasian Journal of Agricultural & Environmental Sciences* 11, no. 2 (2011): 221–227.

Si, Yu, Yakupjan Haxim, and Lei Wang. "Optimum sterilization method for in vitro cultivation of dimorphic seeds of the succulent halophyte Suaedaaralocaspica." *Horticulturae* 8, no. 4 (2022): 289.

Stanisavljević, Aleksandar, Dejan Bošnjak, Ivna Štolfa, Rosemary Vuković, Toni Kujundžić, and Mato Drenjančević. "Sterilizacijarazličitihtipovaeksplantata u mikropropagacijipodlogatrešnje CAB-6p i Gisela 6." *Poljoprivreda* 23, no. 2 (2017): 31–37.

Teixeira da Silva, Jaime A., Dariusz Kulus, Xinhua Zhang, Songjun Zeng, Guohua Ma, and Abel Piqueras. "Disinfection of explants for saffron (*Crocus sativus*) tissue culture." *Environmental & Experimental Biology* 14, no. 4 (2016): 183–198.

Wendling, Ivar, Stephen J. Trueman, and Aloisio Xavier. "Maturation and related aspects in clonal forestry—part II: Reinvigoration, rejuvenation and juvenility maintenance." *New Forests* 45 (2014): 473–486.

7 Gene Transfer in Plants and Their Application

Manish Kumar, Vichitra Kaushik, Manisha Bhatia, Janmejay Pant, A. Pandurangan, and Rupesh Kumar Gautam*
* Corresponding author: vichitra.kaushik@gmail.com

7.1 INTRODUCTION

The genetic manipulation and modification of cells by transferring genetic material (DNA or RNA) to have a therapeutic (Humbert et al., 2012; Uddin et al., 2020) or prophylactic effect is termed as *gene transfer.* Gene transfer in plants, also termed as *plant transformation*, involves two major events, namely, introducing an unknown DNA sequence to the target plant efficiently, followed by the renewal of a whole new plant variety with desirable characteristics from the transformed cells.

Genetic engineering involves the strategic implementation of the genetic material using laboratory-based technologies in a desired, directed, and predetermined way. Also termed as *genetic modification* (Bekana, 2017) and *manipulation*, this method aims at altering DNA makeup by isolating fragments, transferring genetic material (Begna & Okonkwo, 2020), and recombining them with one or more specialized enzymes in a desired combination. These fragments are further restored to a cell for replication and reproduction. When the recipient organism is a microbe, then after allowing cell multiplication, millions of copies of a specific region of DNA are obtained.

With the advent of genetic engineering, the key objective is manipulating and improving the characteristics of plants by introducing the novel concept of foreign gene encoding. These genetically engineered traits include resistance from an attack of herbicides, pesticides, or pests; tolerance to adverse conditions, such as the extremes of cold, heat, drought, and/or salinity; improvement in nutritional content or yield of important plant products; production of economical crops; and many more. Thus, this innovative concept has revolutionized the field of biology, leading to the extensive commercial production of transgenic plants; modified crops, also termed as *bioengineered* or *genetically engineered crops*; monoclonal antibodies; hormones; vaccines; and medicine. Its successful application offers new opportunities in the production of an ample range of therapeutic products, with additional benefits to the science of clinical trials, research, medical genetics, and biomedicine. The terminology refers to modifying genes, which can be employed in plants, animals, and microorganisms to yield medically useful products. The use of this therapy has created a considerable awareness in both fields of applied and fundamental research, leading to the development of varied methods of crop improvement.

7.1.1 Genetic Engineering Techniques: Conventional vs. Modern Breeding Methods

Plant breeding refers to the overall improvement in the genetic makeup of plants, leading to the production of new desired varieties with better properties than the existing one in all aspects. It is an art and a science of crop evolution by changing traits relating to their economic use to produce higher-yielding, pest-resistant, disease-controlled, and drought-resistant plants well adapted to varied environmental conditions. It has been widely tapped in improving the quality and yield of nutrition for humans as well as animals. The main endeavor is to produce varieties that possess unique agronomic characteristics and superior traits, such as wider adaptability to a variety of agricultural applications. The most commonly addressed traits include tolerance to both abiotic and biotic stress,

DOI: 10.1201/9781003326939-8

TABLE 7.1
Differences between Conventional and Modern Breeding Methods Employing Genetic Engineering Techniques

S. No.	Conventional Breeding Methods	Modern Breeding Methods
1	It is mainly focused on selection criteria through sexual and asexual reproduction.	This is concerned with the insertion of genetic material.
2	Exchange is possible either among two same species or those very closely related to each other.	It readily allows the direct transfer of genes between closely related, or even distantly different, species.
3	Variation is possible through the technique of hybridization.	Innovative technologies are followed in improving genetic characteristics.
4	From the crosses generated by conventional breeding methods, there is no confirmation of producing any particular gene combination.	Compared to conventional method, crop improvement output is very specific and precise, which can be achieved in a short time span.
5	In a few cases, along with desirable genes, undesirable characteristics can also be transferred.	It allows the modification of plants by the removal of particular unwanted genes.
6	The major issue with this process is that it is quite lengthy, and the required results may be achieved in a long time.	Desired results can be quickly attained, in a short period of time.
7	It is less effective, depending on subject analysis, making the result less reliable.	It is more successful based on thorough research and observation.
8	The cost involved is moderate due to the usage of local tools.	High-tech machinery is required, making the process quite expensive.

Source: Lamichhane and Thapa (2022).

increase in the quality of grain and biomass yield, and improvement in the basic quality of characteristics such as taste and palatability and in the concentration of proteins, sugars, lipids, vitamins, and fibers. The core aim is the selection of the best quality among all variants in terms of the yield, quality, and nature of edible parts; ease of processing, harvesting, milling, baking, malting, blending, and cultivation; and tolerance to environmental conditions, adaptation, and pest resistance. There is deliberate crossbreeding to introduce genes from one particular variety to a new genetic background. It makes use of genetic diversity as a source of valuable economic traits and produces new varieties with the aid of intra- and interspecific crosses. Miscellaneous approaches and *in vitro* procedures, such as the fusion of the protoplast, embryo liberation, or mutagenesis, can be employed to generate genetic alteration.

The science of breeding plays an immense role in the field of agriculture to improve the accessibility of food throughout the world. Modern plant breeding is a discipline that is firmly rooted in the creation of variation in biological traits. Principles and technologies from several scientific disciplines are applied with a blend of innovative techniques to manipulate plants for specific purposes (Degefa, 2019).

7.1.2 Importance of Plant Breeding

A remarkable contribution has been made in this diverse field of crop improvement in India as of now. Food grain production has improved enormously from 54.92 million tons in the years 1949–1950 to 241 million tons in 2010–2011. This goal has been accomplished with the research work of a number of eminent scientists who have made significant contributions in this context. T. S. Venkatraman, a famous sugarcane breeder, succeeded in transferring thick stem and high sugar contents from Noble cane to North Indian canes, leading to the development of high sugar content.

B. P. Palis is known for breeding superior disease-resistant varieties of wheat. Pushkarnath, a potato breeder, worked on producing high-yielding varieties of potato. N. G. P. Rao, a reputed sorghum breeder, developed several high-yielding hybrids of grain sorghum. K. Ramiah, a well-known rice breeder, led the production of high-yielding variants of rice. Dharmapal Singh employed this technique to release several varieties of oilseeds, such as rapeseed and mustard, from Kanpur. C. T. Patel, a renowned cotton breeder, developed the world's first cotton hybrid from Gujarat Agricultural University, Surat Station. V. Santhanam developed high-yielding varieties of upland and Egyptian cotton employing this technique.

Genetic engineers have successfully designed enormous schemes in carrying out crossbreeding to pass, retain, and manage essential, indispensable, and vital preferred traits. The classical method of plant breeding is most often quite confusing, unpredictable, and sluggish and might take a long time to yield desirable results. A mutual intentional crossing between two selected lines, followed by repetition in the same manner, can induce desirable characteristics. The science of plant breeding is quite cumbersome, which may take a minimum of 15 years before ultimately resulting in an altogether new variety. Nevertheless, its application is limited to a few categories that have the ability to hybridize sexually.

Recombinant DNA is a technology involving artificial fragments of DNA which can be engineered through varied combinations or insertions of suitable strands, thereby combining novel sequences. It involves the insertion of DNA fragments and the transfer or introduction of specific gene(s) from a suitable vector into varied genetic combinations and traits. There is the manipulation and isolation of DNA molecules from different species, resulting in the formation of a recombinant DNA that can replicate itself, bringing together genetic materials from multiple sources. These innovative combinations are of enormous value in the diverse fields of science, medicine, pharmaceutical, agriculture, and industry. They also contribute to improving health conditions through the discovery of lifesaving drugs, vaccines, and biopharmaceuticals. Many of these recombinant pharmaceuticals play a key role in the treatment of deadly, lethal diseases. Treatment strategies are also improved to a great extent by developing commercial products, hormones, vaccines, antigens, crucial pharmaceuticals, kits meant for diagnosis, monitoring instruments, and the latest approaches used therapeutically. The most remarkable examples of recombinant DNA technology include the production of a synthetic form of human insulin and erythropoietin with the aid of bacteria modified genetically. Another important achievement is the production of different variants of mutant mice meant for the purpose of research. It has brought a revolution by solving environmental issues, cleaning the atmosphere, causing the degradation of hazardous wastes into biofuels, cleaning carbon and toxic wastes, and detecting the presence of contaminants like arsenic in drinking water. An ample range of polymeric proteins has been produced which can be employed as enzymes in industries and the medical field (Ma et al., 2003). Due to their widespread usage, more drugs have been approved by the Food and Drug Administration (FDA) (Liu et al., 2013) for the treatment of ailments like anemia, AIDS, cancer, hereditary disorders, Gaucher disease, human growth hormone deficiency, hemophilia A, immunodeficiency syndrome, Turner's disease, foot ulcer in diabetic patients, diphtheria, warts, hepatitis B, and multiple sclerosis. It allows plants to make a replica of specific genes of advantageous traits, with the purpose of overcoming certain constraints, and also examines the distinct biological characteristics of plants. Genetically modified crops using this technology can diminish modern-day challenges in the field of agriculture (Verma et al., 2022). The technology has been successfully integrated into plant breeding and crop improvement programs, as evident in transgenic crops.

Genetic transformation implies the relocation and insertion of a foreign DNA molecule into the host cell using molecular biology procedures, thereby leading to successive generations inheriting the desired characteristics. The inserted sequence of gene is termed as *transgene*, which is acquired either from a completely different species or from a distantly related plant (Cristina et al., 1999; Rani & Usha, 2013).

Thus, gene transformation is a natural phenomenon involving the genetic alteration of one organism to another to induce altogether new changes in biological traits. Plant transformation has nowadays become indispensable in incorporating a particular gene of interest. It has become the most consistent method in integrating genes into nuclear genome to get the desired gene expression. The genetic transformation of cultured cells and tissues has proven successful in the generation of numerous agronomically useful varieties (Mubeen et al., 2016). The best advantage of genetic engineering is the easy method of gene transfer, which can be either intergeneric, interspecific, or even interkingdom (Kavipriya et al., 2019).

Genetic transformation is the most potent tool in analyzing plant functional genomics, gene exploration, and gene regulation and brings new insight of genetically regulated features. Furthermore, it aids in introducing foreign genes into existing species so they can produce expeditiously, creating novel genetically customized varieties with desired features, contributing to an enormous increase in percentage of crop production and quality. Genome sequencing carried out on numerous plants and organisms has resulted in the rapid advancement and development of this science. It is the most powerful tool in studying plant genomics, which includes gene discovery, its function, and its genetically controlled features.

7.1.3 Beneficial Impact of the Technology of Gene Transfer

1. Resistance to deadly viruses
2. Resistance to attack of insecticides
3. Reinforcement in the ability to fight numerous ailments
4. Strengthening of the ability to face adverse conditions and environmental extremes
5. Improvement in nutritional quality
6. Enablement of species of plants to grow nurtured in varied seasons
7. Easy inducement of resistance to herbicides
8. Ability to fight against pathogens
9. Provision of both biotic and abiotic stress tolerance
10. Aid in delayed ripening
11. Potential therapeutic modality in the medical field for the treatment of a range of diseases
12. Delay of leaf senescence; production of stress-resistant plants
13. Improved and sustainable crop yield and nutritional value

7.2 GENETIC TRANSFER TECHNIQUES

Genetic transfer technique implies the procedure followed in the relocation, integration, and at the same time, expression of a specific transgene into host cells. An unknown gene capable of encoding the required trait must be introduced into the plant which needs to be genetically modified. Along with this, an additional genetic material can also be added to incorporate a desirable trait. This includes a sequence of DNA termed as *promoter*, which decides how, where, and when this gene should be articulated in the host, and, in addition, another gene, termed as *marker*, through which breeders decide by selection techniques which plants should retain the inserted gene. With this latest technology, it is possible to introduce numerous innovative features, such as making plants resistant to antibiotics or tolerant of the attack of lethal herbicides, insecticides, etc.

The method of gene transfer technology is categorized as:

1. Indirect method: vector-mediated transfer of genes
2. Direct method: vector-less transfer of genes

These are quite powerful, effective, and potent techniques meant for the relocation of genes, in both plant and animal cells, with possible applications in gene therapy. The selection of method depends on the target cell and on the main aim of gene manipulation. Along with the choice of DNA transfer method, the most vital steps include assortment, choice, isolation of the desired gene, preparation of recombinant DNA, and further selection of the transformed cells.

7.2.1 Indirect Method: Vector-Mediated Gene Transfer

Plant transformation is a versatile method of incorporation of desired characteristics in the selected plant. A vector-mediated gene transfer, also termed as *indirect method*, is carried out in two processes, either by *Agrobacterium*-mediated plant transformation or an alternative, using viruses as selected vectors. In this successful technique, the required gene is initially combined with a biological vector which aids in transferring it to a particular cell for the purpose of integration, involving the introduction of a foreign DNA molecule. The plant gene vector can be used for transmitting genetic information between both plants and some other organism, such as a bacterium, a fungus, a virus, and an animal, to another plant. It entirely depends on the unique regeneration capabilities of the host plant. The gene vectors generally utilized in this technique are plasmids of *Agrobacterium* viruses and varied transposable elements. The *Agrobacterium*-mediated method of genetic transfer is a successful method which forms an efficient delivery system. This gram-negative soil bacteria is able to cause crown gall disease in many plants.

7.2.1.1 *Agrobacterium*-Mediated Transformation

Agrobacterium-mediated transformation is the most extensively opted method for genetic transfer, leading to the production of novel transgenic plants. This system was historically considered the first successful vector-mediated system, marking a breakthrough in the science of genetic engineering in the year 1983. In many countries, it is the first choice in the production of agronomically and horticulturally significant crops, such as corn, soybeans, cotton, wheat, rice, canola, potatoes, and tomatoes. Modern agricultural biotechnology is highly dependent on it to create transgenic plants, and a wide array of crop science research has benefited from it as well. Also, transformation efficiency has increased to a great extent, depending on the strain of *Agrobacterium* used for transformation (Gelvin, 2003; Nyaboga et al., 2014).

Agrobacterium is the most common natural bacterium, with species *A. tumefaciens* and *A. rhizogenes. A tumefaciens* is recognized for inducing tubers termed as crown galls; on the contrary, the other species, *rhizogenes*, causes hairy root disease. Out of the two known species, the maximum contribution for the advancement of genetic transformation is dependent upon the innate exclusive capability of *A. tumefaciens.* It is considered to be the preferably employed method in case of dicot plants, where the system of plant regeneration is quite well-recognized.

The mandatory requirements for this process are that, in particular, the explants chosen from the plant should produce phenolic compounds capable of activating the virulence genes, and the resulting transformed tissues should have the tendency to redevelop into a whole new plant with desired characteristics.

The basic protocol for this plant transformation includes:

1. *Agrobacterium* development bearing the double vector with the required selected gene
2. Precise detection and selection of an appropriate explant, which can be a cell, tissue, section of the protoplast, callus, or specific organ
3. Carrying out of co-culture of selected explants with a strain of *Agrobacterium*
4. The process of killing of an *Agrobacterium* with an appropriate antibiotic with no harm to the tissue of the plant involved
5. Selection of plant cells which have been transformed
6. Finally, regeneration and booming of a complete plant

7.2.1.1.1 Advantages

On of the major advantages of this method is that it is a natural and successful means of transmitting genes with the capability of infecting plant cells, tissues, as well as organs in transferring bulky DNA molecules. The incorporation of T DNA is a relatively accurate, specific, and successful method, with excellent stability of the transferred gene, and transformed plants can be regenerated effectively. Along with advantages, this technique is quite host-specific, because most crop plants, including pulses and cereals, are quite safe from and not infected by it. For this purpose, potent strains of *Agrobacterium* that have the ability to influence a wide variety of plants have been employed.

7.2.1.1.2 Disadvantages

It shows an absence of the capability to transfer a number of multiple genes which are not easy targets. The cells that can regenerate more effectively pose a problem as they are tricky to transform, for example, embryonic cells, which are quite embedded and lie in deep layers within. Other important limitations that restrict its usage are somaclonal variations and slow regeneration.

7.2.2 Direct Method: Vector-Less Gene Transfer

In a few cases, if the DNA strand is inserted directly into the plant genome, the process is called *direct gene transfer*, also termed as *vector-less transmission*. A direct approach of DNA transfer is dependent on an uncovered DNA delivered directly into the desired cells, in contrast with transfer through an additional source, as already mentioned in a vector-mediated process. The majority of the methods employed in vector-less gene transfer are quite easy to perform without any hassle and, at the same time, very effective, and they are thus used to develop numerous transgenic plants. The concept of introducing DNA into selected plants without the help or involvement of any additional biological moiety, leading to a stable and efficient transformation process, is termed as a *direct approach of gene transfer.*

7.2.2.1 Direct Methods of DNA Transfer

The direct method and/or approach for the transfer of DNA is largely divided into the following classes, along with their subsequent subclassifications:

1. Physical methods of gene transfer
   a. Electroporation
   b. Particle bombardment
   c. Microinjection
   d. Liposome-mediated transformation
   e. Silicon carbide fiber–mediated
2. Chemical-mediated gene transfer
   a. Polyethylene glycol–mediated
   b. Diethylaminoethyl dextran–mediated
   c. Calcium phosphate precipitation
3. DNA imbibition by cells/tissues/organs
4. Transformation with the aid of pollen grains
5. Delivery through growing pollen tubes
6. Laser-induced alteration

7.2.2.2 Physical Methods of Gene Transfer

7.2.2.2.1 Electroporation

The electroporation process of physical gene transfer was first established by scientists Wong and Neumann in the year 1982 in carrying out a study of gene transfer in mouse cells. It involves the inclusion of DNA by exposing them to a voltage of high intensity and electrical pulses for quite a short duration, leading to the formation of temporary pores in the plasma lemma. The basic principle is the comparatively fragile hydrophobic or hydrophilic kind of interaction among the phospholipid bilayer and their capability to instinctively arrange themselves. A rapid and fast current shock can lead to a momentary disturbance, allowing the way specifically for polar molecules. The membrane immediately reseals itself, leaving the cell undamaged afterward. It is a widely employed mechanical method of introducing a variety of polar molecules through this membrane and transgene into a variety of viral, bacterial, fungal, plant, as well as animal cells. It requires the use of a high-frequency electric pulse that has the tendency to momentarily create a disturbance in the phospholipid bilayer, thereby allowing the route for the passage of DNA. Both the host cells and the DNA fragments which are supposed to be transported into the cells are initially submerged for a short while in a solution. After the closure of the first switch, the capacitor gets charged up and saves an ample quantity of high voltage which immediately gets discharged upon the closing of the second switch. On average, 10,000–100,000 V/cm of pulse, which may last in a duration of a few microseconds to a millisecond, is required in this technique, which can vary according to the size of the cell. This electric pulse generated tends to disrupt the phospholipid bilayer present in the membrane, leading to the formation of aqueous pores, which are temporary. When the value of potential across the cell membrane is further increased to 0.5–1.0 V, these stimulating molecules move around the membrane with the aid of these pores. The initiation of this process usually happens when the value of the transmembrane voltage reaches the maximum range of 0.5–1.5 V. The membrane of the cell discharges with the successive inflow of ions, molecules bearing charge, and the membrane pores close immediately, followed by the shuffling and reassembling of the phospholipid bilayer.

The plant cell electroporation technique generally makes use of protoplasts, but opposite to that, thick cell walls of the plant can limit the movement of big-sized, bulky macromolecules (Bates, 1999). The plant is initially incubated in a buffer solution with the preferred DNA pieces and later on exposed to high-current electrical impulses, which aids in forming small temporary holes inside the membrane for the passage of DNA. After entering the cell, the foreign DNA tends to be incorporated into the host genome, resulting in the transformation of the genetic material, and these protoplasts can then be cultured for the regeneration of a complete plant. This method is considered best for species of crops with which it is feasible to regenerate from the protoplast. In previous years, only protoplasts were utilized for the phenomenon of gene transfer, but currently, unbroken pieces of cells, callus tissue, cultures, and immature embryos can also be employed after giving both pre- and post-electroporation treatment.

The electroporation method has been effectively employed in producing transgenic plants of a variety of cereals, for example, rice, wheat, and maize. It is mainly used for DNA transfection, as well as transformation, involving the introduction of foreign DNA into various selected cells. There is direct, straightforward relocation of plasmids in between cells involved in bacterial cell incubation with a different strain of plasmid which possesses various other desired features. This method of transfer can be easily performed among species, and a huge number of plasmids can be easily grown in speedily dividing bacterial colonies with the use of electroporation. It is the most preferred choice in gene transfer to a widespread collection of tissues, including those of muscle, brain, lung, skin, kidney, liver, artery, cornea, etc. The best part of this method is that vector-specific immune responses observed in recombinant viral vectors are avoided; thus, it is quite promising and beneficial in a number of clinical applications.

7.2.2.2.1.1 Advantages It is highly diverse, adaptable, and successful in almost every distinct type of cell as well as species. It is considered to be a highly competent way of transfer, as most cells are able to incorporate the target DNA molecule. It can be very easily performed on an even smaller level, and the quantity of DNA molecules required is lesser compared with other methods. This technique is simple, effortless, convenient, cost-effective, and rapid. The resulting altered cells are in an identical physiological state, and their competence can be further improved to a great extent with the optimization of the percent of electrical field strength and by adding agents such as spermidine.

7.2.2.2.1.2 Disadvantages Cell damage and spoilage are among the major disadvantages observed in this method, which may be caused by irregular and unbalanced intensity pulses creating massive sizes of pores, which in a few cases fail to close after discharge of the membrane. One more problem faced is non-specific transport causing the improper functioning, ion imbalance, and in a few cases, mortality of the cell. Under regular conditions, the percentage of DNA which can be delivered into plant cells is meager in amount, and at the same time, its efficiency is quite inconsistent and unpredictable, depending on the type of plant and the conditions provided during treatment. Last but not the least, the regeneration of a complete plant is not feasible, specifically when cell protoplasts are employed.

7.2.2.2.2 Particle Bombardment/Microprojectile/Biolistic/Gene Gun/Particle Acceleration

In 1987, Prof. Sanford at the University of Cornell in the USA developed for the first time the concept of original bombardment and, based upon that, introduced the term "biolistics." It is also known as *particle bombardment, particle gun, microprojectile bombardment, biolistic/gene gun*, and *particle acceleration*. It makes use of high-speed microprojectiles for the release of desired substances into cells. It has been very successful in carrying out the transformation of a variety of cereals, including rice, barley, grains, wheat, pulses, and maize. The first genetically modified crop, which was maize containing the BT toxin gene, used commercially till now, was successfully developed by the process of particle bombardment. Thus, particle bombardment is an innovative technique meant for introducing the selected foreign DNA into cells (Birch & Franks, 1991). It is considered as the most effective and appropriate method of gene transfer in plants, mammalian cells, and microorganisms. It makes use of unfamiliar DNA strands coated evenly with a minute layer of 0.2–0.7 μm gold and tungsten particles, which can be easily delivered to desired plant cells. The mechanism involves loading the particle gun with the coated particles and accelerating it to maximum velocity by using helium gas under great pressure or an intense amount of electrostatic energy released by a droplet of water when exposed to a high voltage. The target could be the miscellaneous suspension of plants, specific cells, culture, callus, and various tissues. The projectiles penetrate the plant's cell wall and membrane, releasing transgenes for further incorporation into the plant DNA.

This process is widely employed in the genetic transformation of numerous organisms that have a comparatively lesser capacity to regenerate, specifically those which do not respond to the *Agrobacterium*-mediated method in crops like rice, maize, pulses, corn, wheat, chickpea, sorghum, and pigeon pea. Tungsten- and DNA-coated gold particles are fully loaded into the gun in this technique, and the trigger is pulled. In addition, a helium pulse of low pressure assists in the delivery of gold- as well as tungsten-coated particles into the target cell, integrating their DNA fragments into the genome, thereby enabling the host cell for both transcribing and translating its genetic material.

7.2.2.2.2.1 Advantages It is considered to be a simple, trouble-free, and convenient method that involves the coating of both DNA and as well as RNA with a gold microcarrier, the loading of the sample, and the followed by the pointing of the nozzle in the required direction and firing the device. There is no requirement to obtain protoplasts, because the wall of the cell is intact and can be very easily and simply penetrated. The manipulation and variation of the genome of organelles can also be performed, and at the same time, it eradicates the usage of potentially damaging viruses, or lethal, toxic chemical treatment. It has the additional advantage of placing DNA or RNA molecules in the exact position where it is actually required by the organism.

7.2.2.2.2.2 Disadvantages The competence of transformation can be comparatively lesser in contrast to *Agrobacterium-mediated* technique. A special advanced equipment is required to improve its efficiency, and the device, apparatus, material, and required consumables are expensive. In a few cases, the damage of associated cells can be a major risk factor. The target tissue should have the capacity to regenerate quickly, and random integration is also a major obstacle. Sometimes, the probability of multiple copy insertions can also lead to gene silencing.

Generally, plant tissues which are frequently employed in carrying out this technique include primary explants, which can be subjected to particle bombardment quite easily, and embryonic tissues in the proliferation stage, which can be initially bombarded in cultures and later on kept for propagation and regeneration. To protect these vulnerable tissues and cells from being spoilt, cultures should be either maintained on a raised level of osmoticum media or subjected to the phenomenon of plasmolysis. It is believed that the transfer of specific genes in bombardment is actually a process involving two stages. The first phase is called the *pre-integration phase*, in which DNA molecules, acting as vector, are spliced together, resulting in multiple copies of gene fragments. The integrative phase involves the inclusion of copies of genes into the selected host genome, which further facilitates the process of integration. The outcome is that the particle bombardment technique is related with the production of a high copy number, which may be beneficial for regeneration.

7.2.2.2.2.3 Factors Affecting Bombardment

1. *Nature of microparticles used.* Neutral metals, which include tungsten, gold, platinum, etc., are employed as microparticles to transfer DNA molecules. These, with moderately high mass, will move speedily when bombarded and can easily infiltrate tissues. Thus, the type and chemical nature of microparticles may play a key role.
2. *Nature of tissues/cells.* The nature and type of target cells competent in carrying the process of division are best selected for alteration.
3. *Quantity of DNA.* The rate of transformation is observed to be quite slow when a lesser quantity of DNA is utilized. On the contrary, a high amount of DNA can lead to the production of much more copy number, along with the reshuffling and organization of transgenes. Thus, the amount of DNA used for this purpose should be well balanced to accomplish a high level of efficiency in the process of transformation.
4. *Environmental factors.* Miscellaneous environmental parameters are observed to affect particle bombardment, which include atmospheric conditions like temperature, humidity, and photoperiod. These can easily manipulate and alter the physiology of the plant material and, thus, affect the level of gene transfer.

7.2.2.2.2.4 Advantages The major advantage is that the transfer of the desired gene can be effectively done in almost all organized tissues, and a varied plant variety can be employed in the development of transgenic plants.

7.2.2.2.2.5 Disadvantages The major complication in this process is the production of a high number of transgenes, leading to instability in transgene expression because of a phenomenon called gene silencing. The worst part is that the target tissue may get spoiled due to a lack of control of the velocity used in bombardment, and in a few cases, chimeric plants may be regenerated, which are undesirable.

7.2.2.2.3 Microinjection

Microinjection was first initiated by Dr. Marshall A. Barber quite early in the 19th century. In this technique, the DNA strand can be delivered directly without any vector into single living cells with the use of very fine, hollow needles in the form of glass micropipettes. The introduced DNA leads to the expression of certain, specific desired genes which can be further employed to recognize the distinguished functions of dominant genes. Experimenters employing this method use a sophisticated and complex equipment consisting of a powerful microscope and quite fragile micromanipulators to visualize the cell, keep it in hold, and further inject a solution containing the desired DNA molecules (Council, 1987).

It is a direct physical approach for the inclusion of the desired DNA into a selected target cell mechanically, which can be from varied sources, including cells, protoplasts, callus culture, meristems, embryos, etc. It is generally preferred both for the transmission of cellular organelles and also the amendment of chromosomes. In this procedure, a solution of DNA is prepared and directly incorporated inside the cell with the help of capillary glass micropipettes with a 0.5–10 pm tip employing micromanipulators of a microinjection assembly. It is observed to be comparatively easier using protoplasts than cells because the cell wall can interfere and obstruct the phenomenon of microinjection. The protoplast is generally immobilized in a solution of agarose on a glass slide initially covered with polylysine, or by holding them under suction using a micropipette. After completion, the transformed cell is carefully cultured and grown, which later on develops into a complete transgenic plant. A variety of transgenic species of *Brassica napus* and tobacco have been effectively prepared by using this approach. The injection containing the plasmid DNA put inside the lumen of a developing inflorescence by employing a hypodermic syringe is termed *microinjection*. An aqueous solution containing DNA is prepared and introduced into the developing floral tillers 14 days before the start of meiosis. After cross-pollination, seeds which are completely transformed are obtained from these injected tillers. The best part is that there is no requirement of a protoplast, and the instrument used is simple, trouble-free, and economical. This method is preferred in the transfer of genes into cereals, which face a problem in regenerating from cultured cells easily. The direct microinjection of DNA strands into the cultured cells' cytoplasm is widely used as a method of transfection which is extremely competent at the level of individual cells. The main noteworthy application is in the direct introduction of DNA into the oocytes, eggs, and embryos of animals to generate transgenic animals. It can be a very cumbersome, time-consuming, lengthy procedure, and the treatment of only a limited number is possible. Initially, it was employed for the transformation of specific cells that showed resistance to other means and whose transfection was challenging. Its main benefit is that it provides for the straightforward delivery of DNA into the nucleus, avoiding any alternative pathway, and it also makes sure that it is delivered complete and intact. Microinjection is the method of choice for the introduction of huge and bulky vectors, such as introducing YACs into the mouse egg pronuclei. But it has to be taken care of that DNA transmitted through this must be very pure and uncontaminated, which may require a lot of training and preparation. There are chances of shearing in the delivery needle, and huge-sized DNA molecules can be easily protected by suspending them in a highly salted buffer or by using protective agents, such as polyamines. At last, the transfection step of cultured cells is completely automatic, with computer-aided micromanipulation and microinjection techniques, the latest injection capillaries, and the prior standardization of cell preparation procedure.

7.2.2.2.3.1 Advantages The best advantage of microinjection is that there is no requirement of a marker gene. It is possible to introduce the desired target gene into a single cell or multiple cells directly. There is easy detection and recognition of transformed cells, with no requirement of selection of the transformed cell. Used in creating transgenic organisms, the direct nuclear delivery of DNA keeps it intact and suitable for the introduction of large vectors.

7.2.2.2.3.2 Disadvantages The major disadvantage is that it works at as low pace, is time-consuming, is high-priced, and is mandatory to be performed by both trained and highly skilled personnel. This procedure is technically quite demanding and takes a lot of time, as only a maximum of 40–50 protoplasts can be microinjected in an hour. Another drawback is that it is less-specific and less-precise. In addition, the frequency and the speed of transformation obtained are quite low.

7.2.2.2.4 Liposome-Mediated Transformation

This method was first explained in 1965 using carriers such as liposomes, artificially prepared spherical phospholipid vesicles utilized for delivery purposes. This method of transformation shows promising results as intracellular delivery systems. Preloading with DNA content by two methods is done, which includes membrane fusion and endocytosis, leading to the formation of a liposome complex with DNA which integrates with the cell membrane of target cells and releases the content into the cell. Liposomes are chosen as an ideal and perfect drug carrier system since their external morphology is quite the same as those of cellular membranes and because of their extraordinary ability to incorporate a range of substances. Liposomal encapsulation technology, also called LET, is the most recent delivery option preferred by medical investigators for the transmission of specific drugs that aid in curing and act as healing promoters to affected body organs (Akbarzadeh et al., 2013). These are artificially created vesicles enriched with lipids containing a phospholipid membrane effectively employed in mammalian cells in the delivery of a mixed range of compounds. Genes carrying liposomes can be easily engaged to fuse with cell protoplasts and transmit the particular gene of interest. Efficiency can be further increased to a great extent when the procedure is performed in combination with polyethylene glycol (PEG). The mechanism behind this involves the adhesion of particles of liposomes to the protoplast surface, its fusion at the attachment site causing instant plasmid release inside the cell.

Because of its unique properties of compatibility, biodegradability, less toxicity, affinity to both hydrophilic and lipophilic drugs (Johnston et al., 2007), and simple drug delivery to tumor tissues (Hofheinz et al., 2005), its usage has shown an increase at an enormous rate as an investigational and commercial system of drug delivery.

7.2.2.2.4.1 Advantages Because of its encapsulated form, DNA fragments are well-protected from adverse environmental attacks and damage, making it stable, constant, and easily stored for a long period before transfer. This method is appropriate for numerous plant cells, with a good rate of reproducibility and minimal toxicity. There is the protection of the nucleic acid without degradation, and it is quite economic. An efficient delivery of nucleic acid and targeting ligands, like peptides, antibodies, aptamers, and folic acid, is possible, with the retention of the enhanced efficiency of drug delivery.

7.2.2.2.4.2 Disadvantages The major trouble with this method are the complications related to the further renewal of plant cells and tissues from selected protoplasts. Thus, this technique has not been commonly employed because of the additional difficulty of constructing lipid vesicles.

7.2.2.2.5 Silicon Carbide Fiber–Mediated Transformation

This is quite a simple, easy, and inexpensive technique making use of special silicon carbide fibers that can be employed to transform reproducibly in monocot and dicot species. The fibers made up of a silicon carbide popularly known under the trade name Whiskers are easily accessible nowadays in the market (Kaeppler et al., 1992). These special silicon carbide fibers have been widely considered for a number of high-tech applications because of its important benefits, which include strong tensile strength and elastic modulus, excellent property of shock absorption, and resistance to degradation (Choi, 1997). They possess immense intrinsic hardness and have the ability to fracture readily to give sharp cutting edges (Greenwood & Earnshaw, 2012). These can be easily obtained by carrying out thermal reduction of silica in a reducing atmosphere (Mutsuddy, 1990).

In this process, the required plant material is introduced into a buffer having both DNA and silicon fibers. The fibers go through the wall of the cell as well as the plasma membrane, thereby allowing the DNA molecule to gain entry into the vicinity of the cells. It can be further used to speedily deliver alien DNA into whole plant cells for the purpose of investigation of transient gene expression (Kaeppler et al., 1990).

The DNA is initially transferred into the cell cytoplasm and nucleus with the help of silicon carbide fibers roughly 10–80 μm in length and 0.3–0.6 μm in diameter, which is quite the same as in microinjection. These special fibers have the capability to pierce deeply inside both the cell wall and the plasma membrane, thereby successfully transporting DNA into the cells. These silicon carbide fibers coated properly with DNA are vortex and thoroughly mixed with the suspension culture of the plant material. During this mixing, DNA molecules, which closely adhere to the fibers, can easily enter particular cells and, at the same time, get stably integrated with the host genome. The method has proven successful and has shown booming results with maize and tobacco suspension cell culture. In later years, various improvements and amendments have been made in the technique silicon carbide fiber–mediated transformation, which has greatly altered a variety of crops, such as rice, jute, wheat, maize, barley, and many more.

7.2.2.2.5.1 Advantages It allows the direct, instant release of selected DNA into intact-walled tissues and cells, thereby avoiding the separation of the protoplast. The procedure is quite uncomplicated, hassle-free, rapid, and inexpensive, without the involvement of any expensive, sophisticated, and multifunctional apparatus.

7.2.2.2.5.2 Disadvantages The major concern with silicon carbide fibers is that they are carcinogenic and thus need utmost care throughout its handling. In addition, embryonic plant cells are comparatively quite rigid, stiff, and condensed and are opposed to the penetration of these fibers, making the whole process tedious and challenging, with low rate of success.

7.2.2.3 Chemical Gene Transfer Procedures

7.2.2.3.1 Polyethylene Glycol–Mediated Transfer

There is no requirement of a special equipment or apparatus, and this technique can be very easily performed in a laboratory simply by employing a sterile hood (Liu & Vidali, 2011). The best part is, it produces a transformed population of cells with quite a high rate of survival, with a quick division capacity (Potrykus, 1991). Another beneficial advantage is that it can be readily adapted to an extensive range of species, further making its application broader. Polyethylene glycol (PEG) has the tendency to disrupt the plasma membrane of protoplasts completely in the presence of divalent cations, such as Ca^{2+}, thereby making it easily permeable for the entry of naked DNA molecules. This protoplastic culture containing molecules of DNA are then poured into a tube, along with the addition of a 40% mixture of PEG 4000 (w/v) dissolved in mannitol. Afterward, a solution of calcium nitrate is added quite gradually and kept for a few minutes for incubation. DNA molecules can forcibly penetrate the host genome is only a few selected cells which lack cell walls. The naked protoplasts of the plants are thoroughly mixed with molecules of DNA strands containing the foreign gene particle. Further, these two contents are combined together in a selective transformation medium rich in Mg^{2+} ions, followed by the addition of a 20% polyethylene glycol (PEG) solution. After this subsequent treatment procedure is carried out, the concentration of polyethylene glycol is comparatively reduced and, on the contrary, Ca^{2+} concentration is enhanced. This further promotes and encourages the speed and rate of the intentional transformation procedure. The chemical method carried out in the presence of polyethylene glycol is considered to be quite convenient, effective, and uncomplicated.

Polyethylene glycol, along with divalent cation, has the tendency to destabilize and weaken the plasma membrane of protoplasts, thereby rendering it permeable for the entry of naked DNA strands. Through this method, the desired DNA can easily make an entry into the nucleus of the protoplast to get itself associated with the genome. This particular procedure involves various vital steps, such as isolation and separation of the suspension of protoplasts, adding up of plasmid DNA, later on followed by the addition of 40% PEG 4000 (w/v) readily dissolved in mannitol slowly, as well as a solution of calcium nitrate. After the incubation of the prepared mixture for a specific period of time, protoplasts are easily transformed.

7.2.2.3.1.1 Advantages A large number of protoplast cells can be concurrently altered by this scheme, and this PEG technique can be effectively employed for a wide variety of plant species with a high success rate.

7.2.2.3.1.2 Disadvantages It has been observed that many cells are quite sensitive and vulnerable, so that the chemical method cannot be processed, leading to their death in the course of treatment. This method is not considered ideal for a few cases because many of the chemically treated cells do not possess any transfer DNA. Many times, the foreign molecule of DNA gets deteriorated in the cytoplasm itself before even reaching the nucleus. DNA is quite prone to effects of degradation and gets rearranged, making it vulnerable. Many a times, the random incorporation of a new strand of DNA into the genome can lead to undesirable and detrimental behaviors. Using this method, the regeneration of a complete plant from transformed protoplasts is a complicated and tedious assignment.

7.2.2.3.2 Diethylaminoethyl Dextran–Mediated Transfer (DEAE)

This technique of employing DEAE was originally presented by Vaheri and Pagano in 1965 in enhancing and improving the viral infectivity of the cell, but after a few years, it was followed up and considered a successful process for the transfer of plasmid DNA. Diethylaminoethyl dextran, also termed DEAE-dextran, is found to be a soluble form of polycationic carbohydrate moiety that can easily promote interaction among DNA and the endocytic machinery existing in the cell. The principle behind this procedure is that the negatively charged DNA and the DEAE-dextran moiety possessing a positive charge complex with each other, leading to the development of clumps through the phenomenon of electrostatic interaction, forming a polyplex. A minor surplus quantity of DEAE-dextran in this mixture results in the development of the overall net positive charge in the formed complex. The prepared complexes, upon the addition of cells, have the capability to easily bind to the plasma membrane bearing a negative charge and make entry by the process of endocytosis. Complexed DNA delivery with molecules of DEAE-dextran can be further enhanced to a great extent through osmotic shock using either DMSO or glycerol. A range of factors including the number of cells, concentration of the polymer, transfected concentration of DNA, and duration meant for transfection, should be optimized and fixed for a particular given cell line.

7.2.2.3.2.1 Advantages This is quite a simple and inexpensive way of gene transfer, but also quite sensitive and susceptible. It can be very efficiently used for a broad range of cells and can be performed in transient transfection quite successfully.

7.2.2.3.2.2 Disadvantages It is highly lethal for a few cells when administered in great concentrations. In this technique, the efficiency of transfection shows an alteration with the cell type. Further, it is not the method of choice for stable transfection. It has the capacity to typically produce less than 10% delivery in certain primary cells. The problem with the DEAE-dextran-mediated method is that it is not particularly competent in producing stable transformed cell lines. Thus, the major limitation observed is that it yields unstable transformants, which limits its usage.

DEAE-dextran was the first transfection reagent discovered and was quite commonly preferred until lipofection reagents came into existence in the 1990s. It is basically a solubilized form of polycationic carbohydrate that can readily form clumps with DNA through the mechanism of electrostatic interaction. It makes the whole complex positively charged, allowing an interaction with the negatively charged cell membrane, thereby promoting uptake through the process of endocytosis. The complexes formed are quite small in comparison to the particles formed in other procedures of transfection. Quite a meager quantity of DNA can be employed in each experimental step. One common similarity is that the reagents used are quite reasonably priced, and the process is again easy to perform and highly effective. The desirable required DNA can be further complexed with the heavy, high-molecular-weight polymer diethylaminoethyl dextran (DEAE-dextran) and easily transferred to the desired cell. This efficiency can be increased to an extent of 80% when DMSO shock is employed.

7.2.2.3.3 Calcium Phosphate Co-Precipitation–Mediated Transfer

This is considered to be the first chemical transfection technique capably utilized for both animal and plant cells. Calcium phosphate co-precipitation–mediated transfer is the most extensively used technique in introducing foreign DNA plasmids into selected cells. It is an effortless, easy-to-perform, and reliable method appropriate for most cultured cell lines with the use of inexpensive reagents. The best part is, it shows the best results both for transient and stable transformation. The main principle involved is that DNA taken in a buffered phosphate solution is mixed smoothly and thoroughly with calcium chloride, leading to the formation of a smooth, fine co-precipitate of DNA and calcium phosphate. This formed precipitate has the tendency to settle nicely, and a few of the required particles are taken up by the process of endocytosis. The most effective transfection occurs in those which are growing in the form of a monolayer, because they are evenly coated with the required precipitate. The procedure was invented long ago, back in 1973, by scientists Graham and Van Der Erb for the purpose of introducing adenovirus DNA into rat cells. Later on, in a report published in 1962 by Szybalska and Szybalski, the occurrence of calcium was predicted to be accountable for carrying out the successful alteration of human cells with genomic DNA. It was observed that the first mammalian cell lines firmly transfected with plasmid DNA were also formed by this procedure in 1978.

The fundamental mechanisms underlying this procedure are still not well defined clearly, and though it has been observed that DNA–calcium phosphate precipitates can easily penetrate cells internally, DNA is quite well expressed in all cell types. The cost factor, efficiency parameter, ease of performing, and simplicity involved allow its effective usage in primary neuronal cultures, though with minor issues, such as neurotoxicity (Kwon & Firestein, 2013).

In a balanced and optimized calcium phosphate technique of plasmid delivery into primary dissociated neuronal cultures, initially the DNA is mixed with a calcium chloride solution and isotonic phosphate buffer, forming a DNA–calcium phosphate precipitate. In the next step, the aggressively dividing cells in the culture are brought into contact with this precipitate and kept for several hours, leading to the transformation of cells quite effectively. Success rate is highly reliable on the excess DNA employed and, at the same time, the protection of the complex precipitate formed. It has been observed that the addition of dimethyl sulfoxide (DMSO) can further increase the efficiency and level of transformation to a great extent.

7.2.2.4 DNA Imbibition by Cells/Tissue/Organs

DNA imbibition by cells/tissue/organs involves the uptake of exogenous DNA, specifically of dehydrated plant tissues devoid of water content. Desiccated tissues of plants can easily uptake DNA molecules through this innovative process of DNA imbibition (Töpfer et al., 1989). Throughout the procedure of desiccation, a number of physiological and substantial alterations are supposed to happen, including cell wall bursting, leakiness, and variations in cell membrane structure, fast expansion, and the development of a huge water flow. The uptake and transitory expression observed in cereal seed embryos of the DNA plasmid employing this process of DNA imbibition by cells, tissues, and organs were confirmed by Töpfer et al. (Abdul Qayyum et al., 2016).

The selective permeability of the cell membrane is a key deciding aspect of this phenomenon and has to be given utmost importance and notice. This was confirmed by a subsequent increase in uptake as well as expression of DNA after the use of 20% DMSO in this process. The additional benefit is that it is the easiest to execute out of all gene transformation techniques, because target plant tissue formation is quite basic and no specialized apparatus is required to perform it. No sophisticated instruments are required in performing it, and even the preparation of the target plant tissue is quite basic and uncomplicated.

Dry isolated embryos of selected crops like wheat, barley, rye, pea, and bean are completely immersed in a solution of DNA, subsequently leading to the absorption of DNA, and eventually will depict the ideal expression of the marker gene. It has been observed that dry seeds, generally with the removal of seed coats, also have the capability of taking up DNA when dipped in the prepared solution. Further, these imbibed seeds are successfully germinated on a suitable selective medium for the isolation of transformed embryos. It has been experimentally observed that DNA is easily taken up by embryos through cells injured during the isolation step, which then move through plasmodesmata to varied cells.

7.2.2.5 Pollen Transformation

The pollen tube transformation (PTT) method featuring exogenous DNA strand transfer into the plant embryo was proved successfully in the cotton plant for the first time (*Gossypium hirsutum* L.) (Zhou et al., 1983). The pollen tube transformation technique involves the exclusion of the stigma of the recipient plant immediately subsequent to the process of pollination and fertilization, with the further application of an exogenous DNA solution to the style part (Luo & Wu, 1988). This specific donor DNA is further delivered via pollen tube growth till it reaches the recipient plant ovary, where it gets integrated with the fertilized recipient zygote. Thus, for the pollen tube transformation technique to be considered successful, foreign genes can be well and easily included into the recipient's genome at the initial stage of embryo formation itself and are thus shown in transformed seeds as well.

In contrary to other methods, it is comparatively simple, effortless, with ease of performance, and at the same time, it is devoid of the traditional regeneration problems generally observed (Ali et al., 2015; Zhang et al., 2005). Recently, the success of pollen tube–mediated transformation in various economically vital crops clearly suggests that it could also additionally be a method of choice in some obstinate plant species (Huang et al., 1999; Bibi et al., 2013).

This innovative method involves the mechanism of gene relocation by initially soaking the pollen grains in a solution of DNA before its use meant for the purpose of pollination. The method is catching a lot of interest due to its simplicity, effectiveness, and applicability to a wide range of varied species, as depicted.

7.2.2.6 DNA Delivery via Growing Pollen Tubes

In this innovative procedure, stigmas are cut after inducing the process of pollination, clearly revealing the pollen tubes; furthermore, DNA is applied onto the sliced, exposed surface most probably diffused through the germinating pollen tube straight away into the ovule. This process of DNA delivery via growing pollen tubes is considered to be very simple and easy to perform, yielding a promising, consistent results and achieving stable transformations. One major issue in carrying out DNA delivery via growing pollen tubes is that the mechanism underlying DNA transfer into the zygote is still not well-established, thereby limiting its usage on a broad scale.

7.2.2.7 Laser-Induced Transformation

Nowadays, a lot of attention has been created in the utilization of laser apparatus, specifically for nonviral targeted gene transfer, because of the special spatial characteristics of laser light, which are quite precise, well-defined, and proved to be successful (Hongbao & Shen, 2007; Ogura et al., 2004).

In this special technique, the plasma membrane of numerous cells can be momentarily, as well as permanently, permeabilized by various optical methods, and the desired exogenous material can be easily introduced into living cell cytoplasm. As of now, only some mechanisms have been exploited and explained for its manipulation, since laser has the ability to directly focus on the target cells for the purposes of optoinjection, photochemical internalization, laser-induced stress waves, and irradiation of desired cells, which are targeted with light-absorbing particles. In previous years, widespread progress and plentiful breakthroughs have been made in this specialized area, making it a successful method of transformation (Yao et al., 2008).

This innovative method employs a pulse which stays for a short period of time through a specialized focused laser beam. In this procedure, DNA is primarily initially mixed with the cells already present in the given culture, and then, further, a fine clear focus of laser beam is passed onto the surface of the cells, leading to the formation of a minute pore which may cause a sufficient uptake of DNA into cells. The formed pore is quite short-lived, existing for only a short period of time, and can be easily repaired very soon. Laser punctures have the excellent ability to create momentary holes in selected cell membranes for the entry of DNA, which can then easily enter into cell cytoplasm.

Thus, this procedure involves the introduction of selected DNA strands into desired plant cells, altering it completely with the aid of a special laser microbeam. Stress waves induced by laser technique are very precise, leading to the transfer of genes in a targeted manner, as the small pores they create allows DNA from the surrounding solution to enter into the cell cytoplasm. Strong pressure waves induced by nanosecond laser pulses can be employed to transport macromolecules to desired cells. It is further ascertained experimentally that a strong pressure wave, termed a photomechanical or laser-induced stress wave (LISW), also accompanies laser-induced plasma. It has its own additional advantages in terms of stability, constancy, specificity of the host, number of copies, etc., leading to its high success rate.

REFERENCES

Abdul Qayyum, R., Muhammad Azam, A., Muhammad Azmat Ullah, K., Kamran Shahzad, B., Adnan, I., Tahir, I., Ahmad Ali, S., Idrees Ahmad, N., & Tayyab, H. (2016). Science behind cotton transformation. In Y. A. Ibrokhim (Ed.), *Cotton Research* (pp. Ch. 10). IntechOpen. https://doi.org/10.5772/64888

Akbarzadeh, A., Rezaei-Sadabady, R., Davaran, S., Joo, S. W., Zarghami, N., Hanifehpour, Y., Samiei, M., Kouhi, M., & Nejati-Koshki, K. (2013). Liposome: Classification, preparation, and applications. *Nanoscale Research Letters, 8*, 1–9.

Ali, A., Bang, S. W., Chung, S.-M., & Staub, J. E. (2015). Plant transformation via pollen tube-mediated gene transfer. *Plant Molecular Biology Reporter, 33*, 742–747.

Bates, G. W. (1999). Plant transformation via protoplast electroporation. *Plant cell Culture Protocols*, 359–366.

Begna, T., & Okonkwo, J. (2020). Role of recombinant DNA technology in agriculture. *International Journal of Research in Agriculture and Forestry, 7*(12), 8–15.

Bekana, G. (2017). Review on application of genetic engineering to crop improvement and its perceptions in Ethiopia. *Advances in Life Science and Technology, 60*, 1–7.

Bibi, N., Fan, K., Yuan, S., Ni, M., Ahmed, I. M., Malik, W., & Wang, X. (2013). An efficient and highly reproducible approach for the selection of upland transgenic cotton produced by pollen tube pathway method. *Australian Journal of Crop Science, 7*(11), 1714–1722.

Birch, R. G., & Franks, T. (1991). Development and optimisation of microprojectile systems for plant genetic transformation. *Functional Plant Biology, 18*(5), 453–469.

Choi, G. J. (1997). Silicon carbide fibers from copolymers of commercial polycarbosilane and silazane. *Journal of Industrial and Engineering Chemistry, 3*(3), 223–228.

Council, N. R. (1987). Gene transfer methods applicable to agricultural organisms. In *Agricultural Biotechnology: Strategies for National Competitiveness*. National Academies Press (US).

Cristina, A., Maria, Q., & Vieira, M.-L. (1999). Plant transformation: Advances and perspectives. *Scientia Agricola, 56*. https://doi.org/10.1590/S0103-90161999000100001

Degefa, I. (2019). Plant breeding methods: In brief for student. *International Journal of Agriculture and Biology, 3*(1), 156–203.

Gelvin, S. B. (2003). Agrobacterium-mediated plant transformation: The biology behind the "gene-jockeying" tool. *Microbiology and Molecular Biology Reviews, 67*(1), 16–37.

Greenwood, N. N., & Earnshaw, A. (2012). *Chemistry of the Elements*. Elsevier.

Hofheinz, R.-D., Gnad-Vogt, S. U., Beyer, U., & Hochhaus, A. (2005). Liposomal encapsulated anti-cancer drugs. *Anti-Cancer Drugs, 16*(7), 691–707.

Hongbao, M., & Shen, C. (2007). Study of gene transfection using laser irradiation and increased temperature techniques. *Nature and Science, 5*(3).

Huang, G., Dong, Y., & Sun, J. (1999). Introduction of exogenous DNA into cotton via the pollen-tube pathway with GFP as a reporter. *Chinese Science Bulletin, 44*, 698–701.

Humbert, O., Davis, L., & Maizels, N. (2012). Targeted gene therapies: Tools, applications, optimization. *Critical Reviews in Biochemistry and Molecular Biology, 47*(3), 264–281. https://doi.org/10.3109/10409238.2012.658112

Johnston, M. J., Semple, S. C., Klimuk, S. K., Ansell, S., Maurer, N., & Cullis, P. R. (2007). Characterization of the drug retention and pharmacokinetic properties of liposomal nanoparticles containing dihydrosphingomyelin. *Biochimica et Biophysica Acta (BBA)-Biomembranes, 1768*(5), 1121–1127.

Kaeppler, H. F., Gu, W., Somers, D. A., Rines, H. W., & Cockburn, A. F. (1990). Silicon carbide fiber-mediated DNA delivery into plant cells. *Plant Cell Reports, 9*, 415–418.

Kaeppler, H. F., Somers, D., Rines, H., & Cockburn, A. (1992). Silicon carbide fiber-mediated stable transformation of plant cells. *Theoretical and Applied Genetics, 84*, 560–566.

Kavipriya, C., Yuvaraja, A., Senthil, K., & Menaka, C. (2019). Genetic transformation methods for crop improvement: A brief review. *Agricultural Reviews, 40*(4), 281–288.

Kwon, M., & Firestein, B. L. (2013). DNA transfection: Calcium phosphate method. *Neural Development: Methods and Protocols*, 107–110.

Lamichhane, S., & Thapa, S. (2022). Advances from conventional to modern plant breeding methodologies. *Plant Breeding and Biotechnology, 10*(1), 1–14.

Liu, W., Yuan, J. S., & Stewart Jr, C. N. (2013). Advanced genetic tools for plant biotechnology. *Nature Reviews Genetics*, *14*(11), 781–793.

Liu, Y.-C., & Vidali, L. (2011). Efficient polyethylene glycol (PEG) mediated transformation of the moss Physcomitrella patens. *Journal of Visualized Experiments: JoVE*, *50*, 2560.

Luo, Z.-x., & Wu, R. (1988). A simple method for the transformation of rice via the pollen-tube pathway. *Plant Molecular Biology Reporter*, *6*, 165–174.

Ma, J. K., Drake, P. M., & Christou, P. (2003). The production of recombinant pharmaceutical proteins in plants. *Nature Reviews Genetics*, *4*(10), 794–805.

Mubeen, H., Naqvi, R. Z., Masood, A., Shoaib, M. W., & Raza, S. (2016). Gene transformation: Methods, uses and applications. *Journal of Pharmaceutical and Biological Sciences*, *4*(2), 54.

Mutsuddy, B. C. (1990). Electrokinetic behavior of aqueous silicon carbide whisker suspensions. *Journal of the American Ceramic Society*, *73*, 2747–2749.

Nyaboga, E., Tripathi, J. N., Manoharan, R., & Tripathi, L. (2014). Agrobacterium-mediated genetic transformation of yam (Dioscorea rotundata): An important tool for functional study of genes and crop improvement. *Frontiers in Plant Science*, *5*, 463.

Ogura, M., Sato, S., Nakanishi, K., Uenoyama, M., Kiyozumi, T., Saitoh, D., Ikeda, T., Ashida, H., & Obara, M. (2004). In vivo targeted gene transfer in skin by the use of laser-induced stress waves. *Lasers in Surgery and Medicine*, *34*(3), 242–248.

Potrykus, I. (1991). Gene transfer to plants: Assessment of published approaches and results. *Annual Review of Plant Physiology and Plant Molecular Biology*, *42*(1), 205–225.

Rani, S. J., & Usha, R. (2013). Transgenic plants: Types, benefits, public concerns and future. *Journal of Pharmacy Research*, *6*(8), 879–883.

Töpfer, R., Gronenborn, B., Schell, J., & Steinbiss, H.-H. (1989). Uptake and transient expression of chimeric genes in seed-derived embryos. *The Plant Cell*, *1*(1), 133–139.

Uddin, F., Rudin, C. M., & Sen, T. (2020). CRISPR gene therapy: Applications, limitations, and implications for the future [review]. *Frontiers in Oncology*, *10*. https://doi.org/10.3389/fonc.2020.01387

Verma, R., Kumari, P., Megha, S., & Rani, L. (2022). Application of recombinant DNA technology in agriculture: A review. *International Journal of Advanced Research in Biological Sciences*, *9*(3), 138–146.

Yao, C.-P., Zhang, Z.-X., Rahmanzadeh, R., & Huettmann, G. (2008). Laser-based gene transfection and gene therapy. *IEEE Transactions on Nanobioscience*, *7*(2), 111–119.

Zhang, Y., Yin, X., Yang, A., Li, G., & Zhang, J. (2005). Stability of inheritance of transgenes in maize (Zea mays L.) lines produced using different transformation methods. *Euphytica*, *144*, 11–22.

Zhou, G.-Y., Weng, J., Zeng, Y., Huang, J., Qian, S., & Liu, G. (1983). [29] Introduction of exogenous DNA into cotton embryos. In *Methods in Enzymology* (Vol. 101, pp. 433–481). Elsevier.

8 Immobilization Techniques of Plant Cells and Their Application on Secondary Metabolite Production

Jashanpreet Singh Sandhu, Vichitra Kaushik, Alakh Niranjan Sahu, Vipin Saini, and Jeevan Jyoti Kaushik*
* Corresponding author: vichitra.kaushik@gmail.com

8.1 PLANT CELLS

Cells have been a structural part of biotechnology, as almost all concepts of biotechnology revolve around cells and their structure. Cells are known to produce valuable stuff in harmony to the body, such as organic materials or products and energy for usage in performing its functions. To enhance this process, cells are modified with biotechnological approaches which help in achieving desired materials and products out of the modifications performed in usual cells. Plant cells are considered to have 40 different types with regards to their shapes and structures, whereas a common pictorial representation can give an idea about the structure of the plant cell (Figure 8.1) (Martin et al., 2001; Morales-Quintana et al., 2021).

Plant cells are the major producers of products like steroids or terpenoids which can be used in the biotransformation processing of complex biological compounds. Plant cells are also involved in the production of alkaloids and anthocyanins, which are delivered through products into the human body to perform functions like anticancer, anti-inflammatory, etc. (Reddy & Goto, 2022; Tesoriere, 2021). The biotechnological era is being used widely for the production of food products, with genetically modified foods. These foods are made through the elimination of unwanted substances from food products (Smith, 2003; Lin et al., 2020; Román et al., 2018). This aspect can be well explained with an instance: glucosinolates in canola meals. *Glucosinolate* is an active compound obtained from vegetables like cabbage, mustard, broccoli, and rapeseed. It belongs to the family Brassicaceae, used for its anti-inflammation, antioxidant, and antimicrobial properties, whereas it is accompanied by a number of side effects that put its advantages in the shade, such as goiter, renal lesions, GIT irritation, and alteration in hepatic functions. Therefore, this compels their removal from food, where this biotechnology comes into the picture (Abudabos et al., 2019; Hussain et al., 2022).

8.1.1 Historical Aspects of Plant Cell Immobilization

In the mid-1970s, numerous papers had pointed to the immobilization of subcellular patches in animals, which ultimately led to the entry of the concept of immobilization of the plant cell, together with chloroplasts, microbodies, peroxisomes, and mitochondria. Arkles et al. (1975) mentioned the immobilization of rat liver mitochondria with the help of face assimilation on alkalinized glass blob. This can be taken in studies including plant cells in which the primary composition is based on the immobilization of subcellular patches (Arkles et al., 1975). Two times latterly, Tanaka et al. (1977) paralyzed incentive peroxisomes containing an alcohol enzyme, a catalase, and a D-amino acid enzyme into matrices of picture-cross-linkable resins (Tanaka et al., 1977). Afterward, Kastle

DOI: 10.1201/9781003326939-9

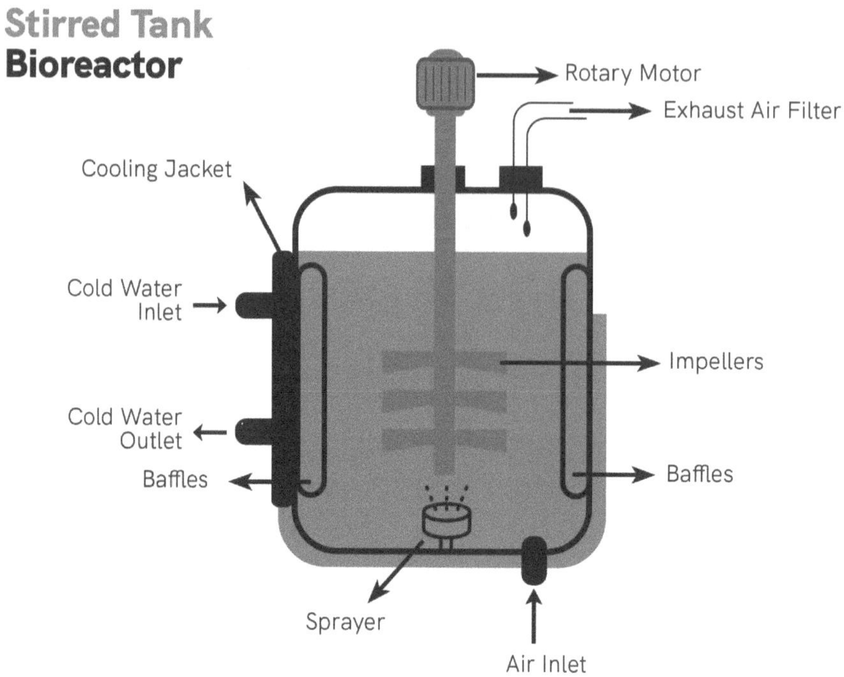

FIGURE 8.1 Plant cell.

et al. (1978) used entangled microsomes attained from rat liver as a medicine detoxifier, and Ochiai (1978) paralyzed plastids to organize a chloroplast captain and get a photocurrent. These paralyzed subcellular medication squares measure terribly thrilling motives for every educational and application study (Ochiai et al., 1983).

Further, in an analysis of the immobilization of microorganisms' complete cells and subcellular patches, papers on the immobilization of factory and beast cells have been published. Paralyzed factory cells have been taken up in studies as a profitable catalyst for drug products.

Brodelius et al. (1979) paralyzed cells of *Morinda*, *Catharanthus*, and *Digitalis* through the metallic element alginate approach and used them for the metamorphosis of natural wares similar to anthraquinones, ajmalicine isomers, and digitalin (Brodelius et al., 1979), whereas in the case of breast cells, Rechnitz et al. (1979) paralyzed a thin slice of porcine excretory organ on the bottom of a membrane captain that detected ammonia energy via a gas-passable membrane and used the device for the determination of L-amino acids (Rechnitz et al., 1979).

8.1.1.1 Plant Tissue Culture

Plant tissue culture is described as a culture medium which consists of plant cells clustered in a petri dish, expected to produce the desired product or an old product with the desired characteristics. Plant tissue culture, or plant tissue culture media, is the laboratory-based creation of a sterile environment to provide exactly the same nutrition fit for the aseptic growth of plant cells (Espinosa-Leal et al., 2018). This can be well understood in correlation to the concept of "totipotency." *Totipotency* is defined as the ability of a cell (plant or animal) to differentiate to produce a number of cells in a favorable environment, such as nutrient medium. Taking plant cell immobilization into consideration, plant tissue culture acts as a foundation for immobilization techniques, as all cells are proliferated artificially in the nutrient media to gain exceptionally good numbers of plants and related products. The composition of the plant tissue culture holds great significance in defining the interrelationship between cells and the related factors on the overall growth of the plants in the latter stages. This concept also helps in determining the problems that may hinder the growth of the plants (Hussain et al., 2012; Norouzi et al., 2022).

8.1.1.2 Historical Aspects of Plant Tissue Culture

Plant tissue culture is a cell-based culture developed artificially in a lab (in vitro). It can also be called a *sterile culture*. Plant tissue culture is an essential and tedious process, and it is a vital part of microbiology as well as the industry of applied science. Explained in other words, plant tissue culture is a technology, or an innovation, to create cells which eventually form tissues, followed by the formation of an organ and related components, in an artificially created environment (in vitro conditions), which helps in the formation of body parts in body-like conditions or in providing predetermined physical and chemical conditions, known as plant tissue culture. Taking a view of the beginnings of plant tissue culture, various scientists have contributed to providing clarity on the existence of plant tissue culture with various proofs, few of which are discussed in brief to help in enlightening on the concept of plant tissue culture and in understanding the position of plant tissue culture in building a whole plant and its origin.

Philip R. White, in a 1939 experiment, made the conclusion that related plant cells exist as individual entities but are capable of making a lot of cells by differentiation. In his publication "Potentially Unlimited Growth of Excised Plant Callus in an Artificial Nutrient," he also said that cells are identical and are worthy enough to influence one another (Thorpe, 2012). He has done multiple experiments, out of which he finally made the assumption that vegetative cells may produce embryos in artificially created conditions, which might help in their effective growth. Also, through these experiments he believed that the concept of totipotency is supremely bland. Through these experiments, he also framed the notion that the technique and the nutrient solution used are the two main factors in the production of the desired plant, and these techniques can be employed to trace and rectify various issues in plant growth. He published these skyrocketing results in a presentation delivered in 1902 with the resulting proof of these experiments. Further investigations presented that root tips can also be used in a culture medium to bring out plant totipotency. The viability of this method was proved through the preparation of tomato root tip cultures in harmony with explants containing meristematic cells. Whereas these root cultures have different origins altogether, these techniques have a vital role in physiological and viral research fields. Another instance to describe the potentiality of this method is the employment of barley embryos in the early 19th century. The main course of action was the use of in vitro culture in the efficient harvest of embryos from non-viable seeds between *Linum perenne* M. and *Linum austriacum* species, which were conveniently cross bred. Another method observation was the Precocious germination. *Acer pseudoplatanus* cambial tissue was harvested by Gautheret to create the first authentic plant tissue cultures. The nutrient media used in the differentiation of roots and shoots were potent enough to give birth to tissues sourced with indole acetic acid and vitamins. The aforementioned results and innovations in the context of plant tissue culture, especially in vitro cultures, have helped in gathering the interest of researchers in the development of better and more potent products from plant parts and a suitable artificially created environment. Hence, in the 21st century as well, scientists are inclined toward results, and considerations of these experiments look forward toward the development of potential products using these techniques, leading to the creation of a different and advanced scenario in the agricultural and microbiological fields (Loyola-Vargas et al., n.d.).

8.2 IMMOBILIZATION OF PLANT CELLS

Immobilization is a technique which is not only limited to the plant cell, as it is a very wide topic to be discussed. Stripping it down to plant cells, immobilization acts as a beginning of culturing cells and producing modified plants with desired characteristics (Pais et al., 1988; Lindsey et al., 1983). Basically, immobilization includes the process of adding a plant cell into the bioreactor or reacting it with an active catalyst, a process that ensures the cell is not in contact with any substrate or product carried in the mobile phase. The immobilization of the plant cell is a promising method meant to produce various types of plants from a single type of cell or culture (Abdelraof et al., 2022). The versatile nature of plant cells is completely known, leading to these cells' getting established as a good carrier of oxygen; they are large in size, so they are prone to breakage, and they harmonize for metabolic production. The immobilization of plant cells can simply uplift the quality of plant tissue cultures by making them easily available to react with product biocatalysts, which makes the process even more efficient. Biotransformation is specifically linked with immobilization for the production of products of higher value from products of lower value (Lapponi et al., 2022).

8.2.1 Advantages of Immobilization

- Immobilization is an uninterrupted process which helps in maintaining the continuity of the process, leading to the removal of all the inhibitors of metabolism.
- It helps in an easy separation of cells from their culture medium.
- The high biomass level is directly proportional to an increased volume of production as it increases the density, which eventually reduces contact time.
- Immobilization helps in the decrease in mechanical damage caused by shear stress caused by the formation of aggregates, thus inhibiting growth and leading to the formation of foam. Therefore, it helps in dealing with issues like aggregation, foaming, and growth.

8.2.2 Disadvantages of Immobilization

- The immobilization of a plant is process-independent, which means that it depends on the product release rate instead of the biosynthesis process rate.
- Immobilization may alter the rate of the biosynthetic process.
- The cell production–favoring microenvironment may produce undesired reactions with the by-products and the metabolites.

8.3 CLASSIFICATION OF IMMOBILIZATION METHODS

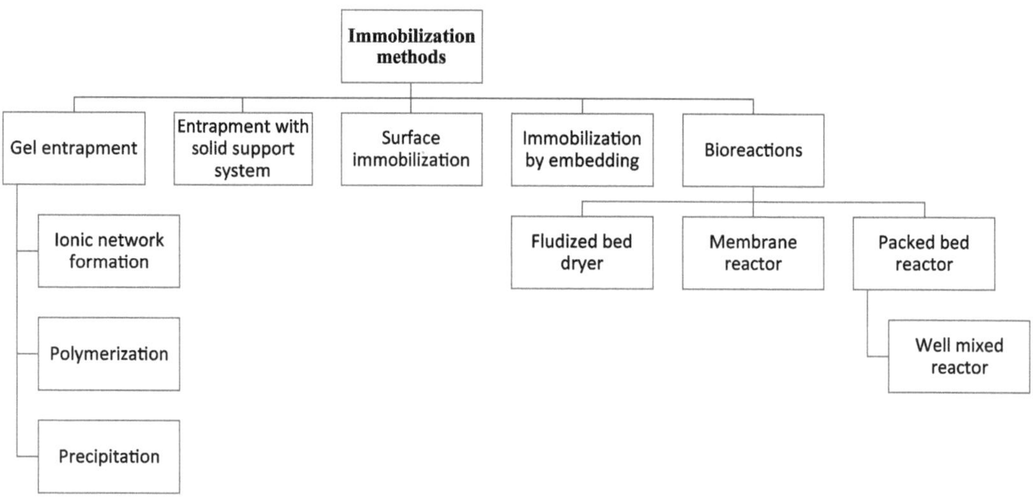

8.3.1 Gel Entrapment by Ionic Network Formation

Gel entrapment is a popular immobilization technique dating back many decades. The gel is employed as a medium to immobilize plant cells. The usual process of immobilization of plant cells consists of the mixing of a suitable gelling agent with a cell-containing medium. Choosing the right medium and gelling agent is an important aspect of or deciding factor in the biochemical properties of the plant cells that are entrapped in the culture medium. Gelling agents can be used as per the suitability of the cells. These agents are of three types:

S. No.	Type of Gelling Agent	Example
1	Inotropic	Alginates, chitosan, carrageenan
2	Thermal	Agarose, agar
3	Chemical	Gelatine, polyacrylamide

Out of these gelling agents, alginates are preferred because they offer a wide range of viability as well as a variety of cultures which are metabolically active. This is a toxicity-free process which acts through the mixing of a counter-ion with a suspension containing cells when poured drop-wise. The other form of this method is that which consists of three agents, namely, a metal, a linker, and a cell. All three parts come together to form a complex. This entrapment combination is a widely employed method globally on both small and large scales (Brodelius et al., 1980; Mandal et al., 2010).

8.3.2 Gel Entrapment by Polymerization

Polymerization through gel entrapment includes the process of free radical polymerization. Monomers (tiny molecules that harmonize to form a bigger molecule with the same chemical

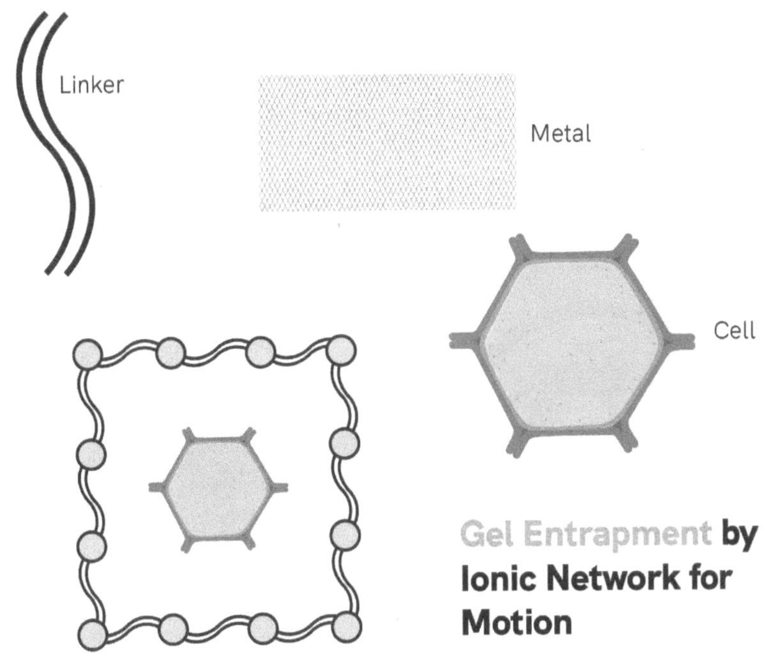

FIGURE 8.2 Gel entrapment by ionic network formation.

and different physical properties) such as epoxide, ethylene, ethyl methacrylate, polyacrylamide, etc. polymerize in the existence of a cell suspension, whereas cell suspension is already embrangled in the lattice of polymers. Therefore, free radical monomers produce a free radical polymer. This process is usually carried out at lower temperatures, such as 10°, or even lower than that, in the presence of a cross-linking agent (commonly used cross-linking agent is N,N'-ethyl bisacrylamide) used to form a lattice or a matrix in the absence of oxygen because the free radical polymer formed as a result of this process will react with oxygen on priority, which will eventually form an oxygen polymer which will rapidly lower down the rate of polymer production.

8.3.3 Gel Entrapment by Precipitation

The method of gel entrapment by precipitation includes a few factors which helps in the precipitation of natural (starch, cellulose, alginate, chitosan, etc.) and synthetic polymers (polyvinyl chloride, polyethylene glycol, etc.). These factors may include the solvent's pH, temperature, and the usage or selection of material for precipitation. This method may also cause some loss of viability, as heat is involved, which will alter the rate of precipitation.

8.3.4 Entrapment with Solid Support System

8.3.4.1 Hollow Fiber Reactor–Mediated Entrapment in Solid Support

Hollow fiber reactors are widely used in large-scale industries for the immobilization of plant cells to produce desired products on a larger scale. This includes the process of repeated circulations of a nutrient medium through mounted fibers. The cells are mounted over the shell side of the reactor. For instance, the usage of polyurethane foam as a solid support system. While using polyurethane foam, when the cell suspension passes through, it gets entrapped in the cavity through adhesion as well as adsorption. This method, being swift, easy, and cost-effective, is widely used (Jose et al., 1983).

8.3.4.2 Immobilization through Surface Binding

This process involves the immobilization of plant cells on the surface of the immobilization matrix. This process can occur on both synthetic and as well as natural matrix. While it is not known

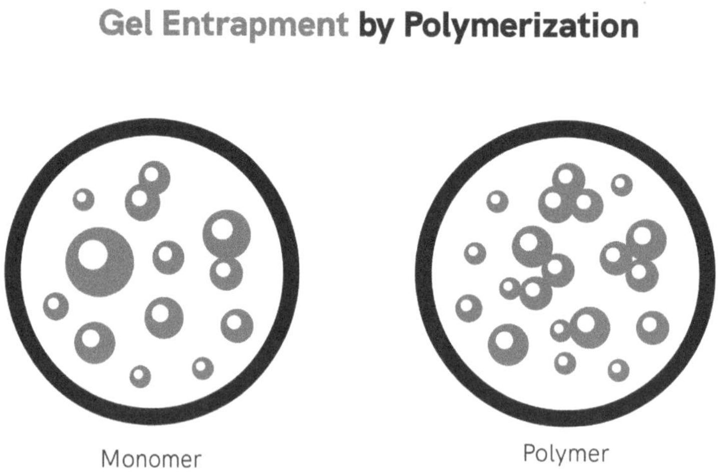

FIGURE 8.3 Gel entrapment by polymerization.

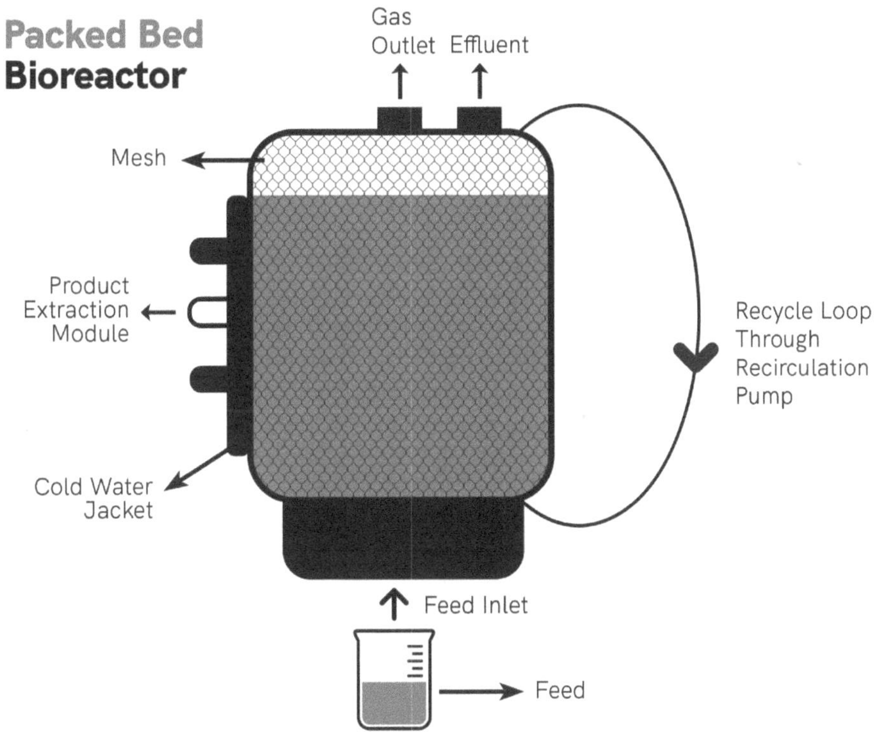

FIGURE 8.4 Gel entrapment by precipitation.

whether this process produces a stable outcome or not, this is still a relatively rapid process. Various studies have been conducted previously to enlighten on the utility of this method. Some instances are given in the following table:

S. No.	Title	Year	Ref. No.
1	Isolation of heavy metal–immobilizing and plant growth–promoting bacteria and their potential in reducing Cd and Pb uptake in water spinach	2022	Wang et al., 2022
2	Process optimization and modeling of Cd^{2+} biosorption onto the free and immobilized *Turbinaria ornata* using Box–Behnken experimental design	2022	Fawzy et al., 2022
3	Enzyme Immobilization and Co-Immobilization: Main Framework, Advances and Some Applications	2022	Bié et al., 2022
4	Immobilization of As and Sb by combined applications Fe–Mn oxides with organic amendments and alleviation their uptake by *Brassica campestris L*	2021	Rong et al., 2021
5	Immobilization of protoplasts by anchoring to microcarriers	1982	Bornman et al., 1982

8.3.5 Immobilization by Embedding

In this process, an embedding of two different materials is done, through which a solution containing cells is passed, and the outcome is monitored. This process hardly alters the viability of the cell. For example, a mixture of organic (silica) and inorganic (PVC) materials is converted into flat beds or slim sheets, and the cell suspension is passed and entrapment is recorded. The structure may have different pore and diameter sizes. The layers are arranged in a non-uniform manner (Zakria et al., 2021).

8.3.6 Bioreactors

8.3.6.1 Fluidized Bed Reactor

A fluidized bed reactor is a complex equipment that is primarily made for the commercial production of desired products formed after the removal of unwanted substances. This machinery was initially designed for the treatment of wastewater, and further research created a model for the immobilization of cells (both plant and human cells) (Hulst et al., 1989). Talking about cell-specific fluidized bed reactors, these reactors use liquid or gaseous energy for the suspension of particles. This phenomenon is called fluidization, which is considered to be more efficient if the particle's surface area is high, which means the preferred size of the particles is small (Godia et al., 1995). The attrition levels of fluidized bed reactors are relatively lower compared to those of other methods, which helps in obtaining better immobilization products. Apart from this, there are a few drawbacks to this process, such as the formation of clogs and the undue pressure imparted by liquid on the solid substrate (Panda et al., 1989; Archambault et al., 1989).

8.3.6.2 Membrane Reactor

These reactors are widely oriented for utility in entrapping plant cells. Membrane reactors primary come in two variants, the hollow fiber reactor and the spiral wound reactor. Out of these two, the hollow fiber reactor is more extensively used. Entrapment of plant cells is done by either the outer region or the inner tube of the membrane reactor. These reactors are primarily considered for cells that are diametrically thick from the center surface or the inner core, as such parts may be substrate-deficient. Whereas this type of membrane reactor may help in the characterization of morphology and the chemical configuration of the cell, another pro of this reactor is that it can be used repeatedly. Considering the demerit, membrane reactors may cost more to install initially (Prenosil et al., 1983).

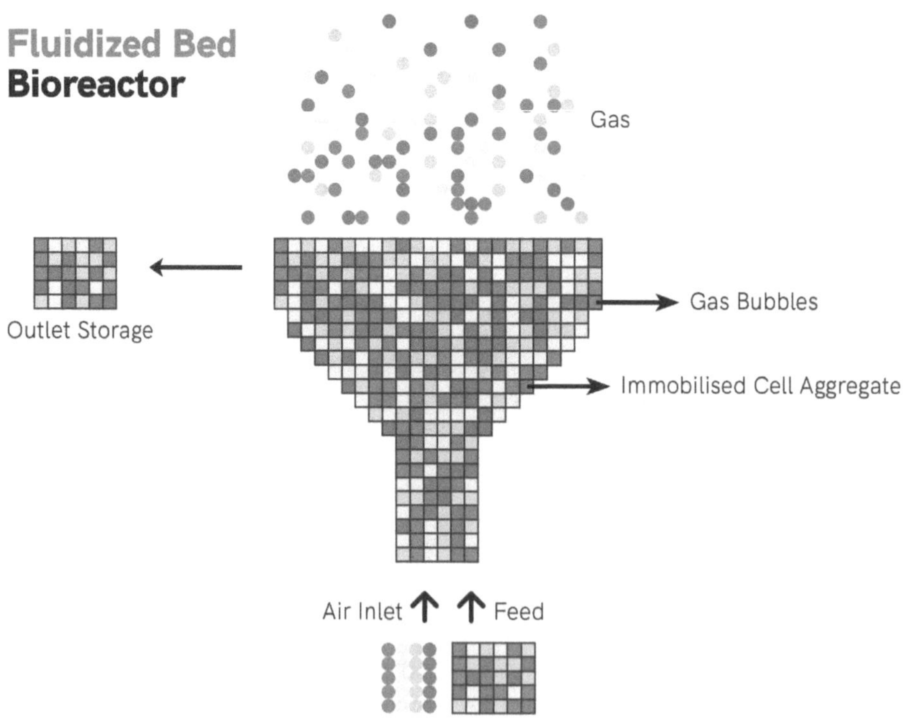

FIGURE 8.5 Fluidized bed reactor.

8.3.6.3 Packed Bed Reactor

This type of reactor also includes two types of sites for the immobilization of cells. Immobilization may be done either on the surface of the cells or within the support system. Once the cells get immobilized, they are checked for viability. Cells which are immobilized within the support system are packed in the form of beds containing a large number of beds. In contrast to this, the packed bed reactor may have several cons, such as that the filters in packed bed reactors may get blocked when the substrate is passed, leading to the blockage of the passage for solvents or fluids. Another point to be discussed is the low degree of mixing; this may alter or even hinder gaseous exchange, primarily of oxygen, carbon dioxide, etc. (Hama et al., 2007).

8.3.6.4 Well-Mixed Reactor

The well-mixed reactor is also called the stirred tank reactor. This technique majorly uses liquid as a solvent. This equipment uses sprayers which spray the cell containing a solute which is later mixed with the solvent. This equipment is capable enough to control pH and temperature. However, shear and centrifugal forces may cause damage to the cells, which may hinder their viability. This equipment consists of a tank and a blade. The blade should be designed in a way that it produces an evenly mixed product (Hama et al., 2007).

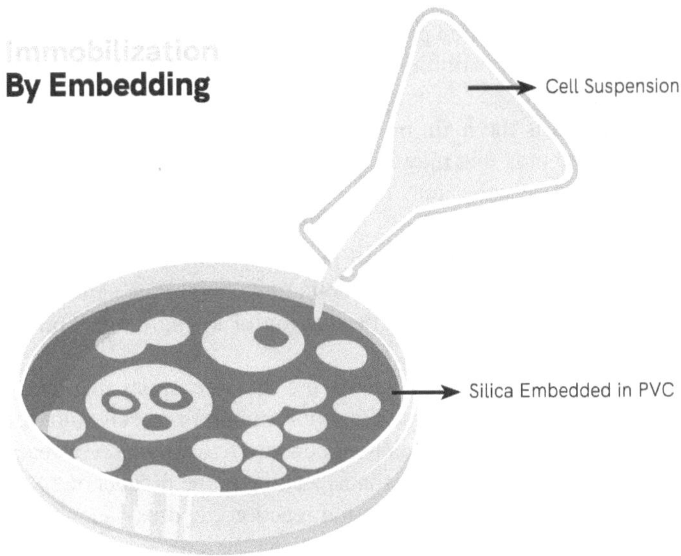

FIGURE 8.6 Packed bed reactor.

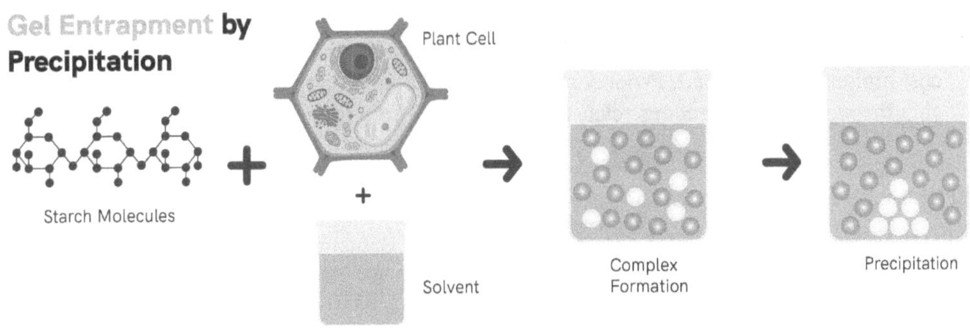

FIGURE 8.7 Well-mixed reactor.

8.4 THE APPLICATION OF PLANT IMMOBILIZATION IN SECONDARY METABOLITE PRODUCTION

The roots of plant biotechnology have spread so wide that each and every customized product is produced by the means of plant biotechnology. Along with the elevated production of the desired biotechnological products, metabolite production has also gained a lot of popularity. The production of secondary metabolites gives rise to a lot of new products out of old ingredients. Pharmaceuticals and nutraceuticals have collectively emphasized the production of secondary metabolites as useful products. The plus point of this technology is that it is not plant-specific—that is, almost all compounds, such as glycosides (rhubarb), alkaloids (morphine, codeine), terpenes (menthol), etc., can be used to produce valuable secondary metabolites. The production of secondary metabolites from plant cells is based on various steps under controlled conditions. The respective process can be repeated as per product requirements.

Choosing the right parent plant. Choice of plant is an initial step, and it is the foundation step of the production of secondary metabolites. Any part of a plant can be used in the production initiation of secondary metabolites, such as seeds, leaves, stems, barks, roots, etc. The chosen part of the plant should be preferably fresh or be a young explant.

Selection of the right culture media. Choosing the right culture medium is an important aspect in the production of desired products and secondary metabolites. The culture media should be confirmed for compatibility with cell cultures and should ensure a wide production band and swift growth.

Metabolism. Metabolism is the main process in the production of secondary metabolites. There are various elements that may influence the production of secondary metabolites. These elements can be divided into two categories, namely, physical (e.g., pH, which is kept at medium; temperature, which is optimized; and sunlight, which is controlled, timely given) and chemical (proportion of growth factors in the culture medium, microelements, macroelements, etc.). This metabolic process is often called "targeted metabolism" as it specifically focuses on the production of a single metabolite at a particular time.

Another important part of this process includes the set of reactions that are made to happen to produce a desired secondary metabolite from immobilized cell culture. Such reactions are also called biotransformation reactions. These are basically a set of reactions that occur between the old products acting as substrates, leading to the production of new compounds. Examples of such reactions are isomerization reactions, hydroxylation reactions, glycosylation reactions, hydrogenation and dehydrogenation reactions, etc. These reactions are capable enough to cause alterations in the actions of elicitors (physical, chemical, and microbial factors), whereas these factors importantly contribute to elevating the production of secondary metabolites.

Extraction of metabolized products. The extraction of metabolized products is a tedious process, as a minute loss of the secondary metabolite or any of its related products may cost higher than expected. Product recovery should be done in a controlled manner, probably through the spontaneous release of the product into the medium.

REFERENCES

Abdelraof, Mohamed, Houssni El Saied, and Mohamed S. Hasanin. 2022. "Green Immobilization of Glucanobacter Xylinum onto Natural Polymers to Sustainable Bacterial Cellulose Production." *Waste and Biomass Valorization* 13 (4): 2053–2069. https://doi.org/10.1007/s12649-021-01666-w.

Abudabos, Alaeldein M., Elsayed O. S. Hussein, Mutahar H. Ali, and Muath Q. Al-Ghadi. 2019. "The Effect of Some Natural Alternative to Antibiotics on Growth and Changes in Intestinal Histology in Broiler Exposed to Salmonella Challenge." *Poultry Science* 98 (3): 1441–1446. https://doi.org/10.3382/ps/pey449.

Archambault, J., B. Volesky, and W. G. W. Kurz. 1989. "Surface Immobilization of Plant Cells." *Biotechnology and Bioengineering* 33 (3): 293–299. https://doi.org/10.1002/bit.260330307.

Arkles, B., and W. S. Brinigar. 1975. "Respiratory Properties of Rat Liver Mitochondria Immobilized on an Alkylsilylated Glass Surface." *Journal of Biological Chemistry* 250 (22): 8856–8862. https://doi.org/10.1016/s0021-9258(19)40752-7

Bié, Joaquim, Bruno Sepodes, Pedro C. B. Fernandes, and Maria H. L. Ribeiro. 2022. "Enzyme Immobilization and Co-Immobilization: Main Framework, Advances and Some Applications." *Processes* 10 (3): 494. https://doi.org/10.3390/pr10030494.

Bornman, Chris H., and Anders Zachrisson. 1982. "Immobilization of Protoplasts by Anchoring to Microcarriers." *Plant Cell Reports* 1 (4): 151–153. https://doi.org/10.1007/bf00269185.

Brodelius, P., B. Deus, K. Mosbach, and M. H. Zenk. 1979. "Immobilized Plant Cells for the Production and Transportation of Natural Products." *FEBS Letters* 103 (1): 93–97. https://doi.org/10.1016/0014-5793(79)81257-0.

Brodelius, Peter, and Kjell Nilsson. 1980. "Entrapment of Plant Cells in Different Matrices." *FEBS Letters* 122 (2): 312–316. https://doi.org/10.1016/0014-5793(80)80463-7.

Espinosa-Leal, Claudia A., César A. Puente-Garza, and Silverio García-Lara. 2018. "In Vitro Plant Tissue Culture: Means for Production of Biological Active Compounds." *Planta* 248 (1): 1–18. https://doi.org/10.1007/s00425-018-2910-1.

Fawzy, Mustafa A., Hadeer Darwish, Sarah Alharthi, Mayasar I. Al-Zaban, Ahmed Noureldeen, and Sedky H. A. Hassan. 2022. "Process Optimization and Modeling of Cd2+ Biosorption onto the Free and Immobilized Turbinaria Ornata Using Box–Behnken Experimental Design." *Scientific Reports* 12 (1). https://doi.org/10.1038/s41598-022-07288-z.

Godia, Francesc, and Carles Sola. 1995. "Fluidized-Bed Bioreactors." *Biotechnology Progress* 11 (5): 479–497. https://doi.org/10.1021/bp00035a001.

Hama, Shinji, Hideki Yamaji, Takahiro Fukumizu, Takao Numata, Sriappareddy Tamalampudi, Akihiko Kondo, Hideo Noda, and Hideki Fukuda. 2007. "Biodiesel-Fuel Production in a Packed-Bed Reactor Using Lipase-Producing Rhizopus Oryzae Cells Immobilized within Biomass Support Particles." *Biochemical Engineering Journal* 34 (3): 273–278. https://doi.org/10.1016/j.bej.2006.12.013.

Hulst, A. C., and J. Tramper. 1989. "Immobilized Plant Cells: A Literature Survey." *Enzyme and Microbial Technology* 11 (9): 546–558. https://doi.org/10.1016/0141-0229(89)90082-3.

Hussain, Altaf, Iqbal Ahmed, Hummera Nazir, and Ikram Ullah. 2012. "Plant Tissue Culture: Current Status and Opportunities." *Recent Advances in Plant in Vitro Culture*, October. https://doi.org/10.5772/50568.

Hussain, S., A. U. Rehman, H. K. Obied, D. J. Luckett, and C. L. Blanchard. 2022. "Extraction, Chemical Characterization, In Vitro Antioxidant, and Antidiabetic Activity of Canola (Brassica napus L.) Meal." *Separations* 9: 38. https://doi.org/10.3390/separations9020038.

Jose, Wilfredo, Henrik Pedersen, and Chee-Kok Chin. 1983. "Immobilization of Plant Cells in a Hollow-Fiber Reactor." *Annals of the New York Academy of Sciences* 413 (1 Biochemical E): 409–412. https://doi.org/10.1111/j.1749-6632.1983.tb47916.x.

Kastle, P. R., W. H. Baricos, R. P. Chambers, and W. Cohen. 1978. "Hollow Fiber-Entrapped Liver Microsomes: A Potential Extracorporeal Drug Detoxifier." *Enzyme Engineering* 4: 199.

Lapponi, María J., Mariana B. Méndez, Jorge A. Trelles, and Cintia W. Rivero. 2022. "Cell Immobilization Strategies for Biotransformations." *Current Opinion in Green and Sustainable Chemistry* 33 (February): 100565. https://doi.org/10.1016/j.cogsc.2021.100565.

Lin, Wenwei, and Zhenbiao Yang. 2020. "Unlocking the Mechanisms Behind the Formation of Interlocking Pavement Cells." *Current Opinion in Plant Biology* 57: 142–154. https://doi.org/10.1016/j.pbi.2020.09.002.

Lindsey, K., M. M. Yeoman, G. M. Black, and F. Mavituna. 1983. "A Novel Method for the Immobilisation and Culture of Plant Cells." *FEBS Letters* 155 (1): 143–149. https://doi.org/10.1016/0014-5793(83)80227-0.

Loyola-Vargas, Victor M., and Felipe Vázquez-Flota. n.d. "An Introduction to Plant Cell Culture: Back to the Future." *Plant Cell Culture Protocols*, 3–8. https://doi.org/10.1385/1-59259-959-1:003.

Mandal, Sanchita, S. Senthil Kumar, Balakrishnam Krishnamoorthy, and Sanat Kumar Basu. 2010. "Development and Evaluation of Calcium Alginate Beads Prepared by Sequential and Simultaneous Methods." *Brazilian Journal of Pharmaceutical Sciences* 46 (4): 785–793. https://doi.org/10.1590/s1984-82502010000400021.

Martin, Cathie, Kiran Bhatt, and Kim Baumann. 2001. "Shaping in Plant Cells." *Current Opinion in Plant Biology* 4 (6): 540–549. https://doi.org/10.1016/s1369-5266(00)00213-2.

Morales-Quintana, Luis, Andrea Barrera, Rasme Hereme, Karla Jara, Claudia Rivera-Mora, Felipe Valenzuela-Riffo, Pedro E. Gundel, Stephan Pollmann, and Patricio Ramos. 2021. "Molecular and Structural Characterization of Expansins Modulated by Fungal Endophytes in the Antarctic Colobanthus Quitensis (Kunth) Bartl. Exposed to Drought Stress." *Plant Physiology and Biochemistry* 168: 465–476. https://doi.org/10.1016/j.plaphy.2021.10.036.

Norouzi, Omid, Mohsen Hesami, Marco Pepe, Animesh Dutta, and Andrew Maxwell P. Jones. 2022. "In Vitro Plant Tissue Culture as the Fifth Generation of Bioenergy." *Scientific Reports* 12 (1). https://doi.org/10.1038/s41598-022-09066-3.

Ochiai, Hideo, Atsuo Tanaka, and Saburo Fukui. 1983. "Immobilized Organelles." *Immobilized Microbial Cells*, 153–187. https://doi.org/10.1016/b978-0-12-041104-7.50010-5.

Pais, S., F. Mavituna, and J. M. Novais. 1988. *Plant Cell Biotechnology*. Berlin, Heidelberg: Springer.

Panda, A. K., Saroj Mishra, V. S. Bisaria, and S. S. Bhojwani. 1989. "Plant Cell Reactors—a Perspective." *Enzyme and Microbial Technology* 11 (7): 386–397. https://doi.org/10.1016/0141-0229(89)90132-4.

Prenosil, Jiri E., and Henrik Pedersen. 1983. "Immobilized Plant Cell Reactors." *Enzyme and Microbial Technology* 5 (5): 323–331. https://doi.org/10.1016/0141-0229(83)90002-9.

Rechnitz, G. A., M. A. Arnold, and M. E. Meyerhoff. 1979. "Bio-Selective Membrane Electrode Using Tissue Slices." *Nature* 278 (5703): 466–467. https://doi.org/10.1038/278466a0.

Reddy, A. V. B., and M. Goto. 2022. *Engineering Perspectives in Biotechnology [Course Notes]*. Course Hero. https://www.coursehero.com/file/100449778/Engineering-Perspectives-in-Biotechnologydocx/.

Román, J., A. Castillo, and A. Mahn. 2018. "Molecular Docking of Potential Inhibitors of Broccoli Myrosinase." *Molecules* 23 (6): 1313. https://doi.org/10.3390/molecules23061313

Rong, Qun, Chaolan Zhang, He Huang, Chuanzhang Li, Xinyu Nong, Hecheng Zhao, Kai Zhong, Xingzi Qin, and Yapeng Yang. 2021. "Immobilization of as and Sb by Combined Applications Fe–Mn Oxides with Organic Amendments and Alleviation Their Uptake by Brassica Campestris L." *Journal of Cleaner Production* 288 (March): 125088. https://doi.org/10.1016/j.jclepro.2020.125088.

Smith, Laurie G. 2003. "Cytoskeletal Control of Plant Cell Shape: Getting the Fine Points." *Current Opinion in Plant Biology* 6 (1): 63–73. https://doi.org/10.1016/s1369-5266(02)00012-2.

Tanaka, Atsuo, Shigeki Yasuhara, Masako Osumi, and Saburo Fukui. 1977. "Immobilization of Yeast Microbodies by Inclusion with Photo-Crosslinkable Resins." *European Journal of Biochemistry* 80 (1): 193–197. https://doi.org/10.1111/j.1432-1033.1977.tb11871.

Tesoriere, Luisa. 2021. "Update on Nutraceuticals." *Nutraceuticals* 1 (1): 1–1. https://doi.org/10.3390/nutraceuticals1010001.

Thorpe, Trevor. 2012. "History of Plant Tissue Culture." *Methods in Molecular Biology (Clifton, N.J.)* 877: 9–27. https://doi.org/10.1007/978-1-61779-818-4_2.

Wang, Xiaoyu, Debao Cai, Mingfei Ji, Zhaojin Chen, Lunguang Yao, and Hui Han. 2022. "Isolation of Heavy Metal-Immobilizing and Plant Growth-Promoting Bacteria and Their Potential in Reducing Cd and Pb Uptake in Water Spinach." *Science of the Total Environment* 819 (May): 153242. https://doi.org/10.1016/j.scitotenv.2022.153242.

Zakria, Hazirah Syahirah, Mohd Hafiz Dzarfan Othman, Roziana Kamaludin, Siti Hamimah Sheikh Abdul Kadir, Tonni Agustiono Kurniawan, and Asim Jilani. 2021. "Immobilization Techniques of a Photocatalyst into and onto a Polymer Membrane for Photocatalytic Activity." *RSC Advances* 11 (12): 6985–7014. https://doi.org/10.1039/D0RA10964A.

9 Cloning of Plant Cells
Different Methods of Cloning and Its Applications—Advantages and Disadvantages of Plant Cell Cloning

Zeeshan Khan, Aqsa Akram,
*Muhammad Irfan, and Khushbukhat Khan**
* Corresponding author: khushi-khan2011@hotmail.com

9.1 INTRODUCTION

Long before the first mammals, plants were the only living things on the planet. Plants' unique reproductive strategies are one reason they can survive in any habitat on the globe. The mixing of gametes from two individuals of opposing sexes is the sole way for the majority of animals to reproduce sexually. The majority of plants, on the other hand, may be cloned to generate offspring asexually from only a single parent (Ramesh, 2023).

The word "clone" comes from the Greek word "klan," which means a slip or twig. Since many years ago, maybe even before agriculture began commercially, plants have been "cloned" through stem cuttings or whole-plant divisions. The intentional creation of individuals that are genetically similar is known as cloning. Every new plant created is a duplicate of the original—natural clones are monozygotic (identical) twins. Every cell in a clone's body has a nucleus, which is where the chromosomes are located, that houses identical sets of genetic material. As a result, the DNA and genes in the nucleus of two cloned cells are identical. The energy-producing "factories" known as mitochondria, which are present in all cells, including eggs, contains DNA. These structures are located in the cytoplasm, which is the part of a cell that is not the nucleus. Since they can reproduce on their own, mitochondria have their own DNA. Although the word "clones" is also used to describe people with identical nucleus DNA but differing mitochondrial DNA, true clones are those in which both the nuclei and the mitochondria have the same DNA (Ramesh, 2023).

Naturally, plants may produce copies of themselves through a variety of processes, including layering, stolon, and tubers. Each year, horticulturists and gardeners use the capacity for plant cloning to produce millions of new plants. The majority of plants you may buy from nurseries and food shops were originally clones. It is advantageous to create plants by cloning since it ensures that the progeny plants will have the same characteristics and attributes as their parent plant. Many times, certain plants have distinctive qualities and attributes that appeal to people, such as sweet-tasting fruits, huge fruits, or beautifully colored flowers. Let us take the scenario where a gardener discovers a rose shrub with a distinctive yellow blossom and wants to produce other roses with the same yellow flowers. Growing the yellow rosebush's seeds will result in progeny with distinctive genetic combinations and, thus, various colored blooms. While some of the progeny may have yellow blooms, others are more likely to have inherited genes for other hues. The only way for the gardener to guarantee that all the progeny produced will have yellow blooms is through cloning (Ramesh, 2023).

DOI: 10.1201/9781003326939-10

9.2 HISTORICAL BACKGROUND

German scientists Matthias Schleiden and Theodor Schwann proposed the cell theory in 1838, which states that all life is composed of cells and that all cells are the result of the evolution of earlier cells (Singh & Bhalla, 2006).

This notion serves as the basis for the idea of totipotency, which holds that since cells must have all the genetic material needed to build an entire multicellular creature, all cells in a multicellular entity retain the capacity to rejuvenate, or recreate, the entire organism (Harold, 2022). This served as the foundation for plant cell culture studies.

Early in the 20th century, Austrian scientist Gottlieb Haberlandt made the first botched effort at cultivating isolated plant tissues. The first effective culture of distinct plant tissue was presented in 1939 by Professor R. J. Gautheret and colleagues as a continually dividing callus tissue. An unstructured mass of proliferating cells, such as during a wound reaction, is what is referred to as a callus. However, it took W. H. Muir and his associates until 1954 to successfully grow the first whole plant from a single adult plant cell. Then, in the 1970s and the 1980s, plant cell and tissue culture methods advanced quickly as a result of a greater knowledge of plant physiology, particularly on the function of plant hormones in plant proliferation and maturation. Numerous plant species have been successfully cloned from a single cell, proving and supporting the theory of totipotency (Singh & Bhalla, 2006).

9.3 TYPES OF PLANT CLONING

Regardless of the species, multiple techniques are used to artificially clone a plant. Some cloning techniques make use of the plant's naturally occurring participants, offspring which can be divided from their parent. Given that the parent plant frequently gives the child with water and nutrients as it grows, this type of cloning is typically quicker and more successful. Other cloning techniques entail plucking a piece or component from the original plant and nurturing it in the appropriate environment for temperature, light, and moisture. Plant cloning techniques that are often used include the following.

9.3.1 Traditional Techniques

9.3.1.1 Cutting

The simplest, most popular, and most straightforward way of cloning plants is undoubtedly with cuttings. *Cuttings* are pieces that have been cut away from the parent plant. The cutting is often collected in the late spring from the mother plants' neo-growth portion. The cutting is placed in a growing substrate within perfect lighting, temperature, and humidity conditions after being treated with a growth accelerator and an antifungal agent. The medium, also known as the growth substrate, can be different, although it is often highly porous and made of a combination of sand and perlite. Plantlets from excised stem cuttings and leaf cuttings that were split off in the ambient habitat have a higher recurrence rate than seedlings from sexual reproduction (Nahar, 2025).

Cuttings have the resources essential for plant establishment—water and carbon—stored and readily available. The plant's detached portion is still physiologically functioning, which enables mitotic exercise and the development of new root systems for the absorption of water and nutrients. The persistence of epigenetic knowledge, heritable sequences of phenotypic variations that are not caused by changes in DNA but instead result from histone alteration and DNA methylation, and the other benefits of asexual reproduction of plants are also favorable from an evolutionary perspective.

9.3.1.1.1 Applications of Cutting

Plants *cutting* is a part of a plant used in horticulture for vegetative (asexual) propagation. A part of the stem or root of the donor plant is placed in a suitable medium, such as moist soil. The plant fragment will start to grow as a distinct plant from the parent if circumstances are right, a process known as striking.

9.3.1.2 Budding

A single bud from a desired plant is inserted into a crack in the bark of a suitable rootstock through the process of budding to produce a beneficial variety (cultivar) and rootstock combination (Lewis & Alexander, 2008). T-budding and chip budding are grafting methods for fruit trees that employ a solitary scion wood bud as opposed to scion wood with many buds. Apples, pears, peaches, and a wide variety of ornamentals are just a few of the plants that might benefit from budding (Iqbal & Singh, 2020). The scion and rootstock must be biocompatible, the scion buds must be substantially formed and quiescent, and finally, the meristematic tissue from the scion and the rootstock must be aligned with excellent contact in order for the bud to take place. The procedure must be carried out with extra caution to keep the buds from withering out (Yadav & Singh, 2018).

The two most popular techniques for growing fruit trees and numerous perennials are the "T" bud and the "chip" bud. One of the simplest methods of propagation is T-budding. The rootstock, also known as the "bark is slipping," must be actively developing in order to use the T-budding approach. This indicates that the bark may be removed from the wood without damaging it. When the bark is firm, chip budding can be done, allowing for usage earlier and afterward in the league season. The chipped bud can recover more quickly and quickly with repetition. The fundamentals of managing scion wood and rootstock are similar; however, each approach requires distinct cuts. A plant's beneficial traits can be maintained through budding, allowing for unaltered propagation. Compared to sexual reproduction, it is much more rapid.

9.3.1.2.1 Applications of Budding

Certain plant species are propagated by horticulturists, botanists, and gardeners by budding. Budding, also known as *bud grafting*, is commonly used in nurseries and greenhouses, although amateurs can perform it with a little experience. Budding can be used to make clones of a single species or to blend cultivars from suitable species (Singh & Kumar, 2021).

9.3.1.3 Grafting

Theoretically, any two plants with a continuous cambium and close botanical kinship can be grafted. *Grafting* is the process of affixing a piece of one plant (bud or scion) to or on a stem, root, or branch of another (stock) so that a union is formed, and the partners grow together. The addition is referred to as the scion, while the component that gives the root is known as the stock. *Interstock* is the term for the middle component when there are more than two sections involved.

Grafts across genera and families are infrequently effective, but those spanning species are almost always unsuccessful. The proximity of botanical relationships within a genus is not a guaranteed indicator of the likelihood of success, but it is the best one at hand in the absence of documented experience. Numerous intricate physiological and environmental elements play a role in determining whether two plants will proceed to grow or be reconcilable after being grafted simultaneously (Graft | Description, 2025).

The most fundamental of the grafting principles is the matching of the scion and the stock cambiums (meristematic tissue, the cells of which are undifferentiated and capable of frequent cell division). In the majority of woody trees and shrubs, the central core of the wood is covered by a distinct single-cell layer called the cambial tissue, which is located directly under the bark. There are various benefits to grafting trees and other plants. By grafting a plant onto a rootstock that possesses these qualities, for instance, you can increase a plant's ability to survive the winter or its resilience to pests and diseases. Without attempting to transplant the tree, grafting may also be used to substitute it with a tree of a distinct variety.

9.3.1.3.1 Applications of Grafting

Grafting is used in modern horticulture for a variety of purposes, including the repair of damaged trees, the production of dwarf trees and shrubs, the extension of plant resistance to specific diseases, the preservation of varietal characteristics, the adaptation of varieties to challenging soil or climatic conditions, the assurance of pollination, the production of multi-fruited or multi-flowered plants, and the propagation of some species (such as hybrid roses) that cannot be propagated in any other way.

9.3.1.4 Layering

Asexual propagation by layering is a tried-and-true technique that promotes roots without the use of specialized equipment or specially created environmental circumstances, like a mist bench. The home gardener can quickly grasp the necessary skills. By applying layering strategies, desirable shrubs, vines, and houseplants can be grown in a smaller area. Compared to when they are reproduced from cuttings, hard-to-root plants grown using layering procedures often establish roots more rapidly and with higher overall success. Daughter plants are available to be divided and transplanted a month or two after layering is started in the spring or early summer (Yadav & Singh, 2018).

The tip, simple, complex (serpentine), mound (stool), and air are the five fundamental forms of layering. In order for new roots to sprout during tip, simple, mound, and complex layering, a portion of the stem must be buried; however, during air layering, new roots develop above the soil's surface (Awotedu et al., 2021). Plants that are challenging to root from a cutting benefit more from layering as a propagation technique. An area of the plant that is rooted in a much larger area may be layering. You can grow a bigger plant more quickly as a consequence.

9.3.1.4.1 Applications of Layering

Plants that do not grow well from cuttings might be layered. Until it grows roots and can sustain itself, the layered stem receives water and nutrients from the parent plant. Layering is used to propagate plants that are difficult to reproduce using conventional methods, such as some huge shrub layering of the ground.

9.4 PLANT MICROPROPAGATION/IN VITRO PROPAGATION

The word "micropropagation" (also known as "in vitro propagation") is most frequently used to refer to the clonal, true-to-type plant propagation techniques that involve various tissue, cell, and organ culture techniques. It suggests the aseptic cultivation of tiny fragments (explants) of tissues and organs in enclosed tubes with predetermined culture medium and in a regulated environment, illustrated in Figure 9.1. Micropropagation, together with genetic modification, is the quickest approach to growing a large number of clonal plants of diverse plant species that are typically virus- or other pathogen-free, as it provides the artificial environment with different ingredients, illustrated in Figure 9.2.

It is based on the concept of cell totipotency, which refers to a cell's capacity to divide, create all the differentiated cells necessary for organ development, and regenerate into a whole plant—illustrated via Figure 9.3. The various ways of cultivating plant tissues may have certain benefits over the conventional methods of propagation.

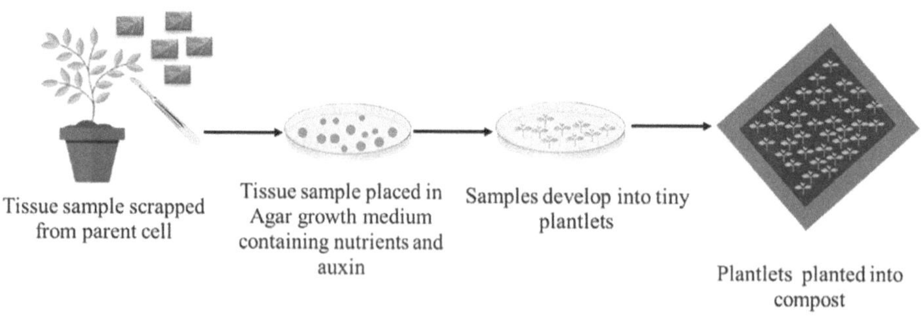

FIGURE 9.1 Diagrammatic representation of the process involved in plant tissue culture.

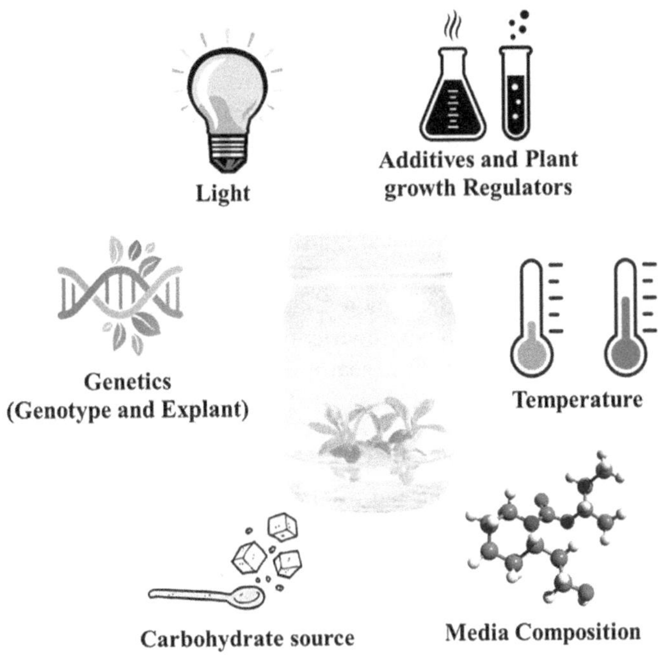

FIGURE 9.2 The constituents required for plant tissue culture.

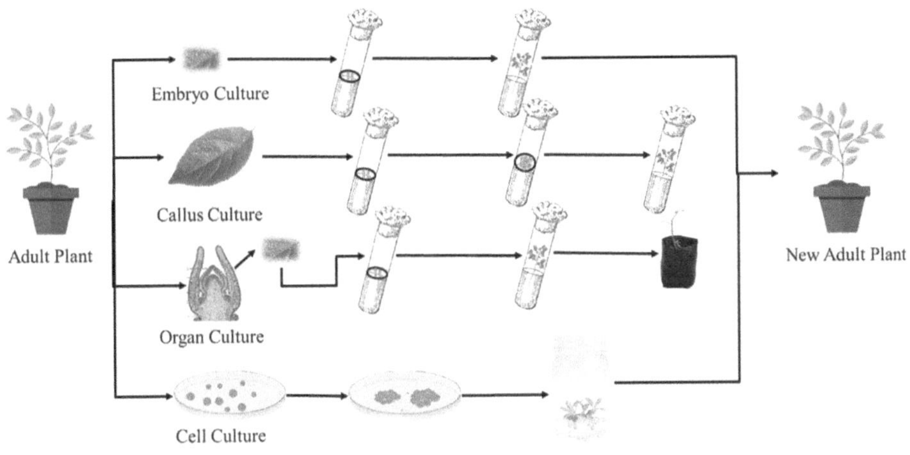

FIGURE 9.3 Types of plant tissue culture.

9.4.1 Different Methods of Micropropagation

9.4.1.1 Meristem Culture

Meristem culture is applied in vitro for shoot apical meristem culture. Morel and Martin invented meristem culture in 1952 to culture dahlia (Chen & Henny, 2008). Morel used meristem culture to micropropagate the orchid *Cymbidium* in 1965 (Arditti & Code, 2007). The meristem culture method works well in eradicating viruses, preserving genes, transforming genetic material, and breeding plants. It has been used to propagate a variety of crops, including sugarcane, strawberries, and others, with success.

In the meristem culture, an apparently pre-existing shoot meristem develops, and lateral roots rejuvenate via these shoots. The primordium meristem is located in the shoot tip beyond the youngest leaf. It may be up to 250 mm long and 100 mm in diameter. A branch tip of 100–500 mm would include one or three leaf primordia in contrast to the apical meristem (Xue et al., 2023). When viral removal is the goal, shoot tips of up to 10 mm are employed to obtain disease-free plants. Shoot tip culture, using explants between 5 and 10 mm, is employed for a quick clonal multiplication. Consequently, the vast preponderances of meristem cultures are basically shoot tip cultures; also used for quick clonal multiplication are nodal explants of various sizes (Lai & Lai, 2019).

When the primary goal is micropropagation, the size of the shoot tip employed for culture is unimportant. However, if the goal is to obtain virus-free stock (or stock free of other diseases), the apical meristem must be excised with as little nearby tissue as possible. If you want to get more than one plantlet from each shoot tip, the shoot tip can be finely sliced into pieces. Certain species, such as cauliflower, employ curd (the inflorescence) pieces.

For plants with subterranean stems, shoot tips or tissue fragments containing buds could be handled. The best explants are those plucked from actively developing plants early in the growth cycle. Meristem explants are then put on Murashige and Skoog (MS) medium, which is thought to work well for the preponderance of species. For some species, a lower salt content is adequate. To get rid of endophytic contamination, the growth medium can be supplemented with fungicides (bavistin) or antibiotics (chloramphenicol/streptomycin). Similar to micropropagation, meristem culture involves initiating the culture, multiplying the shoots, allowing the developing shoots to root, and then transferring the plantlets to pots or soil (Gupta et al., 2022).

9.4.1.2 Shoot Tip Culture

A very effective way of plant propagation is the culture of bigger stem apices or lateral buds (ranging from 5 to 10 mm in diameter to dissected buds). Two types of explants are examined in vitro for size and relative placement in the shoot apex of a typical dicotyledon. Shoot culture has been modified to become node culture. For clonal descendants that proliferate without chromosomal alterations or meiotic abnormalities and are morphologically homogeneous, in vitro shoot tip production is frequently utilized. Between 0.1 and 10 mM, 6-benzylamino purine and kinetin encouraged shoot development, whereas at 5 and 10 mM, 6-benzylamino purine caused tomato apical shoot tips to multiply (*Lycopersicon esculentum* Mill.) (Luna et al. 1994). On *Gossypium* meristem shoot tip culture, a straightforward approach has been described in which optimum shoot proliferation is noted on media containing 0.46 mM kinetin. The previous procedures for cotton meristem tip cultivation have been replaced by this straightforward methodology (Saeed et al., 1997).

9.4.1.3 Single-Node Culture

Explants made from mature plants or seedlings with a single node can be used to study branch formation since nodes include axillary buds that can form branches when exposed to cytokinin in an in vitro environment. Such axillary branches develop roots in the rooting material, resulting in the formation of many plantlets (Barna & Wakhlu, 1995).

9.4.1.3.1 Applications of Meristem and Shoot Culture

One of the most common uses of tissue culture methods is rapid clonal multiplication. By utilizing meristem and shoot tip culture, numerous orchids and ornamental plants have been produced. Compared to meristem tip culture, the rate of propagation is higher during organogenesis and embryogenesis. Meristem culture is still seen as a viable method for mass reproduction amidst the rate since it preserves the genetic integrity of cultures. Meristem culture is helpful in breeding initiatives when hybrid plants fail to develop and generate nonviable or abortive seeds. *Agrobacterium*-mediated genetic modification also uses the meristem and shoot tips.

9.4.1.3.2 Limitations of Meristem and Shoot Culture

The manpower and financial costs of tissue culture can increase. Due to the environment in which they are produced, it is possible that propagated plants will be less resistant to illnesses. Prior to culture, the material must be thoroughly checked; if abnormalities are missed, the young plants may get contaminated. While the overall performance is high provided the proper protocols are implemented, success with tissue culture is not guaranteed. The process could nevertheless result in a secondary metabolic chemical response that inhibits or even causes the demise of the newly formed explants or cells.

9.4.1.4 Embryo Culture

The isolation of immature or mature embryos is known as embryo cultivation. When starting callus cultures in plant tissue culture, for instance, zygotic or seed embryos are frequently employed profitably as explants. When the seed included endosperm throughout maturity, the embryo developed appropriately because it had a nourishing tissue. However, crossing between the two dissimilar species caused the endosperm tissue to deteriorate, impeding the establishment of the embryo and, thus, preventing the growth of a functional plant. Culturing of the embryo was done in order to retrieve such a hybrid embryo. This process of cultivating embryos (hybrids) in vitro is referred to as embryo rescue and is frequently used to enhance crops. In 1904, Hanning successfully acquired the plants by first cultivating an isolated mature embryo from a hybrid between *Cochlearia* and *Raphanus* (Rogo et al., 2023). Laibach isolated zygotic embryos from nonviable *Linum perenne* subsp. *austriacum* (L.) seeds in 1925 and 1929 (Pandey, 2022). It is impossible for these hybrids to grow naturally. For the first time, Van Overbeek et al. used coconut milk in 1941 to aid in the growth of the Datura embryo and callus (Singh & Kumar, 2009). One of the hardest challenges to successfully establishing in vitro cultures is acquiring contamination-free seeds or embryos without harm or oversterilization. Sugii illustrated the numerous disinfestation techniques (ethanol dip and flame, gas sterilization, and bleach sterilization) that are effective in creating in vitro cultures of a number of vulnerable Hawaiian plant species.

9.4.1.4.1 Applications of Embryo Culture

Using embryo culture, it is possible to produce viable embryos from nonviable seeds produced by interspecific and intergeneric crosses, micropropagate hard species (such as *Pinus palustris*), burst dormancy, produce haploid plants (such as *Hordeum vulgare* and *Hordeum bulbosum*), and germinate seeds of obligatory pathogens without the host.

9.4.1.4.2 Limitations of Embryo Culture

Embryo culture requires overcoming multiple technical as well as biological challenges. Embryo early-stage extraction becomes challenging because minimal size and fragility lead to mechanical failure or death of embryos. Due to this obstacle, embryos often fail to initiate the culture procedure. The identification of suitable complex nutritional media required for hybrid embryo development stands as an essential barrier that hinders progress. The specific nutritional needs change considerably because of embryo development stage and species type which demands extensive tests. The dependence of early-stage embryos on external culture conditions becomes more critical because they lack endosperm support thus complicating precise in vitro replication. The combination of these aspects decreases the effectiveness of embryo rescue methods mainly when working with far-related species during hybridization programs.

9.4.1.5 Leaf Culture

Leaf culture is the aseptically growing plucked leaf primordial or an immature young leaf of the shoot apex in a chemically specified media while being carefully monitored to ensure that the development sequences are followed.

In order to understand *phyllotaxy*—the organized arrangement of leaf primordia around the perimeter of the shoot apical meristem—it is vital to understand the developmental phases of the leaf primordia. The primary compounds adequate to create a primordium and set in action all the processes required to make a full leaf are auxin and a localized protein known as expansion.

9.4.1.5.1 Application

The extremely straightforward culture medium needed for the full growth and development of leaf primordia is one of the biggest benefits of this method of cultivation. In order to understand how nutrients, growth factors, and alterations in the leaf's environment affect it, it is helpful to utilize leaf primordium. Additionally, it is helpful for understanding how ferns evolve, first becoming leaves, then later becoming sporangia.

9.4.1.6 Fruit Culture

There are some signs that fruit explant cultures have produced improved outcomes. Einset investigated the impact of orange juice from an "acidless" variety on the in vitro development of citrus fruits, such as citron *(Citrus medica)*, lemon (Citrus limon), grapefruit (*Citrus paradisi*), sweet orange (*Citrus sinensis*), and mandarin (*Citrus reticulata*) explant (juice vesicle or albedo tissues) cultures. He used a variety of in vitro cultured fruits grown on a base medium to demonstrate the substantial growth-stimulating impact of orange juice on these fruits. Additionally, it was shown that adding "acidless" juice had stimulation effects, not that citric acid itself was the cause of this stimulation.

9.4.1.7 Seed Culture

When explants are obtained from in vitro generated plants for the multiplication of orchids, seed culture is a crucial approach. Each explant must be sterilized properly in order to prevent toxicity and damage to plant regeneration. The ideal approach in this situation is to cultivate seeds to grow sterilized seedlings. In contrast, certain plant species have tiny seeds that may not germinate effectively. One of the best instances of this is an orchid. The Orchidaceae family is one of the largest angiosperm families, and many of its members are extremely significant as natural remedies and in the horticulture business. The following criteria impede the proliferation of most orchids: small size and difficulty in using directly in the field, non-endospermous quality with low nutrient content, reliance on a particular fungus for germination, early seedling maturation, and poor germination rate. Only a small percentage of the millions of seeds generated per capsule naturally germinate in the wild. The symbiotic interaction between orchids and fungi, which is necessary for germination, used to be a significant barrier to in vitro production. In vitro asymbiotic orchid germination was demonstrated by Knudson (Kauth et al., 2008).

Asymbiotic germination refers to the process of growing such plants' seeds in vitro, which is reliant on the added growth medium. On the other hand, epiphytic orchids are easier to grow from seed than terrestrial orchids. The parameters for in vitro germination are more demanding for terrestrial orchid seeds because of their rigid seed coat. Lee detailed the pre-processing steps and time of seed collection for various terrestrial Asian orchids to increase in vitro germination (Lee, 2011). In addition, he has listed the critical phases of in vitro germination: (1) seed culture of immature seeds (optimal timing of seed harvest is key to maximizing germination), (2) the choice of appropriate pretreatment conditions (i.e., the length of time and solution concentrations) necessary to enhance germination, and (3) the growth of succeeding seedlings.

For the creation of a large variety of fruitful cultivars, hybridization and cross-pollination have proven to be very reliable processes. Because of the sluggish growth of plantlets, a low multiplication rate, poor roots, and somaclonal diversity, in vitro culture technologies are currently difficult to employ for large clonal replication. In this context, micropropagation through protocorm-like structures obtained from germinating embryos and somatic tissues is a key tactic in getting genetically stable plants and in the enhancement of quality (Neumann et al., 2020). To produce huge numbers of seedlings quickly and efficiently, seed culture is frequently utilized. MS medium enriched with activated carbon to grow *Capparis spinosa* seedlings on a wide scale after sterilizing the seeds with 0.1% $HgCl_2$ for 12 minutes (Khalaf, 2011).

9.4.1.8 Endosperm Culture

The embryonic development consumes the endosperm, which is a transient structure. This tissue provides seeds with nutrition. When two male gametes or nuclei merge with two distinct female nuclei independently during the fertilization phase in angiosperms, this is referred to as double fertilization. One of the male nuclei unites with the egg to generate the zygote (diploid tissue), while the other male nucleus fuses with the two polar nuclei to form the endosperm (triploid tissue) in the embryo sac. For the growing embryo in the seed, this endosperm serves as food. Non-endospermous seeds are those in which the embryo fully uses the seed's embryo, whereas endospermous seeds are those in which the embryo partially utilizes the endosperm (coconut, coffee).

Lamp and Mill first tried cultivating maize endosperm on a nutritional medium in 1933, but they were not entirely successful (LaRue, 1936). LaRue created callus tissue from endosperm culture in 1949, which inspired other workers to continue the work on endosperm culture for several plant species (LaRue, 1949). In general, a mature endosperm is not a desirable substrate for explants because it transforms after pollination, making it appropriate for culture. Thus, endosperm was chosen for this reason promptly following a few days of pollination. Endosperm tissue was removed while the seed or kernel was still sterile, and the entire object was surface-sterilized. The endosperm culture should be kept in the dark for callus generation, and high light is needed for organ maturation. It can also be used to produce triploid plants, which are triploid in their chromosomal makeup, in contrast to serving the purpose of delivering sustenance to the growing embryo. Tetraploid and diploid plants are crossed to create triploid plants for breeding programs.

Typically, these triploids are created by mating diploids with tetraploids that have been colchicine-induced, then saving the resulting triploid embryos. Producing triploids, however, can be challenging due to a strong crossability barrier. According to a recent study, cultures of excised endosperm of immature fruits with zygotic embryos at the early dicotyledonous stage could produce triploid plants of the ornamental *Phlox drummondii*. When contrasted to naturally existing diploid plants, these triploid plants were found to be larger and more robust, and their flowers had a vivid color and a bigger central eye, which increased their aesthetic appeal.

There are two further categories of cultures in this category:

- *Mature culture.* Mature cultures are made from mature endosperm, such as mature *Ricinus communis* embryos grown on White's medium supplemented with 2,4-D, kinetin, and yeast extract.
- *Immature culture.* Immature culture is made from endosperm that is less than 8 days old, such as that found in maize, wheat, and barley. In order to thrive and expand, this kind of culture essentially needs additional ingredients, like yeast extract, vitamins, and casein hydrolysate.

9.4.1.8.1 Applications of Endosperm Culture

The endosperm culture is a useful research platform for physiological and morphogenetic research. The investigation of differentiation and metabolism is greatly advanced by this system. In crops like apple, banana, mulberry, sugar beet, tea, and watermelon, where seeds are not important commercially, triploidy can be used. Triploids outperform diploids in several plants, notably those that are clonally propagated, producing better pulpwood. Sterility in seeds is not a major drawback for these plants, because they may be multiplied vegetatively.

Crosses between auto-tetraploids and auto-diploids are made in the case of the traditional method of triploid creation. These crosses may occasionally fail, which can make it difficult to produce triploids.

9.4.1.9 Isolated Root Culture

Kotte and Robbins first tested the root culture technique on the tips of pea, maize, and cotton roots (White, 1934). These were gently removed from the plant and aseptic conditions cultivated in the nutritional medium for a few weeks before growth stopped. Following the development of White's medium (yeast extract containing inorganic salts, sucrose, and three vitamins that were later substituted), other researchers attempted to alter and establish an appropriate medium for root cultivation (White, 1934). The variations in plant growth factors between healthy and ill plants were investigated at this time by a number of researchers. Stanley contrasted the healthy excised tomato cultured roots with a virus-infected plant under comparable conditions in 1938, and he extracted the protein of the *Aucuba mosaic* virus from ill removed tomato roots cultivated in vitro (Stanley, 1938).

Later, various studies were carried out to keep isolated roots' capacity for regeneration for a longer time. Miguel Raggio and Nora Raggio created a revolutionary technique for the development of isolated roots in 1956 by altering White's medium (Raggio & Raggio, 1956). The majority of experiments were built on modifications to White's medium. Using a modified version of White's medium, *Brassica* isolated culture was created in 1984 (Lazzeri & Dunwell, 1984). It was proven that isolated roots' capacity for regeneration was maintained during the course of five months in a culture. Thus, by incubating on an agar-solidified medium containing cytokinin and auxin, it is feasible to redevelop shoots from pieces of cultivated roots (Lazzeri & Dunwell, 1984).

Root tip cultures are typically kept in an agitated liquid medium with the proper auxin. Tropane alkaloids are more prevalent in the roots of solanaceous plants. The production of alkaloids dramatically increased when such roots are recreated from undifferentiated calli. Therefore, *Datura*, *Hyoscyamus*, and *Atropa* roots are bred to generate a large number of alkaloids. Furthermore, because of their sluggish growth rate and growth-related culturing resistance, many species have sparked little interest as a source of secondary products.

9.4.1.9.1 Applications of Isolated Root Culture

In comparison to other tissue cultures, root cultures have a higher level of genetic stability. As a result, it is appropriate and frequently utilized for the storage of specific species' germplasm. Root culture growth is reportedly inhibited by auxins or gibberellins. Naturally occurring shoots that arise from roots demonstrate the idea that roots have the capacity to differentiate into shoots. Supplementing hormones that cause new growth can do this (cytokinin to the medium). The tomato is the most famous instance of a plant grown from root-derived culture.

9.4.1.10 Callus Culture

The first person to record the production of callus on living plants was Henri-Louis Duhamel du Monceau, who studied the wound healing reactions in elm trees. E. F. Simon was successful in inducing callus from poplar stems in 1908, together with roots and buds. The earliest findings of callus development in vitro were made in 1939 by three different researchers. P. White created a callus from hybrid *Nicotiana glauca* procambial tissues that were undergoing tumor development without the use of hormones.

A *callus* is made up of an amorphous mass of parenchymatous cells with thin walls that are loosely distributed and developed from the mother tissue's growing cells. Nearly every portion of the plant contains calli. A callus frequently develops at the cut end of the stem or root as a result of injury. It is a more or less unorganized tissue that has both differentiated and pluripotent cells, including cells that are starting points for the development of calluses, like leaves, roots, and shoots. As a result, cells with diverse levels of differentiation are included in the word "callus." Unspecialized parenchyma cells make up the callus's mass of cells. The explants' parenchymatous cells typically go through this differentiation. Therefore, a *callus* might be conceptualized as a less differentiated and disorganized parenchymatous tissue. This amorphous tissue typically results from the uncontrolled proliferation of plant cells. Thus, callus cultures are essentially aggregates of cells that have grown as a result of the uncontrolled proliferation of cells from a particular plant organ. In this method, plant cells multiply on proper conditions to create *calluses*, which are clumps of undifferentiated cells. Before cell division may take place, when callus comprises undifferentiated cells that are present in solitary explants, the cells must first undergo dedifferentiation (i.e., transform to become meristematic). The development of callus is illustrated via Figure 9.4.

When explants are isolated and solely include meristematic tissue, cell division occurs without dedifferentiation. Mature adult cells have a brief window of time during the dedifferentiation process to switch from adult to juvenile state. The capacity for growth and division is higher in the revitalized cells. Under specific conditions, these cells have the capacity to regenerate organs and/or embryos. Dedifferentiation is therefore an important stage to take into account while propagating plants in vitro. Suitable callus induction hormones are often supplemented at the time of callus commencement. Its progression and maturation require sterile surroundings and a growth-supporting media. Callus production from the explants is influenced by exogenously administered growth regulators that are present in the nutritional media. Auxin alone, cytokinin alone, and both auxin and cytokinin together comprise the three categories of hormone injections for callus formation.

9.4.1.10.1 Applications of Callus Culture

In addition to acting as a suitable source of resource for protoplast isolation, callus culture offers an insight into the plant's capacity for regeneration. Callus may be retained forever in the in vitro environment for extended lengths of time and can be utilized as inoculums to start suspension cultures. It also serves as a better supply of cells for the genetic modification of plant genomes.

9.4.1.10.2 Limitations of Callus Culture

High biochemical variation and a slow growth rate are some drawbacks of callus culture. These restrictions prevent cultivators from using callus culture techniques at a higher degree.

9.4.1.11 Suspension Culture

The earliest endeavor to isolate individual cells from flowering plant leaves was conducted by Haberlandt (Haberlandt, 1902). In 1958, Muir made it abundantly evident that plant cells could be raised in liquid (suspension) cultures, just like microbes (Muir et al., 1958). Nevertheless, relying

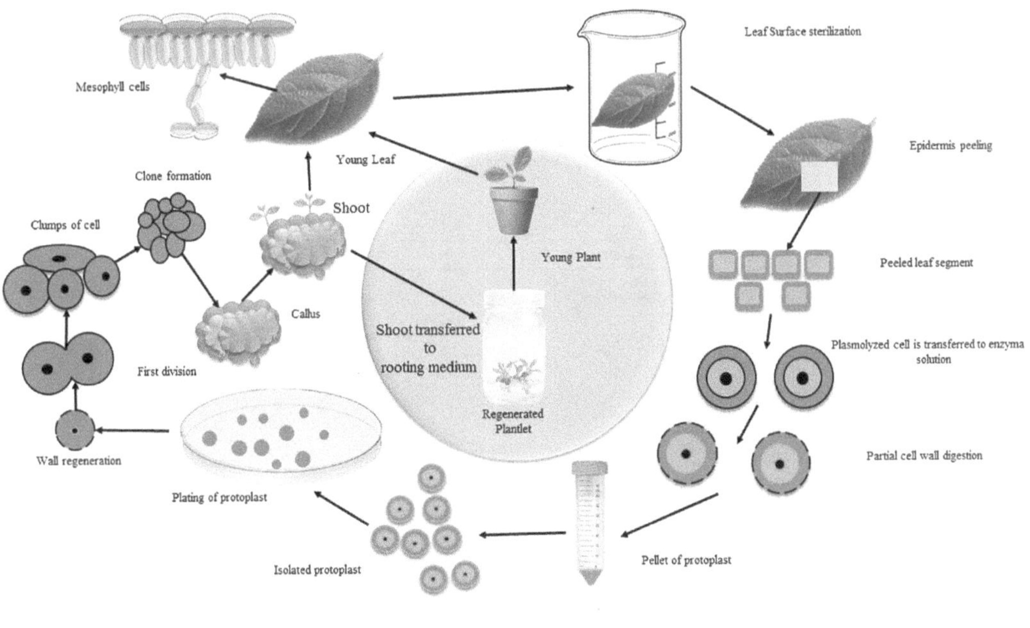

FIGURE 9.4 An illustration of the procedure for isolating, nurturing, and regenerating plantlets from leaf protoplasts.

on the development cycle stage and on culture conditions, particular plant cells within a population of cultured cells usually exhibit cytogenetic and metabolic differences. Spatial heterogeneity is the name given to this type of variation (Lindsey & Yeoman, 1983).

Isolated cells are cultured in liquid media in a process known as suspension culture. Cells can be removed mechanically or via enzyme action from in vivo plant material, and callus can be produced from any explants. Similar to callus culture, cell suspension cultures can be started from almost any region of the plant. The most typical method of starting suspension cultures from calluses that are already developing in a culture is this one. In general, callus cultures are split into two groups: friable and compact. In contrast to *friable* callus, which has loosely connected cells, *compact* callus has firmly packed cells. Suspension cultures are created based on the callus type (friable callus or soft callus easily separates with less force than does compact callus).

It is best to start cell suspensions with comparatively large inoculums so that the liberated concentration rises quickly. Cell cultures are typically used to describe these cell types, which are less organized than tissue or callus cultures. Usually, the liquid media used to develop the callus have properties that are comparable to those of this liquid medium. The hormone and inorganic salt concentrations may need to be changed, though. In order to retain a consistent allocation of cells of different sizes and shapes, maintain cell clumping in the medium, and support cell respiration in the liquid media, agitation is necessary for suspension cultures. Agitation is necessary to break up cell aggregation and establish a homogeneous allocation of cells of different sizes and shapes. The inoculums split apart and produce distinct cell clusters as cell division progresses, which then split apart once more to produce individual cells and other tiny groupings. Growing a suspension of single cells solely is not necessarily achievable since plant cells naturally stick collectively.

Batch culture and continuous culture are the two different kinds of suspension cultures, illustrated via Figure 9.5.

9.4.1.11.1 Batch Cultures

A closed culture technique known as batch culture uses a little quantity of nutrients. Batch culture involves maintaining cells in conical flasks on orbital shakers at a speed of 80–120 rpm as they grow in a finite volume of liquid media. Batch cultures come in a variety of forms, including stirred, shaken, slowly rotating, and spinning cultures.

9.4.1.11.2 Continuous Cultures

Continuous cultures are defined as those that need a constant supply of the medium's component, such as the cell suspension. By removing the old liquid medium and providing a new medium to

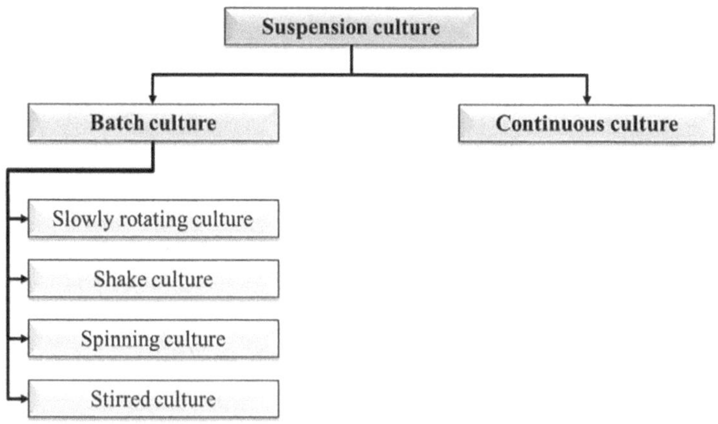

FIGURE 9.5 Types of suspension culture.

stabilize the physiological condition of proliferating cells, this method is kept in a stable state for extended periods of time. Since there is a finite supply of one of the nutrients in the growing media, when the nutrient is used up during exponential development, growth will halt. Before the nutrient runs out, a fresh medium with a small amount of nutrient is applied.

Nutrient depletion is prevented by the constant flow of nutrients and the constant state of the cells' nutrient levels. Based on the diversity of plant cells, many bioreactor layouts and sizes have been developed for continuous culturing (Fowler, 1982; Panda et al., 1989). Two categories of continuous cultures exist: The two systems are (1) closed continuous culture, where cells are removed from the evacuated media and reintroduced back to the suspension culture, and (2) open continuous culture, where the medium is reintroduced and a similar volume of the suspension culture is harvested.

9.4.1.11.3 *Applications of Suspension Culture*

For the generation of secondary metabolites, mutant selection, immobilization, biotransformation, somatic embryogenesis, genetic manipulation, commercial production of cell mass, and induction of mutation of plant cells, as well as for the collection of isolated cells used in protoplast production, single-cell protein hybrids or cybrids, the creation of single-cell lines or cell clones, and the study of the effects of various chemicals.

9.4.1.11.4 *Limitations of Suspension Culture*

Over prolonged subculture periods, suspension culture production declines. The plant cells' sluggish development and limited production. Cells may be harmed by shear forces.

9.4.1.12 Protoplast Culture

Protoplasts are cells that do not have a cell wall (cells with their walls removed and removed from the vicinity of their neighboring cells):

Vasil changed an earlier definition of protoplast: "The protoplast is a part of plant cell which lies within the cell wall and can be plasmolysed and isolated by removing the cell wall by mechanical or enzymatic procedure." An illustration of the procedure for isolating, nurturing, and regenerating plantlets from leaf protoplasts is shown in Figure 9.6.

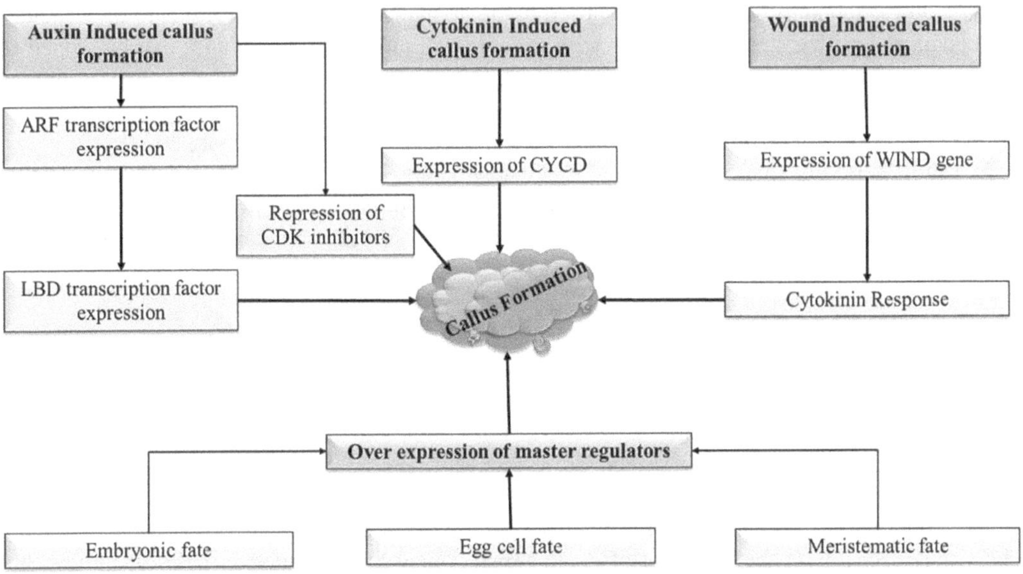

FIGURE 9.6 Mechanisms involved in callus formation.

9.4.1.12.1 Applications of Protoplast Culture

Protoplast culture is employed in the creation of unique interspecific and intergeneric crosses between plants that are difficult or impossible to hybridize traditionally in order to overcome sexual incompatibility. To supplement the selection of green intraspecific hybrids of tobacco, Melchers and Labib utilized albino mutants.

Creating hybrids. A *hybrid* is a cell or plant that combines the cytoplasm of two distinct species with the nucleus of one species. Variable frequencies of these are often generated during protoplast fusion. A hybrid cell may become a hybrid by merging the normal protoplasts of one species with enucleated protoplasts (cytoplasm), by removing the nucleus of one species from a normal heterokaryon, or by gradually losing one species' chromosomes during subsequent mitotic divisions. By irradiating one parenteral protoplast prior to fusion to render the nuclei inactive, or by producing enucleate protoplasts of one species and fusing them with regular protoplasts of different species, cybrids can be created at a high frequency (Melchers & Labib, 1974).

Adsorption of foreign materials. Isolated protoplasts are thought to be particularly suited for genetic and cytoplasmic alterations since they are naked by nature. Plant protoplasts have reportedly been shown to have the unusual ability to absorb isolated nuclei, DNA, chromosomes, chloroplasts, cyanobacteria, nitrogen-fixing bacteria, and virus particles.

Protoplast-based gene transfer is an alternative to employing *Agrobacterium.* Direct gene transfer is accomplished by using isolated protoplasts. In 1974, Jeff Schell and colleagues discovered that the *Agrobacterium* cell's plasmids, not its bacterial chromosomes, are what give it its engineering capabilities. An incredible achievement in recovering the first transgenic plant was made possible by the transformation of the tobacco protoplast by an *Agrobacterium* (Tempé et al., 1977).

9.5 PROS AND CONS OF PLANT CELL CLONING

9.5.1 Advantages

It aids in the quicker and more efficient production of individuals of the same species. It is similar to a rapid and reasonably quick means of meeting the needs of huge quantities of a single-species plant. All cloned plants develop at the same rate; hence, the period between planting and harvesting is more predictable. This may be a significant boost to crop production and earnings for farmers in an agricultural economy. Cloning can be used to create plants that are more resistant to pesticides and chemical fertilizers. A single plant with exceptional nutritional advantages can be cloned to produce identical plants with the same benefits. This can aid in the resolution of difficulties related to poor vegetable and fruit quality. Food difficulties can be overcome by generating vast amounts of crops, since yields can be increased through plant cloning (Christou & Twyman, 2004).

In a short period of time, we can grow many new plants. There should be such conditions that are precisely controlled, we can get the characteristics we want. Hybrid plants are often more vigorous and develop at a faster rate than regular plants. In rapid pace, new genes are discovered (Goulet et al., 2017).

9.5.2 Disadvantages

As some plants have same genetic information, they are susceptible to the same diseases. When a field of plants is genetically similar, a disease or insect that may afflict one plant will most likely affect all of them, inflicting severe economic loss. No new beneficial characteristics can arise (Strange & Scott, 2005).

No variation causes the gene pool to reduce. Cloning has a negative impact on genetic diversity. The same species with the same set of characteristics may be generated in large quantities, but cloned plants lack genetic diversity. This does not occur in nature. Every plant, regardless of species, has its unique level of resistance to pathogens and environmental changes. This is a significant disadvantage in cloned plants. Cloned plants cannot survive shifting environmental variables, and if a disease affects even one cloned plant, the entire harvest can be destroyed. Cloned plants have similar appearances; using them for decoration might be detrimental. Home gardening and landscaping might get boring over time. One of the major concerns of selective breeding is that it might lead to a loss of diversity. This leaves limited room for plant evolutionary development in the future. When damaged plants are cloned and multiplied, the illness can spread far from its original environment, affecting considerably more producers and consumers than previously. Because the rhizomes used to clone banana plants might look free of Panama disease while afflicted, banana cloning contributes to the spread of this fatal disease (Ordonez et al., 2015). If plant cloning is done on a large scale, there is a chance that food will become more commercialized. While many third world nations can be assisted with food shortages, those in power will not pass up the opportunity to commercialize plant cloning in order to acquire more control over food supplies (Ronald, 2011).

Plant cloning, which is thought to ease deforestation problems, poses the risk of producing low biodiversity. We all know that forests are excellent examples of biological variety. If the same species are planted, the richness of diversity will inevitably decline.

9.6 CONCLUSION

Plants are survivors who have been on Earth long before the earliest creatures. Plants' unique reproductive techniques are one of the reasons they have been able to adapt to every habitat on the globe. Most animals can only reproduce sexually by combining gametes from two opposite-sexed individuals. Most plants, on the other hand, may create offspring asexually from a single parent by cloning. Each year, horticulturists and gardeners use the capacity for plant cloning to produce millions of new plants. Most plants you may buy in nurseries and food shops were originally clones. It is advantageous to create plants by cloning since it ensures that the progeny plants will have the same characteristics and attributes as their parent plant. Many times, certain plants have distinctive qualities and attributes that appeal to people, such as sweet-tasting fruits, huge fruits, or beautifully colored flowers.

REFERENCES

Arditti, J. and Code, B.I.M. (2007). Plenary Presentation: History of Orchid Propagation. *Proceedings Asia Pacific Conference on Plant Tissue and Agribiotechnology (APaCPA).*

Awotedu, B., Omolola, T., Akala, A., Awotedu, O. and Olaoti-Laaro, S. (2021). Vegetative Propagation: A Unique Technique of Improving Plants Growth. *World News of Natural Sciences*, *35*, 83–101.

Barna, K. and Wakhlu, A. (1995). Modified Single Node Culture Method—A New Micropropagation Method for Chickpea. *In Vitro Cellular & Developmental Biology-Plant*, *31*, 150–152.

Chen, J. and Henny, R.J. (2008). Role of Micropropagation in the Development of the Foliage Plant Industry. *Floriculture, Ornamental and Plant Biotechnology*, *5*, 206–218.

Christou, P. and Twyman, R.M. (2004). The Potential of Genetically Enhanced Plants to Address Food Insecurity. *Nutrition Research Reviews*, *17*(1), 23–42.

Fowler, M.W. (1981). The Large Scale Cultivation of Plant Cells. *Prog. Ind. Microb.*, *16*, 227–229.

Goulet, B.E., Roda, F. and Hopkins, R. (2017). Hybridization in Plants: Old Ideas, New Techniques. *Plant Physiology*, *173*(1), 65–78.

Graft: Description, Types & Uses (2022a). *Encyclopedia Britannica.* https://www.britannica.com/science/propagation-of-plants

Gupta, S., Singh, A., Yadav, K., Pandey, N. and Kumar, S. (2022). Micropropagation for Multiplication of Disease-free and Genetically Uniform Sugarcane Plantlets. In *Advances in plant tissue culture* (pp. 31–49). Elsevier.

Haberlandt, G. (1902). Kulturversuche mit isolierten Pflanzenzellen. Sitzber. Akad. Wiss. Wien, Math. *Naturwiss. Kl*, *3*, 69–92.

Harold, F.M. (2022). *On life: Cells, genes, and the evolution of complexity*. Oxford University Press.

Iqbal, M. and Singh, K.K. (2020). Propagation of Temperate Fruit Crops. *Innovative Agriculture and Botany*, Victorious Publishers (India), pp. 119–135.

Kauth, P.J., Dutra, D., Johnson, T.R., Stewart, S.L., Kane, M.E. and Vendrame, W. (2008). Techniques and Applications of in vitro Orchid Seed Germination. *Floriculture, Ornamental and Plant Biotechnology: Advances and Topical Issues*, *5*, 375–391.

Khalaf, G.M.H.A. (2011). *Capparis spinosa L.* "*Caper*": *In vitro cultures and bioassay*. Tesis Máster. Hebrón, Palestina. Palestine Polytechnic University.

Lai, R. and Lai, S. (2019). Role of Tissue Culture in Rapid Clonal Propagation and Production of Pathogen–free Plants. In *Crop improvement utilizing biotechnology* (pp. 73–116). CRC Press.

LaRue, C.D. (1936). The Growth of Plant Embryos in Culture. *Bulletin of the Torrey Botanical Club*, *63*(7), 365–382.

LaRue, C.D. (1949). Cultures of the Endosperm of Maize. *American Journal of Botany*, *36*(10), 798–798.

Lazzeri, P. and Dunwell, J. (1984). Establishment of Isolated Root Cultures of Brassica species and Regeneration from Cultured-root Segments of *Brassica oleracea* var. Italica. *Annals of Botany*, *54*(3), 351–361.

Luna, J.M., McGrath, D.G., William, R., Seiter, S., Ten, T. (1994). Collaborative learning and participatory on-farmresearch: cover crops andvegetable production systems. *Hortscience*, *29*(5), 427.

Lee, Y.-I. (2011). In Vitro Culture and Germination of Terrestrial Asian Orchid Seeds. *Methods Mol Biol. 710*, 53–62. doi: 10.1007/978-1-61737-988-8_5.

Lewis, W.J. and Alexander, D. (2008). *Grafting and budding: A practical guide for fruit and nut plants and ornamentals*. Landlinks Press.

Lindsey, K. and Yeoman, M. (1983). The Relationship between Growth Rate, Differentiation and Alkaloid Accumulation in Cell Cultures. *Journal of Experimental Botany*, *34*(8), 1055–1065.

Melchers, G. and Labib, G. (1974). Somatic Hybridisation of Plants by Fusion of Protoplasts: I. Selection of Light Resistant Hybrids of "Haploid" Light Sensitive Varieties of Tobacco. *Molecular and General Genetics MGG*, *135*(4), 277–294.

Muir, W., Hildebrandt, A. and Riker, A. (1958). The Preparation, Isolation, and Growth in Culture of Single Cells from Higher Plants. *American Journal of Botany*, *45*, 589–597.

Nahar, K. (2025). Propagation Technologies: Seedling, Stem Cutting and Grafting. In *Jatropha curcas L: A potential 2G energy crop to produce biofuel in Bangladesh: agronomy, biotechnology, biodiesel and byproducts* (pp. 33–55). Springer.

Neumann, K.-H., Kumar, A., Imani, J.(2020). Plant Propagation: Meristem Cultures, Somatic Embryogenesis Micropropagation, and Transformation of Somatic Embryos in Bioreactors. In *Plant Cell and Tissue Culture–A Tool in Biotechnology* Springer, Cham (pp. 107–183) https://doi.org/10.1007/978-3-030-49098-0_7.

Ordonez, N., Seidl, M.F., Waalwijk, C., Drenth, A., Kilian, A., Thomma, B.P., Ploetz, R.C. and Kema, G.H. (2015). Worse Comes to Worst: Bananas and Panama Disease—When Plant and Pathogen Clones Meet. *PLoS Pathogens*, *11*(11), e1005197.

Panda, A., Mishra, S., Bisaria, V. and Bhojwani, S. (1989). Plant Cell Reactors—A Perspective. *Enzyme and Microbial Technology*, *11*(7), 386–397.

Pandey, A.K. (2022). *Reproductive biology of angiosperms*. CRC Press.

Raggio, M. and Raggio, N. (1956). A New Method for the Cultivation of Isolated Roots. *Physiologia Plantarum*, *9*(3), 466–469.

Ramesh, K. (2023). A Review on Asexual Reproduction System in Plants. *Int. Res. J. Edu. Techno.*, *5*(6), 6.

Rogo, U., Fambrini, M. and Pugliesi, C. (2023). Embryo Rescue in Plant Breeding. *Plants*, *12*(17), 3106.

Ronald, P. (2011). Plant Genetics, Sustainable Agriculture and Global Food Security. *Genetics*, *188*(1), 11–20.

Saeed, N.A., Zafar, Y. and Malik, K.A. (1997). A Simple Procedure of Gossypium meristem Shoot Tip Culture. *Plant Cell, Tissue and Organ Culture*, *51*, 201–207.

Singh, A.K., & Kumar, A. (2021). *Plant propagation and nursery management*. Bhavya Books (BET).

Singh, M. and Kumar, S. (2009). *Plant tissue culture*. APH Publishing.

Singh, M.B. and Bhalla, P.L. (2006). Plant Stem Cells Carve Their Own Niche. *Trends in Plant Science*, *11*(5), 241–246.

Stanley, W. (1938). Aucuba Mosaic Virus Protein Isolated from Diseased, Excised Tomato Roots Grown In Vitro. *Journal of Biological Chemistry*, *126*(1), 125–131.

Strange, R.N. and Scott, P.R. (2005). Plant Disease: A Threat to Global Food Security. *Annual Review of Phytopathology*, *43*(1), 83–116.

Tempé, J., Petit, A., Holsters, M., Montagu, M. v. and Schell, J. (1977). Thermosensitive Step Associated with Transfer of the Ti Plasmid During Conjugation: Possible Relation to Transformation in Crown Gall. *Proceedings of the National Academy of Sciences*, *74*(7), 2848–2849.

White, P.R. (1934). Potentially Unlimited Growth of Excised Tomato Root Tips in a Liquid Medium. *Plant Physiology*, *9*(3), 585.

White, P.R. (1936). Plant Tissue Cultures. *The Botanical Review*, *2*(8), 419–437.

Xue, Y., Hiti-Bandaralage, J.C.A., Jambuthenne, D.T., Zhao, Z. and Mitter, N. (2023). Micropropagation of Duboisia Species via Shoot Tip Meristem. *Horticulturae*, *9*(12), 1313.

Yadav, D. and Singh, S. (2018). Vegetative Methods of Plant Propagation: I-cutting Layering and Budding. *Journal of Pharmacognosy and Phytochemistry*, *7*(2), 3267–3273.

10 Secondary Metabolism in Tissue Cultures with Emphasis on the Production of Medicinal Agents

Precursors and Elicitors on the Production of Secondary Metabolites—Applications and Associated Problems

Gajanand Engla, Love Kumar Soni, Mayur Chaurey, and Arun Patidar*
* Corresponding author: gajanand54@gmail.com

10.1 INTRODUCTION

Plants serve as the main source of crucial secondary metabolites (SMs), which hold medicinal significance in the advancement of drug discovery. Secondary metabolites are frequently employed by plants for diverse functions, such as resistance to herbivory, interspecies interactions, and various forms of stress. These secondary metabolites serve vital functions in the human body as drugs, pigments, flavors, and medications. The pharmaceutical industry's significant faith on medicinal plants and their extraction has resulted in the endangerment of these plants. The utilization of plant cell culture technologies is accessible as a reasonable approach to generate and investigate plant SMs. For the enhancement of plant SM production, a variety of *in vitro* techniques (elicitation, crown gall, hairy root culture system, cell suspension culture system, etc.) have been utilized widely. Although cell suspension culture and elicitation are the most common techniques used to increase SM production, it has been demonstrated that hairy root culture and other organ cultures can also meet the demand for secondary metabolites (Winter & Tang, 2012).

Nowadays, the utilization of *in vitro* methods such as plant cell, tissue, and organ cultures to produce medicinally significant SMs is an effective strategy. Recent years have seen the widespread implementation of in vitro strategies, owing to the understanding and regulation of the SM pathways in commercially valuable plants. Moving forward, these techniques will also offer a beneficial approach to maintaining the viability of medicinal plants as sustainable sources of compounds with medicinal significance, as well as facilitate the effective synthesis of unknown, valuable, and desired compounds. Cultures of plant cells, tissues, and organs are environment-friendly techniques for generating secondary metabolites. Nevertheless, the majority of experiments that aim to generate specific products encounter obstacles in field cultivation. Certain plant species exhibit an increase in

 DOI: 10.1201/9781003326939-11

secondary metabolites when callus cultures are subjected to elicitors, such as heavy metals, salicylic acid, methyl jasmone, and chitosan. Shoot teratoma cultures and hairy root cultures yield substantial quantities of monoterpenes and alkaloids, respectively (Yuan et al., 2016).

Plants derived from cell and tissue culture techniques are independent of seasonal and geographical variation. They generate SMs in varying quantities and characteristics in relation to the mother plants, and these attributes are susceptible to alteration over time. Progressions in plant tissue culture methodologies offer promising avenues for enhancing the production of secondary metabolites (Nasri et al., 2014).

In addition to advantages in cell biology, genetics, and biochemistry research, plant cell cultures are crucial to the commercial production of SMs. Cell suspension culture systems are more rapid than tissue and organ culture methods for industrial application and large-scale production of SMs. They serve as a supreme and trustworthy resource in the manufacturing process of natural products. Suspension culture cells are extremely susceptible to variations in medium composition, callus excellence, and genetic diversity. A variety of techniques are utilized to produce suspension cultures that are relatively homogeneous; these techniques have been applied successfully to numerous plant species. An advanced technique in plant tissue and organ culture, hairy root culture increases the production of stem cells (SMs) by utilizing *Agrobacterium rhizogenes*, a naturally occurring vector system, to transform necessary plant species. By introducing genetic modifications into the genome of dicotyledonous plants, *Agrobacterium rhizogenes* induces hairy root disease, which manifests as the growth of hairy roots at the site of infection on the host plant. Active in the transformation process are *vir* genes that become functional exclusively when acetosyringone is present. Hairy root culture demonstrates the most significant advantage in terms of the biosynthetic capacity of SM production in comparison to their mother plants in plant tissue culture. It enables an unrestricted proliferation in culture media devoid of growth hormones and promotes the synthesis of SMs that are absent in the mother plant. For the production of SM, hairy root cultures have been established in numerous dicotyledonous and monocotyledonous plants (Atanasov et al., 2015).

In addition to the ***Ri-plasmid***, transgenic root systems may facilitate the integration of supplementary genes into the cells of the host plant. In addition to investigating the biological qualities, properties, and gene expression profile of metabolic pathways, hairy root cultures are utilized to identify intermediates and essential enzymes that are involved in the biosynthesis of SMs. Shoot culture, akin to hairy root cultures, is a technique utilized in plants to generate secondary metabolites (SMs). The process entails the introduction of *Agrobacterium tumefaciens* into the aerial portions of plants, resulting in the development of either transgenic shoots (shooty teratomas) or non-transgenic shoots that possess an adequate concentration of hormones. Shooty teratomas are generated through the integration of *Agrobacterium tumefaciens* (***Ti plasmid***) into the plant genome or the transformation of the plant's genetic material with *Agrobacterium tumefaciens* nopaline. Although the precise mechanism underlying the development of shooty teratomas remains unknown, a considerable diversity of plant species exhibits this characteristic. The primary applications of shooty teratomas are biotransformation and terpene production, as in the case of nicotine biotransformation in *Nicotiana tabacum* and mint oil. Although plant cell suspension cultures are effective techniques for producing vital SMs, numerous trials fail to yield the intended SMs (Rosenthal, 1991).

Nevertheless, by means of biotransformation processes, these cell cultures are capable of transforming exogenous compounds into novel compounds that possess distinct properties. *Biotransformation* is an industrial process employed to generate novel active constituents from plant species, including *Catharanthus roseus*, tobacco hyoscyamine, and *Eucalyptus peruviana*, each possessing distinct qualities. The conversion of microorganisms and organic compounds into capsaicin, glycosides, and 8-nordihydrocapsaicin constitutes this process. The production of secondary metabolites is substantially influenced by genetic engineering, given that plants are capable of generating chemical compounds that perform particular functions internally and exert beneficial impacts on the human body. Utilizing genetic engineering of microorganisms and plants and *in vitro* reconstruction of plant metabolic pathways, numerous secondary metabolites, including isoprenoids, flavonoids, and stilbenes, have been produced. Since antiquity, plants have been utilized for their medicinal properties; secondary metabolites in particular are exceptional sources of bioactive compounds with added value. They find application in the cosmetics, dietary supplement, pharmaceutical, fragrance, and dye sectors.

In vitro micropropagation can be an effective technique for the synthesis of numerous secondary metabolites. In addition to being a critical element in the production of secondary metabolites, callus culture is also essential in determining the optimal culture conditions for each plant species. The production of secondary metabolites in plants is substantially influenced by oxidative stress. Phenolic compounds, which are generated via the phenylpropanoid pathway, function as a protective mechanism against a wide range of biotic and abiotic stressors.

10.1.1 Overview of Secondary Metabolism in Tissue Cultures

Since antiquity, plants have been utilized for medicinal purposes, and secondary metabolites are a substantial source of bioactive compounds with added value. Metabolites are synthesized by plants as a reaction to stressors. They serve a multitude of physiological functions encompassing applications in the pharmaceutical, cosmetic, dietary supplement, fragrance, flavor, and pigment sectors. India is a significant producer of valuable medicinal plants which are prized for their comparative health benefits to synthetic medications. The global commerce of medicinal plants and their metabolites amounted to $60 billion in the year 2000. This figure is projected to increase at a 7% average annual rate until it reaches $5 trillion by 2050. Plant tissue culture (PTC) methodologies enable the efficient production of commercially viable secondary metabolites through controlled environmental conditions and without regard to the seasons. This permits bulk propagation. In order to improve the synthesis and production of secondary metabolites, which are influenced by biotic and abiotic variables, including growth and physiology, temperature, humidity, light intensity, and culture media composition, both conventional and metabolic engineering methods have been applied (Ncube & Van Staden, 2015).

Plants generate secondary metabolites in reaction to pathogen invasion or stress, which include flavonoids, alkaloids, terpenoids, and phenylpropanoids. The synthesis of proteins implicated in defense or resistance is induced by signaling compounds, including salicylic acid and methyl jasmonate, which initiate the production of these metabolites via transcriptional stimulation of genes. In addition to fungi, bacteria, viruses, and yeast, secondary metabolic pathways can be initiated by microbes. Due to increased yield and quality, the use of bacterial, fungal, and yeast extracts to induce secondary metabolites is acquiring popularity (Guerriero et al., 2018).

Metabolic engineering is the modification of an organism's metabolic pathways to generate commercially significant metabolites through the use of contemporary biological techniques, such as genomics, proteomics, and metabolomics. Plants possess an extensive array of metabolic pathways that facilitate the biosynthesis of intricate metabolites. These metabolites, which are economically significant, can be reconstituted in heterologous hosts to enable overproduction and isolation. For secondary metabolite production, transcription factors, which bind to specific sites on DNA to regulate the transcription or expression of target genes, have been identified. The study of plant tissue culture has facilitated the large-scale synthesis of various bioactive compounds utilized in pharmaceuticals. These compounds consist of secondary metabolites, such as amino acids, alkaloids, terpenoids, steroids, saponins, phenolics, and alkaloids. The growing need for environmentally friendly and economically viable natural phytochemicals derived from plants necessitates the implementation of plant tissue culture techniques for mass cloning (Sandra & Anabela, 2018).

10.1.2 Importance of Medicinal Agents Derived from Secondary Metabolites

Secondary metabolites are bioactive compounds that confer competitive advantage to plants. These minute molecules influence a diverse array of processes in plants and other living organisms,

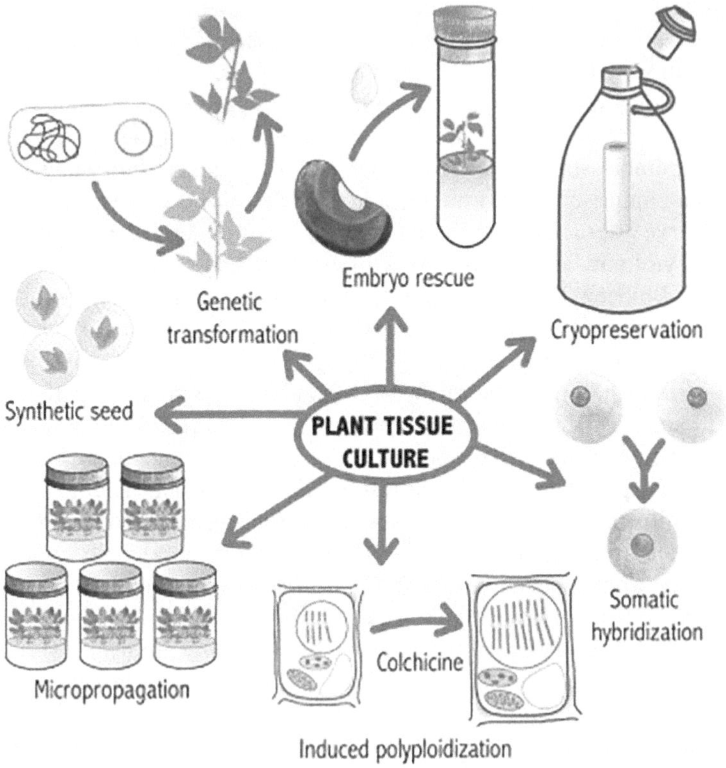

FIGURE 10.1 Tissue cultures.

including, but not limited to, stimulating flowering, fruit set, and abscission; sustaining perennial growth; signaling deciduous behavior; functioning as antimicrobial agents; and serving as attractants or repellents. In the plant kingdom, more than 50,000 secondary metabolites have been identified; numerous contemporary pharmaceuticals derive their effects from secondary plant metabolites. Orchids generate phytoalexins, which are low-molecular-weight compounds that bestow resistance against microorganisms, as a reaction to microbial challenges. Among the principal phytoalexins found in orchids are 9,10-dihydrophenanthrenes; among these, orchinol was the initial to be isolated. Orchids have yielded more than 40 dihydrophenanthrene phytoalexins, of which a number have been synthesized, including the three original phytoalexins. Phytoalexins possess bacteriostatic and fungistatic properties but lack bactericidal and fungicidal activity. Orchids are utilized in Nepalese traditional medicine solely to treat minor ailments, such as boils, acne, and incisions, or as a demulcent. A considerable variety of plants consumed by humans also generate phytoalexins; however, these compounds are typically found in trace amounts and would not pose any health risks unless the vegetable in question is consumed in excessive quantities. It is crucial from a consumer acceptance perspective that the quantity of phytoalexins in agricultural crops be reduced through the use of sound agricultural practices (Hussein & El-Anssary, 2018).

Hydrocarbons, which are composed of hydrogen and carbon, are the most basic compounds. They can exist in the form of linear chains or circular structures. These entities serve as the foundational components of more intricate molecules and are composed of double or triple bonds. Olive oil is composed of numerous saturated hydrocarbons, whereas marsh gas consists of methane, which is a saturated hydrocarbon. An additional unsaturated hydrocarbon, ethylene, is secreted by apples and their hybrids as a plant hormone. An alcohol or ethanol is produced when a hydrocarbon is modified with a hydroxyl group (McMurry, 2015).

Terpenes, which comprise pine oil, floral fragrances, growth inhibitors gibberellic acid and abscisic acid, and certain insecticidal substances, are essential plant metabolites. All 30,000 terpenes share a fundamental quality: They all consist of five-carbon isoprene units that recur. Terpenoids and lignoids have demonstrated antiviral properties, as evidenced by the inhibition of coronaviruses by at least 22 of them, including the perilous SARS-CoV 3CL virus. Despite their comparable structures, triterpenes and steroids are biosynthesized via distinct pathways. Steroids comprise one five-membered ring and three six-membered rings, whereas steroid saponins serve as crucial precursors in the synthesis of steroid medications. Triterpenes, saponins possess cytotoxic, antitussive, expectorant, analgesic, and anti-inflammatory properties. Stilbenoids and bibenzyls are a class of hydrocarbons characterized by an ethane moiety, to which two benzene rings are affixed. Orchids utilize dihydro-p-coumaric acid and acetate or malonate to produce bibenzyls (Kolewe et al., 2008).

Phenanthrenes are angular polynuclear hydrocarbons that are predominantly found in orchids and other higher plants. They are present in the structure of steroids and are related to alkaloids, such as morphine. Three primary categories comprise phenolic compounds: monophenanthrenes, diphenenthrenes, and triphenenthrenes. It was discovered that monophenanthrenes, the most prevalent, exhibited no antitumor activity against cancer cell lines, whereas biphenanthrenes and triphenanthrenes demonstrated such activity. Both lusianthridin, a dihydroxymethoxyphenanthrene, and denbinobin, a phenanthroquinone derived from *Dendrobium nobile*, demonstrate cytotoxic properties both *in vitro* and *in vivo*.

Approximately half of the 210 compounds in the first category are hydroxyl- and/or methoxy-substituted. The remaining compounds are derivatives of 8,10-dihydro or dehydro. Glycosides are uncommon, but one in *Dendrobium chrysanthum* and three in *Bletilla striata* have been identified. Spirochenthraquinones, an additional class of monomeric phenanthrenes, have been identified in the appendicula of *Cremastra sinensis* and *Spiranthes sinensis*. In *Pleione bulbocodioides* and *Pholidota yunnanensis*, bibenzyl derivatives of phenanthrenes were identified (Yue et al., 2016).

Alkaloids are compounds derived from plants that typically contain one or more nitrogen atoms organized in a heterocyclic ring. These compounds exhibit a significant impact on both animals and humans. Optical active elements (OPEs) contribute to plant germination and predator protection. It is believed that alkaloids, which have a bitter flavor, aid in germination and protect plants from predation. A scarcity of quinine and various medicinal alkaloids during World War II prompted an extensive alkaloid screening of plants in the 1950s and 1960s. An early investigation was conducted on secondary metabolites found in orchids, including alkaloids. In 1974, Luning documented the screening of 2,044 species across 281 genera of orchids for the presence of alkaloids. Although not all species belonged to the 56 medicinal orchid genera from Asia that were examined, a majority of them (53.6%) contained alkaloid-positive species. A mere 14.6% of the examined orchid species exhibited positive alkaloids (defined as those present in concentrations of 0.1% or greater) (V. Schulz et al., 2001).

Orchid alkaloids are typically classified into two primary groups: those containing pyrrolizidine and those containing dendrobine. Although dendrobine, the initial alkaloid identified in Chin Shih Hu, is the most abundant in alkaloids, pyrrolizidine compounds are its most significant. Numerous orchid species have been found to contain bibenzyl alkaloids, and to date, over 30 alkaloids have been isolated from the genus *Dendrobium*.

Secondary metabolites consist of a variety of compounds, ranging from simple phenols with a single aromatic ring to complex phenols, such as tannins and lignins. These consist of anthraquinones, coumarins, quinones, and naphthoquinones, which impart color, aroma, and fragrance to vegetation. Certain compounds exert physiological effects on animals, including the flavoring vanillin. It has been discovered that denbinobin, a 1,4-phenanthrenequinone extracted from *Ephermerantha lonchophylla*, suppresses tumor growth by impeding angiogenesis and inhibits HIV-1 replication. Additionally, it could potentially possess an antifibrotic effect on the liver and serve as a viable precursor in the development of compounds that safeguard the liver against cirrhosis. Aromatic compounds with a cyclic carbon ring, including flavonoids, phenols, and tannins, absorb ultraviolet radiation and produce a variety of pigments found in plants. Flavonoids are frequently advocated for their antioxidant properties, as certain varieties exhibit antibacterial, antiviral, antiallergic, anti-inflammatory, antiplatelet, and antineoplastic effects (Ekor, 2014).

Anthocyanins are essential for insect mimicry because they impart the hues yellow, red, mauve, pink, magenta, and purple to flowers. Anthocyanins obtain their parent compound from 2-phenylbenzopyran. A single dominant gene confers the pink coloration to numerous orchid species, whereas hybrids belonging to the Vanda-Aranda-Renanthera group exclusively possess a solitary anthocyanin found in the blossoms. Antioxidant, antimicrobial, anti-inflammatory, antimutagenic, antitumor, antidiabetic, vasorelaxant, immunomodulatory, and possessing both estrogenic and anti-estrogenic properties are among the diverse pharmacological activities of flavonoids.

Bulbophyllum odoratissimum is composed of the flavonoid chrysin, which possesses analgesic and anti-inflammatory properties. Cyclooxygenase-2 (COX2) expression induced by lipopolysaccharide is inhibited by chrysin via the inhibition of the nuclear factor for IL-6 (NF-IL6) DNA-binding activity. Aside from the health supplement industry's promotion of chrysin as an aromatase inhibitor, in vivo research has shown that neither experimental animals nor humans experienced any change in steroid levels when orally administered chrysin.

Bioactive polysaccharides, also known as carbohydrates containing beta 1–3, 1–4, or 1–6 branch chains derived from botanicals, are extensively advocated as tonics and anticancer agents in traditional Chinese medicine (TCM) and Kanpō medicine. While the majority of them are found in fungi, a few are also present in other botanicals. In China and Japan, polysaccharides found in orchids are gaining scientific interest; however, their applicability as adjunctive therapies is hampered by the absence of acceptable controlled trial data and a lack of standardization.

Numerous orchid secondary metabolites, including alkaloids, terpenes, stilbenoids, bibenzyls, phenanthrenes, coumarins, and flavonoids, have been investigated. Presently, under investigation is the use of genetic transformation to enhance medicinal orchids or increase the quantity of the pharmaceutically significant compounds they produce. In order to meet the criteria for clinical testing, a compound must exhibit efficacy at exceedingly low dosages, be non-lethal, and manifest minimal or negligible severe adverse effects (Thomford et al., 2018).

10.1.3 Significance of Precursors and Elicitors in Secondary Metabolite Production

Metabolites, which are organic compounds synthesized by organisms, play a vital role in ecological interactions and defensive mechanisms. They find extensive application in pharmaceuticals, agriculture, and various other sectors. The formation of secondary metabolites is influenced by a variety of factors, such as parental and tertiary influences, environmental conditions, and genetic makeup. The quantity and variety of secondary metabolites produced during the synthesis of these compounds can be significantly impacted by the concentration and accessibility of precursors, which are molecules that serve as substrates or fundamental components. Primary metabolites, such as sugars, amino acids, and fatty acids, are commonly employed as structural components in the synthesis of secondary metabolites. For precursor feeding to enhance the formation of secondary metabolites, specific substances are added to the growth medium or environment. Inducing or enhancing secondary metabolite production in organisms requires the use of vectors, substances, or signals. By acting via a variety of signaling molecules, they can affect the activity of enzymes involved in the synthesis of secondary metabolites and the expression of genes, which can originate from a variety of sources. Gaining an understanding of the roles that precursors and elicitors play in the synthesis of secondary metabolites has significant implications for numerous fields, including biotechnology, pharmacology, and agriculture. *Precursors* are molecular entities that serve as substrates or fundamental components during the synthesis of secondary metabolites. The yield and variety of finalized products may be influenced by them. The availability of precursors can be altered via metabolic engineering, genetic engineering, or supplementation. For instance, supplementing microorganism cultures with particular amino acids can enhance their capacity to synthesize antibiotics. *Elicitors* are exogenous stimuli or environmental agents that induce or enhance the synthesis of secondary metabolites within an organism. Artificial substances such as salicylic acid and natural substances like chitosan and jasmonic acid are examples. For a variety of purposes, precursors and elicitors are essential in optimizing the production of secondary metabolites. Precursors and elicitors have the potential to enhance the manufacturing process of pharmaceutical compounds, thereby facilitating the development of innovative drugs or more effective formulations. Agricultural biotechnology is significantly dependent on secondary metabolites for the functioning of plant defense mechanisms against pathogens, parasites, and environmental challenges. Natural product chemistry encompasses the discovery and synthesis of novel secondary metabolites with diverse biological functions and chemical structures via precursors and elicitors. Bioprocessing and biomanufacturing: By integrating precursor feeding and elicitation techniques, bioprocessing processes can enhance the economics, productivity, and efficiency of secondary metabolite manufacturing processes (Anand et al., 2019).

10.2 PRECURSORS IN SECONDARY METABOLITE BIOSYNTHESIS

Secondary metabolites play an important role in the regulation of physiological processes and facilitate plant development and growth by responding to their surroundings. Hydrocarbons, phytoalexins, alkaloids, flavonoids, polysaccharides, phenolic compounds, and terpenes are some of the classes into which they are classified. Exacerbating environmental pressures are plant hormones, including abscisic acid, jasmonates, salicylic acid, and polyamines. It is believed that elicitation procedures based on *in vitro* culture are more advantageous for the production of industrial secondary metabolites because they are independent of environmental conditions and reduce the danger of microbial contamination. Phosphoric metabolites, which are essential for defense mechanisms, wound healing, electron transport, antioxidants, and wound response, are generated from precursors via the shikimic acid pathway and the Krebs cycle. The accumulation of secondary metabolites can be augmented through the manipulation of plant growth conditions and the implementation of *in vitro* culture techniques that enable sustained synthesis and extraction. Sessile plants are vulnerable to osmotic imbalances, pathogenic or environmental stresses that induce cellular dehydration, and biochemical and physiological changes. The ability of secondary metabolites to adapt to environmental constraints is exemplified by the antioxidant properties exhibited by polyamines and phenyl amides, which are present in tobacco and legumes, respectively (Murthy et al., 2014).

Salinity, temperature, radiation, heavy metals, and drought are examples of abiotic stressors that have a substantial effect on the growth and development of plants. To fight with such obstacles, plants have devised adaptation mechanisms, including the synthesis of essential secondary metabolites, the expression of resistant proteins, controlled metabolism, osmotic adjustment, and stomatal closure. These secondary metabolites regulate ionic balance and enzyme activity, repair oxidative damage, and preserve the relationship between the genotype and the phenotype of a plant. The plant's capacity to photosynthesize is impaired as a result of the reactive oxygen species (ROS) produced and the electron transport chain disrupted by drought stress. In order to withstand arid duress, plants have developed response mechanisms, such as the accumulation of secondary metabolites, such as terpenes, alkaloids, and phenolics. The capacity of plants to synthesize secondary metabolites is also affected by temperature stress. Brassicaceae, for example, demonstrates a reduction in the accumulation of carotenoids and β-carotene when subjected to temperature stress. Conversely, *Silybum marianum*, which grows in acidic pH and high-temperature environments, accumulates an abundance of flavonolignans (Ochoa-Villarreal et al., 2016)

Light is essential for plant metabolism and the synthesis of secondary metabolites; gaseous pollutants have a significant effect on secondary metabolite production. Phosphorus and deaminase oxidase activity are mineral nutrients that contribute to the production of biomass and secondary metabolites. Hormones that mediate growth and immunity against environmental stressors, including gibberellic acid (GA), auxin, jasmonic acid (JA), ethylene, and abscisic acid (ABA), are essential for plant defense mechanisms to function (Hussain et al., 2012).

Animal cell culture technology is indispensable for the genetic modification and the investigation of differentiation, proliferation, and control in the biological sciences. It involves the synthetic cultivation of cells derived from organisms, such as humans, animals, or invertebrates. Early in the 20th century, mammalian cell cultures were first utilized, but polio outbreaks in the 1950s prompted the commencement of industrial-scale research.

Isolated cells are sustained and proliferated in an appropriate artificial environment during tissue culture. The culture medium is the most critical component in animal tissue culture, and sterile conditions are required for the development of cell lines. For the fabrication of synthetic tissues, organotypic and histotypic cultures replicate organs and tissues, respectively. Homogeneous cultures, which are derived from primordial cell cultures, serve as invaluable instruments in the study of cell biology and origin.

Primarily, explant culture, organ culture, and cell culture are the three methods utilized to obtain cell culture. In contrast to organ culture, which utilizes adult organs or entire embryonic organs, the primary explant culture employs animal tissue fragments that adhere to the surface via an extracellular matrix (ECM) constituent. When it comes to quantifying cell proliferation and establishing consistent culture conditions, quantification is vital. Trypsin utilization and cell passage are critical stages in cell culture. Primary cell cultures are obtained directly from tissues and organs because they are more representative of the cell types present in those tissues than cell lines. Secondary cell culture is selected due to its accessibility and ease of growth. Cell lines originate from progenitor cultures and may be finite or continuous in nature (Ogita, 2015).

Mammalian cell lines are commonly employed in the production of pharmaceuticals, vaccines, therapeutic proteins, and anticancer drugs. The growth cycle of a cell line consists of three distinct phases: latency, log, and plateau. Animal cell culture can be employed to conduct a multitude of cell-based assays, encompassing the investigation of morphology, protein expression, cell proliferation, differentiation, apoptosis, and toxicity across diverse environments. Cell proliferation is influenced by the subsequent variables: pH, oxygen, temperature, and cell viability. Due to the fact that toxic substances can induce cell mortality or metabolic alterations, cytotoxicity is a significant issue in cell culture; for example, viability assays consider the protease marker assay, ATP assay, and MTT/MTS/resazurin assay. The activity of marker enzymes can be assessed, or the cytoplasm or nucleus of deceased cells can be colored using vital dyes (Wu et al., 2016).

Culture media are essential components in animal cell culture as they provide vital nutrients, energy, amino acids, and vitamins. Serum-containing media is an indeterminate supplement consisting of proteins, lipids, amino acids, growth factors, vitamins, and minerals, among other complex molecules. Media can be further classified as either natural or artificial. Conversely, there are several disadvantages to consider, such as contamination, variation, cytotoxic and inhibitory substances, and downstream processing. Producing biopharmaceuticals using serum-free media is hazardous to public health and a source of inadvertent contamination.

Organ culture is a technique utilized to regulate the development of plants by utilizing immature components of flowers and fruits, root tips, stem tips, and leaf primordia, among other plant organs. Determinate organ cultures exhibit determinate growth, including that of fruits, flowers, and foliage, and cease growth at maturity. In leaf culture, an embryonic or primordial young leaf of the stalk apex is grown aseptically in a chemically defined medium, from which it is removed.

The investigation of phyllotaxy, which involves the organization of leaf primordia around the perimeter of the shoot apical meristem, is significantly aided by leaf primordia. Their capacity for facilitating the development of somatic embryos has been established through extensive research on ferns, specifically cinnamon. The cultivation of leaf primordia is a valuable method in investigating the impacts of environmental fluctuations, nutrient variations, and growth factors on the leaf. Further research has demonstrated the positive impact of orange juice, specifically that of an "acidless" variety, on the in vitro development of explant cultures originating from a variety of fruits. Antenna development and stamen specification are complex processes that encompass remarkable occurrences, including meristem transition, cell division, differentiation, and intercellular communication.

The cultivation of flowers is a crucial method for researching the initiation, induction, and development of flowers. By exercising control over the environment and media components, it is possible to manipulate various variables that have an impact on these processes. The enhancement of ornamental and horticultural characteristics is contingent upon genetic transformation, while tissue culture is indispensable for the initiation and differentiation of flower buds. Significant quantities of organic additives and iron are necessary to produce flowers with consistent color over the course of many generations *ex vitro*. Parthenogenesis and embryo development dependent on accession are essential for the production of carrot haploid and doubled haploid plants. Sugar concentration and photoperiods both influence flower blossoms. Through the use of in vitro flowering, plant supply and preservation can be enhanced year-round. The formation of flower buds is substantially influenced by plant tissue culture, wherein the cytokinin, medium, and photoperiods are pivotal factors (Wu et al., 2016).

Experimental systems such as ovum culture and seed culture are employed to investigate fruit physiology, seed development, and ovum development. The cultivation of cotton ovules presents several benefits, including enhanced efficacy in embryogenesis, reduced susceptibility to the donor plant genotype, and a substantial yield of haploids and dihaploids. *Androgenesis* is the process by which embryonic pollen divides and develops into tissue in order to generate haploids. For the investigation of embryo formation, nucellus cultures are dependable for virus-free citrus plants. By cultivating the extremity point of the shoot, the meristem culture frees plants from viral infections and indeterminate organs from indefinite growth by adding modules.

10.3 ELICITORS: INDUCERS OF SECONDARY METABOLITE PRODUCTION

Secondary metabolites consist of volatile hydrocarbons, alkaloids, glycosides, flavonoids, tannins, and resins. The majority of these secondary metabolites are obtained through the isolation of wild or cultivated plants, as their chemical synthesis is either impracticable or uneconomical. An appealing alternative, biotechnological production in plant cell cultures has encountered restricted commercial success owing to a lack of comprehension regarding the biosynthesis process of these metabolites. *In vitro* plant specimens and/or plant cells exhibit morphological and physiological reactions to microbial, physical, and/or chemical stimuli referred to as "elicitors" (Efferth, 2019).

Elicitation refers to the mechanism by which plants stimulate or augment the synthesis of secondary metabolites in order to safeguard their viability, endurance, and competitive edge. Decades of research have been devoted to determining the nutritional value of plants, which are an essential component of our daily diet. Pharmaceutical industry–relevant bioactive constituents, or phytopharmaceuticals, include quinine, morphine, codeine, cocaine, and colchicine; anti-cancer substances include belladonna alkaloids, *Catharanthus* alkaloids, and physostigmine, pilocarpine, and reserpine; and steroids include diosgenin, digoxin, and digitoxin. Higher plants are abundant in these substances. Nevertheless, the economic viability of chemically synthesizing compounds derived from plants is frequently hindered by their extraordinarily intricate structures and precise stereochemical demands. As an alternative to the extraction of entire plants, the biotechnological synthesis of valuable secondary metabolites in plant cell or organ cultures is an attractive alternative (Fischer et al., 1999).

Elicitors are substances that, when introduced into a living cell system, initiate or enhance the biosynthesis of particular compounds. The introduction of elicitors in minute quantities induces or enhances the biosynthesis of metabolites; this process is known as elicitation. Elicitors may be categorized according to their origin (exogenous or endogenous) or by their nature (abiotic or biotic). Non-biological substances, including inorganic salts, physical factors (e.g., Cu, Ca, and Cd ions), and elevated pH, are examples of abiotic elicitors. Biological elicitors include glycoproteins and G proteins, as well as polysaccharides derived from plant cell walls and microbes. *Endogenous elicitors* are substances that originate within the cell, as opposed to exogenous elicitors, which originate from outside the cell (Xu et al., 2011).

Although the precise mechanism by which elicitation occurs in plant cells remains indescribable, numerous hypotheses have been advanced. These include messenger Ca^{2+}, factors that influence the integrity of cell membranes, intracellular pathway inhibition or activation, and alterations in osmotic stress. Certain scientists postulate that the binding of the elicitor to a receptor located on the plasma membrane induces elicitation, rapid alterations in protein phosphorylation patterns, activation of protein kinases, stimulation of mitogen-activated protein kinases, production of reactive oxygen species, and a hypersensitive response to cell death at the site of the infection.

The utilization of elicitation to increase the production of secondary metabolites from plant cell cultures has facilitated the exploration of a novel research domain that holds significant economic implications for the pharmaceutical sector. The synthesis of secondary metabolites is influenced by various factors, including the concentration and selectivity of the elicitor, the duration of exposure, the age of the culture, the cell line, growth regulation, nutrient composition, and the quality of the cell wall materials. Diverse elicitors found within plant cells are capable of inducing the synthesis of secondary metabolites. Exogenous elicitors, including fungicides, herbicides, and xenobiochemicals, are generated through the membrane-destroying processes of the cell. Numerous plant species contain endogenous elicitors, which consist of polysaccharides, peptides, enzymes, and fatty acids (Karuppusamy, 2009).

Arachidonic acid, *Capsicum annuum*, *Cicer arietinum*, medicarpin, maackiain, *Ascochyta rabiei*, *Cupressus lusitanica*, *Datura stramonium*, *Dioscorea deltoidea*, steroid (diosgenin), glycine max, glyceollin, *Medicago sativa*, phytoalexins, and glycine max are all examples of biotic elicitors. Arachidonic acid, oxidative stress, and amino acid deprivation are abiotic elicitors that are generated within plant cells. Abiotic elicitors consist of xenobiochemicals, herbicides, and fungicides, among others. By exploiting the chemical structures of these compounds, one can induce secondary metabolites from plant cells. Anthraquinones, alkaloids (tropane), 5'-phosphodiesterase (PDase), acridone epoxide, *Ruta graveolens*, *Pythium* sp., *Trichoderma viride*, *Pythium aphanidermatum*, jasmonic acid, azadirachtin, berberine, capsidiol, debneyol, scopoletin, nicotine, *Nicotiana tabacum*, *Phytophthora cryptogea*, codeine, morphine, and diosgenin are among the plant elicitors that stimulate secondary metabolite production. Research contrasts the elicitation-induced production of secondary metabolites across various species and cell cultures. Multiple variables affect the production of secondary metabolites, including cell culture conditions and elicitors.

By using a variety of parameters, the elicitation of plant cell cultures to produce active compounds has been investigated. The elicitation process is significantly influenced by the nutrient composition, age of the subculture, elicitor concentration, and duration of elicitor exposure. Cells stimulated with increased concentrations of elicitor extracts of *T. viride*, *A. niger*, and *F. moniliforme* exhibited greater accumulation of ajmalicine. An additional 10% concentration nevertheless had an adverse effect on the accumulation of ajmalicine. The age of the subculture is a significant factor in determining the elicitation-based production of bioactive compounds. On elicitation, *C. roseus* cells from cultures aged 20 days produced greater quantities of ajmalicine, according to one study. Ajmalicine accumulation peaked at 166 μg/g DW in 20-day-old cells stimulated with *T. viride* extracts, followed by 88 μg/g DW in cells stimulated with *F. moniliforme* and 90 μg/g DW in cells stimulated with *A. niger*, respectively. In addition to elicitor specificity, environmental conditions, the presence of growth regulators, the nutrient composition of the medium, and the cell line or clones of the microbial elicitor utilized all impact elicitation efficiency. It has been documented that the induction and generation of secondary metabolites through plant cell cultures can increase the accumulation of products and generate active compounds.

Anyhow, with the progress made in the organic synthesis and semi-synthesis of compounds resembling those generated by plants, the commercial significance of secondary metabolite extraction from plants persists. Producing many of these metabolites at economic costs is challenging, if not impossible. Furthermore, consumers tend to be more receptive to naturally occurring products compared to those that are artificially produced. A considerable number of aromas or fragrances derived from mixtures of hundreds of different compounds, such as strawberry and jasmine, and biochemicals with complex molecular structures continue to be extracted naturally.

For the synthesis of plant secondary metabolites, the development of alternatives to the intact plant is of considerable interest. In contrast to the initial emphasis on tissue and cell cultures, contemporary methodologies entail the application of molecular biology techniques to augment the metabolic pathways that generate particular compounds. Plants exhibit a rapid activation of diverse defense mechanisms in response to infection by pathogenic microorganisms. These include hypersensitive cell death, oxidative cross-linking of cell wall proteins, production of phytoalexins and hydrolytic enzymes, and incrustation of cell wall proteins with phenolics. The production of antimicrobial secondary metabolites is induced by microbial invasion of plants in the same manner that stress factors, such as ultraviolet radiation, osmotic shock, fatty acids, inorganic salts, and heavy metal ions, do (Pence, 2011).

The availability of secondary metabolites is frequently constrained, and cultured plant cells are widely recognized as a possible source of phytopharmaceuticals. However, in comparison to intact plants, secondary metabolites are synthesized by an extremely small number of cell cultures. The potential of elicitation in plant cell culture systems is worth considering, given its demonstrated efficacy in the fermentation of antibiotics and other fermented products. Researchers have a close watch and examine the synthesis of indole glucosinolates, sanguinarine accumulation, and rosmarinic acid in plant cell cultures. For example, the effect of biotic stressors and elicitors on the biosynthesis of scopolamine and tryosumine is investigated; biosynthesis of the sesquiterpenic phytoalexin capsidiol in elicited root cultures of chili peppers is investigated; and the impact of elicitor dosage and exposure duration on the biosynthesis of indole alkaloids by hairy root cultures of *Catharanthus roseus* is also investigated. Work was also done on the metabolic and biosynthetic processes of the phytoalexin lubimin and other analogous compounds found in *Datura stramonium*, the accumulation of isoflavonoid conjugates in cell suspension cultures in response to fungal elicitors, and the hydrolysis of phosphatidylinositol 4,5-bisphosphate and the production of inositol 1,4,5-trisphosphate in cultured soybean cells are investigated, and the findings are noteworthy (Chen et al., 2018).

Potential future methodologies in generating secondary natural products in plant cell suspension cultures involve the application of fungal and light elicitors to stimulate the synthesis of enzymes involved in phytoalexins in parsley cell suspension cultures. The many research findings indicate that the implementation of fungal elicitors in suspension cultures of plant cells can result in the generation of a diverse array of secondary natural products.

10.4 STRATEGIES TO ADDRESS CHALLENGES AND FUTURE DIRECTIONS

Due to their susceptibility to pests and animals, plants are also susceptible to secondary metabolites, which are employed in pharmaceuticals, agrochemicals, aromatics, and food additives as well. More than 80% of the human population relies on homeopathic remedies to promote health, and more than 40% of the pharmaceuticals produced in Western countries are derived from natural resources. Globally, between 35,000 and 7,000 plant species are utilized for medicinal and preventative purposes; therefore, standardization and quality assurance are crucial for the herbal market. The production of medicinal products derived from plants grown in soil encounters various obstacles, including inadequate availability, specialized conditions, protracted cultivation periods, low concentrations, hazardous environmental conditions, and time-consuming and costly harvesting processes. These challenges may be surmounted through the use of plant tissue culture (PTC) methods, which involve controlling the plant's biosynthetic pathway in order to increase the yield of beneficial compounds or prevent the accumulation of undesirable substances.

Enhancing the production of bioactive compounds *in vitro* through the use of molecular techniques and gene technology is feasible for cell, organ, and tissue culture, induction of somaclonal variation, somatic embryogenesis, somatic hybridization, genetic transformation, and hairy root development. The propagation of seedless medicinal plants, or those with microscopic or unviable seeds, is a suitable application for PTC techniques. Although biotechnology products are more expensive than those derived from cultivated or untamed plants, large-scale PTC techniques provide a more consistent and controlled supply of secondary metabolites in terms of both quantity and quality. A multitude of artificially devised techniques have been implemented by researchers in an effort to maximize the produce of targeted secondary metabolites by optimizing culture conditions (Palazón et al., 1997).

Under controlled conditions, PTC techniques enable the regeneration of entire plants from organs, minor tissues, or plant cells. In medicinal and other plant species, this technique is utilized more frequently as a supplement to conventional breeding techniques for plant improvement and modification. The regeneration ability of cultivated plant materials is significantly impacted by the physiological condition of the donor plant.

PTC techniques are crucial prerequisites for plant enhancement and propagation, as they enable the year-round cultivation of pathogen-free plants, the rapid production of thousands of plantlets, and the acquisition of such plants. Furthermore, it facilitates comprehension of the molecular biology of plant differentiation, produces novel cultivars of a specific plant, and advances the understanding of the impact of particular biotic and abiotic factors on a tested plant, all while requiring minimal space for propagation.

Phytotherapy is an essential and complementary component of modern medicine and pharmacotherapy; it employs herbal remedies and natural medicinal resources to prevent and treat disease. Herbal medications and remedies comprise active constituents derived from botanical medicinal products. Aerial plant components, such as seeds, leaves, and flowers, frequently accumulate and synthesize secondary metabolites to a greater extent than their subterranean counterparts.

Medicinal plant extracts find widespread application in the treatment of chronic illnesses, such as cancer, diabetes, and hypertension. The antioxidants melatonin and serotonin, which are present in a wide range of plant taxa, have been identified in *Ocimum sanctum* L. plants cultivated in the field and in greenhouses. Antitumor effects have been observed *in vitro* in cultures of *Artemisia japonica*, *Hovenia dulcis*, and *Aegle marmelos*, while methanolic extracts derived from diverse plant species exhibit potent antischistosomal activity.

When we talk about medicinal plants, the plant tissue culture (PTC) is a promising technique that resolves concerns such as low yield and vulnerability to biotic and abiotic stresses. Bioreactor applications, conservation, propagation, induction of polyploidy, and plant engineering are all possible with PTC. By utilizing seedling tissues or shoot meristems, numerous plant species have been able to accomplish high multiplication. For the long-term storage of medicinal plant materials, cryopreservation is advised, particularly for plants with resistant spores. PTC exhibits a higher degree of efficiency in evaluating the impact of various experimental conditions on the synthesis of secondary metabolites derived from medicinal plants when compared to naturally occurring plant materials. It generates novel forms of plant secondary metabolites and paves the way for the production of engineered molecules, both of which may have beneficial effects on strategies involving biological control, food, pharmaceuticals, and others.

A wide range of PTC techniques are effectively utilized to propagate, synthesize, and extract secondary metabolites in vitro. The utilization of root culture is advised due to its consistent metabolite yield and genetic stability. The formation of fibrous roots *in vitro*, independent of other plant organs, enables bioreactors to generate secondary compounds en masse.

The manipulation of *in vitro* conditions to promote the synthesis of secondary metabolites via the regulation of the primary metabolic pathway in plants is an up-front process. The significant uses of compounds produced *in vitro* include agrochemicals, pigments, drugs, cosmetics, flavors, dietary additives, and perfumes.

Under aseptic conditions, bioactive or secondary metabolites, such as flavonoids, alkaloids, phenolics, terpenoids, saponins, steroids, tannins, glycosides, colorants, fragrances, and volatile oils, can be produced from cultured plant materials. Industrial-grade active secondary metabolites have been derived from *Lithospermum erythrorhizon*, *Coptis japonica*, and *Papaver somniferum* cell cultures. The concentration of secondary bioactive metabolites was greater in *in vitro* cultured *Swertia chirayita* than *in vivo* plants. The antimicrobial properties of plant products regenerated *in vitro* were associated with the presence of more bioactive metabolites.

The absence of substantial quantities of pigments in *in vitro* cultivated plant materials facilitates the extraction and purification of pharmaceutical compounds, leading to reduced manufacturing costs. The regulation of secondary metabolite production can be achieved through the utilization of *in vitro* culture methods. Increasing the production of secondary metabolites can be accomplished through the use of techniques such as modifying environmental conditions or selecting an elite variant clone. Improving the yield of antioxidants and other beneficial secondary metabolites requires, as a minimum, the optimization of culture media. Various differentiation pathways and biomass values are observed in media containing PGRs of varying types and concentrations; therefore, it is necessary to investigate the impact of carbon source concentration and type on the productivity of culture biomass and metabolites. Sucrose is frequently employed as a carbon source to promote biomass growth and the biosynthesis of secondary metabolites.

Physical culture conditions, including light and pH, can have a substantial effect on the secondary metabolite production of plants cultured *in vitro.* The ideal pH is 5.8 for standard plant tissue culture processes; it must be modified to facilitate the synthesis of bioactive compounds. In addition to coniferin and lignin, acidification of the cytoplasm may influence the accumulation of secondary metabolites. In cell suspension cultures, alkaloid production increases dramatically as pH rises to between 4.3 and 9.0.

The composition of the culture media is a critical factor in optimizing the production efficacy of a particular cell line within a scaled-up production system. Ascertain the genetic stability of the cultivated plant materials through the utilization of established indicators, including metabolite production, growth parameter index over extended subculture cycles, and molecular markers.

A callus culture is an undifferentiated and disorganized mass that is produced when plant material is cultured on an agar medium and cells divide. Suspension cultures as PTC techniques for the production of secondary metabolites are widely used; however, certain cells might persist in the medium as tissue clusters, impeding growth and diminishing the synthesis of the secondary compounds that were intended.

Suspension cell cultures offer several benefits, including expedited biomass production, straightforward regulation of chemical and physical conditions, sustainable product manufacturing, control over market demand, ability to select plant cell lines, and potential for new product development.

Supplementary cell lines and selective culture result in a greater yield of secondary metabolites compared to the utilization of primary tissues and standard culture conditions. By optimizing cell immobilization and genetic composition, it is possible to enhance the synthesis of secondary metabolite compounds *in vitro* (Zhang et al., 2004).

10.5 APPLICATIONS OF ELICITORS AND PRECURSORS IN MEDICINAL AGENT PRODUCTION

Elicitation serves as a viable approach to enhance the output of *in vitro* cultures through the facilitation of secondary metabolite synthesis. Elicitors are physical, chemical, and/or microbial agents that stimulate or augment the synthesis of secondary metabolites in order to maintain the viability and competitiveness of plants. Several variables affect the efficacy of elicitation, including the type of elicitor, the concentration, and the duration of exposure. When implementing elicitation strategies, the age, cultivated cell line, medium composition, type, and concentration of growth regulators of the cultures are crucial parameters.

Elicitors can be biotic or abiotic in nature; biotic agents consist of exogenous or endogenous microbes, while abiotic agents comprise heavy metals, including methyl jasmonate, salicylic acid, yeast extract, chitosan, inorganic ions, and ultraviolet radiation, among others. Enhancing the bioproduction of secondary metabolites in cultured plant materials is possible through the implementation of these elicitors. As elicitors, abiotic stimuli, including temperature, light parameters, and water potential, may also be employed. The amplification of secondary metabolite yields can be achieved through the combination of elicitors and target gene overexpression; by this means we can stimulate biosynthesis and growth in plant species.

Elicitor purposes may include the co-cultivation of plant cells alongside particular microorganisms. Under controlled conditions, *in vitro* cultured plants exhibit an increased metabolic rate and shortened biosynthesis cycles. For the stimulation of secondary metabolism in cultured plant materials, precursor feeding involves the introduction of bioactive molecule intermediates either at the outset or throughout the *in vitro* culture period. *Precursors* are chemical compounds that have the potential to undergo biosynthetic conversion into secondary metabolites by *in vitro* cultured plant materials. Numerous secondary compounds are generated at a higher rate when organic compounds, such as vitamins or amino acids, are introduced into the culture medium. For example, the introduction of amino acid supplementation resulted in increased yields of asiaticoside and triterpenes in *Centella asiatica* and valine, threonine, and isoleucine in shoot cultures of *Hyraceum perforatum*.

Biodiversity is essential for plant enhancement and the selection of an elite lineage among medicinal plants. The improvement of the productivity of *in vitro* cultivated plant materials is significantly aided by high-yielding cell lines. Striking for the genotype that exhibits high yield and secondary metabolite contents is crucial in order to prevent exorbitant expenses (Yu et al., 2005).

Variability in the quantification and measurement of active metabolites may result from spatial and temporal fluctuations that occur throughout the process. To identify high-yielding cell lines, which can be generated by subjecting the plant material population to toxic inhibitors, biosynthetic precursors, and/or stressful environments and selecting cells that exhibit superior production of the desired components, both quantitative and qualitative methods may be utilized in assessment.

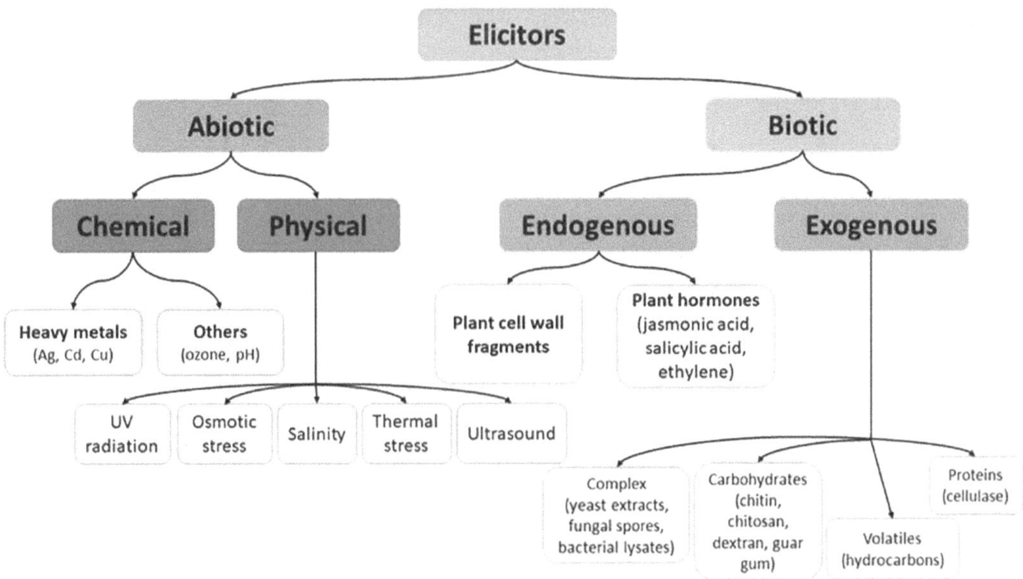

FIGURE 10.2 Applications of precursors.

PTC Techniques and Gene Technology

- Molecular techniques and gene technology can be utilized to increase the yields of bioactive compounds during *in vitro* culture procedures.
- PTC techniques are well-suited for the propagation of medicinal plants that lack seeds or have seeds that are microscopic or unviable.
- PTC techniques on a large scale provide a regulated supply of secondary metabolites in addition to a more uniform product in terms of both quantity and quality.
- Under controlled conditions, PTC techniques enable the regeneration of entire plants from organs, minor tissues, or plant cells.
- They are capable of yielding thousands of plantlets in a brief period of time, produce plants devoid of pathogens, and can be utilized throughout the year.
- Their propagation requires minimal space, they are capable of generating novel cultivars of a specific plant, and they contribute to the comprehension of the molecular biology underlying plant differentiation.
- Herbal pharmaceuticals and remedies consist of active ingredients derived from herbal medicinal products. It is common for aerial plant parts to accumulate and synthesize secondary metabolites to a greater extent than their subterranean counterparts.
- Extracted botanicals are extensively utilized in the treatment of chronic ailments, such as cancer, hypertension, and diabetes.
- A wide range of PTC techniques have been effectively implemented to facilitate the in vitro growth, synthesis, and extraction of secondary metabolites.
- It is simple to manipulate *in vitro* conditions in order to promote the production of secondary metabolites by adjusting the pathway of primary metabolism in plants.

Compounds derived from cultured plants and pharmaceuticals.

- Under aseptic conditions, bioactive or secondary metabolites may be produced from cultured plant materials.
- Pharmaceutical compounds derived from plant materials cultivated *in vitro* can be extracted and purified with greater ease owing to the minimal presence of pigments, which consequently leads to reduced manufacturing costs.

10.6 CONCLUSION

Plants serve as vital sources of secondary metabolites (SMs), which are indispensable in the fields of pharmaceuticals, pigments, flavorings, and drug development. As a consequence of the pharmaceutical industry's substantial reliance on these plants, they are progressively facing extinction. To produce and investigate SMs, plant cell culture technologies have been implemented, with production-enhancing techniques including elicitation, hairy root culture system, and suspension culture systems. Cell suspension culture systems are rapid techniques utilized in industrial settings to generate large quantities of SMs. Genetic engineering is a pivotal factor in the synthesis of secondary metabolites, given that it enables plants to generate chemical compounds with specific physiological functions and beneficial impacts on the human body. Utilizing the genetic engineering of microorganisms and plants and the in vitro reconstruction of plant metabolic pathways, numerous secondary metabolites, including isoprenoids, flavonoids, and stilbenes, have been produced.

India is a significant producer of valuable medicinal plants, with a projected value of $5 trillion by the year 2050. Plant tissue culture methods enable the efficient production of secondary metabolites for commercial purposes, facilitating the propagation of large quantities under precise environmental conditions that are not influenced by seasonal factors. In the plant kingdom, more than 50,000 secondary metabolites have been identified; numerous modern-day pharmaceuticals derive their effects from these metabolites.

Secondary metabolites play an important role in pharmaceuticals, agriculture, and other industries and are essential organic substances produced by organisms. Genetic composition, environmental conditions, precursors, and elicitors all exert an influence on them. *Elicitors* are external agents or environmental stimuli that induce or enhance the production of secondary metabolites by an organism. *Precursors* are molecules that function as substrates or building blocks in the process of secondary metabolite synthesis. Gaining insight into the roles of precursors and elicitors in the synthesis of secondary metabolites has far-reaching implications across numerous fields, such as agriculture, pharmacology, and biotechnology.

Adaptation mechanisms that plants have evolved in response to environmental constraints include the biosynthesis of essential secondary metabolites, the expression of resistant proteins, controlled metabolism, osmotic adjustment, and stomatal closure. In the biological sciences, animal cell culture technology is indispensable for investigations into differentiation, proliferation, regulation, and genetic modification. Utilizing plant cell cultures, which are derived from a variety of plant organs, is vital for nourishing organized plant growth.

Pharmaceutical industry–utilized bioactive constituents, phytopharmaceuticals, and substances such as steroids and anticancer agents are abundant in plants. The utilization of elicitation to increase the production of secondary metabolites from plant cell cultures has eased the exploration of a novel research domain that holds significant economic implications for the pharmaceutical sector.

Notwithstanding the progress made in organic synthesis and semi-synthesis, the commercial significance of extracting secondary metabolites from plants persists significant. By manipulating the plant's biosynthetic pathway, plant tissue culture (PTC) methodologies can conquer the challenges, by providing a regulated provision of secondary metabolites and a product (secondary metabolites) that is more consistently produced in terms of both quantity and quality.

REFERENCES

Anand, U., N. Jacobo-Herrera, A. Altemimi, N. Lakhssassi, A comprehensive review on medicinal plants as antimicrobial therapeutics: Potential avenues of biocompatible drug discovery, Metabolites 9 (2019) 258, doi: http://dx.doi.org/10.3390/metabo9110258.

Atanasov, A.G., B. Waltenberger, E.M. Pferschy-Wenzig, T. Linder, C. Wawrosch, P. Uhrin, V. Temml, L. Wang, S. Schwaiger, E.H. Heiss, J.M. Rollinger, Discovery and resupply of pharmacologically active plant-derived natural products: A review, Biotechnology Advances 33 (8) (2015) 1582–1614, doi: http://dx.doi.org/10.1016/j.biotechadv.2015.08.001.

Chen, L., Y. Cai, X. Liu, Soybean hairy roots produced *invitro* by *Agrobacterium* rhizogenes-mediated transformation, Crop Journal 6 (2018) 162–171.

Efferth, T., Biotechnology applications of plant callus cultures, Engineering 5 (2019) 50–59, doi: http://dx.doi.org/10.1016/j.eng.2018.11.006.

Ekor, M., The growing use of herbal medicines: Issues relating to adverse reactions and challenges in monitoring safety, Frontiers in Pharmacology 4 (2014) 177, doi: http://dx.doi.org/10.3389/fphar.2013.00177.

Fischer, R., N. Emans, F. Schuster, S. Hellwig, J. Drossard, Towards molecular farming in the future: Using plant-cell-suspension cultures as bioreactors, Biotechnology and Applied Biochemistry 30 (2) (1999) 109–112.

Guerriero, G., R. Berni, J.A. Muñoz-Sanchez, F. Apone, E.M. Abdel-Salam, A.A. Qahtan, A.A. Alatar, C. Cantini, G. Cai, J.F. Hausman, K.S. Siddiqui, S.M.T. Hernández-Sotomayor, M. Faisal, Production of plant secondary metabolites: Examples, tips and suggestions for biotechnologists, Genes (Basel) 9 (6) (2018) E309, doi: http://dx.doi.org/10.3390/genes9060309.

Hussain, M.S., S. Fareed, M. Saba Ansari, A. Rahman, I.Z. Ahmad, M. Saeed, Current approaches toward production of secondary plant metabolites, Journal of Pharmacy and Bioallied Sciences 4 (1) (2012) 10–20, doi: http://dx.doi.org/10.4103/0975-7406.92725.

Hussein, R., A.A. El-Anssary, Plants secondary metabolites: The key drivers of the pharmacological actions of medicinal plants, in: P. Builders (Ed.), Herbal Medicine. Intech Open, 2018, pp. 11–30, doi: http://dx.doi.org/10.5772/intechopen.76139.

Karuppusamy, S., A review on trends in production of secondary metabolites from higher plants by *in vitro* tissue, organ and cell cultures, Journal of Medicinal Plants Research 3 (2009) 1222–1239.

Kolewe, M.E., V. Gaurav, S.C. Roberts, Pharmaceutical active natural product synthesis and supply *via* plant cell culture technology, Molecular Pharmaceutics 5 (2008) 243–256, doi: http://dx.doi.org/10.1021/mp7001494.

McMurry, J.E., Organic chemistry with biological applications, Secondary Metabolites: An Introduction to Natural Products Chemistry, (2015), pp. 1016–1046 Stamford, USA.

Murthy, H.N., E.J. Lee, K.Y. Paek, Production of secondary metabolites from cell and organ cultures: Strategies and approaches for biomass improvement and metabolite accumulation, Plant Cell Tissue Organ Culture 118 (2014) 1–16, doi: http://dx.doi.org/10.1007/s11240-014-0467-7.

Nasri, H., A. Baradaran, H. Shirzad, M. Rafieian-Kopaei, New concepts in nutraceuticals as alternative for pharmaceuticals, International Journal of Preventive Medicine 5 (12) (2014) 1487–1499.

Ncube, B., J. Van Staden, Tilting plant metabolism for improve dmetabolite biosynthesis and enhanced human benefit, Molecules 20 (2015) 12698–12731.

Ochoa-Villarreal, M., S. Howat, S. Hong, M.O. Jang, Y.W. Jin, E.K. Lee, G.J. Loake, Plant cell culture strategies for the production of natural products, BMB Reports 49 (3) (2016) 149–158, doi: http://dx.doi.org/10.5483/bmbrep.2016.49.3.264.

Ogita, S., Plant cell, tissue and organ culture: The most flexible foundations for plant metabolic engineering applications, Natural Product Communications 10 (5) (2015) 815–820.

Palazón, J., M.T. Piñol, R.M. Cusido, C. Morales, M. Bonfill, Application of transformed root technology to the production of bioactive metabolites, Recent Research Developments in Plant Physiology 1 (1997) 125–143, doi: http://cabidigitallibrary.org/doi/full/10.5555/19981613963.

Pence, V.C., Evaluating costs for the *invitro* propagation and preservation of endangered plants, *In Vitro* Cellular and Developmental Biology Plant 47 (1) (2011) 176–187.

Rosenthal, G.A., The biochemical basis for the deleterious effects of L-canavanine, Phytochemistry 30 (1991) 1055–1058.

Sandra, G., R. Anabela, Production of plant secondary metabolites by using biotechnological tools, secondary metabolites-Sources and applications, in: R. Vijayakumar (Ed.), Secondary Metabolites Sources and Applications. IntechOpen, 2018, doi: http://dx.doi.org/10.5772/intechopen.79766.

Schulz, V., R. Hänsel, V.E. Tyler, Rational Phytotherapy. A Physician's Guide to Herbal Medicine, Springer-Verlag, Berlin, 2001.

Thomford, N.E., D.A. Senthebane, A. Rowe, D. Munro, P. Seele, A. Maroyi, K. Dzobo, Natural products for drug discovery in the 21st century: Innovations for novel drug discovery, International Journal of Molecular Sciences 19 (6) (2018) 1578, doi: http://dx.doi.org/10.3390/ijms19061578.

Winter, J.M., Y. Tang, Synthetic biological approaches to natural product biosynthesis, Current Opinion in Biotechnology 23 (2012) 736–743, doi: http://dx.doi.org/10.1016/j.copbio.2011.12.016.

Wu, C.F., A. Karioti, D. Rohr, A.R. Bilia, T. Efferth, Production of rosmarinic acid and salvianolic acid B from callus culture of *Salvia miltiorrhiza* with cytotoxicity towards acute lymphoblastic leukemia cells, Food Chemistry 201 (2016) 292–297, doi: http://dx.doi.org/10.1016/j.foodchem.2016.01.054.

Xu, J., X. Ge, M.C. Dolan, Towards high-yield production of pharmaceutical proteins with plant cell suspension cultures, Biotechnology Advances 29 (2011) 278–299.

Yu, K.W., N.M. Hosakatte, E.J. Hahn, K.Y. Paek, Ginsenoside production by hairyroot cultures of *Panax ginseng*: Influence of temperature and light quality, Biochemical Engineering Journal 23 (2005) 53–56.

Yuan, H., Q. Ma, L. Ye, G. Piao, The traditional medicine and modern medicine from natural products, Molecules 21 (5) (2016) 559, doi: http://dx.doi.org/10.3390/molecules21050559.

Yue, W., Q.L. Ming, B. Lin, K. Rahman, C.J. Zheng, T. Han, L.P. Qin, Medicinal plant cell suspension cultures: Pharmaceutical applications and high-yielding strategies for the desired secondary metabolites, Critical Reviews Biotechnology 36 (2016) 215–232, doi: http://dx.doi.org/10.3109/07388551.2014.923986.

Zhang, L., R. Ding, Y. Chai, M. Bonfill, E. Moyano, K.M. Oksman-Caldentey, T. Xu, Y. Pi, Z. Wang, H. Zhang, G. Kai, Z. Liao, X. Sun, K. Tang, Engineering tropane biosynthetic pathway in *Hyoscyamus niger* hairy root cultures, Proc. Natl. Acad. Sci. U.S.A. 101 (17) (2004) 6786–6791, doi: http://dx.doi.org/10.1073/pnas.0401391101.

Biotransformation and Transgenesis

11 Biotransformation

Bioreactors for Pilot- and Large-Scale Cultures of Plant Cells and Retention of Biosynthetic Potential in Cell Cultures

*N. Joshi, G. Karri, and V. Joshi**
* Corresponding author: vinu.jsh@gmail.com

11.1 INTRODUCTION

In recent times, biotechnology methods have been used increasingly for producing secondary metabolites. Secondary metabolites can be obtained by extraction from natural sources. However, there are some disadvantages to this extraction process, such as reliance on geographical and seasonal factors and potential environmental challenges associated with the extraction. Further, many organic solvents possessing toxic and corrosive properties (like pentane, acetone, chloroform, etc.) are used in conventional extraction procedures. Owing to these disadvantages, the extraction of these compounds needs more innovative approaches.

Biotechnological methods use various microorganisms, plant, and animal cells, or purified enzymes, for the production of useful compounds. These are better options because of their economical, imperishable, and inexhaustible properties. One such important technique is biotransformation, which is defined as the process of converting one chemical compound to another by living organisms. Here, a simple, easily available, and low-cost molecule is transformed into a useful and precious end product by biological systems through their enzymes. It involves the enzymatic alteration of the functional groups of organic compounds to produce a chemically distinct end product. Biotransformation examines the special qualities of biocatalysts, such as their regio- and stereoselectivity and their ability to induce reactions at optimal pH and temperature levels.

Plant cell cultures, which are cultivated *in vitro* from isolated plant cells, have emerged as a promising tool in studying and utilizing biotransformation in various applications. This is because plant cells have a rich repository of enzymes, enabling them to perform a variety of chemical reactions on various types of substrates. These cultures can be used to produce useful compounds, such as phenolics, terpenes, alkaloids, and flavonoids, which are of great interest in medicine, agriculture, and the food industry (Krasteva et al., 2020).

The procedure of the biotransformation experiment involves the addition of an exogenous substrate/precursor to the culture media, which is then absorbed by the cultured tissues for enzymatic conversion into another compound (Figure 11.1). The process occurs under a controlled environment, and there is an opportunity to monitor and optimize various parameters, such as nutrient supply, pH, and temperature, which affect cell growth and their conversion ability. The biotransformation rate can be influenced by several factors, such as the dissolvability of precursor compounds, the permeability of cells for precursor uptake and product excretion, the level of enzyme activity, the distribution of enzymes within the cells, the occurrence of undesirable side reactions, and the

DOI: 10.1201/9781003326939-13

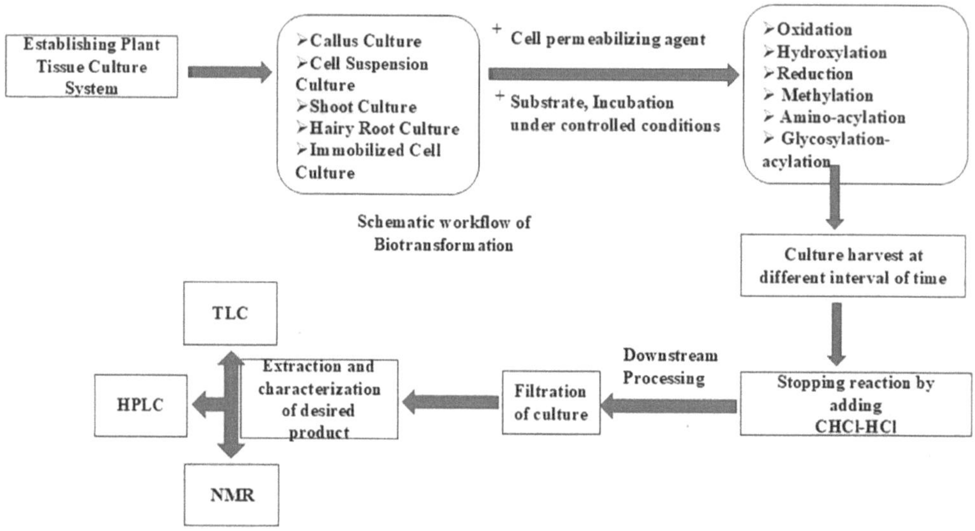

FIGURE 11.1 Schematic workflow of the biotransformation process.

presence of enzymes that break down the desired product and selective extraction of metabolites from media (Dave et al., 2014). Further, pH changes, osmotic influences, and elicitation can also affect the bioconversion ability of cells (Giri et al., 2001).

Without a doubt, biotransformation offers a chance to re-examine the already-studied natural products with the goal of discovering better/novel phytocompounds as well as enhancing the value of well-known compounds for expanded utilization. Plant cell/organ cultures are used in these processes to produce a plethora of compounds with distinct structural alterations. They also guarantee the year-round production of pathogen-free compounds, making them more appropriate for therapeutic and food applications. Furthermore, there is also an opportunity to tailor the biotransformation process by genetic engineering of plant cells to produce specific compounds of interest. Notably, the entire biotransformation process can result in the creation of a completely new set of compounds.

Besides application in metabolite production, biotransformation reactions are also known for the detoxification/degradation of environmental pollutants, thereby proving to be an effective tool in bioremediation. Upscaling of plant cell culture performing biotransformation by bioreactor cultivation can offer sustainable and environmentally friendly solutions for producing valuable compounds and mitigating pollution.

11.2 BIOTRANSFORMATION BY PLANT CELL CULTURES FOR PRODUCING SECONDARY METABOLITES

Plant cell, tissue, and organ cultures are the most extensively utilized systems for producing biologically active secondary compounds for pharmaceutical applications (Eibl et al., 2018). Besides the synthesis of natural compounds, these systems have also been utilized widely for biotransformation reactions (Table 11.1). This section presents a detailed account of biotransformation reactions performed in different plant cell culture platforms.

11.2.1 UNDIFFERENTIATED CALLUS CULTURE SYSTEM

Callus is an unorganized and undifferentiated group of cells which can be obtained from any explant inoculated on plant tissue culture medium supplemented with auxins, mostly 2,4-D, alone or in combination with cytokinins. Callus cultures are known to retain the biochemical totipotency of mother plants. There are numerous examples of commercially available chemicals created through biotransformation in undifferentiated callus cultures. *In vitro* rose root callus culture (liquid) has been used for the biotransformation of a number of precursors, such as phenylalanine, trans-cinnamic acid, cinnamaldehyde, and cinnamyl alcohol, for the synthesis of cinnamoyl alcohol glycosides (CAG), including rosarin, rosavin, and rosin (Mirmazloum et al., 2019). All the investigated precursors, aside from phenylalanine, aided in the production of CAGs. The amount of rosin increased by 80 times after 24 hours, and by 130 times after 96 hours, with the addition of 2 mM trans-cinnamic acid and cinnamyl alcohol, respectively. György et al. (2004) reported the conversion of cinnamyl alcohol into rosin (three- to sixfold accumulation compared to wild plant) in callus culture of *Rhodiola rosea.* To increase the yield of products, glucose was switched for sucrose in the medium of liquid cultures of *Rhodiola rosea* callus, and that resulted in a double accumulation of rosavin and rosin (György et al., 2005). Gil et al. (1995) found that *Myrtillocactus geometrizans*– and *Nicotiana tabacum*–dissociated callus tissues readily convert 2-carene into diastereomeric alcohols carenols and carenones through allylic hydroxylation. Up to 90% of carene was biotransformed by the two plant species into carenols, which were then further oxidized into carenones. Low quantities of podophyllotoxin (0.005% of dry weight) were produced by *Juniperus chinensis* callus cultures (Muranaka et al., 1998). However, by incorporating precursor phenylalanine and elicitor chito-oligosaccharides in the callus culture, the production of podophyllotoxin was enhanced 11- and 15-fold, respectively.

11.2.2 Cell Suspension Culture System

Cell suspension cultures are produced by inoculating friable callus onto the liquid medium, followed by agitation. Cultivating plant cells in suspension offers the benefit of easy scalability in bioreactor systems. In contrast, propagating callus cultures on semisolid medium is a steady, expensive, and labor-intensive process that demands extensive production space.

Furthermore, culture in liquid medium facilitates better absorption of substrate and nutrients and therefore holds significance for biotransformation reactions. Numerous studies have examined the ability of cell suspension cultures to support biotransformation processes. The precursor, costic acid, was bioconverted into the eudesmanic acids 3-oxo-costic acid, 3,5-dihydroxycostic acid, and 3,5-dihydroxycostic acids using *Tessaria absinthioides* cell suspension cultures. The doubling times of cultures introduced with 5% and 10% of inocula were 6.30 days and 6.93 days, respectively. Also, 2.5% of starting cell densities were found to be insufficient to reach comparable rates (Kurina and Donadel, 2003).

Rattan et al. (2022) studied the effects of a phenylalanine precursor on the production of phenylpropanoids in suspension cultures of *Rhodiola imbricata.* It was found that on day 12, the 1 mM precursor increased rosavin and rosarin content, while the 3 mM precursor enhanced p-Coumaric acid amount (2.50 mg/g DW). Also, the total phenolic content increased significantly on the addition of 2 mM phenylalanine.

Mathieu et al. (2009) investigated the capacity of a suspension culture of *Vitis vinifera* to transform two new C13-norisoprenoidic volatiles, ionone and dehydrovomifoliol, in addition to the monoterpenes geraniol and linalool. It was found that β-Ionone was metabolized to 14 norisoprenoidic volatiles oxygenated products, while the transformation of dehydrovomifoliol leads to oxygenated and reduced derivatives.

Suspension cultures of *Lithospermum erythrorhizon* can transform esculetin to esculin (7-hydroxycoumarin-6-O-D-glucoside) via glucosylation (Tabata et al., 1984). When either glucose, maltose, or lactose was used in place of sucrose, tissue growth and esculin production were considerably inhibited. Esculetin levels in the cells decreased over the course of the prolonged culture period, with 56% glucosylating after 24 hours of addition into the medium and 23% decline after 4 days of culture. Despite the significantly slower cell development in the 2,4-D medium than in the NAA or IBA medium, this medium produced the maximum esculin.

In suspension cultures of *Rhodiola rosea*, Furmanowa et al. (1999) noted that trans-cinnamyl alcohol was glycosylated to yield the physiologically active compound rosavin. The glycosylation of the exogenously administered trans-cinnamyl alcohol was assessed using HPLC, confirming the biotransformation capability of *R. rosea* cells at various growth phases. Depending on the strain and bioconversion duration, all *R. rosea* cultures converted the substrate into a variety of products at a range of 92.8–99.8%. Also, *Catharanthus roseus* high-density cultures produced 1.54 g (6% of dry weight) of vanillin glucoside per liter of suspension (Sommer et al., 1997).

According to Van Uden et al. (1990), feeding L-phenylalanine precursors to suspension cultures of *Linum flavum* L. increased the synthesis of 5-methoxypodophyllotoxin (5-MPT) three- to fivefold. The cells began concentrating 5-MPT as soon as they were put in a hormone-free medium, and its accumulation increased by 40–50 times on the 18th day of culture cycle. Van Uden et al. (1993) looked at how *L. flavum*–derived cell cultures converted the deadly lignan podophyllotoxin into podophyllotoxin-D-glucoside. In another study, Van Uden et al. (1995) investigated the conversion of lignan desoxypodophyllotoxin into 5-methoxypodophyllotoxin and 5-methoxypodophyllotoxin-D-glucoside in suspension cultures of *Podophyllum hexandrum* and *L. flavum*.

Zaragoza-Martínez et al. (2016) reported on triterpene synthesis in suspension cultures of *Jatropha curcas* employing precursor (mevalonate) feeding and jasmonic acid (JA) elicitation. *Catharanthus roseus* suspension culture cells underwent the bioconversion of (–) and (+) carvone into 5-hydroxyneodihydrocarveol and 5-hydroxycarvone (Hamada et al., 1997). In order to create the oxygenated products (1S)-3-caren-5-one and (1S)-2-caren-4-one, Dvorakova et al. (2011) studied the transformation of (1S)-2-carene and (1S)-3-carene by suspension culture of *P. abies*. He et al. (2015) discovered a new terpenoid indole alkaloid, 3-hydroxy-4-imino-catharanthine, in suspension cultures of *Catharanthus roseus* via catharanthine biotransformation.

Cell suspension cultures of *D. lanata* synthesized three novel compounds (Moyano et al., 2007). Moreover, these cultures were also reported to produce a selective cell death–triggering cardenolide compound (Pádua et al., 2012; Munkert et al., 2017). The suspension culture of *Catharanthus roseus* was able to convert the anthelminthic drug (–) α-santonin into useful compounds, such as eudesmane and guaiane, and progesterone to 17α-hydroxyprogesterone (Schneider et al., 2018). *Azadirachta indica* cells transformed dydrogesterone (drug for repeated miscarriage) into 20R-hydroxy-9β,10α-pregna-4,6-diene-3-one (Asthana and Asthana, 2015).

Suspension cultures of *Peganum harmala* converted geranyl acetate and linalyl acetate to the increased levels of geraniol and linalool, respectively, when these substrates were concentrated on controlled-release polymer discs (Zhu and Lockwood, 2000). This method accumulated biotransformation products at three- to fivefold higher concentrations compared to the direct incorporation of substrates into the medium. Cormier and Ambid (1987) used *Vitis vinifera* suspension in a biphasic system with a liquid nutrient phase above a lipid (Miglyol 812) layer in the biotransformation of geraniol. This system proved beneficial because it increased the substrate load five times without negatively impacting the culture, enhanced the perseverance of developing end products, and allowed the suspension culture to be recycled or used again in the geraniol bioconversion process.

Despite tremendous exploitation, suspension cultures suffer from the significant drawback of showing somaclonal variation, which can result in unstable biochemical activity (Casas et al., 1998). Therefore, there is a need for regular screening to preserve extremely productive lines to avoid variation and instability issues (Doran, 2009).

11.2.3 Differentiated Shoot Culture System

Differentiated systems such as shoot and root cultures can be obtained from various explants, like seedling parts, shoot tips, internodes, nodal segments, root segments, etc. The shoot cultures are produced either by axillary shoot bud proliferation or by adventitious shoot bud differentiation upon inoculating the explants on a medium supplemented with cytokinins alone or with cytokinin–auxin at 10:1. Adventitious root cultures are produced by inoculating explants on a medium containing auxin alone or with auxin–cytokinin at 10:1. The ability to synthesize and concentrate larger amounts of tissue-specific metabolites, as well as improved genetic and biosynthetic stabilities, distinguish differentiated cells from undifferentiated *in vitro* systems (Espinosa-Leal et al., 2018; Kroll and Brenner, 2020). Additionally, differentiated cultures are also amenable to large-scale cultivation in bioreactors. There are several instances of the production of valuable compounds by biotransformation in differentiated cultures. The microshoots of *Scutellaria lateriflora* showed the synthesis of arbutin from the extrinsic precursors hydroquinone (HQ) and 4-hydroxybenzoic acid (4-HBAc) (Kawka et al., 2020). Szopa et al. (2018) reported the great biosynthetic potential of shoot cultures of *Aronia arbutifolia* and *A. prunifolia* grown in a liquid medium. The effectiveness of the transformation of hydroquinone into arbutin was dependent on the dose and concentration of the precursor used. The best efficiency (74.33–67.37%) was obtained with the smallest doses (100–200 mg/L).

In another study, exogenously given hydroquinone was transformed into the important drug and cosmetic arbutin by agitated cultures of *E. purpurea*, *E. affine*, *M. melissophyllum*, and *R. graveolens* L. The majority of the examined cultures showed 30% biotransformation efficiency (Skrzypczak-Pietraszek et al., 2005). In agitated shoot cultures of *R. graveolens* L. and *H. perforatum* L., Piekoszewska et al. (2010) were able to optimize the factors for the transformation of hydroquinone into its D-glucoside, arbutin, and achieved maximum levels (7.2–7.8%) of this important product.

Da Silva et al. (2013) noticed the biotransformation of limonene and farnesyl cations into carvone and trans-farnesene in the shoot cultures of *Croton floribundus*. Also, in another study, different cardenolides were added to *Digitalis lanata* shoot cultures grown under light and dark conditions. It was noted that the cardenolide digoxigenin transformed into digoxigenin, 3-epidigitoxigenin, and digitoxigen-3-one (Theurer et al., 1998).

11.2.4 Hairy Root Culture System

The infection of explants with *Agrobacterium rhizogenes* results in the development of hairy root culture. These cultures, owing to their biochemical and genetic constancy, fast growth rate, and plant growth regulator–free and affordable culture conditions, provide an appealing framework for implementing biotransformation (Banerjee et al., 2012). They are also known to thrive in liquid environments and are therefore amenable to bioreactor cultivation.

Hydroquinone was converted into arbutin using the hairy root cultures of *Brugmansia candida* (Casas et al., 1998). The usage of 20–40 mg/L doses of hydroquinone led to the best bioconversion and the least cell damage. At concentrations of 30–120 g/L, sugars (glucose, sucrose, sorbitol, and mannitol) increased bioconversion, with the highest rate record for sucrose.

Asada et al. (1993) reported the production of three new products (3-O-[6-O-malonyl-β-D-glucopyranosyl-(1—>2)-β-D-glucopyranosyl]18β-glycyrrhetinic acid, 30-O-[β-D-glucopyranosyl-(1—>2)-β-D-glucopyranosyl] 18 β-glycyrrhetinic acid, and 30-O-(6-O-malonyl-β-D-glucopyranosyl)18 β-glycyrrhetinic acid) in hairy root cultures of *Panax ginseng* fed with the precursor 18-β-glycyrrhetinic acid.

Rhodiola kirilowii hairy root cultures transformed cinnamyl alcohol (2.5 mM) into rosarin and rosavin. Furthermore, rosavin production was elevated when cultures were fed CA on the day of inoculation and additional sucrose on day 14 of the growth cycle. The hairy root culture showed the highest rosavin content, 505 ± 106 mg/L, upon the addition of CA, and extra sucrose was applied on the first day of the culture cycle (Grech-Baran et al., 2014).

11.2.5 Immobilized Cell Culture System

Biotransformation using immobilized cell cultures offers several advantages compared to using free cells, like enhanced stability, improved reusability, simplified downstream processing and product recovery, enhanced productivity, better control over reaction conditions, reduced risk of contamination, and compatibility with continuous processes.

The choice of immobilization method and support material should be carefully considered based on the specific requirements of the biotransformation process to optimize the advantages mentioned above.

Johnson et al. (1996) reported that vanillylamine and ferulic acid were transformed into vanillin and capsaicin in sodium alginate–immobilized cell cultures of *Capsicum frutescens*. The highest levels of vanillin (315 μg/culture) and capsaicin (190 μg/culture) were noted upon treatment with ferulic acid (2.5 mM) and vanillylamine (2.5 mM), respectively, on day 15 of the growth cycle. Furuya et al. (1991) evaluated the conversion of theobromine to caffeine in coffee (*C. arabica* L.) cells immobilized in a reticulate polyurethane foam cube matrix and reported that the phases of immobilized cells had an impact on biotransformation efficiency.

Mentha cells immobilized in polyacrylamide-hydrazide were compared to its suspension culture for their ability to transform (–) menthone and (+) pulegone into (+) neomenthol and (+) isomenthone. It was found that an immobilized culture had the same transformation efficiency as freely suspended cells (Galun et al., 1983). Rao and Ravishankar (2000) converted caffeic acid and protocatechuic aldehyde to capsaicin and vanillin using suspension and immobilized cells of *C. frutescens* Mill. In both cultures, the concentration of capsaicin increased 2.02- and 3.1-fold, respectively, compared to control after 15 days of growth. Immobilized *C. roseus* cells converted racemate into (R)-4-pyridyl-1-ethanol through stereoinversion, with 100% efficiency (Takemoto and Achiwa, 1998).

By using immobilized *Gardenia jasminoides* and *Daucus carota* cells in a stereoselective reduction pathway, it is possible to biotransform acetophenone into (R)- and (S)-α-phenethyl alcohol with an enantiomeric purity of >99% (Akakabe and Naoshima, 1994). Alfermann et al. (1980) studied the transformation of digitoxin and ß-methyldigitoxin by *D. lanata* cells immobilized in alginate gel. The conversion of ß-methyldigitoxin to ß-methyldigoxin involves hydroxylation reaction. *Mucuna pruriens* cells immobilized in alginate gels transform the precursor L-tyrosine to L-dihydroxyphenylalanine (L-DOPA) *in vitro*. In suspension-cultured *M. pruriens* cells, 90–95% of the endogenously produced L-DOPA was accumulated intracellularly, while in alginate-entrapped cells, 90% of the L-DOPA generated by the plant cells was released into the medium (Wichers et al., 1983).

Corchete and Yeoman (1989) investigated the biotransformation of (–) codeinone to (–) codeine that transpired in *Papaver somniferum* cells immobilized in reticulated polyurethane foam. The immobilized cells had a biotransformation rate of 79%, whereas suspended cells had a rate of 57%. Only 12.2% of the total amount of codeine produced was detected intracellularly; majority (87.3%) of the substance was excreted into the medium. A bioconversion ratio of 86% was attained when orthophosphate was removed from the growth medium. Vaněk et al. (1999) examined the production of trans-carveol and carvone from S-(–)-limonene by immobilized cultures of *Dioscora deltoidea* and *Solanum aviculare*. The maximum yield of 6.8 mg carvone was found in polyurethane foam–entrapped cells, while the highest cis- and trans-carveol yields (6 mg and 5 mg, respectively) were recorded in carrageenan gel–immobilized cells.

11.2.6 Biotransformation in Transgenic Lines

Engineering plants with the genes that encode for enzymes catalyzing the desired biosynthetic reaction is another approach to increase the conversion rate. In line with this idea, Hashimoto et al. (1993a) expressed hyoscyamine 6-b-hydroxylase gene in *Atropa belladonna* to convert hyoscyamine to scopolamine. Later, the authors also engineered the hairy roots of *A. belladonna* and noted a higher accumulation of scopolamine from hyoscyamine (Hashimoto et al., 1993b).

A genetic engineering approach to improving biotransformations in *Nicotiana* sp. has been investigated. Suspension cultures established from tobacco (*Nicotiana tabacum* L.) expressing

IFR enzyme (alfalfa) transformed 2′-hydroxyformononetin (2′OHF) into vestitone, with 60% transformation rate. The addition of cadaverine (1,5-diaminopentane) to the hairy root culture of *N. tabacum* showed a decreased content of nicotine and an increased content of the minor alkaloid anabasine (Walton et al., 1988). It could be that the substrates cadaverine and N-methyl putrescine competed in the pathway for nicotine biosynthesis. Fecker et al. (1993) proposed that since cadaverine is produced by lysine decarboxylation, its accumulation could be enhanced by engineering cells with lysine decarboxylase (ldc) gene. Hence, the expression of the ldc gene from the bacterium *Hafnia alvei* in hairy roots of *Nicotiana* showed three and ten times' rise in anabasine and cadaverine synthesis, respectively.

Ichinose et al. (1999) reported the expression of furostanol glycoside 26-O-b-glucosidase cDNA from *Costus speciosus* in *N. tabacum* and found the enzyme's activity in the engineered plant.

Shi and Lindemann (2006) transformed *Cucumis sativus* L. cotyledon explants with *Agrobacterium rhizogenes* 15834 harboring cardenolide 16′-O-glucohydrolase gene (cgh 1) in pBI121. The resultant hairy roots expressing recombinant cgh 1 were used to test the glycolytic conversion of the substrate lanatoside glycosides. Thus, the cgh I–altered cucumber hairy roots could be a helpful illustration of how recombinant enzymes can biotransform natural compounds. Georgiev et al. (2013) reported the production of scopolamine from hyoscyamine in transformed tobacco cell cultures.

11.2.7 Biotransformation for Biodegradation and Detoxification

In addition to the production of commercially interesting compounds, biotransformation reactions are also known to catalyze detoxification/biodegradation, that is, the conversion of an active harmful compound to a less-active form which is safe and reusable for the environment. Plant cell and tissue culture is a useful technique for detoxification (phytoremediation) processes. It has been well demonstrated that plant cells are competent to break down a large number of pollutants, including xenobiotics, independent of the associated microbes (Van Aken and Schnoor, 2002). They have the intrinsic capacity for metal accumulation and toxicity tolerance. Besides remediation of heavy metal–contaminated soils, other applications of plant cell culture–based biotransformation in biodegradation include the removal of organic pollutants and the treatment of wastewater. Generally, callus, cell suspension, and hairy root cultures are used in phytoremediation studies. In comparison to the whole plant, the callus and the suspension culture system enables a better uptake of chemicals from the medium, and they are not dependent on the translocation of the toxicant/pollutant for tissue-specific activity (Camper and McDonald, 1989; Lucero et al., 1999). These culture systems also have the benefit of allowing for the feeding of large amounts of pollutants relative to soil levels, which leads to the accumulation of intermediates and products at levels appropriate for characterization by nuclear magnetic resonance and mass spectrometry techniques (Laurent et al., 2007). These benefits are further increased by upscaling the cell cultures in bioreactors (Knops et al., 1995). Further, the extraction and purification of products from in vitro cultures is easier than with the whole plant due to the low levels of interfering pigments in comparison to whole plants, resulting in higher yield (Schmidt, 2001). Glucose conjugations of exogenously added substances are considered to be detoxification reactions. Glucosylation and hydroxylation seem to be specific biotransformation reactions of higher plant cells. The enzymes for glucose conjugation or hydroxylation are not constitutive and can be inducible in some cases. Glucosyl conjugation, including glucosylation, glucosyl esterification, and hydroxylation, is the most effective reaction among various biotransformations by in vitro plant cultures (Yokoyama, 1996).

Apart from callus and suspension culture, the hairy root culture offers some more benefits for phytoremediation research, as the responses of hairy roots are expected to be realistic because of its structural and functional similarity with natural roots. Moreover, hairy roots are an authentic and consistent experimental platform as they are known to demonstrate genetic and biochemical stability compared to dedifferentiated systems (Aird et al., 1988; Flores et al., 1987). They also do not require exogenous plant regulators for growth and multiplication. Further, genetic engineering for improved detoxification by hairy roots is also simple, as the genes involved in degradation reaction can be inserted into Ri plasmid during the process of infecting explants with *A. rhizogenes* for hairy root production. Such transgenic hairy roots can also be used as explants to produce plants with a phytoremediation capability.

Like hairy roots, shooty teratomas also offer advantages for phytoremediation studies. Shooty teratomas have been used in phytoremediation work (Kučerová et al., 1999); however, the success of the transformation process to produce shooty teratomas is much less over that of hairy roots (Subroto et al., 1997). Further, morphological changes and vitrification of shoots in teratoma can also pose notable practical difficulties.

There are intrinsic impracticality for the utilization of tissue cultures in direct phytoremediation, such as the maintenance of high levels of aseptic conditions, the absence of which can result in difficulty in the remediation of soil or waste stream as they themselves contain lots of microbes, the requirement of sugar in the medium can result in microbial contamination under non-sterile conditions, the costly generation of a large amount of biomass in bioreactors for waste treatment, and the unfeasibility in treating waste in bioreactors. Furthermore, a lot of the circumstances used to cultivate plant tissues in vitro are not very similar to what plants encounter in the wild. Plant cultures' ability for biotransformation can be influenced by growth stage and culture age. Plant tissue culture allows for easier control of experimental variables than whole plants do. On the other hand, it can be challenging to apply in vitro knowledge of pollutant metabolism to mature plants in the field.

In phytoremediation research, plant tissue culture plays a crucial role as a model system in comprehending the metabolic and tolerance mechanisms functioning in entire plants. Scholars possess the ability to discern and examine the inherent enzymatic potential of plant species for phytoremediation, which is unaffected by microbial impacts or constraints resulting from metabolite compartmentation within structured plant architectures. Considering these facts, investigations have been carried out to explore the detoxification potential of cell cultures in several plant species. *Datura innoxia* cell suspension cultures grown on medium supplemented with 2,4,6-trinitrotoluene (TNT) consumed almost 99% of it after 12 hours of incubation (Lucero et al., 1999). Cell lysates and media showed the presence of amino dinitrotoluenes (ADNTs), known TNT metabolites. Within 12 hours, almost all TNTs and ADNTs had been removed. Total ADNTs were never greater than 26% of original TNT, implying that additional biotransformation stages occurred.

Laurent et al. (2007) investigated the breakdown of 2,4-dichlorophenol (DCP) in tobacco (*Nicotiana tabacum* L.) suspension cultures and found the main end products as DCP-(6-O-acetyl)-glucoside, DCP-glycoside conjugates, DCP-(6-O-malonyl)-glucoside, and their precursor, DCP-glucoside.

Alderete et al. (2009) reported that tobacco hairy roots transformed with peroxidase genes from tomatoes (tpx1 and tpx2) detoxified phenols with a higher efficiency in contrast to the control plants.

Podlipná et al. (2013) investigated the ability of *Phragmites australis* to detoxify benzimidazole anthelmintics, namely, albendazole (ABZ) and flubendazole (FLU). The cultures showed an accumulation of ten metabolites of ABZ and five metabolites of FLU, including novel products, such as glucosylglucosides, acetylglucosides, and xylosylglucosides.

Hughes et al. (1996) reported the ability of hairy root cultures of aquatic plants *Myriophyllum spicatum*, *Catharanthus roseus*, and *Myriophyllum aquaticum* to transform 2,4,6- trinitrotoluene (TNT). Both the culture medium and cell lysates showed low levels of aminated nitrotoluenes.

Vila et al. (2005) reported the use of tobacco suspension cultures to detoxify 2,4,6-trinitrotoluene (TNT) into end products formed by attaching a glycose (sugar) residue to the hydroxylamine group of either 4-hydroxylamino-2,6-dinitrotoluene (4-HADNT) or 2-hydroxylamino-4,6-dinitrotoluene (2-HADNT).

Bhadra et al. (2001) investigated the potential of *Myriophyllum aquaticum* and hairy root cultures of *Catharanthus roseus* to disintegrate the explosives RDX and HMX. It was found that HMX was intact in *M. aquaticum* and showed minimal biological activity in the HR cultures. On the other hand, RDX was efficiently eliminated by both plant systems from the medium as well as from the cell mass of the plants.

TABLE 11.1
Biotransformation Reactions Performed by *In Vitro* Plant Cell and Organ Cultures

Sr. No.	Precursor	Plant	Culture System	Method	Product	Factors Influencing Biotransformation	Biotransformation Rate	References
1.	Phenylalanine, trans-cinnamic acid, cinnamaldehyde, and cinnamyl alcohol	*Rhodiola rosea*	Callus culture (liquid)	Shake flask	CAGs (rosin, rosavin, rosarin)	Type of precursor used; except for phenylalanine, all produced CAGs with varying degrees	Trans-cinnamic acid-rosin-80-fold increase Cinnamaldehyde-rosin 130-fold increase	(György et al., 2004)
2.	Cinnamyl alcohol and tyrosol	*Rhodiola rosea*	Liquid culture of compact callus	Shake flask	Rosavin and rosin	Type of sugar source used; glucose more beneficial for CAG production	Biotransformation rate doubled	(György et al., 2005)
3.	Δ2-carene	*Myrtillocactus geometrizans* and *N. tabacum*	Callus culture	Semisolid agar medium	Carenols and Carenones	Presence and absence of terpene and Tween 80	90% biotransformation of carene into carenols	(Gil et al., 1995)
4.	Phenylalanine	*Juniperus chinensis*	Callus culture	Semisolid agar medium	Podophy-llotoxin	Combination of biogenic precursor podophyllotoxin and elicitor chito-oligosaccharides	11–15 fold increase in product formation	(Muranaka et al., 1998)
5.	Hydroquinone and 4-hydroxybenzoic acid	*Scutellaria lateriflora*	Shoot culture	Agitated	Arbutin	Influence of PGR concentration in the MS and LS media	Arbutin production increased from 0.43 g% DW to 5.63 g/100 g DW	(Kawka et al., 2020)
6.	Hydroquinone	*Aronia arbutifolia* and *Aronia* × *prunifolia*	Shoot culture	Agitated	Arbutin	Doses and concentration of precursor administered in the medium	Up to 4 fold increase in total arbutin content	(Szopa et al., 2018)
7.	Hydroquinone	*Ruta graveolens* L. and *Hypericum perforatum* L.	Shoot culture	Agitated	Arbutin	Doses and concentration of precursor added	7.8 and 7.2% (dry weight) arbutin production	(Piekoszewska et al., 2010)
8.	Hydroquinone	*Asteraceae, Gentianacea, Lamiaceae, Tenore, and Rutaceae*	Shoot culture	Agitated	Arbutin	Comparison of different types of cultures and plant species for product accumulation	Maximum from 35.1 % to 65.6 % biotransformation efficiency	(Skrzypczak-Pietraszek et al., 2005)
9.	Limonene and farnesyl	*Croton floribundus* Spreng	Shoot culture		Carvone and trans-β-farnesene	Effect of different plant growth regulators and their combinations	3.5% and 3.57% increase of Carvone and trans-β-farnesene formation	(Da Silva et al., 2013)
10.	Digoxigenin	*Digitalis lanata*	Shoot culture		Digitoxigen-3-one and 3-epidigitoxigenin	Comparison of light-grown and dark-grown shoot cultures for product formation	Product accumulation of 0.3umol g1 d.w.	(Theurer et al., 1998)

(Continued)

TABLE 11.1 (*Continued*)
Biotransformation Reactions Performed by *In Vitro* Plant Cell and Organ Cultures

Sr. No.	Precursor	Plant	Culture System	Method	Product	Factors Influencing Biotransformation	Biotransformation Rate	References
11.	Ilicic acid and γ-costic acid	*Tessaria absinthioides*	Cell suspension culture	Shake flask	Tessaric acid; 3-oxo-γ-costic acid; 3β, 5β-dihydroxycostic	Effect of inoculum size	Total accumulation rate of tessaric acid was 15.10×10^{-2} mg g^{-1} dry wt	(Kurina and Donadel, 2003)
12.	β-ionone and dehydrovomifoliol	*Vitis vinifera* cv. *Gamay*	Cell suspension culture	Shake flask	3-Oxo-a-ionol + 4-oxo-b-ionol; 3-hydroxy-b-ionol and vomifoliol	Addition of monoterpenes, geraniol, and linalool: biogenetically common pathways sharing compounds	75% of β-ionone biotransformed into the generated metabolites	(Mathieu et al., 2009)
13.	Trans-cinnamyl alcohol	*Rhodiola rosea*	Cell suspension culture	Shake flask	Rosavin	Different strain of *R. rosea* (green and yellow), and different growth stages of culture	The biotransformation rate was 92.8–99.8% depending on the strain and bioconversion period	(Furmanowa et al., 1999)
14.	Phenylalanine	*Rhodiola imbricata*	Cell suspension culture	Shake flask	Phenylpropanoids (rosavin, rosarin, and *p*-coumaric acid)	Doses and concentration of added precursor (phenylalanine)	Total phenolic content increased to 43.72 mg/g DW	(Rattan et al., 2022)
15.	Esculetin	*Lithospermum erythrorhizon*	Cell suspension culture	Shake flask	Esculin	Stage of cell growth and media composition (PGR and sugar)	98.6% of esculin was recovered in the culture medium	(Tabata et al., 1984)
16.	Vanillin glucoside	*Catharanthus roseus*	Cell suspension culture	Shake flask	Vanillyl alcohol	Different plant cell suspension cultures of different plant families	Yield of 60% with 1.54 g/L of vanillin glucoside (6% DW)	(Sommer et al., 1997)
17.	L-phenylalanine	*Linum flavum* L.	Cell suspension culture	Shake flask	5-methoxy podophyllotoxin	Presence and absence of PGR	5-MPT production increased by 40–50 fold	(Van Uden et al., 1990)
18.	Podophyllotoxin	*Linum flavum*	Cell suspension culture	Shake flask	Podophyllotoxin-β-D-glucoside			(Van Uden et al., 1995)
19.	Desoxypodophyllotoxin	*Linum flavum and Podophyllum hexandrum*	Cell suspension culture	Shake flask	5-methoxypodophyllotoxin and 5-methoxypodophyllotoxin-β-D-glucoside	Addition of cyclodextrin or cyclodextrin-complexed desoxypodophyllotoxin in the medium	Highest bioconversion percentage of 52.3% was found at day 14	(Van Uden et al., 1995)
20.	Mevalonate	*Jatropha curcas*	Cell suspension culture	Shake flask	Betulin, betulinic acid, and lupeol	Combination of jasmonic acid elicitation and mevalonate as biosynthetic precursor	3-fold increase in formation of products	(Zaragoza-Martínez et al., 2016)
21.	Geraniol and nerol, and (+)- and (−)-carvone	*Catharanthus roseus*	Cell suspension culture	Shake flask	5β-hydroxyneodihydrocarveol and 5α-hydroxycarvone	Different precursors for various product formation	Approximately 40% and 10% of yield respectively	(Hamada et al., 1997)
22.	(1S)-2-carene and (1S)-3-carene	*Picea abies*	Cell suspension culture	Shake flask	(1S)-2-caren-4-one and (1S)-3-caren-5-one	Optimization of biotransformation duration for selective product synthesis	Approximately 80% of (1S)-2-carene biotransformation	(Dvorakova et al., 2011)

23.	Catharanthine	*Catharanthus roseus*	Cell suspension culture	Shake flask	3-hydroxy-4-imino-catharanthine	Effect of precursor dosage	9.75% bioconversion efficiency	(He et al., 2015)
24.	Ferulic acid and vanillylamine	*Capsicum frutescens*	Immobilized cell culture	1.5% (w/v) sodium alginate	Capsaicin and vanillin	Influence of precursors added individually or in combination with each other	315 μg/culture vanillin and 190 μg/culture capsaicin	(Johnson et al., 1996)
25.	Theobromine	*Coffea arabica L.*	Immobilized cell culture	Polyurethane foam cubes	Caffeine	Phases of the immobilized cells	Highest caffeine content was 569 and 568 mg/L	(Furuya et al., 1991)
26.	(−) menthone and (+) pulegone	*Mentha piperita*	Immobilized cell culture	Cross-linked polyacrylamide-hydrazide	(+) neomenthol and (+) isomenthone	Comparison of freely suspended cells and cells entrapped in alginate and PAAH-G	A biotransformation ratio of 46 % to 70 % was recorded	(Galun et al., 1983)
27.	Protocatechuic aldehyde and caffeic acid	*C. frutescens*	Immobilized cell culture	Calcium alginate beads–entrapped cells	Vanillin and capsaicin	Combined, separate, and pulse feeding of precursors	Vanillin and capsaicin maximum buildup was 5.63 mg l−1 and 3.83 mg l−1 respectively	(Rao and Ravishankar, 2000)
28.	4-pyridyl-1-ethanol	*Catharanthus roseus*	Immobilized cell culture	Calcium alginate beads	(R)-4-pyridyl-1-ethanol			(Takemoto and Achiwa, 1998)
29.	Acetophenone	*D. carota, N. tabacum, and Gardenia jasminoides*	Immobilized cell culture	Calcium alginate beads	(R)- and (S)-b-phenethyl alcohol	Comparison of different plant species for the same product formation	A maximum of 88% conversion	(Akakabe and Naoshima, 1994)
30.	ß-methyldigitoxin	*D. lanata*	Immobilized cell culture	Alginate gel entrapment	ß-methyldigoxin	Comparison of light-grown and dark-grown plant cells	Yield of 67% (521.4 mg)	(Alfermann et al., 1980)
31.	L-tyrosine	*Mucuna pruriens*	Immobilized cell culture	Alginate gel entrapment	L-DOPA	Comparison of freely suspended cells and alginate-entrapped cells	90% of L-DOPA released into the medium	(Wichers et al., 1983)
32.	Codeinone		Immobilized cell culture	Reticulated-polyurethane foam	Codeine	Effect of culturing cells in the absence of phosphate or nitrate	79% of biotransformation ratio	(Corchete and Yeoman, 1989)
33.	(S)-(−)- and (R)-(+)-limonene	*S. aviculare and Dioscorea deltoidea*	Immobilized cell culture	Alginate, carrageenan, pectate gels, polyurethane foam	Cis- and trans-carveol carvone	Comparison of product formation in different immobilization techniques	6.8 mg carvone, 6 mg Cis-c and 5 mg Trans-c was the highest yields observed	(Vaněk et al., 1999)
34.	Hydroquinone	*Brugmansia candida*	Hairy root culture	Shake flask	Arbutin	Various concentrations of precursor HQ and different sugars	8 to 12 fold increase in arbutin production	(Casas et al., 1998)
36.	Digitoxigenin	*Panax ginseng*	Hairy root culture		Digitoxigenin stearate		0.6% conversion rate	(Kawaguchi et al., 1990)
37	18-β-glycyrrhetinic acid	*Panax ginseng*	Hairy root culture	Shake flask	30-O-[β-D-glucopyranosyl-(1—>2)-β-D-glucopyranosyl]18 β-glycyrrhetinic acid	Comparison with non-hairy root transformed cells	688 mg of total product formation including all known and unknown compounds	(Asada et al., 1993)

(Continued)

TABLE 11.1 (*Continued*)
Biotransformation Reactions Performed by *In Vitro* Plant Cell and Organ Cultures

Sr. No.	Precursor	Plant	Culture System	Method	Product	Factors Influencing Biotransformation	Biotransformation Rate	References
38.	Cinnamyl alcohol	*Rhodiola kirilowii*	Hairy root culture	Shake flask	Rosarin, rosavin, and rosin	Comparison of non-transformed wild type (NTWT) and hairy root culture	505 ± 106 mg/L of rosavin was produced only in hairy root cultures	(Grech-Baran et al., 2014)
39	Silicone oil and n-hexadecane	*Frangula alnus* Mill.	Immobilized cell culture	Four-phase airlift bioreactor	Anthraquinones	Comparison of bioreactor- and shake flask–immobilized cells	10–30 fold increase in bioreactor immobilized cells	(Sajc et al., 1995)
40.	Geranyl acetate and linalyl acetate	*Peganum harmala*	Cell suspension culture	Controlled release polymer bioreactor	Geraniol, linalool, and a-terpineol	Comparison of cultures grown in p-2-hydroxyethyl methacrylate (p-HEMA) polymer disc method and control (freely suspension cells)	Increase from 10 mg l^{-1} to 55.5 mg l^{-1} for geraniol, from 5 mg l^{-1} to 14 mg l^{-1} for linalool, and from 5 mg l^{-1} to 12 mg l^{-1} α-terpineol	(Zhu and Lockwood, 2000)
41.	Geraniol	*Vitis vinifera*	Cell suspension culture	Two-phase system bioreactor	Nerol	Comparison between one phase and two-phase system	5-fold increase in product formation	(Cormier and Ambid, 1987)
42.	Genistin	Soybean (*Glycine max*)	Cell suspension culture	High-speed counter-current chromatography bioreactor (HSCCC)	Genistein	Optimization of the flow rate for the HSCCC bioreactor (4.0, 3.0, 2.0, and 1.0 mL min^{-1})	2.0 mL min^{-1} flow rate had a conversion rate of over 90%	(Wang et al., 2019)
43.	Digitoxin and digitoxigenin	*Digitalis lanata*	Shoot culture	20 L airlift bioreactors	Deacetyllanatoside C	Comparison of growth parameters in a two-phase system semicontinuous production	Yield of 43.8 g deacetyllanatoside C	Kreis and (Reinhard 1990)
44.	2′-hydroxyformononetin	Transgenic tobacco	Cell suspension culture	*Agrobacterium tumefaciens* transformation	Vestitone	Compared to non-transgenic tobacco plants	Bioconversion percentage reaching up to 60%	(Cooper et al., 2002)
45.	Lanatoside A, lanatoside C, purpurea glycoside, deacetyllanatoside	Transgenic *Cucumis sativus* L.	Hairy root culture	*Agrobacterium rhizogenes* transformation	α-Acetyldigitoxine α-Acetyldigoxine, digitoxin, and digoxin	Comparison with untransformed, rcgh-transformed, and WT-transformed hairy roots	More than 20% turnover compared with WT hairy roots	(Shi and Lindemann, 2006)
46.	(E)-cinnamic acid	*Plantago lanceolata*	Hairy root culture		(E)-p-coumaroyl-1-O-β-D-glucopyranoside			(Fons et al., 1999)
47	Vanillin	*Pharbitis nil*	Hairy root culture		Vanillin-4-O-β-D-glucopyranoside (1.4%)			(Kanho et al., 2004)

Gao et al. (2000) reported the assimilation and conversion of o,p'-DDT and p,p'-DDT by axenic cultures of *Mariophyllum aquaticum*, *Spirodela oligorrhiza*, and *Elodea canadensis*. Within six days, almost all the DDT was removed from the medium and degraded into DDD and DDE, at rates ranging from 1% to 13%. The main degradation pathway was observed to be the reduction of aliphatic chlorine atoms.

Van Aken et al. (2004) reported the transformation of RDX by poplar tissue cultures incubated under different light conditions (light and dark) and found partial reduction of RDX with no significant influence of light on it.

Brazier-Hicks et al. (2007) examined the detoxification of pollutant 3,4-dichloroaniline (DCA) by glycosyltransferase UGT72B1 in root cultures of *Arabidopsis thaliana*. The overexpression of UGT72B1 in the plant revealed an improved sensitivity of *Arabidopsis* seedlings to DCA, while plants with UGT72B1 knocked out showed increased resistance compared to the control plants.

Podlipná et al. (2008) investigated the behavior of nitroglycerine (NG) in regenerants of *Juncus inflexus* and *Phragmites australis* and in flax (*Linum usitatissimum*) suspension cultures. The results showed that flax cells absorbed and transformed NG into dinitroglycerine isomers, which were tested to be less toxic than the parent compound.

Agostini et al. (2003) tested the ability of hairy roots of *Brassica napus* to degrade 2,4-dichlorophenol (2,4-DCP), a hazardous contaminant commonly present in industrial effluents. The cultures showed high removal abilities (91–94%) within 15 minutes, and maximum removal (97–98%) was achieved after 1 hour. The underlying mechanism for 2,4-DCP degradation was identified to be root peroxidase–mediated oxidation.

Teng et al. (2009) studied the suspension cultures of *Hordeum vulgare*, *Oryza sativa*, *Panax quinquefolium*, and *Nicotiana tabacum* for their ability to transform ingenol-3-angelate (anticancer compound). All the cultures showed the presence of products, viz, 16-hydroxy-ingenol-3-angelate and ingenol in the medium, with the highest amount of the former in *H. vulgare*, while ingenol-5-angelate was detected in *H. vulgare* at low levels in both medium and cells. These compounds showed growth-retarding activity against MM96L melanoma cells.

11.3 BIOREACTOR CULTIVATION OF CELLS FOR LARGE-SCALE BIOTRANSFORMATION REACTIONS

A *bioreactor* is a large culture vessel designed to grow cells or tissues in liquid nutrient medium under agitation. It offers controlled environmental parameters, such as temperature, pH, oxygen levels, and nutrient supply, which are crucial for maintaining cell viability, growth, and biosynthetic potential (Verdú-Navarro et al., 2023). Bioreactors serve as indispensable tools for the mass cultivation of cells, which is a pre-requisite for the industrial-level production of secondary metabolites.

11.3.1 Bioreactor Design and Operations for Plant Cell Cultures

Scaling up the cultivation process begins with moving cell cultures from flasks into lab bioreactors. Plant cell cultivation in bioreactors has been a challenge even after using well-growing, biosynthetically potent cell lines. The major reason for this limitation are the peculiar characteristics of plant cells, such as the large size of cells and vacuoles, a rigid cell wall that makes cells susceptible to shear stresses, slow multiplication rate and lengthy cultivation cycles of cell cultures, culture heterogeneity and cell aggregation, excessive foaming and sticking of cell biomass to the bioreactor walls, and the complicated mechanisms of cell growth tuning and target metabolite production (Titova et al., 2024). Such specific behavior of cultured plant cells bears implications on bioreactor configuration, thereby posing complexity in its design and operations (Table 11.2).

By thorough fine-tuning of bioreactor design with respect to plant cell culture requirements, the tricky transition of cell culture from shake flask to a larger scale can be eased to some extent. Optimization includes the creation of low-shear conditions, sufficient mixing, adequate nutrient

TABLE 11.2
Characteristics of Plant Cells Influencing Bioreactor Cultivation

	Characteristics	Implications
1	Large size, rigid cell wall, large vacuole, making cells shear-sensitive	Cultivation under low shear conditions
2	Slow growth rate	Maintenance of aseptic conditions for longer duration
3	Cell aggregation, leading to culture heterogeneity (w.r.t. chemical and physical environment)	Mass transfer limitations, mixing and aeration requirements

and oxygen availability, defoaming methods, etc., generating an ideal environment for excessive biomass accumulation and metabolite synthesis.

According to the published articles, the behavior of cell cultures to parameter alterations is often cell line– or species-specific, with the oxygen levels and shear stress caused by agitation being the major growth-retarding parameters. Constant mixing in plant cell suspension cultivation serves the purposes of a uniform dispersion of cell biomass, even distribution of gases and nutrients to the cell biomass and efficient mass transfers. In shake flasks, mixing is achieved by the constant shaking of the suspension cultures on rotary shakers, while in bioreactors, compressed air or mechanical means are used. However, with the increase in size, mixing efficiency gets compromised; hence, the complexity of the configuration to address problems of temperature gradients, variations in the concentrations of limiting substrates, and the creation of "dead zones" due to biomass settling is increased. At the same time, mixing methods should ensure minimum cell damage due to shear stress. Several investigators have achieved this balance by selecting stirrers (spiral stirrer) (Ulbrich et al., 1985), stirring conditions (stirrer rotation speed and air supply rate) (Pavlov et al., 2005), and gas distribution devices.

Aeration is necessary to supply oxygen to the plant cells during cultivation, necessary for the growth and biosynthesis process, and also to maintain culture homogeneity (Thanh et al., 2014). For this purpose, the level of dissolved oxygen in the culture medium is generally kept as 10–15%. Currently, a variety of gas-distributing devices, such as rings, lattices, and single-point spargers, are used to aerate plant cell suspension cultures grown in bioreactors. Each type of bioreactor has a unique sparger design that is chosen to guarantee optimal airflow while minimizing turbulence and to facilitate mass transfer across the bioreactor's working space. Sparger holes with 1–3 mm diameter facilitate an effective air bubble dispersion and inhibit clumping and aggregation (Marks, 2003). To evaluate the physiological condition of plant cell cultures during bioreactor cultivation, a comprehensive assessment of airflow effectiveness, including a computerized evaluation of oxygen utilization by cells and various methods of analyzing the solubility of gases in the nutrient medium, may be useful (Rechmann et al., 2007).

During the cell proliferation phase in the bioreactor, it is essential to keep an ideal circumstance of mass transfer, which is determined by the supply of energy and the type of bioreactor used (Xia et al., 2015). In this regard, a major parameter to be checked during mass cultivation of cells includes the volumetric coefficient of oxygen transfer (KLα). For instance, the upscaling of suspension cultures of *D. lanata* (Lee and Kim, 2004), *Panax ginseng* (Dong et al., 2013), and *Taxus chinensis* (Zhong et al., 2002) has shown the importance of oxygen transfer, and its drawbacks. However, these parameters are of significance only if mixing conditions are identical to those maintained in a laboratory-scale bioreactor.

11.3.2 Cultivation Regimes in Bioreactors

The slow growth of plant cells necessitates an extended cultivation process. Under such conditions, different cultivation regimes are followed to achieve optimal biomass growth and product yield (Xu et al., 2011). The batch culture method is generally used for mass cultivation but is often expensive as it involves repeated steps of sterilization, medium addition, and cleaning the bioreactor. These activities lead to the evolution of more sophisticated systems, such as continuous and semicontinuous cultures, fed-batch, two-stage, cell immobilization, and perfusion (Ochoa-Villarreal et al., 2016). The two-stage strategy involves a separate growth and production phases. The most famous example of bi-stage culture is the one used for the industrial production of shikonin by the Mitsui Petrochemical Company, Japan. The perfusion culture often provides an optimal approach as it involves the continuous feeding of fresh medium and the removal of waste cell–free medium. As a result, cells become reusable, optimizing batch turnaround times and thus boosting production (Georgiev et al., 2013). The elimination of used media promotes productivity and lessens the chance of retardation due to the buildup of repressing metabolites in the medium. Nevertheless, in contrast to batch system, perfusion systems are typically operated on a lesser scale.

11.3.3 Types of Bioreactors Used for Plant Cell Culture

Many kinds of bioreactors have been used to cultivate plant cells for the production of secondary metabolites, ranging from laboratory (2–20 L) to pilot (75 L) and industrial size (630 L). A pilot-size bioreactor is used for method verification and upscaling, an industrial bioreactor is used for the massive production of cell biomass and their metabolites, and a laboratory bioreactor is used for research and refinement (Figure 11.2). However, for biotransformation, most of the studies have been performed at the lab scale only.

The commonly used plant cell bioreactors with their characteristics are presented here. There are different types of bioreactors based on their construction, such as stirred tank, airlift, bubble column and wave, membrane bioreactors, etc. (Table 11.3).

11.3.3.1 Stirred Tank Bioreactors

Stirred tank bioreactors (STR), also known as continuous stirred tank reactors (CSTRs), feature a stirring mechanism (e.g., impeller) to promote uniform mixing and oxygen transfer throughout the culture medium (Figure 11.3a). Stirred tank bioreactors offer scalability, ease of operation, and flexibility in controlling culture conditions (Ruffoni et al., 2010). These are commonly used to scale up the culture process from laboratory to industrial levels (Tanaka, 2000). Suspension culture of several plant species, such as *Taxus*, *Artemisia annua*, have been grown in STR for secondary metabolite production (Frense, 2007; McCoy and O'Connor, 2008)

11.3.3.2 Airlift Bioreactors

Airlift bioreactors utilize gas lift principles to circulate the culture medium and provide aeration. They consist of a riser and downcomer section, with gas injected at the bottom to create buoyancy and lift the medium (Figure 11.3b). Airlift bioreactors offer gentle agitation and efficient oxygen transfer, making them suitable for shear-sensitive plant cell cultures, such as *Berberis wilsoniae* (Breuling et al., 1985) and *Hypericum perforatum* L. (Cui et al., 2014).

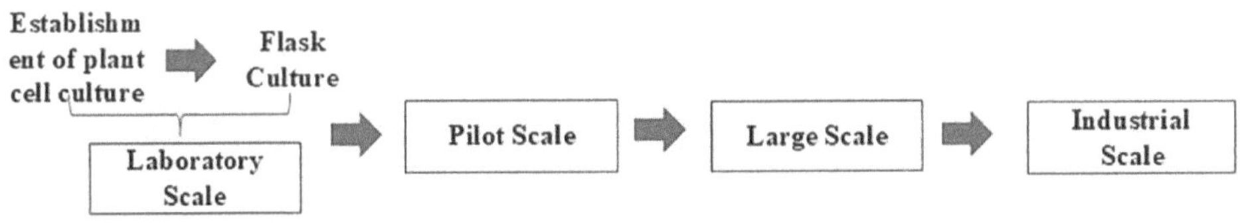

FIGURE 11.2 Steps for scale-up of plant cell culture for biotransformation.

11.3.3.3 Bubble Column Bioreactors

Bubble column bioreactors rely on the injection of gas bubbles at the bottom to generate mixing and aeration. The rising bubbles create upward flow, facilitating nutrient distribution and gas exchange within the culture medium.

Bubble column bioreactors are simple in design and well-suited for large-scale plant cell culture production. They were used for the cultivation of *Eurycoma longifolia* Jack and *Taxus cuspidata* Siebold cells (Shim et al., 2015; Son et al., 2000). According to De Jesus et al. (2017), airlift and bubble-type bioreactors are not advised for cell suspensions with elevated viscosities or significant final cellular biomass contents because of their comparatively limited mass transfer properties.

11.3.3.4 Wave Bioreactors

Wave bioreactors are identified by wave-induced motion, in which mass and energy transfer are mechanically regulated using the 2D rocking, stirring rate, medium volume, and culture tank design (Georgiev et al., 2013). They are seldom used for large-scale cultivation because of their drawbacks, which include substantial loss of energy during medium stirring, technical difficulties in determining the ideal conditions, and the high cost of additional hardware.

TABLE 11.3
Characteristics of Bioreactors Common for Plant Cell Cultivation

Bioreactor Type	Description	Advantages	Disadvantages	Reference
Stirred tank bioreactor	Traditional bioreactor with a stirring mechanism to maintain homogeneity of culture medium and oxygenation.	Good mixing for uniform nutrient distribution and gas transfer.	Mechanical shear may damage fragile plant cells. Requires higher energy input.	(Tervasmäki et al., 2018)
Airlift bioreactor	Utilizes airlift principle for mixing and oxygenation, with separate zones for liquid circulation and aeration.	Gentle mixing, reducing cell damage. Lower energy consumption compared to stirred tank bioreactors.	Limited scalability due to size constraints. May have lower mass transfer rates.	(Kumar et al., 2020)
Bubble column bioreactor	Consists of a column with gas sparging at the bottom to generate bubbles, promoting mixing and oxygen transfer.	Simple design, easy to scale up. Well-suited for cultures with low viscosity.	Limited control over bubble size and distribution. Risk of bubble coalescence, leading to poor oxygen transfer.	(De Jesus et al., 2017)
Perfusion bioreactor	Maintains continuous flow of fresh medium while removing spent medium, allowing for prolonged cell cultivation.	Continuous nutrient supply and waste removal. Mimics physiological conditions more closely.	Complex setup and operation. Higher risk of contamination.	(Georgiev et al., 2013)
Wave bioreactor	Characterized by wave-induced motion, where mass and energy transfer are manually adjusted via the rocking angle, agitation rate, medium level, and culture vessel geometry.	Disposable type of bioreactors. Aeration parameters are achieved easily. Risk of contamination and foaming is low.	Energy losses are high during liquid agitation; complex in design; consume a substantial amount of energy and are rarely used for the large-scale commercial cultivation of plant cells.	(Eibl et al., 2010)

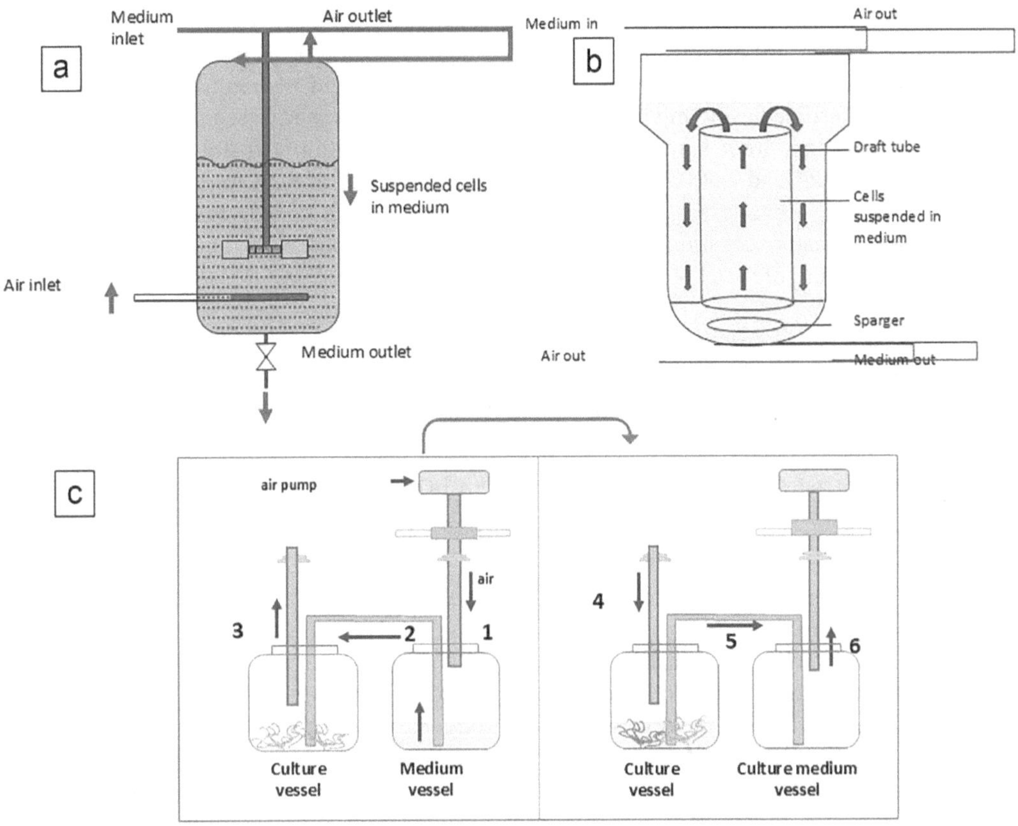

FIGURE 11.3 Types of bioreactors: (a) STR, (b) airlift bioreactor, (c) temporary immersion, (d) two-phase bioreactor.

Source: (a) Redrawn from Furusaki and Takeda (2017). (b) Redrawn from Furusaki and Takeda (2017). (c) Redrawn from Georgiev et al. (2014). (d) Redrawn from Kreis and Reinhard (1990).

11.3.3.5 Other Types

Other bioreactor configurations employed for plant cell cultivation include packed bed reactors, membrane bioreactors, and photobioreactors, each offering unique advantages depending on the specific requirements of the culture system (Georgiev et al., 2013). A temporary immersion bioreactor has also been used to produce secondary compounds alongside micropropagation and hairy root culture (Mirzabe et al., 2022) (Figure 11.3c). Based on the preparation and sterilization methods, these bioreactors can be classified into three categories (Verdú-Navarro et al., 2023).

11.3.3.5.1

Autoclavable bioreactors are typically made of glass and have capacities ranging from 200 mL to 10 L. They can be completely sterilized in an autoclave, along with probes and the culture media. Greater capacities are not feasible as it is difficult to put them in an autoclave and they get heavier to handle and might even shatter while being moved around.

11.3.3.5.2

Sterilization-in-place (SIP) bioreactors are made of stainless steel, having volumes ranging from 15 L to 500 m^3. They are decontaminated using heat and steam pressure.

11.3.3.5.3

Single-use bioreactors (SUB), or disposable bioreactors, are made of plastic and contain probes. These are naturally sterilized by gamma radiation or by treatment with ethylene oxide, and the medium can be sterilized using microfiltration or heat, or it can be purchased pre-sterilized. Their working volumes range from 1 mL to 100 mL in microscales, 100 mL to 10 L in benchtop scale, 10–250 L in pilot size, and 2,500 L in commercial size. Examples of SUB for plant cell cultivation include stirred bag, orbitally shaken, or stirred bioreactors, wave-mixed, plastic-lined, and slug bubble. Wave bioreactors constitute the most efficient type of disposable bioreactor and have been utilized for the culture of many plant species cells, including *Nicotiana tabacum* and *Vitis vinifera* (Eibl et al., 2010).

The most pertinent bioreactor is selected taking into account various variables, such as the selective requirements of the plant cell culture, the properties of the product, scalability aspects, and accessible resources (Merchuk, 1990; Shukla and Gottschalk, 2013).

11.3.4 Pilot-Scale and Industrial-Scale Production Facilities

The conditions of laboratory-scale bioreactor experiments form the basis for the initial experimentations in pilot-scale bioreactors. Different parameters, such as media composition, oxygenation levels, agitation rates, and cell density, are tested before scaling up to larger bioreactors. Pilot-scale systems help in identifying the most efficient conditions for cell growth, product synthesis, and biosynthetic potential retention, ultimately streamlining the transition to industrial scale.

The large-scale production of plant cell cultures represents a pivotal stage in the commercialization of biotransformation processes for the sustainable production of bioactive compounds. These are specialized facilities designed to accommodate the cultivation of large volumes of cells under controlled conditions. These facilities are equipped with state-of-the-art bioreactor systems, downstream processing units, and monitoring/control systems to ensure optimal culture performance and product quality. However, addressing the cultivation challenges is crucial for maximizing productivity and product quality in large-scale bioreactor systems. The specific requirements of plant cell culture consequently lead to high production costs and price of the final product.

Recent developments in plant cell culturing in bioreactors include microcarrier optimization for improved growth, genetic engineering for the generation of metabolites, and 3D culture techniques that imitate natural habitats (Hesami et al., 2021). Many sensors, such as for conductivity or pH, are used for an automated culture condition management and real-time monitoring (Ruffoni et al., 2010). It has long been practiced having precise control over environmental parameters, including temperature, dissolved oxygen, and luminescence regarding photobioreactors (Ruffoni et al., 2010). Moreover, new routes, improved nutrient composition, and the amalgamation of omics technologies with bioreactor facts for a thorough comprehension of biological events are all made possible by synthetic biology (Reski et al., 2018). Together, these technologies are enabling further productive and systematic plant cell cultivation techniques that will be used to produce important chemicals, such as nutraceuticals and medicines.

Research has typically concentrated on upscaling standardized technical procedures from pilot to larger capacities utilizing cell lines with known growth and manufacturing potential. Nevertheless, it is generally quite tough to anticipate the alterations that cells in culture will undergo when they are shifted to bioreactors and to perfectly blend the technological parameters of the bioreactor to the needs of the culture.

For the commercialization of the bioprocess, it is imperative to develop authentic mathematical models, defined industrial designs, and upscaling approaches. Models that take into consideration the intricate relationships between cells and their surroundings can make the process development rapid and cost-effective (Motolinía-Alcántara et al., 2021).

11.3.5 Biotransformation Process in Bioreactors

A literature survey shows that there are very few reports on the biotransformation process performed in bioreactor systems. Details of the process conditions of specific biotransformation reactions carried out by plant cell culture are presented in what follows.

An external-loop airlift bioreactor was configured to produce extracellular products from the calcium alginate–entrapped cells of *Frangula alnus* in a Murashige and Skoog medium. It was a four-phase system that used scattered gas and silicone oil and n-hexadecane as solvents. The continuous extraction of anthraquinones by these solvents in the bioreactor yielded 10- to 30-fold greater cell productivity compared to the cells immobilized in a flask. Based on the blending style, the entrapped cell extraparticle and intraparticle diffusional constraints, and growth kinetics, substrate utilization, and product development, a mathematical model was generated to discern the timeline of a batch culture (Sajc et al., 1995).

Wang et al. (2019) designed a unique bioreactor with high-speed counter-current chromatography (HSCCC) to effectively convert genistin to genistein. Based on the KD values, solvents were selected for the bioreactor. The stationary phase's retention was taken into consideration when choosing the ideal reactor parameters. After experimenting with various flow rates, the HSCCC bioreactor was found to work best at 2.0 mL min^{-1}, resulting in a transformation rate of over 90% in less than 24 hours. This innovative bioreactor made it simple to immobilize and recycle the enzyme, making it suitable to produce genistein. Reinhard et al. (1989) cultivated cell lines of *Digitalis lanata* cell in 1 L shake flasks or 20 L airlift bioreactors for the conversion of beta-methyldigitoxin into beta-methyldigoxin. This process was enhanced in terms of the supply of the substrate and the composition of the culture medium, followed by upscaling in airlift bioreactor (300 L) in a semicontinuous way, leading to the synthesis of 513.3 g beta-methyldigoxin after 12.5 weeks of culture cycle. Later, Kreis and Reinhard (1990) demonstrated the production of deacetyl lanatoside C from digitoxin by cell lines of *Digitalis lanata* shake flasks, which was subsequently upscaled using a two-stage cultivation (growth and production phase) method in a semicontinuous manner in 20 L airlift bioreactors, accumulating 43.8 g deacetyl lanatoside C after 7 days in the production stage.

Kreis and Reinhard (1992) reported the biotransformation of digitoxin to digoxin by cell suspension of *Digitalis lanata* cells in airlift bioreactors (300 L).

Kutney (1993) developed a cell line of *P. peltatum* capable of the recurrent transformation of butanolide to the podophyllotoxin equivalent (50% yield) in bioreactor in a semicontinuous manner for greater than 15 cycles of 24 h.

Kutney et al. (1997) reported the increased production of selected triterpenes by fungal elicitation in the TRP4a cell line of *T. wilfordii* in airlift bioreactors.

11.4 CONCLUSION

Biotransformation is an efficient technology that produces novel compounds with altered molecular structures and superior regio- and stereospecificity and frequently provides a simple, inexpensive, and more eco-friendly outcome than semi-synthetic routes, which are still expensive and produce toxic by-products.

Using this approach on a variety of natural products, a distinct collection of natural compound equivalents has been created. Consequently, the production of new biotransformed metabolites may open up previously unexplored paths for the development of phytochemicals. The ability of the new class of altered natural products to enhance the solubility and pharmacodynamics of pharmaceutical drugs may open up new opportunities in the pharmaceutical sector.

Plant cells harbor a rich repository of enzymes, and therefore, its in vitro culture should be an ideal system for biotransformation reactions with enormous conversion potential. However, slow progress has been made in research in exploiting the biocatalytic capability of in vitro cultures. This technology in the plant system has not been as successful as bacterial biotransformation for several reasons, such as challenges with regard to plant cell cultivation, inconsistent and low yields, complex plant metabolism, and genetic instability. Further, there are difficulties in upscaling this process in bioreactors. More precisely, for deciphering a secondary metabolite pathway, a multidisciplinary approach including determining metabolic intermediates and potential reaction sequences, isolating and identifying particular enzymes, and pinpointing their subcellular localization is required. The limited and scattered information on this subject significantly hinders the use of biotransformation *in vitro*. Full realization of the biotransformation potential of plant cell cultures could be possible with continued research and development in the area of metabolic and genetic engineering to manipulate metabolic pathways for the enhanced production of desired products, as well as improvements in the bioreactor design to optimize culture conditions for higher yields and productivity. Further, employing computational tools to predict biotransformation reactions, model metabolic pathways, design biotransformation experiments, and develop new biocatalysts can contribute to the expansion of this technology's commercial viability.

REFERENCES

Agostini, E., Coniglio, M.S., Milrad, S.R., Tigier, H.A. & Giulietti, A.M. 2003. Phytoremediation of 2, 4-dichlorophenol by *Brassica napus* hairy root cultures. *Biotechnology and Applied Biochemistry* 37(2): 139–144.

Aird, E.L.H., Hamill, J.D. & Rhodes, M.J.C. 1988. Cytogenetic analysis of hairy root cultures from a number of plant species transformed by *Agrobacterium rhizogenes*. *Plant Cell Tissue and Organ Culture* 15: 47–57.

Akakabe, Y. & Naoshima, Y. 1994. Biotransformation of acetophenone with immobilized cells of carrot, tobacco and Gardenia. *Phytochemistry* 35(3): 661–664.

Alderete, L.G.S., Talano, M.A., Ibáñez, S.G., Purro, S., Agostini, E., Milrad, S.R. & Medina, M.I. 2009. Establishment of transgenic tobacco hairy roots expressing basic peroxidases and its application for phenol removal. *Journal of Biotechnology* 139(4): 273–279.

Alfermann, A.W., Schuller, I., & Reinhard, E. 1980. Biotransformation of cardiac glycosides by immobilized cells of *Digitalis lanata*. *Planta Medica* 40(10): 218–223.

Asada, Y., Saito, H., Yoshikawa, T., Sakamoto, K. & Furuya, T. 1993. Biotransformation of 18β-glycyrrhetinic acid by ginseng hairy root culture. *Phytochemistry* 34(4): 1049–1052.

Asthana, G.S. & Asthana, A. 2015. Biotransformation of progesterone to 17a-hydroxyprogesterone by using plant cell suspension culture of *Catharanthus roseus*. *International Journal of Pharmacy Pharmaceutical Sciences* 7: 362–368.

Banerjee, S., Singh, S. & Rahman, L.U. 2012. Biotransformation studies using hairy root cultures—a review. *Biotechnology Advances* 30: 461–468. https://doi.org/10.1016/j.biotechadv.2011.08.010.

Bhadra, R., Wayment, D.G., Williams, R.K., Barman, S.N., Stone, M.B., Hughes, J.B. & Shanks, J.V. 2001. Studies on plant-mediated fate of the explosives RDX and HMX. *Chemosphere* 44(5): 1259–1264.

Brazier-Hicks, M., Edwards, L.A. & Edwards, R. 2007. Selection of plants for roles in phytoremediation: The importance of glucosylation. *Plant Biotechnology Journal* 5(5): 627–635.

Breuling, M., Alfermann, A.W. & Reinhard, E. 1985. Cultivation of cell cultures of *Berberis wilsonae* in 20-L airlift bioreactors. *Plant Cell Reports* 4: 220–223. https://doi.org/10.1007/BF00269294.

Camper, N.D. & McDonald, S.K. 1989. Tissue and cell cultures as model systems in herbicide research. *Reviews of Weed Science* 4: 169–190.

Casas, D.A., Pitta-Alvarez, S.I. & Giulietti, A.M. 1998. Biotransformation of hydroquinone by hairy roots of *Brugmansia candida* and effect of sugars and free-radical scavengers. *Applied Biochemistry and Biotechnology* 69: 127–136.

Cooper, J., Qiu, F. & Paiva, N. 2002. Biotransformation of an exogenously supplied isoflavonoid by transgenic tobacco cells expressing alfalfa isoflavone reductase. *Plant Cell Report* 20: 876–884. https://doi.org/10.1007/s00299-001-0404-1.

Corchete, P. & Yeoman, M.M. 1989. Biotransformation of (–)-codeinone to (–)-codeine by *Papaver somniferum* cells immobilized in reticulate polyurethane foam. *Plant Cell Reports* 8: 128–131.

Cormier, F. & Ambid, C. 1987. Extractive bioconversion of geraniol by a *Vitis vinifera* cell suspension employing a two-phase system. *Plant Cell Reports* 6(6): 427–430.

Cui, X-H., Murthy, H.N. & Paek, K-Y. 2014. Pilot-scale culture of *Hypericum perforatum* L. adventitious roots in airlift bioreactors for the production of bioactive compounds. *Applied Microbiology and Biotechnology* 174: 784–792. https://doi.org/10.1007/s12010-014-1123-8.

Da Silva, B.O., Amaral, A.C.F., Ferreira, J.L.P., Santiago, L.J.M. & Louro, R.P. 2013. Micropropagation and in vitro production of secondary metabolites of *Croton floribundus* Spreng. *In Vitro Cellular & Developmental Biology-Plant* 49: 366–372.

Dave, V., Khirwadkar, P. & Dashora, K. 2014. A review on biotransformation. *Indian Journal of Research in Pharmacy and Biotechnology* 1136.

De Jesus, S.S., Moreira Neto, J. & Maciel Filho, R. 2017. Hydrodynamics and mass transfer in bubble column, conventional airlift, stirred airlift and stirred tank bioreactors, using viscous fluid: A comparative study. *Biochemical Engineering Journal* 118: 70–81. https://doi.org/10.1016/j.bej.2016.11.019.

Dong, Y., Gao, W-Y., Man, S., Zuo, B., Wang, J., Huang, L. *et al.* 2013. Effect of bioreactor angle and aeration rate on growth and hydromechanics parameters in bioreactor culture of Ginseng suspension cells. *Acta Physiologiae Plantarum* 35: 1497–1501.

Doran, P.M. 2009. Application of plant tissue cultures in phytoremediation research: Incentives and limitations. *Biotechnology and Bioengineering* 103(1).

Dvorakova, M., Valterova, I., Saman, D. & Vanek, T. 2011. Biotransformation of (1 S)-2-Carene and (1 S)-3-Carene by *Picea abies* Suspension Culture. *Molecules* 16(12): 10541–10555.

Eibl, R., Kaiser, S., Lombriser, R. & Eibl, D. 2010. Disposable bioreactors: The current state-of-the-art and recommended applications in biotechnology. *Applied Microbiology and Biotechnology* 86: 41–49. https://doi.org/10.1007/s00253-009-2422-9.

Eibl, R., Meier, P., Stutz, I., Schildberger, D., Hühn, T. & Eibl, D. 2018. Plant cell culture technology in the cosmetics and food industries: Current state and future trends. *Applied Microbiology and Biotechnology* 102: 8661–8675.

Espinosa-Leal, C.A., Puente-Garza, C.A. & García-Lara, S. 2018. In vitro plant tissue culture: Means for production of biological active compounds. *Planta* 248: 1–18.

Fecker, L.F., Rugenhagen, C. & Berlin, J. 1993. Increased production of cadaverine and anabasine in hairy root cultures of *Nicotiana tabacum* expressing a bacterial lysine decarboxylase gene. *Plant Molecular Biology* 23: 11–21.

Flores, H.E., Hoy, M.W. & Pickard, J.J. 1987. Secondary metabolites from root cultures. *Trends in Biotechnology* 5 (3): 64–69.

Fons, F., Tousch, D., Rapior, S., Alian, G., Roussel, J.L., Gargadennec, A. *et al.* 1999. Profiles of untransformed and hairy root cultures of Plantago lanceolata. *Plant Physiology and Biochemistry* 37: 291–296.

Frense, D. 2007. Taxanes: Perspectives for biotechnological production. *Applied Microbiology and Biotechnology* 73: 1233–1240. https://doi.org/10.1007/s00253-006-0711-0.

Furmanowa, M., Hartwich, M., Alfermann, A.W., Kozminski, W. & Olejnik, M. 1999. Rosavin as a product of glycosylation by *Rhodiola rosea* (roseroot) cell cultures. *Plant Cell Tissue and Organ Culture* 56: 105–110. https://doi.org/10.1023/A:1006232023274

Furuya, T., Orihara, Y. & Koge, K. 1991. Biotransformation of theobromine to caffeine in suspension and polyurethane foam immobilized coffee (*Coffea arabica* L.) cells. *Plant Cell Reports* 9: 659–662.

Galun, E., Aviv, D., Dantes, A. & Freeman, A. 1983. Biotransformation by plant cells immobilized in cross-linked polyacrylamide-hydrazide. *Planta Medica* 49(9): 9–13. https://doi.org/10.1055/s-2007-969799.

Gao, J., Garrison, A.W., Hoehamer, C., Mazur, C.S. & Wolfe, N.L. 2000. Uptake and phytotransformation of o, p '-DDT and p, p '-DDT by axenically cultivated aquatic plants. *Journal of Agricultural and Food Chemistry* 48(12): 6121–6127.

Georgiev, M.I., Eibl, R. & Zhong, J.J. 2013. Hosting the plant cells in vitro: Recent trends in bioreactors. *Applied Microbiology and Biotechnology* 97: 3787–3800. https://doi.org/10.1007/s00253-013-4817-x.

Georgiev, V., Schumann A., Pavlov, A. & Bley, T. 2014. Temporary immersion systems in plant Biotechnology. *Engineering in Life Sciences* 14: 607–621.

Gil, G., Ferreira dos Santos, P. & Bullard, C. 1995. Biotransformation of Δ2-carene by callus tissues. *Phytochemistry* 38(3): 629–631. https://doi.org/10.1016/0031-9422(94)00680-R

Giri, A., Dhingra, V., Giri, C., Singh, A., Ward, O. & Narasu, M. 2001. Biotransformations using plant cells, organ cultures and enzyme systems: Current trends and future prospects. *Biotechnology Advances* 19: 175–199. https://doi.org/10.1016/S0734-9750(01)00054-4.

Grech-Baran, M., Sykłowska-Baranek, K., Krajewska-Patan, A., Wyrwał, A. & Pietrosiuk, A. 2014. Biotransformation of cinnamyl alcohol to rosavins by non-transformed wild type and hairy root cultures of *Rhodiola kirilowii*. *Biotechnology Letters* 36: 649–656.

György, Z., Tolonen, A., Neubauer, P. & Hohtola A.2005. Enhanced biotransformation capacity of *Rhodiola rosea* callus cultures for glycoside production. *Plant Cell Tissue Organ Culture* 83: 129–135. https://doi.org/10.1007/s11240-005-4010-8

György, Z., Tolonen, A., Pakonen, M., Neubauer, P. & Hohtola, A. 2004. Enhancing the production of cinnamyl glycosides in compact callus aggregate culture of *Rhodiola rosea* by biotransformation of cinnamyl alcohol. *Plant Science* 166: 229–236. https://doi.org/10.1016/j.plantsci.2003.09.011

Hamada, H., Yasumune, H., Fuchikami, Y., Hirata, T., Sattler, I., Williams, H.J. & Scott, A.I. 1997. Biotransformation of geraniol, nerol and (+)-and (−)-carvone by suspension cultured cells of *Catharanthus roseus*. *Phytochemistry* 44(4): 615–621.

Hashimoto, T., Matsuda, J. & Yamada, Y. 1993a. Two step epoxidation of hyoscyamine to scopolamine is catalyzed by bifunctional hyoscyamine 6-hydroxylase. *FEBS Letters* 329: 35–39.

Hashimoto, T., Yun, D.-J. & Yamada, Y. 1993b. Production of tropane alkaloids in genetically engineered root cultures. *Phytochemistry* 32: 713–718.

He, S., Zhu, J., Zi, J., Zhou, P., Liang, J. & Yu, R. 2015. A novel terpenoid indole alkaloid derived from catharanthine via biotransformation by suspension-cultured cells of *Catharanthus roseus*. *Biotechnology Letters* 37: 2481–2487.

Hesami, M., Baiton, A., Alizadeh, M., Pepe, M., Torkamaneh, D. & Jones, A.M.P. 2021. Advances and perspectives in tissue culture and genetic engineering of cannabis. *International Journal of Molecular Sciences* 22: 5671. https://doi.org/10.3390/ijms22115671.

Hughes, J.B., Shanks, J., Vanderford, M., Lauritzen, J. & Bhadra, R. 1996. Transformation of TNT by aquatic plants and plant tissue cultures. *Environmental Science & Technology* 31(1): 266–271.

Ichinose, K., You, S., Kawano, N., Hayashi, K., Yao, X.-S. & Ebizuka, Y. 1999. Heterologous expression of furostanol glycoside 26-O-β-glucosidase of *Costus speciosus* in *Nicotiana tabacum*. *Phytochemistry* 51(5): 599–603.

Johnson, S.T., Ravishankar, G.A. & Venkataraman, L.V. 1996. Biotransformation of ferulic acid and vanillylamine to capsaicin and vanillin in immobilized cell cultures of *Capsicum frutescens*. *Plant Cell Tissue and Organ Culture* 44: 117–121.

Kanho, H., Yaoya, S., Itani, T., Nakane, T., Kawahara, N., Takase, Y. *et al.* 2004. Glucosylation of phenolic compounds by *Pharbatis nil* hairy roots. *Bioscience Biotechnology and Biochemistry* 68: 2932–2939.

Kawaguchi, K., Hirotani, M., Yoshikawa, T. & Furuya, T. 1990. Biotransformation of digitoxigenin by ginseng hairy root cultures. *Phytochemistry* 29: 837–843.

Kawka, B., Kwiecień, I. & Ekiert, H. 2020. Endogenous production of specific flavonoids and verbascoside in agar and agitated microshoot cultures of *Scutellarialateriflora* L. and biotransformation potential. *Plant Cell Tissue and Organ Culture* 142: 471–482.

Knops, M., Schuphan, I. & Schmidt, B. 1995. Biotransformation of 4-nitrophenol by a fermenter grown cell suspension culture of soybean (*Glycine max*): Isolation and identification of conjugates. *Plant Science* 109: 215–224. https://doi.org/10.1016/0168-9452(95)04175-T.

Krasteva, G., Georgiev, V. & Pavlov, A. 2020. Recent applications of plant cell culture technology in cosmetics and foods. *Engineering in Life Sciences* 21: 68–76. https://doi.org/10.1002/elsc.202000078.

Kreis, W. & Reinhard, E. 1990. Two-stage cultivation of *Digitalis lanata* cells: Semicontinuous production of deacetylla-natoside C in 20 litre airlift bioreactor. *Journal of Biotechnology* 16: 123–136.

Kreis, W. & Reinhard, E. 1992. 12 beta-Hydroxylation of digitoxin by suspension-cultured *Digitalis lanata* cells: Production of digoxin in 20-litre and 300-litre air-lift bioreactors. *Journal of Biotechnology* 26(2–3): 257–273. https://doi.org/10.1016/0168-1656(92)90011-w.

Kroll, C.K. & Brenner, W.G. 2020. Cytokinin signaling downstream of the his-asp phosphorelay network: Cytokinin-regulated genes and their functions. *Frontiers in Plant Science* 11: 604489.

Kučerová, P., Macková, M., Poláchová, L., Burkhard, J., Demnerová, K., Pazlarová, J. & Macek, T. 1999. Correlation of PCB transformation by plant tissue cultures with their morphology and peroxidase activity changes. *Collection of Czechoslovak Chemical Communications* 64: 1497–1509.

Kumar, N., Gupta, R. & Bansal, A. 2020. Effect of surface tension on hydrodynamics and mass transfer coefficient in airlift reactors. *Chemical Engineering and Technology* 43: 995–1004.

Kurina Sanz, M.B. & Donadel, O.J. 2003. Biotransformation of eudesmanes by *Tessaria absinthioides* cell suspension cultures. *Plant Cell Tissue and Organ Culture* 73: 123–129. https://doi.org/10.1023/A:1022850219300.

Kutney, J.P. 1993. Plant cell culture combined with chemistry: A powerful route to complex natural products. *Journal of Chemical Research* 26: 559–566.

Kutney, J.P., Han, K., Kuri-Brena, F., Milanova, R.K. & Roberts, M. 1997. Studies with plant cell cultures of the Chinese herbal plant, *Tripterygium wilfordii* synthesis and biotransformation of diterpene analogues. *Heterocycles* 44: 95–104.

Laurent, F., Canlet, C., Debrauwer, L. & Pascal-Lorber, S. 2007. Metabolic fate of [14C]-2, 4-dichlorophenol in tobacco cell suspension cultures. *Environmental Toxicology and Chemistry* 26: 2299–2307.

Lee, S. & Kim, D. 2004. Effects of Pluronic F-68 on cell growth of *Digitalis lanata* in aqueous Wo-Phase systems. *Journal of Industrial Microbiology and Biotechnology* 14: 1129–1133.

Lucero, M.E., Mueller, W., Hubstenberger, J., Phillips, G.C. & O'Connell, M.A. 1999. Tolerance to nitrogenous explosives and metabolism of TNT by cell suspensions of *Datura innoxia*. *In Vitro Cellular and Developmental Biology of Plant* 35: 480–486.

Marks, D.M. 2003. Equipment design considerations for large scale cell culture. *Cytotechnology* 42: 21–33. https://doi.org/10.1023/A:1026103405618.

Mathieu, S., Wirth, J., Sauvage, F-X., Lepoutre, J-P., Baumes, R. & Gunata, Z. 2009. Biotransformation of C13-norisoprenoids and monoterpenes by a cell suspension culture of cv. Gamay (*Vitis vinifera*). *Plant Cell Tissue and Organ Culture* 97: 203–213. https://doi.org/10.1007/s11240-009-9516-z.

McCoy, E. & O'Connor, S.E. 2008. Natural products from plant cell cultures. *Progress in Drug Research* 65: 330–370. https://doi.org/10.1007/978-3-7643-8117-2_9.

Merchuk, J. 1990. Why use air-lift bioreactors? *Trends in Biotechnology* 8: 66–71. https://doi.org/10.1016/0167-7799(90)90138-N.

Mirmazloum, I., Kiss, A., Ladányi, M. & György, Z. 2019. Production of cinnamyl alcohol glycosides by biotransformation in roseroot callus cells. *Plant Cell Tissue and Organ Culture* 139: 29–37. https://doi.org/10.1007/s11240-019-01659-7.

Mirzabe, A.H., Hajiahmad, A., Fadavi, A. & Rafiee, S. 2022. Temporary Immersion Systems (TISs): A comprehensive review. *Journal of Biotechnology* 357: 56–83. https://doi.org/10.1016/j.jbiotec.2022.08.003

Motolinía-Alcántara, E.A., Castillo-Araiza, C.O., Rodríguez-Monroy, M., Román-Guerrero, A. & Cruz-Sosa, F. 2021. Engineering considerations to produce bioactive compounds from plant cell suspension culture in bioreactors. *Plants* 10: 2762. https://doi.org/10.3390/plants10122762.

Moyano, E., Palazón, J., Bonfill, M., Osuna, L., Cusidó, R.M., Oksman Caldentey, K.M. *et al.* 2007. Biotransformation of hyoscyamine into scopolamine in transgenic tobacco cell cultures. *Journal Plant Physiology* 164: 521–524.

Munkert, J., Franco, M.S., Nolte, E., Silva, I.T., Castilho, R.O., Ottoni, F.M. *et al.* 2017. Production of the cytotoxic cardenolide glucoevatromonoside by semisynthesis and biotransformation of evatromonoside by a *Digitalis lanata* cell culture. *Planta Medica* 83: 1035–1043.

Muranaka, T., Miyata, M., Ito, K. & Tachibana, S. 1998. Production of podophyllotoxin in Juniperus chinensis's callus cultures treated with oligosaccharides and a biogenetic precursor. *Phytochemistry* 49: 491–496. https://doi.org/10.1016/S0031-9422(98)00295-7.

Ochoa-Villarreal, M., Howat, S., Hong, S., Jang, M.O., Jin, Y.W., Lee, E.K. *et al.* 2016. Plant cell culture strategies for the production of natural products. *BMB Reports* 49(3): 149–158. https://doi.org/10.5483/bmbrep.2016.49.3.264.

Pádua, R.M., Meitinger, N., Dias De Souza, J.F., Waibel, R., Gmeiner, P., Braga, F.C. & Kreis, W. 2012. Biotransformation of 21-O-acetyl-deoxycorticosterone by cell suspension cultures of *Digitalis lanata* (strain W.1.4). *Steroids* 77: 1373–1380.

Pavlov, A.I., Georgiev, M.I. & Ilieva, M.P. 2005. Production of rosmarinic acid by *Lavandula vera* MM cell suspension in bioreactor: Effect of dissolved oxygen concentration and agitation. *World Journal of Microbiology and Biotechnology* 21: 389–392. https://doi.org/10.1007/s11274-004-3982-6.

Piekoszewska, A., Ekiert, H. & Zubek, S. 2010. Arbutin production in *Ruta graveolens* L. and *Hypericum perforatum* L. in vitro cultures. *Acta Physiologiae Plantarum* 32: 223–229.

Podlipná, R., Fialová, Z. & Vaněk, T. 2008. Toxic effect of nitroesters on plant tissue cultures. *Plant Cell Tissue and Organ Culture* 94(3): 305–311.

Podlipná, R., Skálová, L., Seidlová, H., Szotáková, B., Kubíček, V., Stuchlíková, L. & Vokřál, I. 2013. Biotransformation of benzimidazole anthelmintics in reed (*Phragmites australis*) as a potential tool for their detoxification in environment. *Bioresource Technology* 144: 216–224.

Rao, S.R. & Ravishankar, G.A. 2000. Biotransformation of protocatechuic aldehyde and caffeic acid to vanillin and capsaicin in freely suspended and immobilized cell cultures of *Capsicum frutescens*. *Journal of Biotechnology* 76(2–3): 137–146.

Rattan, S., Kumar, D. & Warghat, A.R. 2022. The influence of phenylalanine feeding on cell growth, antioxidant activity, phenylpropanoids content, and yield in cell suspension culture of *Rhodiola imbricata* (Edgew.). *Plant Cell Tissue and Organ Culture*. https://doi.org/10.1007/s11240-022-02356-8.

Rechmann, H., Friedrich, A., Forouzan, D., Barth, S., Schnabl, H., Biselli, M. *et al.* 2007. Characterization of photosynthetically active Duckweed (*Wolffia ausraliana*) in Vitro culture by respiration activity monitoring system (RAMOS). *Biotechnology Letters* 29: 971–977. https://doi.org/10.1007/s10529-007-9330-9.

Reinhard, E., Kreis, W., Barthlen, U. & Helmbold, U. 1989. Semicontinuous cultivation of *Digitalis lanata* cells: Production of beta-methyldigoxin in a 300-L airlift bioreactor. *Biotechnology and Bioengineering* 34(4): 502–508. https://doi.org/10.1002/bit.260340410.

Reski, R., Bae, H. & Simonsen, H.T. 2018. Physcomitrella patens, a versatile synthetic biology chassis. *Plant Cell Reports* 37: 1409–1417. https://doi.org/10.1007/s00299-018-2293-6.

Ruffoni, B., Pistelli, L., Bertoli, A. & Pistelli, L. 2010. Plant cell cultures: Bioreactors for industrial production In: *Bio-farms for nutraceuticals: Functional food and safety control by biosensors*. Springer. pp. 203–221. https://doi.org/10.1007/978-1-4419-7347-4_15.

Sajc, L., Vunjak-Novakovic, G., Grubisic, D., Kovačević, N., Vuković, D. & Bugarski, B. 1995. Production of anthraquinones by immobilized *Frangula alnus* Mill. plant cells in a four-phase air-lift bioreactor. *Applied Microbiology and Biotechnology* 43: 416–423.

Schmidt, B. 2001. Metabolic profiling using plant cell suspension cultures. In: Hall, J.C., Hoagland, R.E. & Zablotowicz, R.M., editors. *Pesticide biotransformation in plants and microorganisms*. ACS Symp Ser 777. Washington, DC: American Chemical Society. pp. 40–56.

Schneider, N.F.Z., Cerella, C., Lee, J.Y., Mazumder, A., Kim, K.R., De Carvalho, A. *et al.* 2018. Cardiac glycoside glucoevatromonoside induces cancer type-specific cell death. *Frontiers in Pharmacology* 9: 70.

Shi, H.P. & Lindemann, P. 2006. Expression of recombinant *Digitalis lanata* EHRH. cardenolide 16′-O-glucohydrolase in *Cucumis sativus* L. hairy roots. *Plant Cell Reports* 25: 1193–1198.

Shim, K-M., Murthy, H.N., Park, S-Y., Rusli, I. & Paek, K.-Y. 2015. Production of biomass and bioactive compounds from cell suspension cultures of *Eurycoma longifolia* in balloon type bubble bioreactors. *Horticultural Science and Technology* 33: 251–258.

Shukla, A.A. & Gottschalk, U. 2013. Single-use disposable technologies for biopharmaceutical manufacturing. *Trends in Biotechnology* 31: 147–154. https://doi.org/10.1016/j.tibtech.2012.10.004.

Skrzypczak-Pietraszek, E., Szewczyk, A., Piekoszewska, A. & Ekiert, H. 2005. Biotransformation of hydroquinone to arbutin in plant in vitro cultures—Preliminary results. *Acta Physiologiae Plantarum* 27: 79–87.

Sommer, J., Schroeder, C. & Stöckigt, J. 1997. In vivo formation of vanillin glucoside. *Plant Cell Tissue and Organ Culture* 50:119–123.

Son, S.H., Choi, S.M., Lee, Y.H., Choi, K.B., Yun, S.R., Kim, J.K. *et al.* 2000. Large-scale growth and taxane production in cell cultures of *Taxus cuspidata* (Japanese Yew) using a novel bioreactor. *Plant Cell Reports* 19: 628–633.

Subroto, M.A., Mahagamasekera, M.G.P., Kwok, K.H., Hamill, J.D. & Doran, P.M. 1997. Co-culture of hairy roots and shooty teratomas. In: Doran, P.M., editor. *Hairy roots: Culture and applications*. Amsterdam: Harwood Academic. pp. 81–88.

Szopa, A., Kubica, P., Snoch, A. & Ekiert, H. 2018. High production of bioactive depsides in shoot and callus cultures of *Aronia arbutifolia* and *Aronia*× *prunifolia*. *Acta Physiologiae Plantarum* 40: 1–11.

Tabata, M., Umetani, Y., Shima, K. & Tanaka, S. 1984. Glucosylation of esculetin by plant cell suspension cultures. *Plant Cell Tissue and Organ Culture* 3: 3–9. https://doi.org/10.1007/BF00035916.

Takemoto, M. & Achiwa, K. 1998. Deracemization of racemic 4-pyridyl-1-ethanol by *Catharanthus roseus* cell cultures. *Phytochemistry* 49: 1627–1629. https://doi.org/10.1016/S0031-9422(98)00241-6.

Tanaka, H. 2000.Technological problems in cultivation of plant cells at high density. *Biotechnology and Bioengineering* 67: 775–790.

Teng, R.W., McManus, D., Aylward, J., Ogbourne, S.M., Armstrong, D., Mau, S-L. *et al.* 2009. Biotransformation of ingenol-3-angelate in four plant cell suspension cultures. *Biocatalysis and Biotransformation* 27(3): 186–194.

Tervasmäki, P., Latva-Kokko, M., Taskila, S. & Tanskanen, J. 2018. Effect of oxygen transfer on yeast growth—Growth kinetic and reactor model to estimate scale-up effects in bioreactors. *Food and Bioproducts Processing* 111: 129–140.

Thanh, N.T. Murthy, H.N. & Paek, K.Y. 2014. Optimization of ginseng cell culture in airlift bioreactors and developing the large-scale production system. *Industrial Crops and Products* 60, 343–348.

Theurer, C., Kreis, W. & Reinhard, E. 1998. Effects of digitoxigenin, digoxigenin, and various cardiac glycosides on cardenolide accumulation in shoot cultures of *Digitalislanata*. *Planta Medica* 64(08): 705–710.

Titova, M., Popova, E. & Nosov, A. 2024. Bioreactor systems for plant cell cultivation at the institute of plant physiology of the Russian academy of sciences: 50 years of technology evolution from laboratory to industrial implications. *Plants (Basel)* 13(3): 430. https://doi.org/10.3390/plants13030430. PMID: 38337964.

Ulbrich, B., Wiesner, W. & Arens, H. 1985. Large-scale production of rosmarinic acid from plant cell cultures of *Coleus blumei* Benth. In: *Primary and secondary metabolism of plant cell cultures*. Berlin/Heidelberg, Germany: Springer. pp. 293–303.

Van Aken, B. & Schnoor, J.L. 2002. Evidence of perchlorate (ClO4-) reduction in plant tissues (poplar tree) using radio-labeled 36ClO4. *Environmental Science and Technology* 36(12): 2783–2788.

Van Aken, B., Yoon, J.M., Just, C.L. & Schnoor, J.L. 2004. Metabolism and mineralization of hexahydro-1, 3, 5-trinitro-1, 3, 5-triazine inside poplar tissues (*Populus deltoides*× *nigra* DN-34). *Environmental Science & Technology* 38(17): 4572–4579.

Vaněk, T., Valterová, I., Vaňková, R. & Vaisar, T. 1999. Biotransformation of (−)-limonene using *Solanum aviculare* and *Dioscorea deltoidea* immobilized plant cells. *Biotechnology Letters* 21: 625–628.

Van Uden, W., Bouma, A.S., Bracht Waker, J.F., Middel, O., Wichers, H.J., de Waard, P. *et al.* 1995. The production of podophyllotoxin and its 5-methoxy derivative through bioconversion of cyclodextrin-complexed desoxypodophyllotoxin by plant cell cultures. *Plant Cell Tissue and Organ Culture* 42: 73–79. https://doi.org/10.1007/BF00037684.

Van Uden, W., Oeij, H., Woerdenbag, H.J. & Pras, N. 1993. Glucosylation of cyclodextrin-complexed podophyllotoxin by cell cultures of *Linum flavum* L. *Plant Cell Tissue and Organ Culture* 34: 169–175. https://doi.org/10.1007/BF00036098.

Van Uden, W., Pras, N., Vossebeld, E.M., Mol, J.N.M. & Malingre, T.M. 1990. Production of 5-methoxypodophyllotoxin in cell suspension cultures of *Linum flavum* L. *Plant Cell Tissue and Organ Culture* 20: 81–87. https://doi.org/10.1007/BF00114704.

Verdú-Navarro, F., Moreno-Cid, J.A., Weiss, J., & Egea-Cortines, M. 2023. The advent of plant cells in bioreactors. *Frontiers in Plant Science* 14: 1310405. https://doi.org/10.3389/fpls.2023.1310405.

Vila, M., Pascal-Lorber, S., Rathahao, E., Debrauwer, L., Canlet, C. & Laurent, F. 2005. Metabolism of [14C]-2, 4, 6-trinitrotoluene in tobacco cell suspension cultures. *Environmental Science & Technology* 39(2): 663–672.

Walton, N.J., Robins, R.J. & Rhodes, M.J.C. 1988. Perturbation of alkaloid production by cadaverine in hairy root cultures of Nicotiana rustica. *Plant Science* 54: 125.

Wang, D., Khan, M.S., Cui, L., Song, X., Zhu, H., Ma, T. *et al.* 2019. A novel method for the highly efficient biotransformation of genistein from genistin using a high-speed counter-current chromatography bioreactor. *RSC Advances* 9(9): 4892–4899. https://doi.org/10.1039/c8ra10629k.

Wichers, H.J., Malingré, T.M. & Huizing, H.J. 1983. The effect of some environmental factors on the production of L-DOPA by alginate-entrapped cells of *Mucuna pruriens*. *Planta* 158(6): 482–486.

Xia, J., Wang, G., Lin, J., Wang, Y., Chu, J., Zhuang, Y. *et al.* 2015. Advances and practices of bioprocess scale-up. In: *Bioreactor engineering research and industrial applications II*. Berlin/Heidelberg, Germany: Springer. pp. 137–151.

Xu, J., Ge, X. & Dolan, M.C. 2011. Towards high-yield production of pharmaceutical proteins with plantcell suspension cultures. *Biotechnology Advances* 9: 278–299. https://doi.org/10.1016/j.biotechadv.01.002.

Yokoyama, M. 1996. 4 Industrial application of biotransformations using plant cell cultures. *Plant Cell Culture Secondary Metabolism Towards Industrial Application*, 79.

Zaragoza-Martínez, F., Lucho-Constantino, G.G., Ponce-Noyola, T. Esparza-García, F., Poggi-Varaldo, H., Cerda-García-Rojas, C. *et al.* 2016. Jasmonic acid stimulates the oxidative responses and triterpene production in *Jatropha curcas* cell suspension cultures through mevalonate as biosynthetic precursor. *Plant Cell Tissue and Organ Culture* 127: 47–56.

Zhong, J-J., Pan, Z-W., Wang, Z-Y., Wu, J., Chen, F., Takagi, M. *et al.* 2002.Effect of mixing time on taxoid production using suspension cultures of *Taxus chinensis* in a centrifugal impeller bioreactor. *Journal of Bioscience and Bioengineering* 94: 244–250.

Zhu, W. & Lockwood, G.B. 2000. Enhanced biotransformation of terpenes in plant cell suspensions using controlled release polymer. *Biotechnology Letters* 22: 659–662.

12 Biotransformation
Bioreactor Strategies for Scaling Plant Cell Cultures and Metabolic Stability

*Devyani Rajput, Rahul Singh Thakur, Akanksha Dwivedi, G. N. Darwhekar, and Rakhi Khabiya**
* Corresponding author: rakhikhabiya@gmail.com

12.1 INTRODUCTION

In the realm of modern biotechnology, the utilization of plant cells for the production of valuable compounds has gained substantial momentum. Plant cells, with their inherent ability to synthesize a diverse array of secondary metabolites and bioactive compounds, have emerged as a promising resource for various industries, including pharmaceuticals, cosmetics, and agriculture (Chen & Hu, 2006). The process of harnessing the biosynthetic potential of these cells on a larger scale necessitates a profound understanding of their growth dynamics and metabolic pathways, coupled with the development of efficient cultivation strategies (Wendt et al., 2009).

This chapter delves into the crucial domain of biotransformation, focusing on the cultivation of plant cells in bioreactors at both pilot and large scales. Biotransformation, encompassing the conversion of raw materials by biological systems into valuable products, stands as a cornerstone in the production of pharmaceuticals, nutraceuticals, and other high-value compounds. At the heart of this transformation lies the cultivation of plant cells, which demands the meticulous orchestration of environmental parameters, nutrient provision, and reactor design (Hansmann et al., 2013).

In the initial sections of this chapter, we embark on a journey through the fundamental principles of biotransformation and its significance in modern bioprocessing. As we navigate the intricate pathways of plant cell metabolism, we uncover the diverse range of compounds that can be synthesized, encompassing alkaloids, flavonoids, terpenoids, and more. These compounds hold immense therapeutic and commercial potential, motivating researchers and industries to explore innovative approaches for their production (Martin et al., 2004).

The spotlight then shifts to bioreactors, the pivotal platforms where the intricate dance of plant cell cultivation unfolds. A bioreactor's design, configuration, and operation significantly impact the yield, quality, and scalability of the biotransformation process (Lawrence et al., 2009). The transition from laboratory-scale experiments to pilot- and large-scale cultures poses unique challenges, including mass transfer limitations, shear stress, and optimal resource utilization. This chapter elucidates these challenges while presenting state-of-the-art strategies and technologies that have been developed to overcome them (Plunkett & O'Brien, 2011).

A core theme of this chapter revolves around the retention of biosynthetic potential in plant cell cultures. As cells transition from the controlled environment of a laboratory flask to the dynamic conditions of a bioreactor, maintaining their ability to produce target compounds becomes

 DOI: 10.1201/9781003326939-14

paramount (Eibl & Eibl, 2009). Strategies to enhance and sustain secondary metabolite synthesis are discussed, including elicitation techniques, genetic engineering, and innovative bioreactor configurations. These approaches not only safeguard the biosynthetic capacity of plant cells but also open avenues for the production of tailored compounds with improved yields and qualities (Castro et al., 2020).

In conclusion, the chapter "Biotransformation: Bioreactor Strategies for Scaling Plant Cell Cultures and Metabolic Stability" embarks on a comprehensive exploration of the intricate interplay between plant cell biology and bioprocessing engineering. It underscores the evolving landscape of biotechnology, where the fusion of biological insights and engineering innovations paves the way for the sustainable and economically viable production of bioactive compounds. As we venture deeper into this chapter, we unravel the strategies and techniques that propel us closer to unlocking the full potential of plant cells as miniature bioreactors for a myriad applications (Gelinsky et al., 2015).

12.2 FUNDAMENTALS OF PLANT CELL CULTURES

Plant cell culture, a branch of biotechnology, involves the propagation and manipulation of plant cells in controlled environments outside their natural context. This technique provides valuable insights into plant biology, enables the production of secondary metabolites, and facilitates genetic modification, as mentioned in Table 12.1 (Reinert & Bajaj, 2013). Here are the key fundamentals of plant cell culture:

1. **Cell types and sources.** Plant cells can be derived from various tissues, including leaves, stems, roots, and embryos. Explants (small tissue sections) are often used to initiate cultures. Cells can be totipotent—meaning, they have the ability to regenerate into a whole plant under suitable conditions (Shahzad et al., 2017).
2. **Culture medium**. A nutrient-rich culture medium is crucial for cell growth and differentiation. It typically contains macronutrients (nitrogen, phosphorus, potassium), micronutrients (iron, zinc, copper), vitamins, amino acids, and growth regulators (auxins, cytokinins) (Gamborg et al., 1976).
3. **Sterile techniques.** Since contamination can easily affect cultures, sterile techniques are essential. Properly sterilized equipment, culture vessels, and media are used to minimize unwanted microbial growth (Evans et al., 2003).
4. **Initiation and maintenance.** Explants are introduced into the culture medium for initiation. Cells multiply and form a callus tissue. Regular subculturing involves transferring a portion of the growing culture to a fresh medium to maintain active growth (Shahzad et al., 2017).
5. **Callus formation and regeneration.** *Callus* is a mass of undifferentiated cells. By altering the concentration of growth regulators in the medium, callus can be induced to differentiate into roots, shoots, and ultimately, whole plants (Reinert & Bajaj, 2013).
6. **Hormone regulation.** Auxins and cytokinins are essential plant hormones that regulate cell growth and differentiation. Their balance in the medium influences the type of tissue that develops (Long et al., 2022).
7. **Embryogenesis and organogenesis.** Manipulating hormone concentrations can induce embryogenesis (formation of embryos) or organogenesis (formation of organs, such as roots or shoots) from callus tissue (Reinert & Bajaj, 2013).
8. **Somatic hybridization and genetic engineering.** Plant cell cultures are used to develop somatic hybrids by fusing cells from different species. Genetic engineering techniques can also introduce foreign genes into plant cells to produce genetically modified plants (Schum et al., 2001).

9. **Secondary metabolite production.** Plant cells can synthesize various secondary metabolites, such as alkaloids, flavonoids, and terpenoids. By optimizing culture conditions, these metabolites can be produced in higher quantities (Constabel, 1988).
10. **Bioreactors.** For large-scale production, bioreactors provide controlled environments to cultivate plant cells. They regulate factors like temperature, pH, aeration, and nutrient supply (Curtis, 2005).
11. **Applications.** Plant cell culture finds application in plant breeding, conservation, pharmaceuticals, and agriculture. It enables the production of disease-resistant plants, the preservation of endangered species, and the large-scale synthesis of bioactive compounds (Imani et al., 2009).
12. **Challenges.** Plant cell culture can be challenging due to the specific requirements of different plant species, the risk of genetic instability during long-term cultivation, and the difficulties in scaling up cultures to commercial levels (Gupta & Ibaraki, 2006).

Plant cell culture has revolutionized the field of plant biotechnology, offering a platform to study plant growth and development, create genetically modified plants, and produce valuable compounds for various industries. As techniques and technologies continue to advance, the applications and potential of plant cell culture continue to expand.

TABLE 12.1
Advantages of Using Plant Cells for Biotransformation

1. **Biosynthetic potential.** Plant cells possess a diverse array of biosynthetic pathways capable of producing a wide range of secondary metabolites, including pharmaceuticals, flavors, fragrances, and nutraceuticals.
2. **Controlled environment.** Plant cell cultures can be maintained in controlled environments, allowing for the precise manipulation of growth conditions to enhance metabolite production.
3. **Sustainability.** Plant cell cultures offer a renewable and sustainable source of bioactive compounds, reducing the need for resource-intensive plant cultivation and harvest.
4. **Faster production.** Compared to whole plants, plant cell cultures often exhibit faster growth and higher metabolite production rates, leading to quicker product yields.
5. **Genetic modification.** Plant cells can be genetically engineered to enhance the expression of specific biosynthetic pathways or introduce novel pathways, resulting in the higher yields of desired compounds.
6. **Elimination of seasonal variation.** In contrast to plants that are subject to seasonal changes, plant cell cultures provide a consistent and year-round source of bioactive compounds.
7. **Smaller space requirements.** Plant cell cultures require less physical space than traditional agriculture, making them suitable for urban environments or areas with limited arable land.
8. **Reduction of pest and disease concerns.** Growing plant cells in sterile conditions minimizes the risk of pests, diseases, and contamination, ensuring higher product quality.
9. **Preservation of endangered species.** Plant cell cultures offer a means to propagate and preserve endangered plant species without depleting wild populations.
10. **Enhanced compound purity.** Controlled growth conditions in bioreactors can lead to improved purity and consistency of produced compounds, important for pharmaceutical and high-value applications.
11. **Scalability.** Plant cell cultures can be scaled up in bioreactors to meet commercial production demands, offering a feasible route for large-scale metabolite synthesis.
12. **Exploration of non-cultivable species.** Plant cell cultures enable the production of compounds from plant species that are difficult to cultivate or rare in nature.
13. **Tailored compound production.** By manipulating culture conditions and genetics, it is possible to optimize the production for specific compounds, increasing yields and quality.

Source: Ishihara et al. (2003).

12.3 BIOREACTOR SYSTEMS AND DESIGN

In the realm of biotechnology, the efficient cultivation of microorganisms, plant cells, and animal cells is essential for the production of diverse bioproducts. Bioreactors stand at the heart of this endeavor, providing controlled environments where biological systems can flourish and transform raw materials into valuable compounds. The design and operation of bioreactor systems play a pivotal role in optimizing growth, enhancing productivity, and ensuring the desired quality of the final products (Asenjo, 1994). Bioreactors are essential in tissue engineering, not only because they provide an in vitro environment mimicking in vivo conditions for the growth of tissue substitutes, but also because they enable systematic studies of the responses of living tissues to various mechanical and biochemical cues. The basic principles of bioreactor design are reviewed, shown in Figure 12.1 and 12.2, and the bioreactors commonly used for the tissue engineering of cartilage, bone, and cardiovascular systems are assessed in terms of their performance and usefulness (Pörtner et al., 2005). Each type of bioreactor offers distinct advantages and is suitable for specific plant cell culture applications, biosynthetic pathways, and metabolite production, as summarized in Table 12.2. The choice depends on factors such as culture requirements, shear sensitivity, scalability, and desired metabolite production. Table 12.3 mentions several types of bioreactors that can be used for plant cell cultures, each with its own advantages and suitability for different applications (Margaritis & Wallace, 1984).

12.3.1 Parameters Influencing Bioreactor Design

Bioreactor design is influenced by various parameters (Table 12.4). These include temperature, traces of chemical constituents, dissolved oxygen, pH, and the need for a cooling mechanism to maintain appropriate temperature conditions during fermentation. The surrounding parameters associated with the bioreactor, such as temperature, dissolved oxygen, and pH, have an impact on

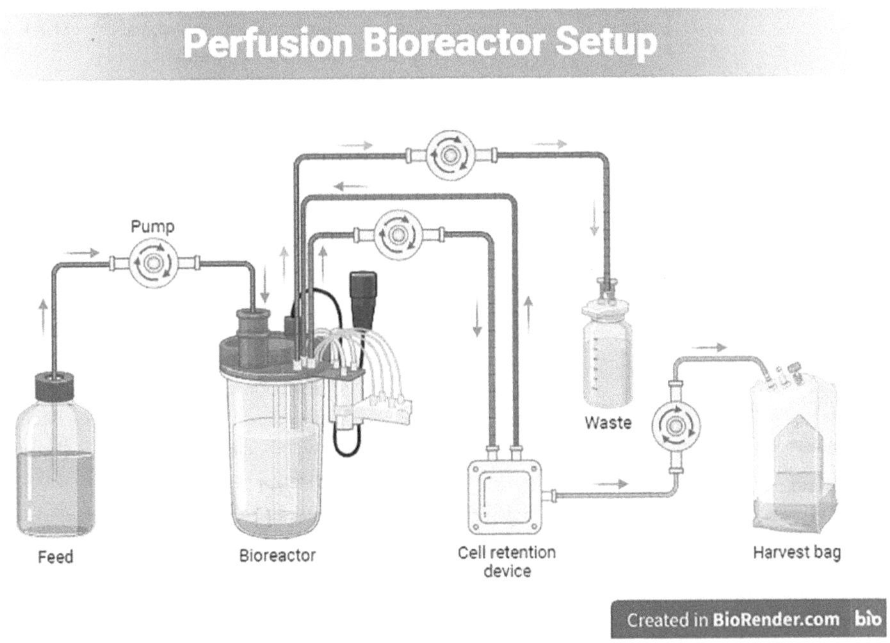

FIGURE 12.1 Stirred tank reactor design.

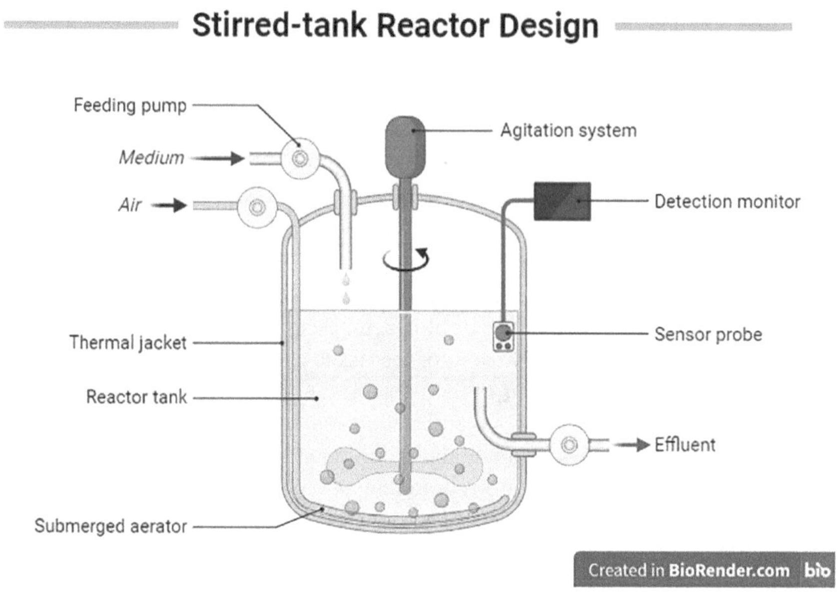

FIGURE 12.2 Perfusion bioreactor setup.

TABLE 12.2
Biosynthetic Pathways and Metabolite Production

Biosynthetic Pathway	Metabolite Produced	Organism/ System	Application/Importance	Ref.
Photosynthesis	Glucose	Plants, algae	Primary energy source, carbon fixation	(Lai & Lan, 2015)
Glycolysis	ATP, pyruvate	Most organisms	Energy production, precursor for various pathways	(Karki et al., 2013)
Citric acid cycle (Krebs cycle)	ATP, NADH, FADH2	Aerobic organisms	Energy generation, precursor for electron transport chain	(Akram, 2014)
Fatty acid biosynthesis	Fatty acids	Plants, animals	Lipid production, energy storage	(Macfarlane & Macfarlane, 2003)
Terpenoid biosynthesis	Terpenes	Plants, microbes	Production of essential oils, pharmaceuticals	(Bian et al., 2017)
Amino acid biosynthesis	Various amino acids	All organisms	Protein synthesis, precursor for numerous molecules	(Park & Lee, 2008)
Polyketide biosynthesis	Polyketides	Microbes, plants	Antibiotics, antifungals, anticancer drugs	(Li et al., 2012)
Flavonoid biosynthesis	Flavonoids	Plants	Pigmentation, antioxidants, plant defense	(Verpoorte & Memelink, 2002)
Secondary metabolite biosynthesis	Various compounds	Microbes, plants	Antibiotics, alkaloids, toxins, pigments	(Verpoorte et al., 2000)

the growth and productivity of microorganisms or cells. The design principles of bioreactors aim to mimic the native tissue environment in tissue engineering and regenerative medicine. In enzymatic hydrolysis of lignocellulose, the bioreactor plays a crucial role as the reaction holder and the place for microorganism growth. The amount of light intensity received by individual algal cells is a key

TABLE 12.3
Some Common Types of Bioreactors Used for Plant Cell Cultures

Type of Bioreactor	Description	Advantages	Suitability	Ref.
Stirred tank bioreactors	Agitated tanks with impellers for mixing and aeration.	• Uniform mixing and oxygenation • Scalable for large-scale production	• General plant cell cultures • High-density cultures	(Reuss, 1994)
Bubble column bioreactors	Relies on rising gas bubbles for mixing and aeration.	• Simple design • Suitable for low shear-sensitive cultures	• Low-shear cultures • Cultures requiring less mechanical agitation	(Pérez et al., 2006)
Airlift bioreactors	Utilizes rising gas bubbles and a draft tube for circulation.	• Gentle mixing • Suitable for shear-sensitive cultures	• Shear-sensitive cultures	(Merchuk, 2003)
Wave bioreactors	Rocking motion creates wavelike agitation.	• Gentle agitation • Suitable for adherent cultures and microcarriers	• Adherent cultures • Microcarrier cultures	(Eibl & Eibl, 2008)
Rotary drum bioreactors	Rotating drum exposes cells to alternating aeration and submersion.	• Intermittent aeration • Suitable for cultures needing air exposure	• Cultures needing intermittent air exposure	(Alam et al., 2009)
Perfusion bioreactors	Continuous addition and removal of culture medium.	• Steady-state conditions • Enhanced cell growth	• Long-term cultures • High-density growth	(Gaspar et al., 2012)
Single-use bioreactors	Designed for one-time use, reducing cross-contamination risks.	• Convenient setup • Reduced cleaning and sterilization	• Various plant cell cultures	(Jossen et al., 2019)
Solid-state bioreactors	Cells grow on a solid support or substrate.	• Suitable for immobilized cultures • Mimics natural conditions	• Immobilized cell cultures • Soil-mimicking cultures	(Ashok et al., 2017)
Temporary immersion bioreactors	Cell aggregates are immersed and submerged at intervals.	• Enhanced gas exchange • Minimized shear forces	• Suspension cultures • Minimized shear stress cultures	(Afreen, 2008)

TABLE 12.4
Different Aspects of a Bioreactor

Aspect	Oxygenation Management	Shear Stress Management
Aeration strategies	• Sparging (gas bubbling). • Diffused aeration. • Membrane aeration.	• Appropriate agitation methods. • Gentle, fluid movement designs.
Oxygen delivery rate	• Adjust based on cell needs. • Monitor uptake rates.	• Choose agitation intensity carefully.
Oxygen sensors	• Integrate real-time sensors. • Monitor dissolved oxygen levels.	• Monitor power input and impeller torque.
Mixing and agitation	• Promote uniform oxygen distribution. • Prevent oxygen gradients.	• Ensure efficient mixing without excessive shear.
Oxygen transfer efficiency	• Optimize impeller design and size. • Adjust agitation rate.	• Consider low-shear impeller designs.
Oxygen limitation prevention	• Avoid oxygen depletion. • Adjust aeration rates as needed.	• Recognize shear-sensitive growth stages.
Shear stress effects	• Note how shear stress damages cells and aggregates.	• Note how high shear affects cell viability and metabolite production.
Agitation intensity	• Balance mixing needs with shear stress risks.	• Avoid abrupt changes in agitation.
Impeller design	• Choose designs that minimize shear.	• Use gentle, fluid movement methods.
Sparger design	• Select spargers with minimal bubble-induced shear.	• Choose designs compatible with shear-sensitive cultures.
Bioreactor geometry	• Opt for even shear distribution. • Minimize localized high-stress areas.	• Consider shear effects during cell division.
Monitoring shear	• Remember that direct measurement is challenging. • Monitor power input and impeller torque.	• Consider indirect indicators of shear stress.
Controlled elicitation	• Choose compatible elicitation methods. • Avoid additional shear stress.	• Plan elicitation to minimize shear effects.

Source: Pörtner et al. (2005).

factor in optimizing microalgae production in photobioreactors. Scale-up and scale-down techniques are important considerations in bioreactor design to ensure efficient and controlled operations (Ntwampe et al., 2010; Vasumathi et al., 2012).

12.4 SCALE-UP STRATEGIES FOR PLANT CELL CULTURES

Plant cell cultures have been explored as a potential alternative source for the production of pharmaceutical agents. However, the scale-up of plant cell cultures for large-scale production is still limited. Strategies to increase product yields from medicinal plant cell suspension cultures (MPCSC) are being investigated, including the optimization of reactor systems and the use of environmental factors to enhance productivity. The establishment of high-yielding cell cultures and the use of immobilized plant cells are also being explored as methods to improve productivity. The challenges associated with scaling up plant cell cultures, such as the need for morphological differentiation and the influence of environmental factors, are being addressed through process-based strategies (Taticek et al., 1991). The development of large-scale fermentation processes for plant cell cultures has been realized, with the aim of commercial production of biochemicals. The definition of large-scale culture varies, but it is clear that both commercial and process development aims can be achieved with different bioreactor sizes (Appelhagen et al., 2018).

12.4.1 Challenges in Transitioning from Laboratory to Pilot and Large Scales

Transitioning from laboratory-scale plant cell cultures to pilot and large scales presents several challenges that must be addressed to ensure a successful and consistent production. Here are some of the key challenges:

1. **Scale effects.** Physiological and metabolic differences between small laboratory cultures and larger-scale cultures can lead to variations in growth, metabolism, and product yield (Kwade et al., 2018).
2. **Mixing and mass transfer.** Achieving uniform mixing and efficient mass transfer becomes more challenging as culture volumes increase, potentially leading to nutrient and oxygen gradients within the bioreactor (Zondag et al., 2023).
3. **Oxygen transfer.** Larger cultures require more efficient oxygen transfer. Ensuring adequate oxygen levels throughout the culture is essential for maintaining growth and productivity (Hirsch et al., 2018).
4. **Shear stress.** The shear forces generated during mixing can cause cell damage, affecting cell viability and product quality. Balancing shear stress is crucial (Yadav et al., 2023).
5. **Contamination control.** The risk of contamination increases with larger volumes. Developing stringent aseptic techniques and maintaining sterility become critical (Huesemann et al., 2022).
6. **Nutrient heterogeneity.** In larger bioreactors, achieving uniform nutrient distribution becomes challenging, potentially leading to uneven growth and metabolite production.
7. **Inoculum preparation.** Scaling up the inoculum from laboratory to pilot or large scale requires maintaining consistency and reproducibility, which can be complex.
8. **Process monitoring and control.** Real-time monitoring and control become more demanding at larger scales. Advanced sensing and automation systems are needed to ensure optimal conditions.
9. **Equipment compatibility.** Transitioning to larger equipment may require adjustments to equipment design, materials, and configurations to accommodate increased volumes.
10. **Feeding strategies.** Developing appropriate feeding strategies for nutrients, growth regulators, and inducers becomes crucial to sustain cell growth and maintain metabolite production.
11. **Kinetics and lag phase.** The kinetics of cell growth and metabolite production can change with scale. Longer lag phases and altered growth rates may impact overall process dynamics.
12. **Cost considerations.** Scaling up incurs higher operational and resource costs, necessitating careful budgeting and cost optimization strategies.
13. **Process validation.** Ensuring that the scaled-up process consistently meets quality and yield standards requires thorough validation and testing.
14. **Regulatory compliance.** As production scales increase, compliance with regulatory requirements becomes more stringent, requiring meticulous documentation and adherence to guidelines.
15. **Environmental impact.** Larger-scale production may have greater environmental implications, necessitating sustainable practices and waste management strategies.
16. **Learning curves and expertise.** Scaling up requires specialized knowledge in bioprocess engineering, bioreactor design, and large-scale operations. Accessing this expertise is crucial.

Addressing these challenges requires a combination of scientific understanding, engineering expertise, and practical experience. Pilot studies that bridge the gap between laboratory- and large-scale production can help identify and mitigate potential issues before full-scale implementation. Collaboration between biologists, engineers, and process experts is essential for a successful transition and the sustained production of desired plant cell culture products (Schliwa, 2013).

12.4.2 Mass Transfer Limitations and Nutrient Requirements

Mass transfer limitations and nutrient requirements are important considerations in plant cell culture. Plant cells are shear-sensitive and require adequate mixing and homogenous conditions in the reactor to achieve optimal growth and product formation. The sticky nature of late exponential phase cells due to excretion of polysaccharides can hinder nutrient and oxygen diffusion to the cells, but this can be minimized by optimizing the concentration of calcium ions or adding pectinase/cellulase (Yeo et al., 2015). In addition, the use of nutrient supplements and a medium cycling strategy shows promise in increasing and sustaining product formation in plant cell culture. The development of novel membrane reactors with double mass transfer has also been explored to improve nutrient supply and remove depleted medium in cell culture. Overall, understanding and addressing mass transfer limitations and nutrient requirements are crucial for optimizing plant cell culture processes (Bosma et al., 1996).

12.4.3 Oxygenation and Shear Stress Management

Oxygenation and shear stress management are important factors in plant tissue culture. The requirement for sterility and avoidance of dehydration in tissue cultures can limit gas exchange, primarily driven by diffusion. Factors such as static liquid media, gelling agents, large tissue mass, and warm temperatures impose severe limits on diffusive flux for gases like O_2, CO_2, and ethylene. The wall of the enclosing vessel, typically made of a gas-impermeable material, is the main barrier to diffusive gas exchange. Shear stress can affect culture viability, cell lysis, and the accumulation of secondary metabolites in plant cell suspension cultures. Increasing shear rates can lead to increased cell death, lysis, and phenolics secretion. The amount of phenolics secreted is directly related to the intensity of shear, possibly due to increased cell permeabilization. Agitation and aeration are crucial for product formation in plant tissue cultures, and the magnitude and distribution of shear stress, as well as the control of gas exchange, are essential for reactor design and scaling-up procedures (Kalbhenn et al., 2018; Peterson, 1999).

12.5 MEDIA FORMULATION AND NUTRIENT OPTIMIZATION

Media formulation and nutrient optimization in plant tissue culture is a critical aspect of efficient development and optimization of new media. Different algorithms, such as gene expression programming (GEP) and M5' model tree, have been used to predict the effects of media components on various growth parameters. Multi-objective evolutionary optimization algorithms, like genetic algorithm (GA) and particle swarm optimization (PSO), have been compared to mono-objective GA optimization technique for precise multi-optimization. Response surface methodology has been used to screen different media factors and determine optimal media formulations for upregulating the accumulation of cartilaginous extracellular matrix (ECM) constituents in chondrocytes (Jayme, 1991). Neuro-fuzzy logic technology has been employed to uncover and integrate knowledge hidden in multiple data from plant tissue culture media formulations, providing a useful approach for media optimization. Macronutrient factors, such as media volume and the amount of tissue, sucrose, nitrogen, and potassium, have been simultaneously altered to optimize tissue culture media and improve plantlet quality (Galbraith et al., 2018).

12.5.1 Designing Culture Media for Large-Scale Cultures

Designing culture media for large-scale cultures involves several considerations. One approach is to develop a chemically defined media that optimizes the growth of specific organisms, such as *Agrobacterium tumefaciens*, for the production of recombinant proteins in plants. Another method is to create a large-scale cell culture system that provides a stable environment for cell growth. This system includes an incubator with supports for cell culture, a medium supply part, and a pump for circulating the medium (Fowler, 1984). Additionally, systems with vessels and connection tools can be used to transfer cell liquid between vessels, disrupting cell aggregates and facilitating cell growth. Methods for preparing high-precision culture media involve accurately weighing chemicals and mixing them according to specific quantity demands, resulting in a medium with high precision and small errors between batches. These approaches offer convenience, quickness, and simplicity in large-scale culture media preparation, making them suitable for application in plant tissue culture and biotechnology (Chen et al., 2022).

12.5.2 Nutrient Supplementation and Metabolic Regulation

Nutrient supplementation and metabolic regulation are important factors in cell culture. Glucose, as a key nutrient, has a significant impact on bioprocesses and affects productivity and product quality. Hydrolysates, such as cottonseed hydrolysates, can be used as media supplements to enhance cell growth, immunoglobulin titers, and productivities. Different carbon sources, including gluconate, glycerol, and acetate, can also affect metabolic regulation in *Escherichia coli* batch cultures. Controlled nutrient feeding can reduce metabolite production and alter cell metabolism, leading to higher cell concentrations in continuous cultures. Regulating nutrient levels, such as glucose, can improve cell yields in cell culture, as demonstrated in the case of spheroids formed from the β-TC6 insulinoma cell line. Overall, understanding and optimizing nutrient supplementation and metabolic regulation are crucial for enhancing cell culture productivity and product quality (Albaugh et al., 2017).

12.6 GENETIC AND EPIGENETIC STABILITY

Genetic and epigenetic stability are crucial considerations in cell culture, as they directly impact the reliability and reproducibility of experimental results and the safety of cell-based therapies (Lund et al., 2012; Mohn & Schübeler, 2009). Here, we will explore these two aspects in more detail:

Genetic stability. *Genetic stability* refers to the preservation of an organism's or cell's genetic information over time and through cell division. In cell culture, maintaining genetic stability is essential to ensure that the cells used in experiments or therapies remain consistent and free from undesirable mutations. Several factors influence genetic stability in cell culture:

- **Avoiding contamination.** Contamination by other cell lines or microorganisms can introduce genetic variations. Strict aseptic techniques and regular testing for contaminants are essential.
- **Monitoring cell line identity.** Regularly verifying the identity of cell lines through techniques like DNA profiling (short tandem repeat analysis) helps prevent cross-contamination and misidentification.
- **Minimizing passages.** Frequent passages of cells can increase the risk of genetic mutations. Keeping passages to a minimum and using low-passage cells whenever possible are advisable.
- **Using frozen stocks.** Cryopreserving cell lines as frozen stocks can help maintain their genetic stability over extended periods.

Epigenetic stability. *Epigenetics* involves changes in gene expression patterns that do not involve alterations to the underlying DNA sequence. Epigenetic stability is crucial in cell culture because it can affect cell behavior, differentiation, and response to stimuli. Factors that influence epigenetic stability include:

Cell culture conditions. The choice of culture media, growth factors, and supplements can influence epigenetic modifications. Maintaining consistent culture conditions is essential for stability.

Cell passage effects. Repeated passages can lead to changes in epigenetic marks. Techniques like single-cell cloning can help preserve epigenetic stability.

Differentiation protocols. When differentiating cells in culture, specific protocols must be followed to ensure that epigenetic marks associated with cell identity are retained.

Epigenetic modifiers. Researchers may use epigenetic modifiers (e.g., DNA methyltransferase inhibitors or histone deacetylase inhibitors) to manipulate cell behavior, but these should be applied with care to avoid unintended consequences.

Epigenetic analysis. Regularly assessing the epigenetic status of cultured cells through techniques like DNA methylation analysis or chromatin immunoprecipitation can help detect and address any changes.

Maintaining genetic and epigenetic stability in cell culture is a complex and ongoing process that requires careful handling, monitoring, and adherence to best practices. Ensuring stability not only contributes to the scientific rigor of research but also holds great importance in the development of safe and effective cell-based therapies.

12.6.1 Maintaining Biosynthetic Potential over Generations

Microbes maintain their biosynthetic potential over generations through various mechanisms. One mechanism is the replication of genetic elements, such as plasmids, which allow the microbes to maintain their genomes extrachromosomally. Another mechanism involves the clustering of viral genomes, which enables rapid increase in copy numbers of genomes per cell. Additionally, the microbiome plays a crucial role in maintaining metabolic pathways that have been present in ancestors for millennia. The disruption of the microbiome can lead to the extinction of key taxa and deprive individuals of essential synthesis and toxin clearance processes. Understanding the transcriptional mechanisms that regulate biosynthesis and transport of essential nutrients, such as biotin, can also contribute to the maintenance of biosynthetic potential in bacteria. These mechanisms ensure the survival and metabolic adaptation of microbes over generations (Johnson & Hill, 2013; Ongley et al., 2013).

12.6.2 Influence of Subculture Methods on Stability

Subculture methods play a significant role in determining the genetic and epigenetic stability of cells in a culture. The choice of subculture techniques and frequency can have both direct and indirect influences on the stability of cultured cells. Here is how subculture methods can impact stability:

1. **Direct impact on genetic and epigenetic stability (Mohn & Schübeler, 2009).**
    - **Passage number.** The frequency of subculture, often measured in passage numbers, can directly affect genetic and epigenetic stability. Frequent passaging increases the risk of accumulating genetic mutations and can lead to changes in epigenetic marks.
    - **Cell density and split ratios.** Maintaining appropriate cell densities and split ratios during subculture is crucial. Overcrowding or underseeding can stress cells and affect their genetic and epigenetic stability.
    - **Subculture techniques.** The method used for subculture, such as trypsinization, scraping, or enzymatic dissociation, can impact cell viability and stress levels, potentially influencing stability.

2. **Indirect impact on genetic and epigenetic stability (Thorstensen et al., 2005).**
    - **Selection pressures.** The process of subculturing can inadvertently select for specific cell subpopulations with altered genetic or epigenetic profiles. This can result in clonal drift, where a particular cell population with different characteristics becomes dominant over time.
    - **Culturing conditions.** The conditions provided during subculture, including changes in culture media, growth factors, and supplements, can affect epigenetic modifications and gene expression patterns.
    - **Cell line cross-contamination.** Improper subculture techniques or insufficient quality control measures can lead to cross-contamination with other cell lines, introducing foreign genetic and epigenetic elements into the culture.

To mitigate the negative impact of subculture methods on genetic and epigenetic stability, it is essential to (Sadikovic et al., 2008):

- **Use low-passage cells.** Whenever possible, start experiments or new cultures with low-passage cells to minimize the risk of accumulated mutations or epigenetic changes.
- **Regularly authenticate cell lines.** Verify the identity of cell lines through DNA profiling or other authentication methods to prevent cross-contamination.
- **Optimize subculture techniques.** Choose subculture methods that are gentle on the cells to minimize stress and maintain stability. Techniques like single-cell cloning or limiting dilution can be employed to ensure uniformity.
- **Monitor cell behavior.** Continuously monitor cell behavior, morphology, and growth rates. Any unusual changes may signal genetic or epigenetic instability.
- **Document subculture protocols.** Maintain detailed records of subculture protocols, including passage numbers, split ratios, and culture conditions, to ensure consistency.

12.6.3 Epigenetic Modifications and Their Implications

Epigenetic modifications play a crucial role in various physiological processes and diseases, including cancer. They involve alterations in DNA methylation, histone modifications, and non-coding RNA regulation. These modifications can regulate gene expression and contribute to the development and progression of diseases, such as diabetes and cancer. In cancer, epigenetic modifications can trigger tumor formation by altering host epigenetic patterns. Understanding the role of epigenetic mechanisms in disease development can lead to the identification of novel biomarkers and therapeutic targets. Additionally, epigenetic therapies have the potential to prevent or reverse drug resistance in ovarian cancer. Furthermore, epigenetic modifications are being explored as therapeutic targets for treating diabetes and its complications. Overall, epigenetic modifications have significant implications in disease pathogenesis and offer opportunities for developing new treatment strategies (Jiménez-Chillarón et al., 2012).

12.7 ELICITATION TECHNIQUES FOR ENHANCED BIOSYNTHESIS

The most effective elicitation techniques for enhanced biosynthesis include priming with reactive oxygen species (ROS) prior to elicitation with *Rhizopus oligosporus/oryzae* (R). Another effective technique is the application of small-molecule elicitors to specifically affect transcription of secondary metabolite gene clusters. In the case of *Centella asiatica*, treatment with methyl jasmonate (MeJA) was found to enhance the biosynthesis of triterpenoids. In suspension cultures of *Panax ginseng*, the addition of vanadate was demonstrated to be the most efficient elicitor for ginsenoside biosynthesis. Chitosan supplementation in cell cultures of *Orthosiphon stamineus* also enhanced the antioxidant activity and increased the total phenolic content. These findings suggest

TABLE 12.5
Elicitors and Their Modes of Action

Elicitor	Source/Type	Mode of Action
Salicylic acid	Plant-derived	Activates defense genes, inducing systemic acquired resistance (SAR)
Jasmonic acid	Plant-derived	Induces the expression of defense-related genes and triggers the jasmonate pathway
Ethylene	Plant-derived	Regulates plant responses to stress, including pathogen defense
Yeast extract	Microbial-derived	Contains cell wall components that stimulate microbial responses
Chitosan	Derived from chitin	Activates plant defenses and induces the production of phytoalexins
Methyl jasmonate	Synthetic analog	Mimics the action of jasmonic acid, inducing defense responses
Elicitin proteins	Microbial-derived	Induces hypersensitive responses in plants, leading to defense mechanisms
Beta-glucans	Microbial-derived	Activate immune responses in animals and promote defense mechanisms
Lipopolysaccharides	Microbial-derived	Trigger immune responses in animals and induce the production of cytokines
Flagellin	Microbial-derived	Recognized by plant receptors, leading to immune responses
Polyinosinic–Polycytidylic Acid (Poly I:C)	Synthetic analog	Stimulates antiviral responses in animals by mimicking viral RNA

Source: Yoshikawa et al. (1993).

that ROS priming, small-molecule elicitors, MeJA, vanadate, and chitosan are effective techniques for enhancing biosynthesis in different plant species (Sivanandhan et al., 2014).

Elicitation techniques can be used to improve the efficiency of biosynthesis processes (Table 12.5). Priming with reactive oxygen species (ROS) prior to elicitation with *Rhizopus oligosporus/oryzae* (R) increased glyceollin production in soybean seedlings by more than fourfold compared to elicitation without priming. Salicylic acid treatment enhanced biomass production and alkaloid accumulation in *Arthrospira platensis*, a microalga with medicinal properties. Treatment with precursors and elicitors such as methyl jasmonate (MeJA) and salicylic acid (SA) promoted the biosynthesis of triterpenoids in *Centella asiatica* hairy root cultures, resulting in rapid biomass accumulation and increased triterpenoid production. Jasmonates, including pentafluoropropyl jasmonate (PFPJA), 2-hydroxyethyl jasmonate (HEJA), and 2-hydroxyethoxyethyl jasmonate (HEEJA), enhance the ginsenoside content and the ratio of Rb group to Rg group in *Panax notoginseng* cells, suggesting that suitable esterification of methyl jasmonate (MJA) with fluorine or hydroxyl groups can increase elicitation activity. Elicitor treatments inhibited the growth of Panax ginseng hairy roots but enhanced ginseng saponin biosynthesis, with selenium and sodium chloride showing positive effects on saponin content and productivity (Narayani & Srivastava, 2017; Zhang et al., 2009).

12.7.1 Strategies to Stimulate Secondary Metabolite Production

Stimulating the production of secondary metabolites in microorganisms, plants, or other biological systems often involves manipulating the environmental or genetic factors that influence these pathways (Narayani & Srivastava, 2017; Wang & Wu, 2013). Here are several strategies to enhance secondary metabolite production:

1. **Optimize nutrient availability.**
    - Ensure that the culture medium contains the appropriate nutrients required for secondary metabolite biosynthesis, including carbon sources, nitrogen sources, minerals, and vitamins.

- Nutrient limitation or specific nutrient supplementation can sometimes trigger the production of secondary metabolites. This strategy is known as "nutrient stress."

2. **Control environmental conditions.**
   - Adjust temperature, pH, and oxygen levels to create conditions favorable for secondary metabolite production. Different metabolites may require specific environmental conditions for optimal synthesis.
   - Employ bioreactor systems that allow precise control of environmental parameters to enhance metabolite production.
3. **Add precursor molecules.**
   - Supplementation with precursor molecules or intermediates of the target secondary metabolite pathway can promote its biosynthesis. This approach is often used in microbial fermentation processes.
   - Providing exogenous precursors can bypass rate-limiting steps and increase metabolite production.
4. **Use elicitation techniques.**
   - Apply physical, chemical, or biological elicitors, as mentioned in a previous response, to trigger stress responses in the organisms. This stress can lead to increased secondary metabolite production.
   - Elicitation can include adding signaling molecules, elicitor peptides, or physical stimuli like UV radiation or mechanical agitation.
5. **Genetic manipulation.**
   - Genetic engineering can be employed to overexpress key genes in the secondary metabolite biosynthesis pathway.
   - Techniques like gene knockout, gene silencing, or the introduction of heterologous genes can be used to enhance the production of specific metabolites.
6. **Stress conditions.**
   - Subjecting the organisms to mild stress conditions, such as changes in osmolarity, pH, or temperature, can stimulate secondary metabolite production as a stress response.
   - Stress-induced metabolite production is often seen in plants and microorganisms.
7. **Fermentation strategies.**
   - Use specialized fermentation strategies, such as fed-batch or two-stage fermentation, to control nutrient feeding and maintain optimal conditions for metabolite production.
   - Continuous monitoring and adjustment of culture parameters during fermentation can maximize yields.
8. **Metabolic pathway engineering.**
   - Modify metabolic pathways by enhancing the availability of precursors or by redirecting carbon flux to favor the production of specific metabolites.
   - Employ synthetic biology techniques to design and engineer pathways for improved metabolite synthesis.
9. **Co-culturing and microbiome manipulation.**
   - Cultivating microorganisms in consortia or manipulating the microbial community can stimulate the production of secondary metabolites.
   - Interactions between different microorganisms can trigger the production of metabolites not typically synthesized by individual strains.
10. **Natural variation screening.**
    - Screen and select naturally occurring strains or mutants that exhibit higher secondary metabolite production.
    - Genetic diversity within a species can lead to variations in metabolite production capabilities.

Selecting the most appropriate strategy for stimulating secondary metabolite production depends on the specific organism, metabolite, and production scale. Combining multiple approaches, such as genetic manipulation and optimized fermentation conditions, often yields the best results in enhancing secondary metabolite yields for industrial or research purposes.

12.7.2 Metabolic Engineering Approaches

Metabolic engineering is a multidisciplinary field that involves the modification of cellular metabolic pathways to enhance the production of desired compounds, improve cellular properties, or develop new metabolic capabilities. It plays a crucial role in biotechnology, synthetic biology, and bioprocess optimization (Jang et al., 2012; Lee & Wendisch, 2017). Here are some common metabolic engineering approaches:

1. **Overexpression of pathway genes**
   - **Application:** Increasing the expression of genes involved in the target metabolic pathway to boost the production of specific compounds.
   - **Mechanism:** Overexpression increases the availability of enzymes and substrates in the pathway, increasing flux toward the desired product.
2. **Gene knockout or silencing**
   - **Application:** Deleting or silencing genes that compete with or divert metabolic flux away from the target pathway.
   - **Mechanism:** This reduces the consumption of precursors and resources by unwanted pathways, redirecting them toward the desired product.
3. **Optimizing regulatory elements**
   - **Application:** Fine-tuning the regulation of genes in the target pathway by modifying promoter regions, ribosome binding sites, or transcription factors.
   - **Mechanism:** Enhanced control over gene expression levels allows for more precise metabolic regulation.
4. **Pathway balancing**
   - **Application:** Adjusting enzyme levels within a pathway to balance flux and avoid the accumulation of intermediates or bottlenecks.
   - **Mechanism:** Ensuring that enzyme activities are coordinated optimally prevents the buildup of toxic intermediates and maximizes product yield.
5. **Heterologous pathway engineering**
   - **Application:** Introducing genes from other organisms (heterologous genes) into the host organism to enable the production of compounds not naturally synthesized.
   - **Mechanism:** The heterologous pathway integrates with the host's metabolism, allowing for the production of the desired compound.
6. **Metabolic flux analysis**
   - **Application:** Using mathematical models and experimental data to analyze and optimize metabolic flux distribution within the cell.
   - **Mechanism:** By understanding how metabolites flow through the metabolic network, researchers can identify bottlenecks and optimize pathways accordingly.
7. **Cofactor engineering**
   - **Application:** Manipulating the availability of cofactors (e.g., NADH, NADPH) to improve the efficiency of specific enzymatic reactions.
   - **Mechanism:** By providing an excess of cofactors or engineering enzymes to preferentially use specific cofactors, enzymatic reactions can be enhanced.
8. **Dynamic control strategies**
   - **Application:** Implementing dynamic control of gene expression or enzyme activity through inducible promoters or feedback regulation.
   - **Mechanism:** This approach allows for the precise timing of metabolic changes, optimizing productivity under different conditions.

9. **Optimized media and fermentation conditions**
    - **Application:** Fine-tuning the growth medium and fermentation parameters to support the metabolic requirements of the target product.
    - **Mechanism:** Adjusting pH, temperature, aeration, and nutrient availability can impact cell growth and metabolite production.
10. **Evolutionary engineering**
    - **Application:** Using selective pressure, such as continuous culturing or mutagenesis, to evolve strains with improved metabolic traits.
    - **Mechanism**: Over time, cells with desirable metabolic traits are selected and enriched in the population.

Metabolic engineering approaches are highly versatile and adaptable to a wide range of organisms, from bacteria and yeast to plants and mammalian cells. The choice of approach depends on the specific goals of the metabolic engineering project and the characteristics of the host organism. These strategies are instrumental in the development of bioprocesses for the production of biofuels, pharmaceuticals, chemicals, and a wide array of bio-based products.

12.7.3 Integration with Large-Scale Cultures

Integrating small-scale cultures with large-scale cultures can be achieved through various methods. One approach is to use a benchtop multi-gas incubator to regulate oxygen concentration in the media, enabling faster formation of cellular foci. Another method involves screening different suspension cell lines for growth, metabolism, and virus productivity to identify promising candidates for the large-scale production of oncolytic viruses. Additionally, a high-throughput design of experiments (DoE) approach in a 96 deep-well plate system can be used to explore the impact of media and feed components on the quality attributes of a monoclonal antibody, providing valuable insights for cell culture development. Furthermore, an automated multi-scale 3D imaging platform allows for the integration of high-density organoid cultures with rapid and live 3D imaging, enabling the characterization of organoids at multiple scales (Beers et al., 2015).

Integrating small-scale cultures with large-scale cultures requires the consideration of best practices. These practices include the cultivation of new business models and political supports, as well as the engagement of local communities and cultures. Additionally, it is important to address the challenge of bridging the gap between the technical and socio-cultural aspects of industrial development. Furthermore, the integration of different models used in social science can be challenging due to the diverse requirements and the abstract nature of the concept of culture. Therefore, a more limited approach is advocated, focusing on specific aspects and influences of culture on society. By following these best practices, it is possible to effectively integrate small-scale cultures with large-scale cultures (van de Vijver, 2018).

12.7.4 Genetic Manipulation for Improved Metabolite Yield

Genetic manipulation techniques have been developed to improve metabolite yield in cell culture. These techniques include gene manipulation, metabolic engineering, and genetic transformation. By modifying the expression of specific genes, such as citrate export gene A (cexA), host cells can be engineered to produce fermentation products other than citric acid. In addition, strategies such as precursor feeding, elicitation, and in situ product removal can be used to enhance metabolite production. Furthermore, the use of a hybrid culture of metabolism-complementary plant cell lines has been shown to improve the yield of secondary metabolites. By combining multiple strategies, synergistic interactions can be achieved, leading to significant increases in metabolite yield. These genetic manipulation techniques offer promising approaches for increasing metabolite yield in cell culture systems (Kumar & Prasad, 2011).

12.7.5 Pathway Engineering and Gene Expression Optimization

Pathway engineering and gene expression optimization in cell culture have been the focus of several studies (Table 12.6). Combinatorial engineering strategies have been used to achieve optimal pathway performance by manipulating regulatory elements and open reading frames. Dynamic regulatory networks have also been explored to maximize production by balancing cell growth and product formation. Design of experiment (DoE) models have been employed to correlate gene expression with strain performance, enabling the identification of optimal gene expression patterns for metabolic pathway efficiency. CRISPR interference has been used to generate expression diversity on a combinatorial scale, allowing for the tuning of multiple genes to achieve the desired metabolic resource distribution. *Yarrowia lipolytica* has been investigated as a cell factory for the production of plant pharmaceutical components, with gene copy number manipulation resulting in increased production of naringenin. These studies demonstrate the importance of pathway engineering and gene expression optimization in cell culture for improving metabolic pathway efficiency and the production of desired compounds (Shen et al., 2019; Vick et al., 2011).

12.7.6 Balancing Growth and Biosynthesis in Large-Scale Cultures

Balancing growth and biosynthesis in large-scale cultures requires optimization of various factors, such as cell density, cell viability, growth stage, cell nutrition, composition of media, dissolved oxygen concentration, temperature, pH, and osmotic pressure of the media. Strategies to optimize culture conditions and promote taxol production in large-scale cultures of *Taxus chinensis* cells have been proposed. Fed-batch culture processes using a serum-free, glutamine-free medium have been developed to extend cell growth and culture lifetime, resulting in increased cell concentration

TABLE 12.6
Metabolic Engineering Approaches

Metabolic Engineering Approach	Application	Mechanism
Overexpression of pathway genes	Increase production of specific compounds.	Enhances enzyme and substrate availability.
Gene knockout or silencing	Divert metabolic flux away from unwanted pathways.	Reduces gene expression to minimize competition.
Optimizing regulatory elements	Fine-tune gene expression in target pathways.	Modifies promoter regions, ribosome binding sites, etc.
Pathway balancing	Prevent accumulation of intermediates and bottlenecks.	Adjusts enzyme levels to optimize pathway flux.
Heterologous pathway engineering	Enable production of non-native compounds.	Integrates foreign genes into host metabolism.
Metabolic flux analysis	Analyze and optimize metabolic flux distribution.	Uses mathematical models to identify bottlenecks.
Cofactor engineering	Enhance efficiency of enzymatic reactions.	Manipulates cofactor availability or enzyme preference.
Dynamic control strategies	Precisely time metabolic changes.	Utilizes inducible promoters or feedback regulation.
Optimized media and fermentation conditions	Support target product production.	Adjusts pH, temperature, aeration, and nutrient availability.
Evolutionary engineering	Evolve strains with improved metabolic traits.	Applies selective pressure for desired traits.

Source: Jang et al. (2012).

and protein titers in mouse and rat growth hormone production. The metabolic engineering of essential amino acids in Chinese hamster ovary (CHO) cells has been shown to reduce the production of inhibitory by-products and enhance cell growth and productivities in fed-batch cultures. Understanding the allocation of intracellular resources over biosynthetic processes, such as protein synthesis, can provide insight into balancing growth parameters and metabolic strategies in microbial growth (Fischer & Sauer, 2005; Ren et al., 2021).

12.8 CASE STUDIES IN BIOTRANSFORMATION AND SCALE-UP

CASE STUDY 1: BIOTRANSFORMATION OF CITRIC ACID (MAMMA & CHRISTAKOPOULOS, 2014)

Background: Citric acid is a widely used organic acid in the food and beverage industry, with application in soft drinks, in jams, and as a pH regulator. Traditionally, it was produced through chemical methods, but there has been a growing interest in using microbial fermentation for citric acid production.

Scale-up challenges:

1. *Microbial strain selection.* Identifying or developing a microbial strain capable of efficiently converting carbohydrates into citric acid.
2. *Optimization of fermentation conditions.* Scaling up the biotransformation process while maintaining optimal temperature, pH, and nutrient levels for citric acid production.
3. *Productivity and yield.* Ensuring high productivity and yield of citric acid on an industrial scale to compete with chemical production methods.
4. *Downstream processing.* Developing scalable methods for the recovery and purification of citric acid from the fermentation broth.

Outcome: Researchers successfully scaled up citric acid production using the fungus *Aspergillus niger*. By optimizing fermentation conditions, including pH control and oxygen supply, they achieved high yields and productivity. Additionally, efficient downstream processing techniques were implemented for citric acid recovery. This biotransformation approach is now widely used in the industry for citric acid production, providing a more sustainable and cost-effective alternative to chemical synthesis.

CASE STUDY 2: ENZYMATIC BIOTRANSFORMATION FOR PHARMACEUTICAL INTERMEDIATES (PATEL, 2002)

Background: The pharmaceutical industry often relies on chemical synthesis to produce complex intermediates. However, environmental concerns and the need for more sustainable practices have led to the exploration of enzymatic biotransformation as an alternative.

Scale-Up Challenges:

1. *Enzyme selection and stability.* Identifying enzymes that are stable and active under industrial conditions.
2. *Substrate compatibility.* Ensuring that the chosen enzymes are compatible with the substrates and reaction conditions required for the synthesis of pharmaceutical intermediates.

3. *Optimizing reaction conditions.* Scaling up enzymatic reactions while maintaining optimal temperature, pH, and reaction kinetics.
4. *Product purity.* Achieving high product purity through efficient downstream processing techniques, such as chromatography or crystallization.

Outcome: Researchers successfully scaled up the enzymatic biotransformation of a key pharmaceutical intermediate. They selected a robust enzyme, optimized reaction conditions, and implemented efficient purification processes. This approach not only reduced the environmental impact of the synthesis but also improved the overall yield and quality of the intermediate. The enzymatic biotransformation method is now an integral part of the pharmaceutical production process, demonstrating the feasibility of sustainable and cost-effective manufacturing.

These case studies illustrate how biotransformation and scale-up can offer sustainable and efficient alternatives to traditional chemical processes in various industries, including food and pharmaceuticals. Careful consideration of strain selection, fermentation conditions, and downstream processing is crucial to the success of such endeavors.

12.9 FUTURE PERSPECTIVES AND CHALLENGES

Bioreactors have emerged as a promising tool for pilot- and large-scale cultures of plant cells, offering advantages, such as controllability, reproducibility, scalability, and low-cost production. However, challenges remain in retaining the biosynthetic potential of cell cultures during scale-up. One study demonstrated the successful scaling up of rice recombinant butyrylcholinesterase production in a pilot-scale bioreactor, achieving consistent high yields (Singh et al., 2021). Another study investigated the biotransformation capabilities of cell suspension cultures derived from *Decalepis hamiltonii* and *Nicotiana tabacum* leaves, showing the conversion of racemic carvedilol and the specific enzyme activity of the cultures. Fungi have also been explored for the bioconversion of tetracyclines in wastewater and manure treatment, highlighting the potential of fungal enzymes to degrade or derivatize these antibiotics. Additionally, the use of bioreactors in plant tissue and cell culture has advanced over time, leading to improved stability, operational ease, and large-scale biomass production. However, challenges remain in scaling up microbial endophyte production and transitioning from laboratory scale to manufacturing (Bergen et al., 2021; Volarevic et al., 2014).

12.10 CONCLUSION

In conclusion, the field of biotransformation using plant cell cultures, particularly in the context of bioreactors for pilot- and large-scale cultures, presents a promising avenue for sustainable and tailored production of valuable compounds. This technology offers a bridge between traditional agriculture and chemical synthesis, addressing the growing need for environmentally friendly and economically viable methods of compound production.

The ability to harness the biosynthetic potential of plant cells in controlled bioreactor environments has paved the way for the production of high-value compounds, including pharmaceuticals, natural flavors, fragrances, and nutraceuticals. This approach offers several advantages, such as reduced land and resource requirements, year-round production, and the possibility of producing rare or difficult-to-cultivate plants.

However, challenges remain, particularly in scaling up these cultures to industrial levels while maintaining biosynthetic potential and product quality. Issues related to genetic stability, metabolite productivity, and cost-effectiveness must be addressed through ongoing research, innovation, and process optimization.

As the fields of metabolic engineering, bioprocessing, and bioreactor technology continue to advance, we can expect to see increased integration of artificial intelligence and omics approaches to monitor and optimize large-scale plant cell cultures. Moreover, sustainability and environmental considerations will drive the adoption of biotransformation methods as industries seek to reduce their ecological footprint.

In summary, the future of biotransformation using plant cell cultures in bioreactors is promising, with the potential to revolutionize the production of valuable compounds. While challenges persist, ongoing research and technological advancements will likely overcome these hurdles, making large-scale biotransformation a vital component of sustainable and customizable compound production for various industries.

REFERENCES

Afreen, F. (2008). Temporary immersion bioreactor. *Plant tissue culture engineering*, Springer, 187–201.

Akram, M. (2014). Citric acid cycle and role of its intermediates in metabolism. *Cell Biochemistry and Biophysics*, *68*(3), 475–478.

Alam, M. Z., Mamun, A. A., Qudsieh, I. Y., Muyibi, S. A., Salleh, H. M., & Omar, N. M. (2009). Solid state bioconversion of oil palm empty fruit bunches for cellulase enzyme production using a rotary drum bioreactor. *Biochemical Engineering Journal*, *46*(1), 61–64.

Albaugh, V. L., Mukherjee, K., & Barbul, A. (2017). Proline precursors and collagen synthesis: Biochemical challenges of nutrient supplementation and wound healing. *The Journal of nutrition*, *147*(11), 2011–2017.

Appelhagen, I., Wulff-Vester, A. K., Wendell, M., Hvoslef-Eide, A.-K., Russell, J., Oertel, A., Martens, S., Mock, H.-P., Martin, C., & Matros, A. (2018). Colour bio-factories: Towards scale-up production of anthocyanins in plant cell cultures. *Metabolic Engineering*, *48*, 218–232.

Asenjo, J. A. (1994). *Bioreactor system design*. CRC Press.

Ashok, A., Doriya, K., Rao, D. R. M., & Kumar, D. S. (2017). Design of solid state bioreactor for industrial applications: An overview to conventional bioreactors. *Biocatalysis and Agricultural Biotechnology*, *9*, 11–18.

Beers, J., Linask, K. L., Chen, J. A., Siniscalchi, L. I., Lin, Y., Zheng, W., Rao, M., & Chen, G. (2015). A cost-effective and efficient reprogramming platform for large-scale production of integration-free human induced pluripotent stem cells in chemically defined culture. *Scientific Reports*, *5*(1), 11319.

Bergen, V., Soldatov, R. A., Kharchenko, P. V., & Theis, F. J. (2021). RNA velocity—Current challenges and future perspectives. *Molecular systems biology*, *17*(8), e10282.

Bian, G., Deng, Z., & Liu, T. (2017). Strategies for terpenoid overproduction and new terpenoid discovery. *Current Opinion in Biotechnology*, *48*, 234–241.

Bosma, T. N., Middeldorp, P. J., Schraa, G., & Zehnder, A. J. (1996). Mass transfer limitation of biotransformation: Quantifying bioavailability. *Environmental Science & Technology*, *31*(1), 248–252.

Castro, N., Ribeiro, S., Fernandes, M., Ribeiro, C., Cardoso, V., Correia, V., Minguez, R., & Lanceros-Mendez, S. (2020). Physically active bioreactors for tissue engineering applications. *Advanced Biosystems*, *4*(10), 2000125.

Chen, H.-C., & Hu, Y.-C. (2006). Bioreactors for tissue engineering. *Biotechnology Letters*, *28*, 1415–1423.

Chen, L., Guttieres, D., Koenigsberg, A., Barone, P. W., Sinskey, A. J., & Springs, S. L. (2022). Large-scale cultured meat production: Trends, challenges and promising biomanufacturing technologies. *Biomaterials*, *280*, 121274.

Constabel, F. (1988). Principles underlying the use of plant cell fermentation for secondary metabolite production. *Biochemistry and Cell Biology*, *66*(6), 658–664.

Curtis, W. R. (2005). Application of bioreactor design principles to plant micropropagation. *Plant Cell, Tissue and Organ Culture*, *81*, 255–264.

Eibl, R., & Eibl, D. (2008). Design and use of the wave bioreactor for plant cell culture. *Plant tissue culture engineering*, Springer, 203–227.

Eibl, R., & Eibl, D. (2009). Application of disposable bag bioreactors in tissue engineering and for the production of therapeutic agents. *Advances in Biochemical Engineering Biotechnology, 112,* 183–207. 10.1007/978-3-540-69357-4_8.

Evans, D. E., Coleman, J. O., & Kearns, A. (2003). *Plant cell culture*. Garland Science.

Fischer, E., & Sauer, U. (2005). Large-scale in vivo flux analysis shows rigidity and suboptimal performance of Bacillus subtilis metabolism. *Nature genetics*, *37*(6), 636–640.

Fowler, M. (1984). Large-scale cultures of cells in suspension. *Laboratory procedures and their applications.* In Methods in Cell Biology. Academic Press, *10*, 167–174.

Galbraith, S. C., Bhatia, H., Liu, H., & Yoon, S. (2018). Media formulation optimization: Current and future opportunities. *Current Opinion in Chemical Engineering*, *22*, 42–47.

Gamborg, O., Murashige, T., Thorpe, T., & Vasil, I. (1976). Plant tissue culture media. *In Vitro*, *12*(7), 473–478.

Gaspar, D. A., Gomide, V., & Monteiro, F. J. (2012). The role of perfusion bioreactors in bone tissue engineering. *Biomatter*, *2*(4), 167–175.

Gelinsky, M., Bernhardt, A., & Milan, F. (2015). Bioreactors in tissue engineering: Advances in stem cell culture and three-dimensional tissue constructs. *Engineering in Life Sciences*, *15*(7), 670–677.

Gupta, S. D., & Ibaraki, Y. (2006). *Plant tissue culture engineering* (Vol. 6). Springer.

Hansmann, J., Groeber, F., Kahlig, A., Kleinhans, C., & Walles, H. (2013). Bioreactors in tissue engineering—Principles, applications and commercial constraints. *Biotechnology Journal*, *8*(3), 298–307.

Hirsch, A., Parag, Y., & Guerrero, J. (2018). Microgrids: A review of technologies, key drivers, and outstanding issues. *Renewable and Sustainable Energy Reviews*, *90*, 402–411.

Huesemann, M. H., Knoshaug, E. P., Laurens, L. M., Dale, T., Lane, T. W., & McGowen, J. (2022). Development of integrated screening, cultivar optimization, and verification research (DISCOVR): A coordinated research-driven approach to improve microalgal productivity, composition, and culture stability for commercially viable biofuels production. *Algal Research*, 102961.

Imani, J., Kumar, A., & Neumann, K.-H. (2009). *Plant cell and tissue culture-A tool in biotechnology: Basics and application.* Springer.

Ishihara, K., Hamada, H., Hirata, T., & Nakajima, N. (2003). Biotransformation using plant cultured cells. *Journal of Molecular Catalysis B: Enzymatic*, *23*(2–6), 145–170.

Jang, Y.-S., Park, J. M., Choi, S., Choi, Y. J., Cho, J. H., & Lee, S. Y. (2012). Engineering of microorganisms for the production of biofuels and perspectives based on systems metabolic engineering approaches. *Biotechnology advances*, *30*(5), 989–1000.

Jayme, D. W. (1991). Nutrient optimization for high density biological production applications. *Cytotechnology*, *5*, 15–30.

Jiménez-Chillarón, J. C., Díaz, R., Martínez, D., Pentinat, T., Ramón-Krauel, M., Ribó, S., & Plösch, T. (2012). The role of nutrition on epigenetic modifications and their implications on health. *Biochimie*, *94*(11), 2242–2263.

Johnson, J. D., & Hill, G. E. (2013). Is carotenoid ornamentation linked to the inner mitochondria membrane potential? A hypothesis for the maintenance of signal honesty. *Biochimie*, *95*(2), 436–444.

Jossen, V., Eibl, R., & Eibl, D. (2019). Single-use bioreactors—An overview. *Single-use technology in biopharmaceutical manufacture*, Wiley, 37–52.

Kalbhenn, J., Schlagenhauf, A., Rosenfelder, S., Schmutz, A., & Zieger, B. (2018). Acquired von Willebrand syndrome and impaired platelet function during venovenous extracorporeal membrane oxygenation: Rapid onset and fast recovery. *The Journal of Heart and Lung Transplantation*, *37*(8), 985–991.

Karki, S., Rizal, G., & Quick, W. P. (2013). Improvement of photosynthesis in rice (Oryza sativa L.) by inserting the C4 pathway. *Rice*, *6*(1), 1–8.

Kumar, R. R., & Prasad, S. (2011). Metabolic engineering of bacteria. *Indian Journal of Microbiology*, *51*(3), 403–409.

Kwade, A., Haselrieder, W., Leithoff, R., Modlinger, A., Dietrich, F., & Droeder, K. (2018). Current status and challenges for automotive battery production technologies. *Nature Energy*, *3*(4), 290–300.

Lai, M. C., & Lan, E. I. (2015). Advances in metabolic engineering of cyanobacteria for photosynthetic biochemical production. *Metabolites*, *5*(4), 636–658.

Lawrence, B. J., Devarapalli, M., & Madihally, S. V. (2009). Flow dynamics in bioreactors containing tissue engineering scaffolds. *Biotechnology and Bioengineering*, *102*(3), 935–947.

Lee, J.-H., & Wendisch, V. F. (2017). Production of amino acids–genetic and metabolic engineering approaches. *Bioresource Technology*, *245*, 1575–1587.

Li, Y.-P., Xu, Y., & Huang, Z.-B. (2012). Isolation and characterization of the citrinin biosynthetic gene cluster from Monascus aurantiacus. *Biotechnology Letters*, *34*, 131–136.

Long, Y., Yang, Y., Pan, G., & Shen, Y. (2022). New insights into tissue culture plant-regeneration mechanisms. *Frontiers in Plant Science*, *13*, 926752.

Lund, R. J., Närvä, E., & Lahesmaa, R. (2012). Genetic and epigenetic stability of human pluripotent stem cells. *Nature Reviews Genetics*, *13*(10), 732–744.

Macfarlane, S., & Macfarlane, G. T. (2003). Regulation of short-chain fatty acid production. *Proceedings of the Nutrition Society*, *62*(1), 67–72.

Mamma, D., & Christakopoulos, P. (2014). Biotransformation of citrus by-products into value added products. *Waste and Biomass Valorization*, *5*, 529–549.
Margaritis, A., & Wallace, J. B. (1984). Novel bioreactor systems and their applications. *Bio/Technology*, *2*(5), 447–453.
Martin, I., Wendt, D., & Heberer, M. (2004). The role of bioreactors in tissue engineering. *TRENDS in Biotechnology*, *22*(2), 80–86.
Merchuk, J. C. (2003). Airlift bioreactors: Review of recent advances. *The Canadian Journal of Chemical Engineering*, *81*(3–4), 324–337.
Mohn, F., & Schübeler, D. (2009). Genetics and epigenetics: Stability and plasticity during cellular differentiation. *Trends in Genetics*, *25*(3), 129–136.
Narayani, M., & Srivastava, S. (2017). Elicitation: A stimulation of stress in in vitro plant cell/tissue cultures for enhancement of secondary metabolite production. *Phytochemistry Reviews*, *16*, 1227–1252.
Ntwampe, S., Chowdhury, F., Sheldon, M., & Volschenk, H. (2010). Overview of parameters influencing biomass and bioreactor performance used for extracellular ligninase production from Phanerochaete chrysosporium. *Brazilian Archives of Biology and Technology*, *53*, 1057–1066.
Ongley, S. E., Bian, X., Neilan, B. A., & Müller, R. (2013). Recent advances in the heterologous expression of microbial natural product biosynthetic pathways. *Natural Product Reports*, *30*(8), 1121–1138.
Park, J. H., & Lee, S. Y. (2008). Towards systems metabolic engineering of microorganisms for amino acid production. *Current Opinion in Biotechnology*, *19*(5), 454–460.
Patel, R. N. (2002). Microbial/enzymatic synthesis of chiral intermediates for pharmaceuticals. *Enzyme and Microbial Technology*, *31*(6), 804–826.
Pérez, J. S., Porcel, E. R., López, J. C., Sevilla, J. F., & Chisti, Y. (2006). Shear rate in stirred tank and bubble column bioreactors. *Chemical Engineering Journal*, *124*(1–3), 1–5.
Peterson, E. L. (1999). Benthic shear stress and sediment condition. *Aquacultural Engineering*, *21*(2), 85–111.
Plunkett, N., & O'Brien, F. J. (2011). Bioreactors in tissue engineering. *Technology and Health Care*, *19*(1), 55–69.
Pörtner, R., Nagel-Heyer, S., Goepfert, C., Adamietz, P., & Meenen, N. M. (2005). Bioreactor design for tissue engineering. *Journal of Bioscience and Bioengineering*, *100*(3), 235–245.
Ren, Y., Sun, H., Deng, J., Huang, J., & Chen, F. (2021). Carotenoid production from microalgae: Biosynthesis, salinity responses and novel biotechnologies. *Marine Drugs*, *19*(12), 713.
Reuss, M. (1994). Stirred tank bioreactors. *Bioreactor system design*. Taylor & Francis.
Sadikovic, B., Al-Romaih, K., Squire, J., & Zielenska, M. (2008). Cause and consequences of genetic and epigenetic alterations in human cancer. *Current Genomics*, *9*(6), 394–408.
Schliwa, G. (2013). Exploring living labs through transition management-challenges and opportunities for sustainable urban transitions. *IIIEE Master thesis*.
Schum, A., Hofmann, K., & Felten, R. (2001). Fundamentals for integration of somatic hybridization in rose breeding. XX International Eucarpia Symposium, Section Ornamentals, Strategies for New Ornamentals-Part II 572.
Shahzad, A., Sharma, S., Parveen, S., Saeed, T., Shaheen, A., Akhtar, R., Yadav, V., Upadhyay, A., & Ahmad, Z. (2017). Historical perspective and basic principles of plant tissue culture. *Plant biotechnology: Principles and applications*, Springer Nature, 1–36.
Shen, X., Wang, J., Li, C., Yuan, Q., & Yan, Y. (2019). Dynamic gene expression engineering as a tool in pathway engineering. *Current Opinion in Biotechnology*, *59*, 122–129.
Singh, M., Zappa, D., & Comini, E. (2021). Solid oxide fuel cell: Decade of progress, future perspectives and challenges. *International Journal of Hydrogen Energy*, *46*(54), 27643–27674.
Sivanandhan, G., Selvaraj, N., Ganapathi, A., & Manickavasagam, M. (2014). Enhanced biosynthesis of withanolides by elicitation and precursor feeding in cell suspension culture of Withania somnifera (L.) Dunal in shake-flask culture and bioreactor. *PLoS One*, *9*(8), e104005.
Taticek, R. A., Moo-Young, M., & Legge, R. L. (1991). The scale-up of plant cell culture: Engineering considerations. *Plant Cell, Tissue and Organ Culture*, *24*, 139–158.
Thorstensen, L., Lind, G. E., Løvig, T., Diep, C. B., Meling, G. I., Rognum, T. O., & Lothe, R. A. (2005). Genetic and epigenetic changes of components affecting the WNT pathway in colorectal carcinomas stratified by microsatellite instability. *Neoplasia*, *7*(2), 99–108.
van de Vijver, F. J. (2018). Towards an integrated framework of bias in noncognitive assessment in international large-scale studies: Challenges and prospects. *Educational Measurement: Issues and Practice*, *37*(4), 49–56.
Vasumathi, K., Premalatha, M., & Subramanian, P. (2012). Parameters influencing the design of photobioreactor for the growth of microalgae. *Renewable and Sustainable Energy Reviews*, *16*(7), 5443–5450.

Verpoorte, R., & Memelink, J. (2002). Engineering secondary metabolite production in plants. *Current Opinion in Biotechnology*, *13*(2), 181–187.

Verpoorte, R., van der Heijden, R., & Memelink, J. (2000). Engineering the plant cell factory for secondary metabolite production. *Transgenic Research*, *9*, 323–343.

Vick, J. E., Johnson, E. T., Choudhary, S., Bloch, S. E., Lopez-Gallego, F., Srivastava, P., Tikh, I. B., Wawrzyn, G. T., & Schmidt-Dannert, C. (2011). Optimized compatible set of BioBrick™ vectors for metabolic pathway engineering. *Applied Microbiology and Biotechnology*, *92*, 1275–1286.

Volarevic, V., Bojic, S., Nurkovic, J., Volarevic, A., Ljujic, B., Arsenijevic, N., Lako, M., & Stojkovic, M. (2014). Stem cells as new agents for the treatment of infertility: Current and future perspectives and challenges. *BioMed Research International*, *2014*.

Wang, J. W., & Wu, J. Y. (2013). Effective elicitors and process strategies for enhancement of secondary metabolite production in hairy root cultures. *Biotechnology of hairy root systems*, Springer Nature, 55–89.

Wendt, D., Riboldi, S., Cioffi, M., & Martin, I. (2009). Bioreactors in tissue engineering: Scientific challenges and clinical perspectives. *Bioreactor systems for tissue engineering*, Springer Nature, 1–27.

Yadav, R. K., Das, S., & Patil, S. A. (2022). Are integrated bioelectrochemical technologies feasible for wastewater management? *TRENDS in Biotechnology*, *41*(4), 484–496. https://doi.org/10.1016/j.tibtech.2022.09.001

Yeo, H., An, J., Reid, R., Rittmann, B. E., & Lee, H.-S. (2015). Contribution of liquid/gas mass-transfer limitations to dissolved methane oversaturation in anaerobic treatment of dilute wastewater. *Environmental Science & Technology*, *49*(17), 10366–10372.

Yoshikawa, M., Yamaoka, N., & Takeuchi, Y. (1993). Elicitors: Their significance and primary modes of action in the induction of plant defense reactions. *Plant and Cell Physiology*, *34*(8), 1163–1173.

Zhang, H.-C., Liu, J.-M., Lu, H.-Y., & Gao, S.-L. (2009). Enhanced flavonoid production in hairy root cultures of Glycyrrhiza uralensis Fisch by combining the over-expression of chalcone isomerase gene with the elicitation treatment. *Plant Cell Reports*, *28*, 1205–1213.

Zondag, S. D., Mazzarella, D., & Noël, T. (2023). Scale-up of photochemical reactions: Transitioning from lab scale to industrial production. *Annual Review of Chemical and Biomolecular Engineering*, *14*, 283–300.

13 Methods Used in Gene Identification, Localization, and Sequencing

Applications of PCR in Plant Genome Analysis

*Ram Baran Singh, Ramachandra Ramanjinappa, and Santosh Anand**
* Corresponding author: sanand.biotech@gmail.com

13.1 INTRODUCTION: GENE

The term "gene" (Greek: *génos*, signifying generation, birth, or gender) can be used in many various ways, depending on how it is defined in terms of heredity, selection, biological function, or molecular structure, although a majority of these can be classified as either Mendelian or molecular genes (Gericke and Hagberg, 2006; Orgogozo et al., 2016; Roth, 2019; Meunier, 2022). Any elementary heritable trait is referred to as the Mendelian gene, which is the classical *gene* in genetics. The definition of a *molecular gene*, which is one that is expressed in terms of nucleotide sequence, is more widely employed in biochemistry, molecular biology, and majority of genetics (Orgogozo et al., 2016). A *gene* is a segment of DNA, which is a long, double-stranded molecule that resembles a twisted ladder or helix. DNA is made up of nucleotides, each consisting of a sugar (deoxyribose), a phosphate group, and one of four nitrogenous bases: adenine (A), thymine (T), cytosine (C), and guanine (G). Genes are located on chromosomes, which are thread-like structures found in the cell nucleus.

The *Arabidopsis* genome contains approximately 125 megabases of sequence and encodes around 25,500 genes. Analysis of the *Arabidopsis* genome sequence reveals that it is similar in many ways to the genomes of other sequenced multicellular organisms but also has some important differences reflecting its plant-specific biology (The Arabidopsis Genome Initiative, 2000). *Arabidopsis* has a small genome size and a haploid chromosome number of five, which makes it a useful model organism for plant genetics and molecular biology research. Humans have 23 pairs of chromosomes, and genes are distributed along these chromosomes. Genes consist of coding regions (exons) and non-coding regions (introns). Coding regions contain the information necessary for synthesizing proteins, which are essential for various cellular functions. Non-coding regions may have regulatory functions, affecting when and how the gene is expressed.

The process of gene expression begins with transcription. In transcription, a specific segment of DNA is used as a template to synthesize a complementary RNA molecule, called messenger RNA (mRNA). The initial mRNA molecule undergoes processing, including the removal of introns and the addition of a 5' cap and a poly-A tail. This processed mRNA is the mature transcript ready for translation. The mature mRNA exits the cell nucleus and enters the cytoplasm, where translation occurs. Translation involves the synthesis of a polypeptide chain (protein) based on the information encoded in the mRNA sequence. The genetic code, a set of three-nucleotide sequences called

DOI: 10.1201/9781003326939-15

codons, specifies the amino acids that make up the protein. Transfer RNA (tRNA) molecules carry specific amino acids to the ribosomes, where they align with the mRNA codons, facilitating the assembly of the protein. The synthesized protein folds into its three-dimensional structure, determining its function within the cell. Proteins play crucial roles in various cellular processes, such as enzyme catalysis, cell structure, signaling, and regulation. Gene expression is tightly regulated to ensure that the right genes are active at the right times and in the right cells. Regulatory elements, such as promoters and enhancers, influence the rate and timing of gene transcription. Genes are passed from one generation to the next during reproduction. Offspring inherit a combination of genes from both parents, contributing to genetic diversity. Gaining knowledge about the structure and functions of genes is crucial for deciphering the intricacies of genetics and molecular biology, hence offering perspectives on growth, evolution, and many hereditary illnesses.

13.1.1 Gene Identification

Gene identification is a crucial initial step toward understanding the genetic basis of various traits, diseases, and biological processes. Many techniques have been created alongside advancements in technology for the purpose of gene identification. A few common methods for identification and localization of the genes in organisms are as follows.

13.1.1.1 Bioinformatics and Computational Approaches

13.1.1.1.1 Sequence Homology Search

One of the most instructive and the initial steps in any examination of recently identified sequences is the use of sequence homology search to find homologous sequences. Comparing DNA or protein sequences to known sequences can help identify genes by detecting homologous regions (Pearson, 2013). It is carried out by comparing the DNA or protein sequence of interest against databases (e.g., NCBI, Ensembl) using tools like the Basic Local Alignment Search Tool. BLAST is a bioinformatics tool that compares a query sequence against a database of known sequences to identify similar sequences. It is widely used in finding homologous genes in different organisms and in identifying homologous sequences that may represent genes with similar functions. Popular similarity searching programs, including SSEARCH (Smith and Waterman, 1981), FASTA (Pearson and Lipman, 1988), PSI-BLAST, BLAST (Altschul et al., 1997), and the HMMER3 (Johnson et al., 2010), calculate precise statistical estimates, guaranteeing that protein sequences that exhibit significant similarity also share similar structures. Sequences with a high degree of similarity can be assumed to be homologous, meaning they have a common ancestor, which makes similarity searches efficient and trustworthy.

13.1.1.1.2 Gene Prediction Algorithms

The process of finding genes along the genome is known as gene prediction, and the process of determining the genomic DNA areas that encode genes is also referred to as "gene finding." In computational biology, *gene prediction*, often known as *gene discovery*, is the process of locating the genomic DNA sections that encode genes. Not only do RNA and protein-coding genes fall under this category, but regulatory sequences and other functional elements may also be predicted. One of the first and most crucial tasks in comprehending a species' genome after it has been sequenced is locating its genes (Yandell and Ence, 2012). Computational tools like GeneMark, Augustus, and Glimmer are used to predict potential genes based on statistical models and sequence features. Gene prediction algorithms utilize these tools to analyze genomic features like open reading frames (ORFs), splice sites, branch points, polypyrimidine tracts, start codons, stop codons, and promoter regions to predict potential gene locations and structures based on these features (Wang et al., 2004). The challenging task of predicting entire gene architectures was undertaken by the programs in the next generation. Numerous programs have been created, such as FGENEH (Solovyev, 1994), GeneID (Guigó, 1992), GeneParser (Snyder and Stormo, 1993), and GenLang (Dong and Searls, 1994).

13.1.1.1.3 Comparative Genomics

Comparative genomic hybridization (CGH). This method compares the genomic DNA of two different organisms to identify variations, including gene duplications, deletions, and mutations. A comparative genomics technique is a promising area in current gene-finding research, as the whole genomes of many different species are being sequenced. This is predicated on the idea that, because mutations in functional parts are more likely to have a deleterious effect on the organism than mutations elsewhere, the forces of natural selection induce genes and other functional elements to undergo mutation at a slower rate than the rest of the genome. The underlying premise of this method is that exons, which are functional sections, have undergone greater evolutionary conservation than non-functional parts, such as intronic or intergenic regions. Thus, this evolutionary pressure for conservation can be found through gene detection, which involves comparing the genomes of related species. Initially, this method was used on the human and mouse genomes, utilizing TWINSCAN/N-SCAN, CONTRAST, SLAM, and SGP (Gross et al., 2007). Comparing genomes of different species can reveal conserved regions, aiding in the identification of functional genes. The comparative genomic hybridization (CGH) technique is used to compare genomic DNA of two different organisms to identify similarities and differences, helping in the identification of genes that are conserved or unique.

Phylogenetic footprinting. By comparing the genomes of related species, conserved non-coding regions can be identified. These regions often correspond to regulatory elements associated with genes.

13.1.2 Genetic Mapping

The techniques used to determine the position and order of a gene underlying phenotypes on a chromosome based on correlation with genetic variation are referred to as "gene mapping" or "genome mapping." Gene mapping also provides information on the relative distance between various gene sites. Genetic mapping is performed by constructing a genetic map using linkage which is based on genotyping data generated by PCR- or sequence-derived DNA markers (Srivastava et al., 2019). Molecular marker defined as a unique DNA sequence presents in close proximity of the desired gene or locus that can be recognized using a variety of molecular techniques, which reveal sites of variation in genomic DNA (Mackill and Ni, 2000). Molecular markers can determine genetic polymorphism/variability, which may be employed to examine genetic variation among diverse genotypes or alleles of a gene for a definite piece of DNA within a population or gene pool (Winter and Kahl, 1995).

A genetic map, or linkage map, is a genomic tool which illustrates the relative location of genetic markers (reflecting sites of genomic variants) on a chromosome. A genetic map construction is based on the concept of genetic linkage: The closer two markers are to each other on a chromosome, the greater the probability that they will be inherited together. By studying inheritance patterns, the relative order and location of genetic markers along a chromosome can be established.

13.1.2.1 Linkage Analysis

This method involves studying the inheritance patterns of genetic markers in families to identify the location of a gene on a chromosome. The determination of linkage between markers and target genes/QTLs controlling important traits is the key step involved in developing a linkage map using data of every marker on each plant of the population. However, linkage analysis can be conducted manually for a limited number of markers, but it is unfeasible to manually evaluate and determine linkages between lots of markers exploited in constructing genetic maps; computer software are required to perform the task. The degree of linkage among markers and traits is typically estimated exploiting odds ratios (ratio of linkage vs. no linkage). Odds ratio is easily articulated as the logarithm of the ratio, which is referred to as logarithm of odds (LOD) value, or LOD score (Risch,

1992), in QTL analysis. The generally used LOD values of >3 are considered in the construction of linkage maps. The mostly used computer software programs for linkage analysis include QTL IciMapping, Mapmaker/EXP, and MapManager QTX (Meng et al., 2015; Manly et al., 2001; Lincoln et al., 1993), which are publicly available on the internet/websites. JoinMap and QTL cartographer are other frequently used tools for constructing linkage maps.

13.1.2.2 Association Studies

Genome-wide association study (GWAS) is used to study genetic variations within populations, to identify the associations between specific genetic markers and phenotypic traits. GWAS is used to identify associations between specific genetic markers and the presence of a particular trait or disease in a population. An alternative approach to QTL mapping is the genome-wide study (GWAS) or association mapping (AM) approach (Gómez et al., 2011), based on the principle of linkage disequilibrium (LD) to detect substantial association between DNA marker/QTL and target (Gupta et al., 2005). The potential advantage of association mapping is the likelihood for mapping with a superior resolution utilizing mass recombination events from a large number of meiosis throughout germplasm evolutionary history. GWAS has the power to evaluate and characterize several alleles concurrently in diploid (Zhao et al., 2007) as well as in polyploid crops (Breseghello and Sorrells, 2006). Association mapping offers a great deal of benefit over linkage mapping since it provides better mapping resolution on accounts of historical mutations and recombinations in genetic lineages, leading to the identification of markers in the vicinity of governing genes (Liu et al., 2016).

13.1.2.3 Experimental Methods

There are several experimental methods employed for gene identification, each with its own strengths and limitations. Here are some commonly used methods:

1. DNA sequencing
   - *Whole genome sequencing (WGS).* In WGS, the entire genome of an organism is sequenced, allowing for the identification of all genes present. This method is comprehensive but can be expensive and computationally intensive.
   - *Exome sequencing.* Exome sequencing focuses on the protein-coding regions of the genome (exons). Since exons represent a small percentage of the genome, exome sequencing is more cost-effective than WGS while still capturing important functional elements.
2. RNA sequencing (RNA-seq)
   - *Transcriptome sequencing.* RNA-seq is used for profiling the entire transcriptome or all the RNA molecules (including mRNA, non-coding RNA, and alternatively spliced transcripts) to identify genes expressed in a particular tissue or under specific conditions. The RNA-seq technique allows the identification and quantification of RNA molecules in a sample, providing information about gene expression levels and alternative splicing events.
3. Functional genomics
   - *Gene knockout.* This involves the targeted inactivation or deletion of a specific gene in a model organism to observe the resulting phenotype. The analysis of the phenotype helps in understanding the function of the gene.
   - *RNA interference (RNAi).* This technique involves introducing small interfering RNAs (siRNAs) or short hairpin RNAs (shRNAs) to selectively silence the expression of a target gene. The resulting phenotype provides insights into the gene's function.
4. Chromatin immunoprecipitation (ChIP)
   - *ChIP-seq.* ChIP-seq combines chromatin immunoprecipitation with high-throughput sequencing to identify DNA binding sites of proteins, such as transcription factors. This helps in understanding the regulation of gene expression.

5. Mass spectrometry
    - *Proteomics.* Mass spectrometry can be used to identify and quantify proteins present in a biological sample. This information is valuable in understanding the translation of genes into proteins.
6. Functional screens
    - *Insertional mutagenesis.* Inserting a foreign DNA sequence into the genome can disrupt the function of a gene. This method is often used in model organisms to identify genes essential for specific biological processes.
    - *Microarray analysis.* Examining the expression levels of thousands of genes simultaneously to identify genes associated with specific conditions or treatments. Differential gene expression patterns can be correlated with specific cellular functions or conditions. High-throughput genotyping arrays are used to identify single-nucleotide polymorphisms (SNPs) associated with traits or diseases.
8. Transposon Mutagenesis
    - *Insertional mutagenesis.* Inserting transposons (jumping genes) randomly into a genome can disrupt gene function. By analyzing the phenotypic effects of these insertions, researchers can identify the location and function of genes. With the use of this method, random gene disruption is found, and the genes impacted are subsequently determined by phenotypic change analysis.
9. PCR-based techniques
    - *Polymerase chain reaction (PCR).* Genes or genetic variants can be identified by PCR, which amplifies particular DNA sequences. It is possible to utilize PCR to amplify particular DNA fragments, particularly those that match relevant genes.
    - *Quantitative PCR (qPCR).* Quantifying gene expression levels in real time.
10. Functional screens
    - *Loss-of-function screens.* Systematically inhibiting gene function to observe the resulting phenotype.
    - *Gain-of-function screens.* Introducing exogenous genes to observe their effects on cellular processes.
11. Protein structure and function analysis
    - *Protein structure prediction.* Predicting protein structures to infer potential functions.
    - *Functional domains identification.* Analyzing protein domains to understand their roles in cellular processes.
12. Epigenetic approaches
    - *DNA methylation analysis.* Studying DNA methylation patterns to identify genes involved in epigenetic regulation.

The choice of method depends on the specific research question, the available resources, and the nature of the genetic information being sought. Often, a combination of these methods is used for comprehensive gene identification and functional analysis.

13.1.3 Gene Localization

Gene localization involves identifying the specific chromosomal location of a particular gene within an organism's genome. Several methods have been developed to facilitate gene localization, and these techniques have evolved over time. There are several methods employed for gene localization, and they can be broadly categorized into the following approaches:

1. **Genetic mapping.** Genetic mapping relies on the observation of inheritance patterns among individuals in a population. These methods are based on the assumption that genes located close to each other on a chromosome tend to be inherited together.

Genetic mapping is typically divided into two types: linkage mapping and association mapping.

a. **Linkage mapping**
    - *Family studies.* Linkage analysis involves studying the inheritance patterns of genetic markers or traits within families. By analyzing the co-segregation of a genetic marker with a trait of interest, researchers can infer the approximate location of the gene responsible for that trait.
    - *Pedigree construction.* Pedigree charts are created to track the inheritance of genetic markers or traits through multiple generations within a family. Statistical methods are then applied to identify chromosomal regions linked to the gene of interest.
    - *Crossing over and recombination frequency.* The frequency of recombination events between two genes on a chromosome is directly proportional to the physical distance between them. By studying the inheritance patterns of linked genes in families, one can estimate the recombination frequency and, consequently, the distance between genes.
    - *Genetic markers.* Genetic markers, such as restriction fragment length polymorphisms (RFLPs) or microsatellites, are used as reference points on a chromosome. The co-inheritance of a genetic marker with a gene of interest helps in estimating the distance between the marker and the gene.

b. **Association mapping**
    - *Population-based studies.* Association mapping involves studying the association between genetic variations (polymorphisms) and the presence or absence of a trait in a population. This is particularly useful for complex traits influenced by multiple genes.
    - *Single-nucleotide polymorphisms (SNPs).* SNPs are variations at a single nucleotide position in the DNA sequence. By identifying SNPs associated with a particular trait, researchers can infer the genomic region harboring the gene of interest.
    - *Case control studies.* Association studies compare the frequency of specific genetic markers or alleles in individuals with a particular trait (cases) and those without the trait (controls). Significant associations can indicate the location of a gene associated with the trait.
    - *Genome-wide association studies (GWAS).* GWAS involve scanning the entire genome for associations between genetic markers and traits. These studies use large cohorts and high-throughput genotyping technologies to identify common variants associated with diseases or traits.

2. **Physical mapping**

Physical mapping involves the direct examination and measurement of DNA sequences and their positions within the genome. This provides information about the physical distances between genes.

- **Fluorescence in situ hybridization (FISH).** FISH is a cytogenetic technique that uses fluorescently labeled DNA probes to bind to specific chromosomal regions. This method allows for the direct visualization of the chromosomal location of a gene.
    - *DNA probes.* FISH uses fluorescently labeled DNA probes that bind specifically to the target gene or DNA sequence. By visualizing the fluorescence pattern under a microscope, researchers can determine the gene's location on a chromosome.
- **Contig assembly.** Physical maps are created by assembling overlapping fragments of DNA (contigs) that span a particular chromosomal region. This process helps researchers identify the relative positions of genes.
    - *Cloning and sequencing.* Large DNA fragments are cloned, and their sequences are determined. Overlapping sequences are identified to assemble contigs, representing continuous stretches of the genome.

- **Restriction fragment mapping.**
  - *Restriction enzymes.* This method involves digesting genomic DNA with restriction enzymes and analyzing the resulting fragments. By comparing fragment sizes and patterns, researchers can create a physical map of the genome.
- **Radiation hybrid mapping.**
  - *Radiation-induced fragmentation.* Ionizing radiation is used to fragment the DNA. The resulting fragments are then mapped to a reference genome using genetic markers, creating a physical map.

3. **Expressed sequence tags (ESTs) mapping**
   - **EST mapping.** ESTs are short sequences derived from cDNA (complementary DNA) libraries, representing expressed genes. Mapping ESTs to genomic DNA helps identify the location of transcribed genes.
4. **Comparative genomics**
   - **Synteny analysis.** Comparative genomics involves comparing the genome organization and gene order between different species. Conserved synteny (gene order) across species can aid in identifying the location of genes in the genome of interest.
5. **Bioinformatics and computational methods**
   - ***In silico* mapping.** Computational methods use algorithms and bioinformatics tools to predict gene locations based on features such as sequence similarities, conserved motifs, and other genomic characteristics.
6. **Transcription mapping**
   - **RNA-seq.** This method involves sequencing the RNA transcripts in a sample. By mapping these sequences back to the genome, researchers can identify the locations of expressed genes.

These methods are often used in combination to refine the localization of genes and enhance the accuracy of the results. Advances in technology and the integration of multiple data types continue to improve the efficiency of gene localization efforts.

13.1.4 Methods of Gene Sequencing

Gene sequencing, also known as DNA sequencing, is the process of determining the exact order of the nucleotides in a DNA fragment, genome, or microbiome. The sequence of the nucleotides encodes the genetic information that cells use to develop and function. Determining the sequence of a gene or the entire genome is key to understanding the function of genes and other parts of the genome. Gene sequencing is essential for biological research, medical diagnosis, biotechnology, forensic biology, virology, and systematics. Gene sequencing is used in multiple things, including diagnosing diseases, characterizing antibody repertoire, guiding patient treatment, and aiding in organism identification. The main methods of gene sequencing are described in what follows:

1. **Sanger sequencing (or chain termination method).** This was the first widely used DNA sequencing method, developed in 1977. It involves making copies of the DNA with fluorescently labeled chain-terminating nucleotides and then separating the fragments by size using gel electrophoresis to determine the sequence (Heather and Chain, 2016).
2. **Next-generation sequencing (NGS) methods.** These high-throughput techniques emerged in the mid- to late 1990s and can sequence millions of DNA fragments in parallel. Common NGS methods include Illumina sequencing, which uses a "sequencing by synthesis" approach, and PacBio and nanopore sequencing, which can sequence long DNA fragments without the need for fragmentation and assembly.
3. **Shotgun sequencing.** This method involves randomly fragmenting the target DNA, sequencing the fragments, and then computationally assembling the full sequence by identifying overlapping regions between the fragments.

4. **Capillary electrophoresis and fragment analysis.** This technique can be used for both sequencing and sizing/genotyping of DNA fragments, with application in areas like cell line authentication and aneuploidy detection (Heather and Chain, 2016).

To summarize, the primary techniques for gene sequencing are shotgun sequencing, capillary electrophoresis–based fragment analysis, Sanger sequencing, and next-generation sequencing methods, such as Illumina and PacBio.

13.2 POLYMERASE CHAIN REACTION (PCR): PRINCIPLES AND APPLICATIONS

An *in vitro* technique called polymerase chain reaction (PCR) is used for the quick and accurate amplification of millions to billions of copies of a certain DNA fragment, allowing for further in-depth genetic analysis. PCR was devised by an American biochemist, Kary Mullis, in the 1980s and received the 1993 Nobel Prize for Chemistry for this transforming technology. The invention of the technique has revolutionized all the molecular genetic research through rapid, accurate, and cost-efficient genomic operations globally. Currently, PCR is widely utilized by research professionals and clinicians for diagnosing pathogens, cloning and gene sequencing, and quickly and sensitively performing complex genomic and quantitative assays. The enzymatic replication of deoxyribonucleotide (DNA) over the template molecule is the basis of the PCR technique. Using primer-mediated enzymes, a small DNA fragment (whose nucleotide sequence is known) is amplified during the PCR process. New DNA strands that are complementary to the template DNA are created by DNA polymerase. Only the pre-existing 3'-OH group can bind a nucleotide added to it by the polymerase enzyme. Consequently, a primer is needed. A *primer* is a brief sequence of nucleic acids that serves as the basis for the production of DNA. Primers are short RNA strands found in living organisms. Before DNA replication can start, an enzyme known as primase, a subtype of RNA polymerase, needs to synthesize a primer. The DNA polymerase's 3' prime end receives more nucleotides.

13.2.1 Basic Principles of PCR

- *Denaturation.* The DNA template is heated to high temperatures (around 95°C) to separate the double-stranded DNA into single strands.
- *Annealing.* The temperature is lowered to allow short DNA sequences called primers to bind to the complementary sequences on the single-stranded DNA templates. The annealing temperature is typically 3–5°C below the melting temperature (*Tm*) of the primers.
- *Extension.* The temperature is raised again to the optimal temperature for a DNA polymerase enzyme to extend the primers by adding complementary nucleotides, synthesizing new DNA strands complementary to the original templates.

These three steps, *viz*, denaturation, annealing, and extension, comprise one PCR cycle. The cycle is repeated multiple times (typically 20–40 times), allowing an exponential amplification of the target DNA sequence. The key components of PCR are the DNA template, Taq polymerase enzyme, primers, and nucleotides (dNTPs). The selection of appropriate primers and the optimization of the annealing temperature are critical for the specificity and efficiency of the PCR process.

13.2.2 Applications of PCR in Plant Genome Analysis

PCR has many applications in plant genome analysis, and it is used to detect and quantify specific DNA sequences in plant samples. This allows researchers to:

1. **Identify genetic diseases or mutations in plants by detecting disease-causing genes.** To identify genetic diseases or mutations in plants by detecting disease-causing genes, researchers utilize various genetic approaches. One common method involves studying plant disease susceptibility genes, which are genes required for susceptibility to certain pathogens (Vogel et al., 2002). Additionally, genetic engineering plays a crucial role in developing disease-resistant plants. Techniques like gene-expression atlases and genomic analysis are used to identify candidate genes for disease resistance in plants, particularly in legumes (Kankanala et al., 2019). Furthermore, genetic studies have shown that plant health is influenced by interactions with disease-suppressive microflora. Research has focused on understanding the genetic basis of these interactions to enhance plant health in natural environments. For instance, a study revealed a genetic analysis in plants to support disease suppression by microbial biocontrol agents, shedding light on the role of plant genes in establishing associations with beneficial microbes (Smith et al., 1999). In brief, the identification of genetic diseases or mutations in plants involves studying susceptibility genes, utilizing genetic engineering for disease resistance, and exploring the genetic basis of interactions with disease-suppressive microflora. These genetic approaches provide valuable insights into enhancing plant health and developing strategies for disease control in agricultural settings.
2. **Monitor gene expression changes in plants in response to environmental stresses.** In order to strengthen their resilience, plants have the ability to detect, interpret, and respond to environmental stress by triggering the expression of specific genes. These stress-inducible genes can be divided into two main types: (a) genes whose protein products directly confer stress resistance functions, such as late embryogenesis-abundant (LEA) proteins, antifreezing proteins, and enzymes for synthesizing osmoregulators, and (b) genes whose products regulate gene expression and signal transduction, such as transcription factors and protein kinases. Research professionals have studied the gene regulatory networks underlying plant stress responses in model organisms like *Arabidopsis*, rice, and yeast, but the full complexity of the gene networks is poorly understood, especially for plants facing multiple stresses (Ni et al., 2009). Recent studies have highlighted the role of epigenetic and chromatin-based mechanisms in plant stress adaptation and stress memory. Plants may be primed to respond more effectively to recurrent stressors, even over generations, by DNA methylation and histone modifications. Transcription factors, microRNAs, and protein stability pathways also contribute to stress memory (Lämke and Bäurle, 2017). Researchers have used various molecular, genomic, and biotechnological approaches to identify key stress-responsive genes and engineer stress-tolerant crop plants. For example, genes involved in osmolyte synthesis, antioxidant defense, and ABA signaling have been successfully introduced into crops to enhance drought, salt, and temperature stress tolerance (Ahanger et al., 2017). Overall, analysis of the gene expression is crucial in understanding the molecular mechanisms underlying plant stress responses and adaptation. Advances in this area can inform the development of more stress-tolerant crop varieties through genetic engineering and breeding (Hiroshima University Press, 2024).
3. **Analyze genetic variation between plant species.** *Genetic variation* refers to the differences in the nucleotide sequences in the genome of the individuals within a species or between different species. Genetic variation between plant species is significant and can be determined at multiple levels. At the DNA sequence level, different plant species will have differences in their genetic code, with varying numbers and sequences of genes (Bhandari et al., 2017). Genetic diversity allows plants to adapt to different environments and develop unique phenotypic and physiological traits. Evolution is a major source of genetic variation among plant species. New alleles and genes arise as a result of the accumulation of mutations and other genetic alterations throughout time (Singh et al., 2020). After then, this genetic variety is subject to natural selection, which favors some features

and eliminates others. Plant species diverge as a result, acquiring unique genetic profiles. Another major contributor to genetic variation between plant species is the process of hybridization. When closely related plant species interbreed, their genetic material can combine, creating new genetic combinations. Hybridization can introduce novel traits and increase the overall genetic diversity within a plant population. Factors like geographic isolation, mating systems, and population size can also influence the degree of genetic variation between plant species. Isolated populations tend to accumulate unique genetic changes, while self-pollinating plants generally exhibit lower genetic diversity compared to outcrossing species. Techniques like DNA sequencing and molecular markers have enabled researchers to quantify and compare the genetic diversity across different plant species in great detail (Singh et al., 2019). This information is crucial in understanding plant evolution, identifying unique genetic resources, and developing strategies for plant conservation and breeding.

4. **Phylogenetic analysis to study the evolutionary relationships between plant species.** PCR plays a crucial role in phylogenetic analysis to study the evolutionary relationships between plant species. PCR is utilized in amplifying specific DNA sequences, enabling researchers to generate large amounts of sequence data suitable for phylogenetic reconstruction (Uribe-Convers et al., 2016). By using PCR in combination with high-throughput sequencing (HTS), researchers can obtain affordable, large, multi-locus datasets for phylogenetic and population genetic studies. This method allows for the analysis of multiple genetic markers simultaneously, providing a more comprehensive understanding of the evolutionary history and relationships between plant species. Additionally, PCR is essential in generating DNA barcodes like *rbcL*, *matK*, *trnH-psbA*, and *ITS*, which are used for species identification and phylogenetic construction in tree species (Kang et al., 2017). PCR's ability to amplify and analyze specific DNA regions is fundamental in molecular phylogenetics, contributing significantly to the study of plant evolution and biodiversity.
5. **Detection and identification of plant pathogens.** Polymerase chain reaction plays a key role in the detection and identification of plant pathogens by providing fast, sensitive, and specific methods for diagnosing plant diseases (Mirmajlessi et al., 2015). PCR-based techniques utilize the total DNA from diseased plant tissue to detect target organisms, allowing for the quantification of pathogen load with high specificity. Real-time PCR (rtPCR) is particularly promising for diagnosing and quantifying pathogen populations in plants, offering specific, reliable, and high-throughput detection of target DNA in real time (Deepak et al., 2007). PCR techniques, especially rtPCR, enable the detection of amplification products during each PCR cycle, enhancing the accuracy of pathogen quantification. Additionally, quantitative (qPCR) is more sensitive than conventional PCR, allowing for early pathogen detection, even before disease symptoms appear (Venbrux et al., 2023). The quantification capability of qPCR enables the determination of action thresholds for disease management, leading to more efficient and sustainable strategies. Overall, PCR-based methods, including rtPCR and qPCR, are essential tools in plant pathology for detecting, quantifying, and characterizing plant pathogens, contributing significantly to disease management and crop protection (Lau and Botella, 2017).

13.2.3 Other Applications of PCR

1. **Identification of transgenes.** In genetically modified (GM) plants, PCR is often used to identify and confirm the presence of transgenes. This is essential for quality control in the production of genetically modified crops.
2. **Quantitative PCR (qPCR).** This variant of PCR allows for the quantification of DNA. In plant genome analysis, qPCR is frequently used to measure gene expression levels,

providing insights into the regulation of genes under different conditions or in response to various treatments.

3. **Comparative genomic studies.** PCR-based techniques, such as random amplified polymorphic DNA (RAPD) and amplified fragment length polymorphism (AFLP), are utilized in comparative genomic studies. These methods help identify differences in the DNA sequences between different plant varieties or species.
4. **DNA sequencing template preparation.** PCR is often used to generate templates for DNA sequencing. This is a critical step in obtaining the sequence information of specific genes or genomic regions in plants.
5. **Epigenetic studies.** PCR can be adapted to study epigenetic modifications, such as DNA methylation. Understanding epigenetic regulation is important in unraveling the complexities of plant development and environmental responses.
6. **Functional genomics.** PCR is employed in functional genomics studies to validate the function of specific genes. This can involve the amplification of target genes for further analysis, such as gene knockout or overexpression experiments.

Overall, PCR plays a pivotal role in advancing our understanding of plant genomes, enabling researchers to explore various aspects of plant biology, genetics, and biotechnology. The key advantages of PCR for plant genome analysis are its sensitivity, specificity, and speed compared to traditional methods. PCR can detect and amplify target DNA sequences from very small samples, making it useful for analyzing limited plant material. The use of species-specific primers allows for a highly selective detection of target genes or organisms. And PCR can generate millions of copies of a DNA sequence in just a few hours, enabling rapid analysis. Overall, PCR has revolutionized plant genomics and biotechnology by providing a powerful tool in detecting, quantifying, and characterizing plant genetic material with high precision and efficiency.

ACKNOWLEDGMENTS

The authors acknowledge the Department of Biotechnology, School of Applied Sciences, REVA University for providing the necessary facilities.

REFERENCES

Ahanger, M. A., Akram, N. A., Ashraf, M., Alyemeni, M. N., Wijaya, L. & Ahmad, P. (June 27, 2017). Plant responses to environmental stresses-from gene to biotechnology. AoB Plants 9(4): plx025.

Altschul, S. F., Madden, T. L., Schäffer, A. A., Zhang, J., Zhang, Z., Miller, W. & Lipman, D. J. (1997). Gapped BLAST and PSI-BLAST: A new generation of protein database search programs. Nucleic Acids Research 25(17): 3389–3402.

Bhandari, H. R., Bhanu, A. N., Srivastava, K., et al. (2017). Assessment of genetic diversity in crop plants—An overview. Advances in Plants & Agriculture Research 7(3): 279–286.

Breseghello, F. & Sorrells, M. E. (2006). Association mapping of kernel size and milling quality in wheat (*Triticum aestivum* L.) cultivars. Genetics 172: 1165–1177.

Deepak, S., Kottapalli, K., Rakwal, R., Oros, G., Rangappa, K., Iwahashi, H., Masuo, Y. & Agrawal, G. (June 2007). Real-time PCR: Revolutionizing detection and expression analysis of genes. Current Genomics 8(4): 234–251.

Dong, S. & Searls, D. B. (1994). Gene structure prediction by linguistic methods. Genomics 23: 540–551.

Gericke, N. & Hagberg, M. (2006). Definition of historical models of gene function and their relation to students' understanding of genetics. Science & Education 16(7–8): 849–881. Bibcode:2007Sc&Ed.16.849G.

Gómez, G., Álvarez M. F. & Mosquera T. (2011). Association mapping, a method to detect quantitative trait loci: Statistical bases. Agronomía Colombiana 29: 367–376.

Gross, S. S., Do, C. B., Sirota, M. & Batzoglou, S. (2007). Contrast: A discriminative, phylogeny-free approach to multiple informant de novo gene prediction. Genome Biology 8(12): R269.

Guigó, R. (1992). Prediction of gene structure. Journal of Molecular Biology 226: 141–157.

Gupta, P. K., Rustgi, S. & Kulwal, P. L. (2005). Linkage disequilibrium and association studies in higher plants: Present status and future prospects. Plant Molecular Biology 57: 461–485.

Heather, J. M. & Chain, B. (2016). The sequence of sequencers: The history of sequencing DNA. Genomics 107(1): 1–8.

Hiroshima University. (2024). Genetic underpinnings of environmental stress identified in model plant. ScienceDaily. www.sciencedaily.com/releases/2024/04/240411130122.htm.

Johnson, L. S., Eddy, S. R. & Portugaly, E. (2010). Hidden Markov model speed heuristic and iterative hmm search procedure. BMC Bioinformatics 11: 431.

Kang, Y., Deng, Z., Zang, R., et al. (2017). DNA barcoding analysis and phylogenetic relationships of tree species in tropical cloud forests. Scientific Reports 7: 12564.

Kankanala, P., Nandety, R. S. & Mysore, K. S. (2019). Genomics of plant disease resistance in legumes. Frontiers in Plant Science 10: 469336.

Lämke, J. & Bäurle, I. (2017). Epigenetic and chromatin-based mechanisms in environmental stress adaptation and stress memory in plants. Genome Biology 18: 124.

Lau, H. Y. & Botella, J. R. (2017). Advanced DNA-based point-of-care diagnostic methods for plant diseases detection. Frontiers in Plant Science 8: 302813.

Lincoln, S., Daly, M. J., & Lander, E. (1993). Constructing genetic linkage maps with MAPMAKER/EXP. Version 3.0. Whitehead Institute for Biomedical Research Technical Report, 3rd Edn.

Liu, N., Xue, Y., Guo, Z., Li, W. & Tang, J. (2016). Genome-wide association study identifies candidate genes for starch content regulation in maize kernels. Frontiers in Plant Science 7: 1046.

Mackill, D. J. & Ni, J. (2000). Molecular mapping and marker assisted selection for major-gene traits in rice. In Proceedings of the Fourth International Rice Genetics Symposium (Eds. G. S. Khush, D. S. Brar & B. Hardy), Los Banos, The Philippines: International Rice Research Institute, pp. 137–151.

Manly, K. F., Cudmore Jr, R. H. & Meer, J. M. (2001). Map manager QTX, cross-platform software for genetic mapping. Mammalian Genome 12(12): 930–932.

Meng, L., Li, H., Zhang, L., & Wang, J. (2015). QTL IciMapping: Integrated software for genetic linkage map construction and quantitative trait locus mapping in biparental populations. The Crop Journal 3(3): 269–283.

Meunier, R. (2022). Stanford encyclopedia of philosophy: Gene. Stanford Encyclopedia of Philosophy. Retrieved 26 January 2024.

Mirmajlessi, S. M., Destefanis, M., Gottsberger, R. A., Mänd, M. & Loit, E. (January 15, 2015). PCR-based specific techniques used for detecting the most important pathogens on strawberry: A systematic review. Systematic Reviews 4(1): 9.

Ni, F. T., Chu, L. Y., Shao, H. B. & Liu, Z. H. (June 2009). Gene expression and regulation of higher plants under soil water stress. Current Genomics 10(4): 269–280.

Orgogozo, V., Peluffo, A. E. & Morizot, B. (2016). The "Mendelian gene" and the "Molecular gene": Two relevant concepts of genetic units. Current Topics in Developmental Biology 119: 1–26.

Pearson, W. R. (2013). An introduction to sequence similarity ("homology") searching. Current Protocols in Bioinformatics 42(1): 1–8.

Pearson, W. R. & Lipman, D. J. (1988). Improved tools for biological sequence comparison. Proceedings of the National Academy of Sciences, USA 85: 2444–2448.

Risch, N. (1992). Genetic linkage: Interpreting LOD scores. Science 255(5046): 803–805.

Roth, S. C. (2019). "What is genomic medicine?" Journal of the Medical Library Association. University Library System, University of Pittsburgh 107(3): 442–448.

Singh, R. B., Mahenderakar, M. D., Jugran, A. K., Singh, R. K., & Srivastava, R. K. (2020). Assessing genetic diversity and population structure of sugarcane cultivars, progenitor species and genera using microsatellite (SSR) markers. Gene 753: 144800.

Singh, R. B., Singh, B., & Singh, R. K. (2019). Development of potential dbEST-derived microsatellite markers for genetic evaluation of sugarcane and related cereal grasses. Industrial Crops and Products 128: 38–47.

Smith, K. P., Handelsman, J. & Goodman, R. M. (1999). Genetic basis in plants for interactions with disease-suppressive bacteria. Proceedings of the National Academy of Sciences, USA 96(9): 4786–4790.

Smith, T. F. & Waterman, M. S. (1981). Identification of common molecular subsequences. Journal of Molecular Biology 147: 195–197.

Srivastava, R. K., Singh, R. B., Srikanth, B., Satyavathi, C. T., Yadav, R. & Gupta, R. (2019). Genome-wide association studies (GWAS) and genomic selection (GS) in pearl millet: Advances and prospects. Frontiers in Genetics 10: 1389.

Snyder, E. E. & Stormo, G. D. (1993). Identification of coding regions in genomic DNA sequences: An application of dynamic programming and neural networks. Nucleic Acids Research 21: 607–613.

Solovyev, V. V. (1994). Predicting internal exons by oligonucleotide composition and discriminant analysis of spliceable open reading frames. Nucleic Acids Research 22: 5156–5163.

The Arabidopsis Genome Initiative. (2000). Analysis of the genome sequence of the flowering plant *Arabidopsis thaliana*. Nature 408: 796–815.

Uribe-Convers, S., Settles, M. L. & Tank, D. C. (February 1, 2016). A phylogenomic approach based on PCR target enrichment and high throughput sequencing: Resolving the diversity within the South American species of Bartsia L. (*Orobanchaceae*). PLoS One 11(2): e0148203.

Venbrux, M., Crauwels, S. & Rediers, H. (May 8, 2013). Current and emerging trends in techniques for plant pathogen detection. Frontiers in Plant Science 14: 1120968.

Vogel, J., Raab, T. K., Schiff, C. & Somerville, S. C. (2002). *PMR6*, a pectate lyase–like gene required for powdery mildew susceptibility in Arabidopsis. Plant Cell 14: 2095–2106.

Wang, Z., Chen, Y., & Li, Y. (2004). A brief review of computational gene prediction methods. Genomics, Proteomics & Bioinformatics 2(4): 216–221.

Winter, P. & Kahl, G. (1995). Molecular marker technologies for plant improvement. World Journal of Microbiology and Biotechnology 11: 438–448.

Yandell, M. & Ence, D. (April 2012). A beginner's guide to eukaryotic genome annotation. Nature Reviews Genetics 13(5): 329–342.

Zhao, K., Aranzana, M. J., Kim, S., Lister, C., Shindo, C., Tang, C., et al. (2007). An Arabidopsis example of association mapping in structured samples. PLoS Genetics 3: e4.

Fermentation Technology

14 Fermentation Technology for the Production of Ergot Alkaloids

Single-Cell Proteins and Enzymes of Pharmaceutical Interest

Akash Jain, Jasmine Chaudhary, Vichitra Kaushik, Pankaj Dagur, Jashanpreet Singh Sandhu, and Manik Ghosh*
* Corresponding author: vichitra.kaushik@gmail.com

14.1 INTRODUCTION

Fermentation is an ancient technology that relies on traditional knowledge. Although this technology's contribution to the pharmaceutical field is not new, its utilization has picked up steam recently as a result of breakthroughs and discoveries (Stanbury et al., 2013). It is a method that turns microorganisms or other organic materials into vital medicines, including insulin, medicinal proteins, enzymes, and antibiotics. In order to enhance the productivity of the target organism, experimentation is often carried out in temperature-controlled tanks (fermenters), which need the addition of nutrients at the right concentration.

The two broad categories of fermentation at industrial scale are submerged (SmF) and solid-state (SSF) fermentation. Submerged fermentation (SmF) employs microorganisms in a liquid nutrient environment for enzyme production, whereas in solid-state fermentation (SSF), microbes thrive on mostly insoluble substrates with little moisture (no free liquid) (Anupuma & Ravindra, 2001; Ravindra, 2001). These low-energy methods have been essential in the utilization of microbes to produce valuable and commercial pharmaceuticals. It has been effectively utilized in changing agricultural wastes, such as yam peels (Aderiye & Ogunjobi, 1998), and it has been used to produce feed from carbohydrate substrates (wastes), which has significant potential for use as animal and cattle feed (Aderiye & Ogunjobi, 1998).

Claviceps fusiformis produces total ergot alkaloids about four times more in SSF than SmF. In the instance of *Claviceps purpurea*, production is comparable, but solid-state fermentation improves the spectrum of alkaloids. The market for ergot alkaloids and their derivatives has increased recently owing to its potential therapeutic uses. However, the production of lysergic acid and other ergot alkaloids via the chemical pathway is not feasible due to the complexity of the entire chemical synthesis, which entails 15 stages and 140 potential intermediates (Hernandez et al., 1992).

The conventional method, which involves growing plants in the field and which has been in existence as a livelihood for many years across nations, has some drawbacks, including an intensive labor requirement, higher reliability on weather conditions, loss of conidia during rainy weather leading to disturbance in the asexual cycle, and damage to sclerotia during harvest. The fermentation technique, which also has benefits for production efficiency, is used to produce ergot alkaloids saprophytically, which substantially overcomes most of these restrictions (Řeháček, 1984; Hernández et al., 1993).

DOI: 10.1201/9781003326939-17

Enzymes, also known as *biological catalysts*, are essential in the medical field because they start biochemical events without being consumed (Robinson, 2015). Due to their numerous uses as pharmaceuticals, industrial catalysts, analytical or diagnostic instruments, and manipulative tools, they are of great interest to the pharmaceutical industry (Meghwanshi et al., 2020; Vellard, 2003). These enzymes can be produced from microorganisms (bacteria, yeasts, fungi, etc.), plants, and animals using the fermentation technique, and with advancements in technology, the commercial production of enzymes has been augmented (Deckers et al., 2020).

As the initial by-product of the fermentation process, single-cell proteins (SCPs) have gained enormous relevance as a superior protein alternative. SCPs can be synthesized by both solid and submerged type of fermentation, each having their own advantages and limitations. The microbial mass produced through fermentation is the major product which can be harvested or used in the synthesis of single-cell protein (Junaid et al., 2020). Nutritional characteristics supplied in a medium, namely, carbon, nitrogen, minerals, etc., which a microbe can use are of utmost importance for the development of SCPs (Aggelopoulos et al., 2014). Agricultural waste products can also be used in the production of SCPs via fermentation due to its low cost. These SCPs were first used and discovered as a new protein source in World War I, and subsequently, its commercial production was initiated (Ukaegbu-Obi, 2016).

To enhance the quality and quantity of pharmaceuticals, as well as to make them more cost-effective, extensive research is being done in various interrelated fields and disciplines. Microbial strains, fermentation techniques, and downstream processes need to be modified and improvised in order to accomplish this goal. The different facets of fermentation technology are highlighted in this book chapter, along with how these techniques are used to produce single-cell proteins, medicinal enzymes, and ergot alkaloids.

14.2 FERMENTATION TECHNOLOGY

Fermentation is an interdisciplinary scientific technique which uses concepts of biological and engineering sciences, particularly microbiology and biochemistry. Industrial fermentation, including both submerged and solid-state, is essentially linked to the extensive commercial use of microorganisms characteristically in large-scale production processes or chemical reactions in various sectors, like microbiology, food, pharmaceuticals, etc. It aims to enhance the biochemical or physiological functions that bacteria may carry out while producing a specific product in the greatest quality and quantity possible. Knowledge from different fields, like microbiology, genetics, chemistry, chemical engineering, etc., can help in designing a suitable fermentation process. Fungi (like yeast) and bacteria are the main microorganisms employed in industrial fermentation (McNeil & Harvey, 2008). Large fermenters or other bioreactors, sometimes with a capacity of several thousand liters, are used for fermentation.

Depending upon the mode of operation, the four processes can be categorized as batch, continuous, perfusion, and fed-batch culture (Ren et al., 2008; Sadh et al., 2018)

Batch culture. The simplest method, which utilizes a predetermined volume of sterilized medium and nutrients inoculated with specific strains for fermentation, after which no further additions are made to the medium as the process starts. However, acid, alkali, and antifoam are added throughout the process. Then the contents are sent for downstream processes, followed by cleaning of the reactor, refilling, and repeating the same procedure for the next batch.

Continuous culture. It involves the instigation of fermentation process in batch manner. After a specified period, a fresh medium containing a sterilized nutrient-rich broth is introduced into the fermenter based upon the growth rate of the microorganism utilized. Along with it, the medium, containing product and biomass (bioreactor fluid), is continuously withdrawn from the overflow line of the vessel. This process is more efficient than batch culture as it can be operated for a long period without any pause, and also, bacterial growth rate could be easily monitored and improvised by controlling feed flow rate.

Perfusion culture. It is comparable to the continuous culture process in incorporating new media regularly, after a specified duration, with the removal of the used media and products simultaneously. However, a significant percentage of viable cells are retained in the fermenter through the use of alternating or conventional tangential flow filtration, or by adhering the cells to substrates, such as capillary fibers, membranes, microcarriers in fixed beds, etc.

Fed-batch culture. It involves the addition of nutrients to the fermenter in either small batches at regular intervals or depending upon the requirement of the process.

14.2.1 Methods of Fermentation

14.2.1.1 Submerged Fermentation (SmF)

The process of growing microorganisms in a nutrient-rich medium wherein the components are either dissolved or suspended is known as *submerged fermentation*, primarily appropriate for microorganisms, like bacteria, which need a higher moisture content to flourish. Closed vessels or stirred tank reactors containing nutrient-rich media can be used preferentially for bacteria and fungi at high oxygen concentration that may facilitate the production of both extracellular and intracellular enzymes (in case of recombinant *E. coli* or other strains).

These selective microorganisms carry out fermentation in reactors having the capacity of 1,000 m^3, causing the metabolization of nutrients and the production of desirable enzymes for the breakdown of nutrients.

Submerged fermentation is manly carried out in four ways for the industrial production of enzymes. All parameters, viz, temperature, pH, oxygen consumption, and carbon dioxide formation, should be controlled to optimize fermentation (Bhargav et al., 2008; Pérez-Guerra et al., 2003; Singhania et al., 2010).

14.2.1.2 Solid-State Fermentation (SSF)

It involves developing microorganisms for enzyme production on a solid substrate, such as rice bran, or straw and saw dust, oil cakes, or an agro-industrial residue, like sugarcane bagasse, banana or cassava waste, etc. It is mainly suitable for those microorganisms that need less moisture content for their development. The substrate can be selected depending upon its availability, cost, moisture level, etc. Particle size is another important criterion for substrate selection, as smaller substrate particles offer larger surface areas for the proliferation of microorganisms. However, too small a surface area impedes respiration efficiency and growth, and larger particles lead to less surface area and thus poor enzyme production. This method is generally more appropriate for those processes where instead of isolated enzymes, crude fermented products themselves are used as main product. Many enzymes, including pectinase, protease, glucoamylase, cellulose, etc., are obtained through this process (Knob et al., 2014; Vidyalakshmi et al., 2009; Favela-Torres et al., 2006).

14.3 PRODUCTION OF ERGOT ALKALOIDS

Ergot alkaloids are basically indole derivatives, containing the tetracyclic ergoline ring system (Figure 14.1), and are grouped according to their structure into clavine alkaloids (Figure 14.2) and lysergic acid alkaloids (Figure 14.3) (Hernandez et al., 1992; Hernández et al., 1993). Ergot alkaloids have received much interest from pharmacologists due to their diverse actions, including effects on uterine and vascular smooth muscle, as well as its future scope in studies based on the sympathetic nervous system. More recently discovered alkaloids and synthetic derivatives have been shown to inhibit prolactin release and stimulate dopamine receptors and have found scope in the treatment of Parkinsonism.

Due to their immense pharmacological importance, it was desirable to produce large quantities of ergot alkaloids economically. Commercially desirable alkaloids can be obtained in three ways:

1. By isolation from *Claviceps purpurea* grown on cereal hosts
2. By extraction of submerged fermentations of *C. paspali* and *C. purpurea*
3. By synthetic derivatization of lysergic acid

Although the parasitic mode accounts for 95% of the world production of ergot alkaloids, it has many disadvantages. In 1936, Von Bekesy experienced this variability, when, using a newly developed inoculation technique, he found his first crop to be alkaloid free. Except the peptide alkaloids,

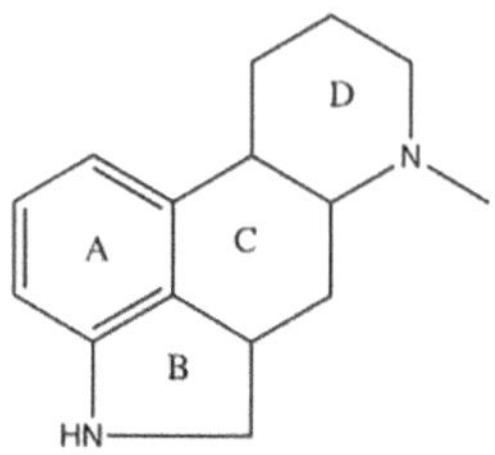

FIGURE 14.1 Ergoline ring system.

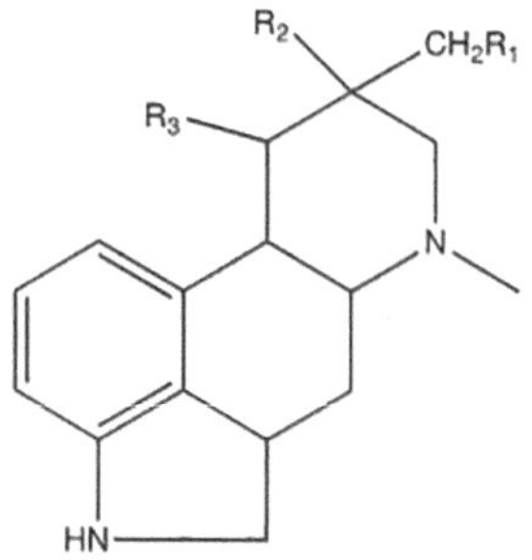

FIGURE 14.2 General formula of the clavinet alkaloids (R_1, R_2, R_3 = H or OH).

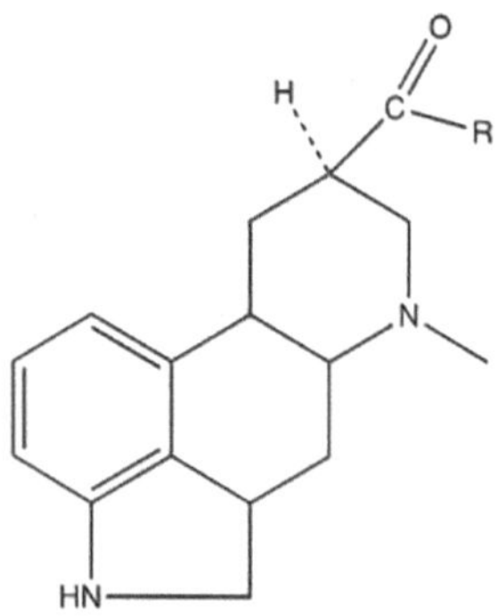

FIGURE 14.3 General formula of the lysergic acid alkaloids: lysergic acid (R = H), lysergic acid amide (R = NH_2), R = residue of an amino alcohol.

the majority of medicinally used ergolines are produced commercially via partial synthesis from lysergic acid. *Paspali* is fermented to produce lysergic acid. Total synthesis is not economical. The saprophytic cultivation of ergot to give high yields of pharmacologically useful peptide alkaloids therefore became a desirable objective. The first *in vitro* production of ergot alkaloids was demonstrated by Abe (1951). Subsequently, many other workers reported the in vitro production of ergot alkaloids, most of which were of the clavine type.

Perhaps one of the biggest breakthroughs in the saprophytic culture of ergot fungi came when Arcamone et al. (1961) discovered a method for the large-scale production of lysergic acid hydroxyethylamide by *Claviceps paspali* (Arcamone et al., 1961). This pharmacologically uninteresting alkaloid is normally produced in trace amounts in parasitic culture, but an industrial fermentation process was developed in which the fungus yielded an alkaloid equivalent to half of the dry weight of the mycelium. Following this, other workers reported successful *C. paspali* fermentation. The *C. paspali* strain used by Kobel et al. produced 8,9 lysergic acid, in which the double bond could be rearranged when treated with dilute alkali to give 9,10 lysergic acid. All *C. paspali* strains, with the exception of that of Kobel et al. (1964), which produced high yields of alkaloid, did not produce conidia and were purple-pigmented. In parasitic culture, *C. paspali* produces abundant conidia and the sclerotia are light-brown in color. These high-alkaloid-producing strains were unstable, and Kobel et al. (1964) reported a progressive decrease in alkaloid production from 1,940 mg/L to 20 mg/L over ten successive routine subcultures (Groger & Tyler, 1963; Kobel et al., 1964; Pacifici et al., 1962).

Currently, industrial-scale ergot alkaloids are produced using SmF and SSF fermentation methods. Other fermentation methods being researched to increase productivity include protoplast utilization, cell-free systems using crude enzymes for biosynthesis, and semicontinuous fermentation with immobilized cells. Liquid surface fermentation and immobilized cells have been partially used to replicate the parasitic lifestyle in a fermentation system, which is widely recognized as a method for generating high yields.

Previous comparisons of *Claviceps fusiformis* ATCC 26019–based production of ergot alkaloids in SmF and SSF processes showed that the latter system produced 3.9 times more total alkaloids. Similar amounts of *Claviceps purpureli* 1029c were produced, but the SSF technique produced an advantageous spectrum of alkaloids (Hernandez et al., 1992).

14.3.1 Submerged Fermentation (SmF) Method for Ergot Production (Amici et al., 1967)

14.3.1.1 Microorganism and Inoculum Development

Microorganisms of the strain 275 Fl, described by Amici et al., have been reported to produce a mixture of 1,100–1,500 μg/mL peptide alkaloids, majorly ergotamine (80%), in this category. By the isolation of certain sectors of their colonies, strains V, C, and W could be observed in medium T2, wherein only strain V produces alkaloids ranging up to 100 μg/mL, whereas others show no such activity.

14.3.1.2 Composition of Liquid Nutrient Medium

The nutrient medium typically includes:

- **Carbon source:** glucose, sucrose, or lactose (20–30 g/L)
- **Nitrogen source:** peptone, yeast extract, or ammonium salts (5–10 g/L)
- **Phosphates:** KH_2PO_4 (1–2 g/L)
- **Magnesium:** $MgSO_4$ (0.5–1 g/L)
- **Trace elements:** $FeSO_4$, $ZnSO_4$, $MnSO_4$ (in trace amounts)
- **pH adjusters:** HCl or NaOH to maintain pH around 5.5–6.5

14.3.1.3 SmF Method

Transfer the prepared inoculum into the fermentation medium. Provide adequate aeration and agitation to ensure proper oxygen transfer and nutrient mixing. This can be achieved using mechanical stirrers or air spargers. Maintain temperature around 24–28°C, depending on the specific strain requirements. Continuously monitor and adjust pH to maintain an optimal range for fungal growth and alkaloid production. The fermentation process typically lasts 7–10 days, during which the fungal mycelium grows and produces ergot alkaloids. At the end of the fermentation period, separate the fungal biomass from the fermentation broth. Extract ergot alkaloids from the biomass using appropriate solvent extraction methods.

14.3.2 Solid-State Fermentation Method (SSF) for Ergot Production (Hernández et al., 1993)

14.3.2.1 Microorganism and Inoculum Development

Sanglier's methodology can be utilized for the development of spore inoculum culture in petri dishes (13 cm internal diameter) containing agar medium or by preparing a PDA slant. After inoculation, cultures are left for incubation at 26°C for 5 days. The agar surface containing spores are then collected with the help of sterile distilled water (50 mL) that has a drop of Tween 80.

14.3.2.2 Composition of Nutrient Medium

The nutrient medium (per liter) is composed of 0.03% KCl, 0.0035% $FeSO_4.7H_2O$, and 0.01% $ZnSO_4.7H_2O$, along with a mineral salt solution. Its pH is maintained at 5.2 using 1 M NaOH. The sterilization conditions of the nutrient medium, sucrose, and their mixing are set at 121°C. The medium is then inoculated with the spore suspension to provide 0.6×10^7 spores/mL of the final liquid medium.

14.3.2.3 SSF Method

Bagasse derived from sugarcane pith is utilized as a solid, inert support after being cleaned, sieved, autoclaved, and dried, as per *Saucedo-Castaeda et al.*'s processing method. After processing, the bagasse is impregnated at a ratio of 30:70 (w/v) with an inoculated liquid medium. Static column fermenters (20 × 4 cm) are filled loosely with a well-mixed moist solid medium (60 g). These column fermenters are kept at 26°C for 120 hours of fermentation, and they are aerated at the required rates using humidified air.

14.4 PHARMACEUTICAL ENZYMES

Enzymes (biocatalysts) are the complex functional proteins required in very small amounts to catalyze biochemical reactions at a high rate under favorable conditions (temperature, pressure, pH, etc.) necessary for any living system to function normally and grow and proliferate without being consumed in the process. They generally create their own specific molecular functioning required to catalyze various biochemical processes involved in metabolic pathways and thus play a very vital and crucial element of every living entity. German physiologist Wilhelm Kuhne first used the term "enzyme" in 1878, and it originated from the Greek words *en*, meaning "within," and *zume*, meaning "yeast," due to the discovery of yeast's ability to generate alcohol from sugars (Robinson, 2015; Meghwanshi et al., 2020).

Enzymes play a key role in various processes involved in the production of food and beverages, meat processing, cheese production, cleaning supplies, clothing, paper products, transportation fuels, biodiesel production, and monitoring devices. The chiral selectivity of enzymes is exploited to synthesize enantiomerically pure pharmaceuticals, agrochemicals, food additives, etc.

Mainstream, currently used pharmaceutical enzymes are hydrolytic in nature and have regional stereospecificity, stability, and selectivity properties, used in asymmetric synthesis and racemic resolution. These enzymes are incredibly useful biodegradable catalysts which work at a high reaction rate, low temperature, and moderate pH. Various enzymes, like amylase, asparaginase, streptokinase, urokinase, collagenase, uricase, hyaluronidase, hydrolase, etc., are of pharmaceutical application as they target a particular biological reaction (reversible or irreversible) to treat a particular condition (Vellard, 2003; Deckers et al., 2020).

14.4.1 α-Amylase

Amylases (α-amylase, β-amylase, and γ-amylase) are significant hydrolase enzymes used widely as catalyst to cleave the internal α-1,4-glycosidic linkages in starch molecules in producing dextrins and other oligosaccharides, like glucose and maltose. However, their activity is pH- and metal cofactor–dependent, so they are also known as calcium metalloenzyme. Amylase enzymes are also used in the bakery industry to convert starch to fermentable sugars, in the detergent industry to convert starch to water-soluble dextrin, in the textile industry to remove starch from woven fabric, in alcohol production, etc. (Almanaa et al., 2020).

Numerous bacterial, fungal, and other genetically engineered microorganism species can produce α-amylase. Thermostable α-amylase can be manufactured from bacterial species like *B. subtilis, B. licheniformis, B. stearothermophilus, B. amyloliquefaciens*, etc. using agricultural and industrial residues, viz, gram husk, mustard oilseed cake, wheat, and rice bran, at a temperature of 45–55°C and pH 6–7 under an incubation period of 48–72 hours, whereas halophilic amylases can be obtained from *Halomonas meridiana* and *Chromohalobacter* and *Halobacillus* species using a carbon source, like maltose, glucose, sucrose, and oil cakes.

Additionally, *Penicillium* species (maintained at 30–55°C and pH 6–7), such as P. *fellutanum, P. roqueforti*, and *P. chrysogenum* (with incubation period of 72 hours), as well as fungal species (maintained at 30–90°C with pH 5–6), like *Aspergillus oryzae, Aspergillus kawachii*, and *Aspergillus niger* (with incubation period of 96 hours), can be used to produce it. The inorganic nitrogen bases used include ammonium chloride, sulphate, and hydrogen phosphate, and the organic nitrogen sources include yeast, peptone, and soybean meals (Saxena & Singh, 2011; Renge et al., 2012).

14.4.2 Protease

Proteases are widely distributed specialized proteolytic enzymes that break down protein molecules through peptide bond hydrolysis. Based on the source, they can be classified as plant proteases, animal proteases, and microbial proteases. Microbial proteases are further classified as serine, cysteine, aspartic, glutamic, threonine, and metalloproteases, depending on the catalytic residue present in the active site. They possess widespread applications in pharmaceutical industries for the synthesis of numerous therapeutic agents, like anticancer, antimicrobial, anti-inflammatory, etc. Various microorganisms can be employed for its synthesis using fermentation techniques (Muthulakshmi et al., 2011).

14.4.3 Collagenase

Collagenase is another protease enzyme which cleaves peptide bonds in collagen with high specificity and is used in the treatment of skin ulcers, burns. This enzyme is also used to produce bioactive collagen peptides, used in the pharmaceutical, food, and cosmetic industries. Various bacterial and fungal strains can be employed to obtain collagenase through fermentation, viz, *Penicillium aurantiogriseum*; *Pseudoalteromonas* spp.; *Bacillus* spp., like *B. tequilensis* and *B. subtilis*; *Pseudomonas* spp.; *Clostridium histolyticum*; *Candida albicans*; etc. (Ferreira et al., 2017).

14.4.4 Asparaginase

L-asparaginase, a tetrameric enzyme, responsible for the breakdown of L-asparagine to L-aspartic acid and ammonia, has gained importance due to its potential therapeutic value in cancer treatment, specifically in acute lymphoblastic leukemia. It can be obtained easily by fermentation owing to its wide-ranging microbial sources, like fungi, yeast, bacteria, and actinomycetes. The submerged fermentation process is used in the commercial synthesis of L-asparaginase from *Escherichia coli* and *Erwinia chrysanthemi* (Muso Cachumba et al., 2019).

14.4.5 Glutaminase

Another important enzyme that catalyzes the conversion of L-glutamine to L-glutamic acid and ammonium anions is L-glutaminase, also known as L-glutamine amidohydrolase. It is in great demand due to its various therapeutic applications in cancer therapy for its potent antileukemic and antiretroviral activity (Jambulingam & Saranya, 2020).

14.4.6 Thrombolytic Enzymes

Thrombolytic enzymes, like streptokinase and urokinase, are used in thrombolytic therapy to treat several cardiovascular diseases. They cause alteration of plasminogen to plasmin and thus helps in the lysis of fibrin clots. Both SmF and SSF are used to produce these enzymes from various microbial sources commercially.

14.4.7 Hyaluronidase

Hyaluronic acid, a high-molecular-weight non-sulfated linear glycosaminoglycan made up of repeating disaccharide units, D-glucuronic acid, and N-acetyl glucosamine, is hydrolyzed by this enzyme. It can function as an adjuvant, which hastens the absorption and dispersion of injected drugs. It can be produced from different strains of *Streptococcus mitis* in a nutrient broth using submerged fermentation (Sahoo et al., 2008).

14.4.8 Uricase

Uricase is another important pharmaceutical enzyme helpful in gout treatment due to its ability to convert urate into allantoin. It can be produced from *Bacillus cereus* by submerged fermentation, or from *Bacillus subtilis* by solid-state fermentation, of shrimp shell wastes (Kotb, 2016).

14.4.9 Lipase

Lipases, or triacylglycerol acyl hydrolases, are hydrolyzing enzymes that catalyze the breakage of ester bonds in triglycerides, resulting in the production of free fatty acids and glycerol at aqueous–organic interfaces. They also catalyze other reactions, like esterification, aminolysis, and alcoholysis, which are helpful in various processes. They are of utmost pharmaceutical importance due to their ability to enantiomerically catalyze the resolution of racemic pharmaceuticals, like antihypertensive and anti-inflammatory agents (Cavalcanti et al., 2005).

14.4.9.1 Production of Pharmaceutical Enzymes

Since most therapeutically used enzymes are easily handled, are economically feasible, have fast growth rates, and can be easily scaled up in large vessels, these are produced by fermentation in large closed fermentation tanks using appropriate and well-defined microbial strains, such as bacteria and fungi (Table 14.4). The selection of suitable microorganism is a very significant aspect in

TABLE 14.4
Fermentation Techniques Used in the Production of Various Pharmaceutical Enzymes

Microorganism Used	Method Employed	Substrate Used	Reference
		Amylase Enzyme	
Aspergillus flavus	SSF	Potato peels	(Ayinde et al., 2023)
	SmF	Mandarin peel (*Citrus reticulata*)	(Ali et al., 2017)
		Cocos nucifera meal	(Arunsasi et al., 2010)
Aspergillus niger	SmF	Jackfruit rag	(Weerasooriya & Piyarathne, 2019)
		Nutrient broth	(Behailu & Abebe, 2018)
Aspergillus oryzae	SSF	Wheat flour	(Farid & Mohamed, 2011)
Aspergillus terreus	SSF	Pearl millet	(Sethi et al., 2016)
Bacillus amyloliquefaciens	SmF	Nutrient broth	(Sharma et al., 2012)
Bacillus cereus	SmF	Soil sample	(Abo-Kamer et al., 2023)
Bacillus cereus and *Bacillus licheniformis*	SSF	Dairy sludge	(Issac & Prince, 2015)
Bacillus subtilis	SSF	Apple residue	(Singh et al., 2022)
		Agro-residue	(Almanaa et al., 2020)
		Banana waste	(Unakal et al., 2012)
Bacillus species	SSF	Potato peel	(Obi et al., 2019)
Citrobacter portucalensis	SmF	Agro-industrial waste	(Liaqat et al., 2024)
Geotrichum candidum	SSF	Natural substrates (olive pomace, potato peels, bread residues)	(Chaib et al., 2024)
Penicillium chrysogenum	SSF	Agricultural wastes (wheat bran, sunflower oil meal)	(Ertan et al., 2006)
Rhizopus oryzae	SSF	Cassava solid waste	(Martgrita et al., 2023)
		Protease	
Aspergillus flavus	SSF	Wheat bran	(Muthulakshmi et al., 2011)
Aspergillus niger	SSF	Agro-industrial waste	(Colla et al., 2023)
		Agro-industrial waste, like rice and wheat bran	(Dhaliwal et al., 2018)
Bacillus circulans	SSF	Agro-industrial waste	(Prakasham et al., 2005)
Bacillus subtilis	SmF	Agricultural waste	(Elumalai et al., 2020)
		Gelatin	(Pant et al., 2015)
Bacillus spp.	SSF	Agricultural residues	(Rajkumar et al., 2011)
	SmF	Soil	(Okpalla et al., 2022)
Rhizopus microsporus	SmF	Nutrient broth	(Var, 2010)
		Asparaginase	
Actinomycetes (Marine)	SSF and SmF	Soybean meal and tryptone glucose, yeast extract	(Basha et al., 2009)
Aspergillus niveus	SSF	Agro-industrial waste	(Amatto & Guimarães, 2019)
Aspergillus niger	SSF	Agricultural wastes from bran of *Cajanuscajan, Phaseolusmungo, Glycine max*	(Mishra, 2006)
Aspergillus terreus	SSF	Sugarcane bagasse	(Muso Cachumba et al., 2019)
Bacillus firmus	SmF	Mangrove sample	(Rudrapati & Audipudi, 2017)
Beauveria bassiana	SSF	Wheat bran	(Nageswara et al., 2014)
Citrobacter spp.	Multiple tubes lactose fermentation method	Water sample	(Shah et al., 2010)

(Continued)

TABLE 14.4 (*Continued*)
Fermentation Techniques Used in the Production of Various Pharmaceutical Enzymes

Microorganism Used	Method Employed	Substrate Used	Reference
Cladosporium spp.	SSF	Agricultural residues, like wheat bran, rice bran	(Kumar et al., 2013)
Enterobacter aerogenes	SmF		(Erva et al., 2017)
Fusarium culmorum	SSF	Agro-based waste, like soybean meal	(Meghavarnam & Janakiraman, 2017)
Pseudomonas aeruginosa	SSF	Soybean meal	(El-Bessoumy et al., 2004)
Streptomyces brollosae	SmF	Dextrose, starch	(El-Naggar et al., 2019)
Serratia marcescens	SSF	Coconut oil cake	(Ghosh et al., 2013)
		Collagenase	
Aspergillus terreus, *Aspergillu sparasiticus*, and *Aspergillus japonicus*	SmF	Chicken feather industrial residue	(Ferreira et al., 2017)
Aspergillus tamari	SSF	Agro-industrial Substrates	(da Silva et al., 2016)
Nocardiopsis dassonvillei	SSF	Chitin waste	(Abdel-Fattah AM, 2013)
		Glutaminase	
Aspergillus flavus	SSF	Wheat bran and Bengal gram husk	(Sathish et al., 2016)
Aspergillus niger	SSF	Agricultural waste	(Nagamani, 2018)
Bacillus subtilis	SmF	Nutrient broth	(Jambulingam & Murugesan, 2020)
Beauveria spp.	SSF	Seawater-based medium containing glutamine	(Sabu et al., 2000)
Fusarium solani-melongenae	SSF as well as SmF	Wheat bran	(Vineetha et al., 2024)
Pseudomonas spp.	SmF	Nutrient broth	(Jyothi et al., 2011)
Streptomyces pseudogriseolus	SSF	Wheat bran, sesame oil cake	(Wardah et al., 2023)
Streptomyces spp.	SmF	Nutrient broth	(Aly et al., 2017; Desai et al., 2016)
Trichoderma koningii	SSF	Sesamum oil cake	(Pallem et al., 2010)
		Agro-industrial by-products, like wheat bran	(El-Sayed, 2009)
Zygosaccharomyces rouxii	SSF	Agro-industrial substrates, like wheat bran and sesamum oil cake	(Kashyap et al., 2002)
		Thrombolytic Enzymes (Streptokinase and Urokinase)	
Bacillus subtilis	SmF	Nutrient broth	(Pan et al., 2019)
Bacillus licheniformis	SSF	Soybean	(Siva Ajithkumar et al., 2015)
Cladosporium spp.	SSF	Agro-based waste, like silkworm pupa, banana peel	(Hariharan et al., 2014)
Enterococcus gallinarum	SmF	Soya flour	(Nawaz et al., 2020)
Streptococcus equisimilis	SmF	Corn steep liquor	(Faran et al., 2015)
		Lipase	
Aspergillus flavus	SSF as well as SmF	Wheat bran	(Colla et al., 2015)
Aspergillus niger	SSF	Rice husk, cottonseed cake, and red gram husk mixture	(Nema et al., 2019)
Bacillus coagulans	SSF	Melon waste	(Alkan et al., 2007)
Penicillium spp.	SSF	Soybean meal	(Rigo et al., 2010)
Penicillium simplicissimum	SSF	Babassu cake	(Cavalcanti et al., 2005)
Penicillium chrysogenum, *Trichoderma harzianum*, and *Aspergillus flavus*	SSF	Agro-industrial residue	(Toscano et al., 2013)
Rhizopus homothallicus	SSF	Perlite	(Velasco-Lozano et al., 2012)
Yarrowia lipolytica	SSF	Cupuaçu fat cake	(Carvalho et al., 2023)

enzyme production using fermentation. Bacteria (*E. coli*, *B. subtilis*), fungi (*A. oryzae*, *A. niger*), yeasts (*S. cerevisiae*, *Pichia pastoris*), etc. are mostly microorganisms. Furthermore, fermentation allows for the acquisition of enzymes that are specifically targeted to perform special functions under specific conditions. Both processes, namely, SmF and SSF, can be employed for enzyme production. While each approach has advantages and disadvantages of its own, submerged fermentation is primarily used because of its many benefits, such as ensuring a high production yield and high quality.

To isolate the crude extracellular enzyme sample from the fermented mass, the process involves filtration, followed by centrifugation. The enzyme is then precipitated and concentrated using ammonium sulfate or organic solvents. After the sample has precipitated, it is treated with water or a buffer in order to concentrate it further, using chromatographic techniques, such as affinity chromatography, gel filtration, ion exchange, and so on. This allows the enzyme to be further separated and purified (Ajithkumar et al., 2015; Unakal et al., 2012).

14.4.9.2 Applications of Pharmaceutical Enzymes

Enzymes are being exploited in the pharmaceutical industry in catalyzing various industrial processes due to their ability to remain viable and perform different chemical transformations, like group protection, deprotection, hydrolysis, acylation, deacylation, esterification, deracemization, kinetic resolution, etc., in situ in both free and immobilized forms with high efficiency. They are also used in the production of various drugs, agrochemicals, biopolymers, etc. Asparaginase, streptokinase, urokinase, deoxyribonuclease, hyaluronidase, etc. are among the major enzymes employed in pharmaceutical applications. They are obtained in a lesser quantity but have high manufacturing cost as compared to industrially significant enzymes, like α-amylase, protease, and lipase, which are needed in enormous amounts and have significantly less manufacturing cost.

Although some enzymes are derived from animal and plant sources, microorganisms are the chief and most convenient source of pharmaceutically important enzymes because of the following:

- Since microbial growth is quick, enzyme production is also rapid. In addition, compared to plants and animals, microbes are simple to handle. They are a more cost-effective source and take up less space.
- Microorganisms can be easily manipulated for strain improvement, mutations, etc. using genetic engineering or other ways; thus, their production can be optimized (Robinson, 2015) (Meghwanshi et al., 2020).

14.5 SINGLE-CELL PROTEIN

Single-cell proteins (SCPs), derived as a by-product of the fermentation process, offer significant potential as an alternative protein source. As the global population continues to grow, competition for food resources intensifies, making SCPs a promising solution to address protein shortages. Given their expanding role not only in therapeutic applications but also as animal feed, numerous pharmaceutical companies in the United States and the UK are increasingly investing in SCPs.

SCPs consist of dried microbial biomass or the total protein content extracted from pure or mixed cultures of bacteria, yeasts, fungi, and microscopic algae (Spalvins et al., 2018). While a wide variety of SCPs are used as animal feed supplements, only a limited number are approved for human consumption. For human use, the substrates for SCP production should preferably be of food-grade quality (Ritala et al., 2017).

Agro-wastes suitable for SCP production include those rich in proteins or lipids, structural polysaccharides, monosaccharides/disaccharides, and starch. These different waste types are evaluated to determine the best substrate for industrial SCP production (Spalvins et al., 2018). SCPs can be produced through both solid-state and submerged fermentation, each with its own advantages and disadvantages. Both methods are utilized and optimized for SCP production (Spalvins et al., 2018; Ritala et al., 2017)

14.5.1 Production of SCPs

14.5.1.1 Cultivation Methods

SCP can be produced by fermenting selective microorganisms on a suitable host, followed by separation processes, with the aim of increasing the yield of cell mass.

The process begins with microbial screening, where an appropriate strain is identified from various sources, such as soil swabs from inorganic or biological materials. This strain is then optimized through selection, mutation, or other genetic techniques. Following optimization, the cultivation conditions for these strains are established, and all metabolic pathways and cell structures are thoroughly analyzed. For large-scale production, the technical aspects of the process need to be combined with engineering aspects and equipment technology to make ends meet. And this is where economic factors, like energy expenditure and other expenses, along with safety factors and environmental considerations, in relation to production come into play. Lastly, safety, innovation safeguards related to legal issues, operating licenses, product licensing for specific purposes, and legal aspects related to novel processes and microbial strains need to be tackled.

14.5.1.2 Submerged Fermentation

In case of submerged fermentation, the substrate is always in a liquid state containing growth-promoting nutrients. The continuously operating fermenter that contains the substrate harvests the product biomass continuously using a variety of methods before the product is filtered or centrifuged and finally dried. To combat the heat produced during cultivation, aeration is used as a cooling device. There are several ways to harvest the microbial biomass. Centrifugation is used to retrieve single-cell organisms, like yeast and bacteria, while filtration is employed for filamentous fungi. Water should be recovered before it dries up, as a final step.

14.5.1.3 Semisolid Fermentation

Although semisolid fermentation uses more solid substrates, such as cassava waste, its preparation method is not clear. Comparatively, fermentation in submerged cultures has a higher start-up costs and higher operating expenses. The process of cultivation involves a variety of tasks, such as stirring and combining a multiphase system, transporting oxygen via liquid phase to microorganisms, and releasing heat generated to the environment. For detecting this mass and energy transfer, a specialized bioreactor, the U-loop fermenter, is used.

The fundamental processes involved in the production of single-cell proteins start with the preparation of an appropriate medium with an adequate carbon source, along with preventative measures to reduce the chances of contamination, followed by the production of desirable microorganisms and, lastly, the separation of the synthesized biomass from its processing.

14.5.1.4 Solid-State Fermentation

The extensive and elaborate texts of previous studies suggest a range of bioreactors, suitable conditions, and microbes required for the production of products like SCPs, feeds, enzymes, ethanol, organic acids, B-complex vitamins, pigments, and flavors. According to those texts, the process starts with seeding the solid substrate, such as rice or wheat bran, with microorganisms, followed by its deposition on flatbeds. The substrate is then allowed to rest several days in a temperature-monitored area. Cultivating unicellular organisms like bacteria or yeast in a liquid medium is optimal. At larger scale, this process requires tanks reaching up to 2,500 m^2 (26,910 ft^2).

It is equally important to continuously provide the microbe with oxygen, along with monitoring other factors, like temperature, pH, and nutrients, to ensure liquid aerobic fermentation, which is often accomplished by agitating the fermentation medium (Wiebe, 2002; Yadav et al., 2014).

14.5.2 Applications of SCPs (Junaid et al., 2020)

The remarkable applications of single-cell proteins evince its significance to be produced on a commercial scale.

They have a variety of applications due to their high nutritional value and versatility. Here are some key applications:

1. **Animal feed.** SCPs are widely used as supplements in animal feed for livestock, poultry, and aquaculture. They provide a rich source of protein, essential amino acids, vitamins, and minerals, enhancing growth and health in animals.
2. **Human nutrition.** Although still limited, SCPs are used as protein supplements in human diets. They can be incorporated into various food products like protein bars, shakes, and dietary supplements, providing a sustainable alternative to traditional protein sources.
3. **Agricultural uses.** SCPs can be used as biofertilizers and soil conditioners, enhancing soil fertility and promoting plant growth. Their use can help reduce dependence on chemical fertilizers.
4. **Bioremediation.** SCP-producing microorganisms can be employed in bioremediation to clean up environmental pollutants. They can degrade toxic substances in soil and water, contributing to environmental sustainability.
5. **Pharmaceuticals.** SCPs have potential applications in pharmaceuticals, including in the production of vaccines, enzymes, and other therapeutic agents. They can be genetically engineered to produce specific proteins or bioactive compounds.
6. **Industrial enzymes.** SCPs can serve as sources for industrial enzymes used in various processes, such as brewing, baking, and biofuel production. These enzymes can improve efficiency and reduce costs in industrial operations.
7. **Aquaculture.** SCPs are used as feed in aquaculture to promote the growth and health of fish and other aquatic organisms. They provide a sustainable and high-quality protein source for aquafeeds.
8. **Space missions.** Due to their high protein content and efficient production methods, SCPs are being considered as a potential food source for long-duration space missions. They offer a compact and sustainable option for meeting the nutritional needs of astronauts. (Ali et al., 2017)

14.6 CONCLUSION

A vast variety of medications which can be used to treat almost any medical condition are produced using fermentation as their base. Examples include hormone imbalance therapy, antibiotics and vaccines, anticancer/cytotoxic medicines, enzymes, SCPs, etc. Numerous multidisciplinary fields are collaborating for research purposes to improvise the manufacturing of pharmaceutical products, both in terms of efficacy and yield. Improvements are being made to microbial strains, fermentation processes, and cutting-edge downstreaming techniques in order to attain the ultimate objective. This will facilitate the global adoption and use of this concept in a better way.

Ergot alkaloids have attracted the attention of pharmacologists because of their diverse impacts, including those on uterine and vascular smooth muscle, as well as their usefulness as a potential resource for digging deeper into the sympathetic nervous system. Due to their immense pharmacological importance, it is desirable to produce large quantities of ergot alkaloids economically. Fermentation technology helps the industries achieve those goals. Similar to this, pharmaceutical enzymes are appealing to the pharmaceutical industry since they focus on a particular biological reaction (either reversible or irreversible) to treat a certain ailment. The production of such enzymes by fermentation is mainly carried out using two processes: submerged fermentation (SmF) and solid-state fermentation (SSF).

Given the prevailing high costs associated with the production of recombinant products and monoclonal antibodies, the relative cost of the entire process will be a significant factor, in addition to product quality and quantity. The ultimate solution to this issue could be provided by studies emphasizing more on the use of affordable processes for plant or microbial expression, hence providing greater medical benefit to the global population.

REFERENCES

Abdel-Fattah, A. M. (2013). Production and partial characterization of collagenase from marine Nocardiopsis dassonvillei NRC2aza using chitin wastes. *Egyptian Pharmaceutical Journal*, *12*(2), 109.

Abe, M. (1951). Isolation and properties of an alkaloid "agroclavine" Produced by Ergot fungus in saprophytic culture. *Annual Report of Takeda Research Laboratories*, *10*, 145.

Abo-Kamer, A. M., Abd-El-Salam, I. S., Mostafa, F. A., Abotaleb, N. S., & Aboshanab, K. M. A. (2023). A promising microbial α-amylase production, and purification from *Bacillus cereus* and its assessment as antibiofilm agent against *Pseudomonas aeruginosa* pathogen. *Microbial Cell Factories*, *22*(1), 141.

Aderiye, B. I., & Ogunjobi, A. A. (1998). Fermentation of yam: Microbiology and sensory evaluation of cooked fermented yam tissues. *Plant Foods for Human Nutrition*, *52*(1), 49–54.

Aggelopoulos, T., Katsieris, K., Bekatorou, A., Pandey, A., Banat, I. M., & Koutinas, A. A. (2014). Solid state fermentation of food waste mixtures for single cell protein, aroma volatiles and fat production. *Food Chemistry*, *145*, 710–716. https://doi.org/10.1016/j.foodchem.2013.07.105.

Ajithkumar, D. S., Vidhya, A. K., Ragunathan, R., & Johney, J. (2015). Production and purification and characterization of streptokinase using Bacillus licheniformis under solid state fermentation. *Journal of Global Biosciences*, *4*(7), 2703–2712.

Ali, E. H., El-Nagdy, M. A., Al-Garni, S. M., Ahmed, M. S., & Rawaa, A. M. (2017). Enhancement of alpha amylase production by Aspergillus flavus AUMC 11685 on mandarin (Citrus reticulata) peel using submerged fermentation. *European Journal of Biological Research*, *7*(3), 154–164.

Ali, S., Mushtaq, J., Nazir, F., & Sarfraz, H. (2017). Production and processing of Single Cell Protein (SCP)—A review. *European Journal of Pharmaceutical and Medical Research*, *4*(7), 86–94.

Alkan, H., Baysal, Z., Uyar, F., & Dogru, M. (2007). Production of lipase by a newly isolated Bacillus coagulans under solid-state fermentation using melon wastes. *Applied Biochemistry and Biotechnology*, *136*(2), 183–192.

Almanaa, T. N., Vijayaraghavan, P., Alharbi, N. S., Kadaikunnan, S., Khaled, J. M., & Alyahya, S. A. (2020). Solid state fermentation of amylase production from Bacillus subtilis D19 using agro-residues. *Journal of King Saud University-Science*, *32*(2), 1555–1561.

Aly, M. M., Kadi, R. H., Aldahlawi, A. M., Alkhatib, M. H., & Wali, A. N. (2017). Production of the antitumor L-glutaminase enzyme from thermotolerant Streptomyces sp. D214, under submerged fermentation conditions. *Journal of Experimental Biology and Agricultural Sciences*, *5*(6), 878–885.

Amatto, I. V. da S., & Guimarães, L. H. S. (2019). Production of L-asparaginase by Aspergillus niveus under solid-state fermentation using agroindustrial byproducts. *International Journal of Scientific Reports*, *5*(9), 561–566.

Amici, A. M., Minghetti, A., Scotti, T., Spalla, C., & Tognoli, L. (1967). Ergotamine production in submerged culture and physiology of Claviceps purpurea. *Applied Microbiology*, *15*(3), 597–602.

Anupama, R., & Ravindra, P. (2000). Value-added food: Single cell protein. *Biotechnology Advances*, *18*(6), 459–479.

Arcamone, F., Chain, E. B., Ferretti, A., Minghetti, A., Pennella, P., Tonolo, A., & Vero, L. (1961). Production of a new lysergic acid derivative in submerged culture by a strain of Claviceps paspali Stevens & Hall. *Proceedings of the Royal Society of London: Series B. Biological Sciences*, *155*(958), 26–54.

Arunsasi, M., Kani, S. M., Jegadeesh, G., & Ravikumar, M. (2010). Submerged fermentation of amylase enzyme by Aspergillus flavus using Cocos nucifera meal. *Kathmandu University Journal of Science, Engineering and Technology*, *6*(2), 75–87.

Ayinde, E. O., Adejumo, T. O., & Adeoyo, O. R. (2023). Production, optimization and characterization of amylase using solid state fermentation method. *Frontiers in Environmental Microbiology*, *9*(1), 9–17.

Basha, N. S., Rekha, R., Komala, M., & Ruby, S. (2009). Production of extracellular anti-leukaemic enzyme lasparaginase from marine actinomycetes by solidstate and submerged fermentation: Purification and characterisation. *Tropical Journal of Pharmaceutical Research*, *8*(4).

Behailu, A., & Abebe, G. (2018). Isolation, production and characterization of amylase enzyme using the isolate Aspergillus Niger FAB-211. *International Journal of Biotechnology and Molecular Biology Research*, *9*(2), 7–14.

Bhargav, S., Panda, B. P., Ali, M., & Javed, S. (2008). Solid-state fermentation: An overview. *Chemical and Biochemical Engineering Quarterly*, *22*(1), 49–70.

Carvalho, A. S. S., Rocha, R. C. P., Sales, J. C. S., Souza, C. E. C., Lemes, A. C., & Coelho, M. A. Z. (2023). Biotechnological valorization of Cupuaçu by-products: Solid-state fermentation for lipase production by *Yarrowialipolytica*. *Fermentation*, *9*, 989.

Cavalcanti, E. d'Avila C., Gutarra, M. L. E., Freire, D. M. G., Castilho, L. dos R., & Sant'Anna Júnior, G. L. (2005). Lipase production by solid-state fermentation in fixed-bed bioreactors. *Brazilian Archives of Biology and Technology*, *48*, 79–84.

Chaib, I., Dakhmouche-Djekrif, S., Bennamoun, L., & Nouadri, T. (2024). Extracellular enzymes producing yeasts study: Cost-effective production of α-amylase by a newly isolated thermophilic yeast *Geotrichumcandidum* PO27. *AIMS Microbiology*, *10*(1), 83–106.

Colla, L. M., Ficanha, A. M. M., Rizzardi, J., Bertolin, T. E., Reinehr, C. O., & Costa, J. A. V. (2015). Production and characterization of lipases by two new isolates of Aspergillus through solid-state and submerged fermentation. *BioMed Research International*, *2015*.

Colla, L. M., Reinehr, C. O., Manfredini, P. G., Cavanhi, V. A. F. & Jorge Alberto Vieira Costa, J. A. V. (2023). Simultaneous production of proteases and antioxidant biopeptides by solid-state fermentation. *Sustainable Food Technology*, *1*, 874–885

da Silva, O. S., de Oliveira, R. L., Souza-Motta, C. M., Porto, A. L. F., & Porto, T. S. (2016). Novel protease from Aspergillus tamarii URM4634: Production and characterization using inexpensive agroindustrial substrates by solid-state fermentation. *Advances in Enzyme Research*, *4*(4), 125.

Deckers, M., Deforce, D., Fraiture, M.-A., & Roosens, N. H. C. (2020). Genetically modified micro-organisms for industrial food enzyme production: An overview. *Foods*, *9*(3), 326.

Desai, S. S., Chopra, S. J., & Hungund, B. S. (2016). Production, purification and characterization of L-Glutaminase from Streptomyces sp. isolated from soil. *Journal of Applied Pharmaceutical Science*, *6*(7), 100–105.

Dhaliwal, M. K., Patil, R. N., & Sambare, S. (2018). Protease production using solid state fermentation. In *Proceedings of the National Conference on Advanced Analytical Tools for Materials Characterization (AATMC-2018). International Journal of Scientific Research in Science, Engineering and Technology*, *5*(4), 80–88.

El-Bessoumy, A. A., Sarhan, M., & Mansour, J. (2004). Production, isolation, and purification of L-asparaginase from Pseudomonas aeruginosa 50071 using solid-state fermentation. *BMB Reports*, *37*(4), 387–393.

El-Naggar, N. E.-A., Moawad, H., El-Shweihy, N. M., El-Ewasy, S. M., Elsehemy, I. A., & Abdelwahed, N. A. M. (2019). Process development for scale-up production of a therapeutic L-asparaginase by Streptomyces brollosae NEAE-115 from shake flasks to bioreactor. *Scientific Reports*, *9*(1), 1–18.

El-Sayed, A. S. A. (2009). L-glutaminase production by Trichoderma koningii under solid-state fermentation. *Indian Journal of Microbiology*, *49*(3), 243–250.

Elumalai, P., Lim, J.-M., Park, Y.-J., Cho, M., Shea, P. J., & Oh, B.-T. (2020). Agricultural waste materials enhance protease production by Bacillus subtilis B22 in submerged fermentation under blue light-emitting diodes. *Bioprocess and Biosystems Engineering*, *43*(5), 821–830.

Ertan, F., Balkan, B., Balkan, S., & Aktac, T. (2006). Solid state fermentation for the production of α-amylase from Penicillium chrysogenum using mixed agricultural by-products as substrate. *Biologia*, *61*(6), 657–661.

Erva, R. R., Goswami, A. N., Suman, P., Vedanabhatla, R., & Rajulapati, S. B. (2017). Optimization of L-asparaginase production from novel Enterobacter sp., by submerged fermentation using response surface methodology. *Preparative Biochemistry and Biotechnology*, *47*(3), 219–228.

Faran, G. E., Zia, M. A., Shahid, M., & Abdullah, S. (2015). Improved streptokinase production: UV irradiation of Streptococcus equisimilis. *Professional Medical Journal*, *22*(5), 656–663.

Farid, M. A. F., & Mohamed, A. H. S. H. (2011). *Amylase production from Aspergillus oryzae LS1 by solid-state fermentation and its use for the hydrolysis of wheat flour*.

Favela-Torres, E., Volke-Sepúlveda, T., & Viniegra-González, G. (2006). Production of hydrolytic depolymerising pectinases. *Food Technology & Biotechnology*, *44*(2).

Ferreira, C. M. O., Correia, P. C., Brandão-Costa, R. M. P., Albuquerque, W. W. C., Lin Liu, T. P. S., Campos-Takaki, G. M., & Porto, A. L. F. (2017). Collagenase produced from Aspergillus sp.(UCP 1276) using chicken feather industrial residue. *Biomedical Chromatography*, *31*(5), e3882.

Ghosh, S., Murthy, S., Govindasamy, S., & Chandrasekaran, M. (2013). Optimization of L-asparaginase production by Serratia marcescens (NCIM 2919) under solid state fermentation using coconut oil cake. *Sustainable Chemical Processes*, *1*(1), 1–8.

Groger, D., & Tyler, V. E. (1963). Alkaloid production by Claviceps paspali in submerged culture. *Lloydia*, *26*(3), 174–+.

Hernández, M. R. T., Lonsane, B. K., Raimbault, M., & Roussos, S. (1993). Spectra of ergot alkaloids produced by Claviceps purpurea 1029c in solid-state fermentation system: Influence of the composition of liquid medium used for impregnating sugar-cane pith bagasse. *Process Biochemistry*, *28*(1), 23–27.

Hernandez, M. R. T., Raimbault, M., Roussos, S., & Lonsane, B. K. (1992). Potential of solid state fermentation for production of ergot alkaloids. *Letters in Applied Microbiology*, *15*(4), 156–159.

Issac, R., & Prince, R. (2015). Production of alpha-amylase by solid state fermentation using Bacillus cereus MTCC 7524 and Bacillus licheniformis MTCC 7445 from dairy sludge-A comparative study. *International Journal of Pharmtech Research*, *8*(9), 111–117.

Jambulingam, K., & Saranya, M. (2020). Studies on optimization of L-glutaminase production under submerged fermentation from marine Bacillus subtilis JK-79. *African Journal of Microbiology Research*, *14*(1), 16–24.

Junaid, F., Khawaja, L. A., & Ali, S. (2020). Single cell proteins as a potential meat substitute: A critical review. *World Journal of Pharmaceutical Research*, *9*, 141–161.

Jyothi, H., Kumar, S., & Vandana, R. (2011). *Production of l-glutaminase by Pseudomonas VJ-6.*

Kashyap, P., Sabu, A., Pandey, A., Szakacs, G., & Soccol, C. R. (2002). Extra-cellular L-glutaminase production by Zygosaccharomyces rouxii under solid-state fermentation. *Process Biochemistry*, *38*(3), 307–312.

Knob, A., Fortkamp, D., Prolo, T., Izidoro, S. C., & Almeida, J. M. (2014). Agro-residues as alternative for xylanase production by filamentous fungi. *BioResources*, *9*(3), 5738–5773.

Kobel, H., Schreier, E., & Rutschmann, J. (1964). 6-Methyl-Δ8, 9-ergolen-8-carbonsäure, ein neues Ergolinderivat aus Kulturen eines Stammes von Claviceps paspali Stevenset Hall. 60. Mitteilung über Mutterkornalkaloide. *Helvetica Chimica Acta*, *47*(4), 1052–1064.

Kotb, E. (2016). Improvement of uricase production from Bacillus subtilis RNZ-79 by solid state fermentation of shrimp shell wastes. *Biologia*, *71*(3), 229–238.

Kumar, N. S. M., Ramasamy, R., & Manonmani, H. K. (2013). Production and optimization of L-asparaginase from Cladosporium sp. using agricultural residues in solid state fermentation. *Industrial Crops and Products*, *43*, 150–158.

Liaqat, U. S., Naz, S., Hussain, H. I. (2024). Optimization and production of highly stable amylase isolated from *Citrobacterportucalensis* using low-cost agro-industrial wastes as substrates: Prospective therapeutics. *Kuwait Journal of Science*, *51*(1), 100156.

Martgrita, M. M., Kembaren, R. F., Hutapea, H. N., Sitepu, I., & Evy Enjelina Simanjuntak, E. (2023). Amylase production by *Rhizopus oryzae* using solid state fermentation with Cassava solid waste as substrate. *Advances in Biological Sciences Research*, *27*, 257–265.

McNeil, B., & Harvey, L. (2008). *Practical Fermentation Technology*. John Wiley & Sons.

Meghavarnam, A. K., & Janakiraman, S. (2017). Solid state fermentation: An effective fermentation strategy for the production of L-asparaginase by Fusarium culmorum (ASP-87). *Biocatalysis and Agricultural Biotechnology*, *11*, 124–130.

Meghwanshi, G. K., Kaur, N., Verma, S., Dabi, N. K., Vashishtha, A., Charan, P. D., Purohit, P., Bhandari, H. S., Bhojak, N., & Kumar, R. (2020). Enzymes for pharmaceutical and therapeutic applications. *Biotechnology and Applied Biochemistry*, *67*(4), 586–601.

Mishra, A. (2006). Production of L-asparaginase, an anticancer agent, from Aspergillus Niger using agricultural waste in solid state fermentation. *Applied Biochemistry and Biotechnology*, *135*(1), 33–42.

Muso Cachumba, J. J., Terán Hilares, R., Brumano, L. P., Marcelino, P. R. F., Antunes, F. A. F., Santos, J. C., & da Silva, S. S. (2019). Extracellular L-asparaginase production in solid-state fermentation by using sugarcane bagasse as support material. *Preparative Biochemistry & Biotechnology*, *49*(4), 328–333. https://doi.org/10.1080/10826068.2019.1566152.

Muthulakshmi, C., Gomathi, D., Kumar, D. G., Ravikumar, G., Kalaiselvi, M., & Uma, C. (2011). Production, purification and characterization of protease by aspergillus flavus under solid state fermentation. *Jordan Journal of Biological Sciences*, *4*(3), 137–148.

Nagamani, J. E. (2018). Production, partial purification and assay of L-glutaminase enzyme from Aspergillus Niger by solid state fermentation. *International Journal of Research and Review*, *5*, 177–181.

Nageswara, S., Guntuku, G., & Tadimalla, P. (2014). Production of L-asparaginase by solid state fermentation using marine fungus. *BioMed Res [Internet]*, 1–9.

Nawaz, A., Gillani, S., Tahir, S., Shah, K., Ashraf, S., Mukhtar, H., & Haq, I. (2020). Biosynthesis of fibrinolytic agent urokinase by Enterococcus gallinarum isolated from sardine. *Revista Mexicana de Ingeniería Química*, *19*(1), 213–225.

Nema, A., Patnala, S. H., Mandari, V., Kota, S., & Devarai, S. K. (2019). Production and optimization of lipase using Aspergillus Niger MTCC 872 by solid-state fermentation. *Bulletin of the National Research Centre*, *43*(1), 1–8.

Obi, C. N., Okezie, O., & Ezugwu, A. N. (2019). Amylase production by solid state fermentation of agro-industrial wastes using bacillus species. *European Journal of Nutrition & Food Safety*, *4*, 408–414.

Okpalla, J., Onyekuru, D. A., Duru, I. E., & Mba, T. O. (2022). Protease production by submerged fermentation in shake flasks using Bacillus sp. isolated from the soil. *Microbiology Research Journal International*, *32*(10), 42–51.

Pacifici, L. R., Kelleher, W. J., & Schwarting, A. E. (1962). Production of lysergic acid derivatives in submerged culture. 1. Fermentation studies. *Lloydia*, *25*(1), 37–45.

Pallem, C., Manipati, S., & Somalanka, S. R. (2010). Process optimization of L-glutaminase production by Trichoderma koningii under solid state fermentation (SSF). *International Journal of Applied Biology and Pharmaceutical Technology*, *1*(3), 1168–1175.

Pant, G., Prakash, A., Pavani, J. V. P., Bera, S., Deviram, G., Kumar, A., Panchpuri, M., & Prasuna, R. G. (2015). Production, optimization and partial purification of protease from Bacillus subtilis. *Journal of Taibah University for Science*, *9*(1), 50–55.

Pérez-Guerra, N., Torrado-Agrasar, A., López-Macias, C., & Pastrana, L. (2003). Main characteristics and applications of solid substrate fermentation. *Electronic Journal of Environmental, Agricultural and Food Chemistry*, *2*(3).

Prakasham, R. S., Rao, C. S., Rao, R. S., & Sarma, P. N. (2005). Alkaline protease production by an isolated bacilluscirculans under solid-state fermentation using agro-industrial waste: Process parameters optimization. *Biotechnology Progress*, *21*(5), 1380–1388.

Rajkumar, R., Ranishree, J. K., & Ramasamy, R. (2011). Production and characterization of a novel protease from Bacillus sp. RRM1 under solid state fermentation. *Journal of Microbiology and Biotechnology*, *21*(6), 627–636.

Ravindra, P. (2001). Studies on production of single cell protein by Aspergillus Niger in solid state fermentation of rice bran. *Brazilian Archives of Biology and Technology*, *44*, 79–88.

Řeháček, Z. (1984). Biotechnology of ergot alkaloids. *Trends in Biotechnology*, *2*(6), 166–172.

Ren, N.-Q., Guo, W.-Q., Wang, X.-J., Xiang, W.-S., Liu, B.-F., Wang, X.-Z., Ding, J., & Chen, Z.-B. (2008). Effects of different pretreatment methods on fermentation types and dominant bacteria for hydrogen production. *International Journal of Hydrogen Energy*, *33*(16), 4318–4324.

Renge, V. C., Khedkar, S. V, & Nandurkar, N. R. (2012). Enzyme synthesis by fermentation method: A review. *Scientific Reviews and Chemical Communications*, *2*(4), 585e90.

Rigo, E., Ninow, J. L., Di Luccio, M., Oliveira, J. V., Polloni, A. E., Remonatto, D., Arbter, F., Vardanega, R., de Oliveira, D., & Treichel, H. (2010). Lipase production by solid fermentation of soybean meal with different supplements. *LWT-Food Science and Technology*, *43*(7), 1132–1137.

Ritala, A., Häkkinen, S. T., Toivari, M., & Wiebe, M. G. (2017). Single cell protein—State-of-the-art, industrial landscape and patents 2001–2016. *Frontiers in Microbiology*, *8*, 2009.

Robinson, P. K. (2015). Enzymes: Principles and biotechnological applications. *Essays in Biochemistry*, *59*, 1.

Rudrapati, P., & Audipudi, A. V. (2017). Production and purification of anticancer enzyme l-asparaginase from Bacillus firmus AVP 18 of mangrove sample through submerged fermentation. *International Journal of Current Microbiology and Applied Sciences*, *5*, 1–18.

Sabu, A., Keerthi, T. R., Kumar, S. R., & Chandrasekaran, M. (2000). L-Glutaminase production by marine Beauveria sp. under solid state fermentation. *Process Biochemistry*, *35*(7), 705–710.

Sadh, P. K., Kumar, S., Chawla, P., & Duhan, J. S. (2018). Fermentation: A boon for production of bioactive compounds by processing of food industries wastes (by-products). *Molecules*, *23*(10), 2560.

Sahoo, S., Panda, P. K., Mishra, S. R., Nayak, A., Dash, S. K., & Ellaiah, P. (2008). Optimization of physical and nutritional parameters for hyaluronidase production by Streptococcus mitis. *Indian Journal of Pharmaceutical Sciences*, *70*(5), 661.

Sathish, T., Uppuluri, K. B., Chari, P. V. B., & Kezia, D. (2016). Sequential optimization methods for augmentation of marine enzymes production in solid-state fermentation: L-glutaminase production a case study. *Advances in Food and Nutrition Research*, *78*, 95–114.

Saxena, R., & Singh, R. (2011). Amylase production by solid-state fermentation of agro-industrial wastes using Bacillus sp. *Brazilian Journal of Microbiology*, *42*, 1334–1342.

Sethi, B. K., Jana, A., Nanda, P. K., DasMohapatra, P. K., Sahoo, S. L., & Patra, J. K. (2016). Production of α-amylase by Aspergillus terreus NCFT 4269.10 using pearl millet and its structural characterization. *Frontiers in Plant Science*, *7*, 639.

Shah, A. J., Karadi, R. V., & Parekh, P. P. (2010). Isolation, optimization and production of L-asparaginase from coliform bacteria. Asian Journal of Biotechnology, *2*(3), 169–177.

Sharma, N., Vamil, R., Agarawal, R., & Bhooshan, S. (2012). An investigation of αamylase production in submerged fermentation by Bacillus amyloliquefaciens. *Journal of Pharmacy Research*, *5*(7), 3548–3550.

Singh, R., Langyan, S., Sangwan, S., Gaur, P., Khan, F. N., Yadav, P., Rohatgi, B., Shrivastava, M., Khandelwal, A., Darjee, S., & Sahu, P. K. (2022). Optimization and production of alpha-amylase using *Bacillus subtilis* from apple peel: Comparison with alternate feedstock. *Food Bioscience*, *49*, 101978.

Singhania, R. R., Sukumaran, R. K., Patel, A. K., Larroche, C., & Pandey, A. (2010). Advancement and comparative profiles in the production technologies using solid-state and submerged fermentation for microbial cellulases. *Enzyme and Microbial Technology*, *46*(7), 541–549.

Siva Ajithkumar, D., Vidhya, A. K., Raghunathan, R., & Jesteena, J. (2015). Production, purification and characterization of streptokinase using Bacillus licheniformis under solid state fermentation. *Journal of Global Biosciences*, *4*(7), 2703–2712.

Spalvins, K., Ivanovs, K., & Blumberga, D. (2018). *Single cell protein production from waste biomass: Review of various agricultural by-products.*

Stanbury, P. F., Whitaker, A., & Hall, S. J. (2013). *Principles of Fermentation Technology*. Elsevier.

Toscano, L., Montero, G., Stoytcheva, M., Gochev, V., Cervantes, L., Campbell, H., Zlatev, R., Valdez, B., Pérez, C., & Gil-Samaniego, M. (2013). Lipase production through solid-state fermentation using agro-industrial residues as substrates and newly isolated fungal strains. *Biotechnology & Biotechnological Equipment*, *27*(5), 4074–4077.

Ukaegbu-Obi, K. M. (2016). Single cell protein: A resort to global protein challenge and waste management. *Journal of Microbial and Biochemical Technology*, *1*(1), 5.

Unakal, C., Kallur, R. I., & Kaliwal, B. B. (2012). Production of α-amylase using banana waste by Bacillus subtilis under solid state fermentation. *European Journal of Experimental Biology*, *2*, 1044–1052.

Var, S. F. U. R. M. (2010). Production of protease by submerged fermentation using Rhizopus microsporus var oligospous. *The Internet Journal of Microbiology*, *9*(1).

Velasco-Lozano, S., Volke-Sepulveda, T., & Favela-Torres, E. (2012). Lipases production by solid-state fermentation: The case of Rhizopus homothallic us in perlite. In *Lipases and Phospholipases* (pp. 227–237). Springer.

Vellard, M. (2003). The enzyme as drug: Application of enzymes as pharmaceuticals. *Current Opinion in Biotechnology*, *14*(4), 444–450.

Vidyalakshmi, R., Paranthaman, R., & Indhumathi, J. (2009). Amylase production on submerged fermentation by Bacillus spp. *World Journal of Chemistry*, *4*(1), 89–91.

Vineetha, M. S., Aldabaan, N. A., Sunil, S., More, S. S., Mahnashi, M. H., Shaikh, I. A., Samdani, M. S., Swami, R., Yadav, A., Rohith, N., Bhavya, J., Mannasaheb, B. A., Alharbi, M. S., Khan, A. A., Dafalla, S. E., & Shakeel Iqubal, S. M. S. (2024). Production and optimization of L-glutaminase from Halophilic *Fusarium solani-melongenae* strain CRI 24 under submerged and solid state fermentation. *Journal of Pure and Applied Microbiology*, *18*(1), 593–604.

Wardah, Z. H., Chaudhari, H. G., & Prajapati, V. (2023). Application of statistical methodology for the optimization of L-glutaminase enzyme production from Streptomyces pseudogriseolus ZHG20 under solid-state fermentation. *Journal of Genetic Engineering and Biotechnology*, *21*, 138.

Weerasooriya, M. K. B., & Piyarathne, S. (2019). Production of extracellular amylase by Aspergillus Niger under submerged fermentation using jack fruit rag as the carbon source. *Indian Journal of Traditional Knowledge (IJTK)*, *19*(1), 158–163.

Wiebe, M. (2002). Myco-protein from Fusarium venenatum: A well-established product for human consumption. *Applied Microbiology and Biotechnology*, *58*(4), 421–427.

Yadav, J. S. S., Bezawada, J., Ajila, C. M., Yan, S., Tyagi, R. D., & Surampalli, R. Y. (2014). Mixed culture of Kluyveromyces marxianus and Candida krusei for single-cell protein production and organic load removal from whey. *Bioresource Technology*, *164*, 119–127.

Newer Paradigm in Medicinal Plant Biotechnology

15 Computational Techniques for Drug Discovery from Medicinal Plants

Ajay Kumar Shukla, Joohee Pradhan,*
Manish Kumar, Sunita Panchawat, and C. P. Jain
* Corresponding author: juhipradhan123@gmail.com

15.1 INTRODUCTION

Computer-aided drug design discovery (CADDD) is a powerful and promising approach to developing drugs more quickly, inexpensively, and effectively. Modern technologies and methodologies have been integrated into many drug discovery programs to enhance drug development productivity and analytical techniques. Ligand screening and structural virtual screening methods, ranging from hit detection to optimization, are widely utilized in these programs. Machine learning, through rapid and intelligent data processing, holds significant potential for improving personalized therapy in basic and translational drug research. Ensuring reliable treatments for all patients is crucial in combating cancer as we know it today.

Advancements in computational drug discovery approaches have substantially contributed to novel drug design and provided valuable insights into drug therapy. Key bioinformatics tools used to predict drug efficacy include transcriptomics, toxicogenomics, and functional genomics. Existing databases and computational techniques are vital for developing new therapeutic strategies (Rahman et al., 2022; Madariaga-Mazón et al., 2021; Clark et al., 2006).

The information that is currently available about the target structure determines which CADDD method is used. Target structures must be thoroughly understood to use designing drugs based on structure (DDBS) techniques. This information is usually gathered by nuclear magnetic resonance (NMR) or X-ray crystallography. When these methods are unavailable, computational techniques, like homology modeling, can predict three-dimensional target structures. Understanding the structure allows for the application of direct docking techniques and virtual high-throughput screening on targets and potential therapeutic compounds. Binding free energy estimations can determine the affinities of compounds to specific targets, followed by further filtering and optimization of potential therapeutic compounds (Kapoor et al., 2022).

Design of drugs based on structure (DDBS) techniques necessitate knowledge of the target's three-dimensional (3D) structure. In contrast, ligand-based methods are employed when this structural information is unavailable. These methods leverage knowledge of molecules that bind to the target to identify, filter, and optimize hits, aiming to discover potential drug candidates for in vitro experimental evaluation. When target structures cannot be determined experimentally or computationally, ligand-based methods are used. These approaches depend on the recognition of the target's known active binders. The two key forms of CDDD approaches are ligand-based and structure-based, and the general pathways for CDDD are shown in Figure 15.1.

Drug development can be accelerated through computational techniques, which are valuable for interpreting and guiding research. Utilizing virtual screening and network pharmacology tools in the process of creating drug candidates from herbal sources can elucidate the molecular basis and mode of action of pharmacologically active phytoconstituents. Computational methods and

DOI: 10.1201/9781003326939-19

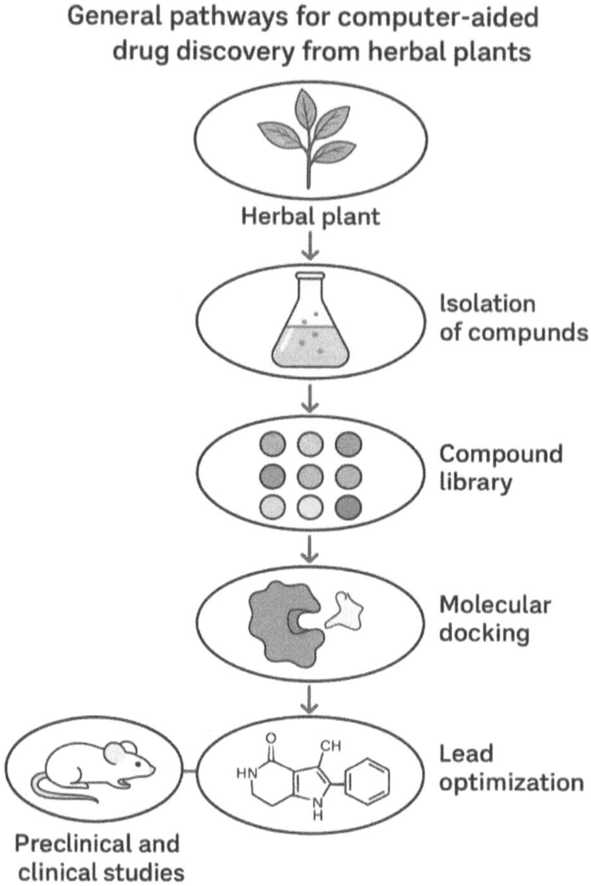

FIGURE 15.1 General pathways for computer-aided drug design discovery from herbal plants.

cheminformatics have played a major role in building and growing freely available compound databases, particularly those pertaining to natural products. The creation of medications derived from natural ingredients depends heavily on these computational techniques. The identification of the genetic variables and processes underlying the onset of disease has been made possible by genome-wide association studies (GWAS), which have aided in the creation of novel natural product medications. DNA barcoding has been employed in biodiversity inventories and in ensuring the authenticity of herbal products. Advanced transcriptomic techniques, like microarray analysis of gene expression, enable the rapid and efficient study of numerous transcripts, providing insights into the molecular processes and biological networks that underlie the pharmacological effects of natural medicinal substances. Proteomics and metabolomics approaches are also pivotal in novel drug development from natural materials, as they can clarify metabolic and biosynthetic pathways, protein expression, and protein function based on therapeutic benefits. The metabolism of natural products and their components can be effectively studied using proteomics and imaging techniques, which help elucidate their therapeutic effects (Kibble et al., 2015; Chen et al., 2016).

Computer-aided drug design discovery (CADDD) is a rapidly advancing field, continually developing new techniques and resources. It shows great promise for drug development and repurposing, significantly accelerating and reducing the costs associated with drug research and development. The increased availability of biological macromolecule and small molecule data has broadened the application of computational drug discovery, integrating it into nearly every stage of the drug discovery and development process, including target identification and validation, lead discovery and optimization, and preclinical testing (Talele et al., 2010; Clark, 2006; Kitchen et al., 2004). Various techniques of computer-aided drug design discovery, mentioned in Figure 15.1, as well as newer introductions, including network pharmacology, cheminformatics, and bioinformatics, are discussed in the following sections.

15.2 TECHNIQUES FOR SCREENING BIOACTIVE COMPOUNDS THROUGH CADDD

The goal of *in silico* toxicity, chemistry, and physical prediction of bioactive chemicals is achieved by algorithmic analysis of toxicity data, quantitative structure–activity relationships (QSARs), and computational modeling. Also known as *computational pharmacology* or *computational therapeutics*, this rapidly developing subject uses software-based methods to collect, process, and synthesize biological and medical data from many sources. Various approaches for the screening of bioactive compounds through CADDD are discussed in what follows.

15.2.1 Designing Drugs Based on Structure (DDBS)/ Structure-Based Drug Design (SBDD)

The availability of gene data and the finding of 3D protein structures of bioactive compounds have both increased throughout the last ten years. The use of computer approaches in drug design has been made easier by developments in computational science. Technologies that accurately ascertain ligand binding to target proteins and their spatial orientation within the protein's structure are utilized in structure-based drug design (SBDD). The Protein Data Bank has precise structural data that is being used to characterize more and more proteins that are being targeted for therapy. Using computational chemistry techniques, SBDD finds or creates chemical compounds to determine the binding site of a protein and the surrounding amino acids. In order to find the ideal "docking position," "molecular docking" inserts molecules into the binding site computationally. Stable interactions are predicted by a variety of scoring techniques, which help in designing or choosing interesting compounds to test experimentally against the target protein. The structure–activity link, or how structural alterations impact biological activity, is revealed by experimental data, which guide subsequent iterations of the design of drugs based on structure (DDBS) (Zhu et al., 2011; Tan et al., 2013).

The workflow of the structure-based drug design process is depicted in Figure 15.2.

15.2.1.1 Protein Structure Determination

The first step of the procedure is to ascertain the target protein's three-dimensional structure by methods such as X-ray crystallography, NMR spectroscopy, or homology modeling in the event that the structure is unknown.

15.2.1.2 Finding the Active Site of the Bioactive Compound

Potential binding sites, where tiny molecules (ligands) can interact with proteins, are found after the protein structure is determined. Since these sites are essential to the biological activity of the protein, medications frequently target them to alter that function.

A target macromolecule's binding location is a crucial factor to consider while refining the pharmacophores of novel ligands. Understanding protein–ligand interactions, post-docking dynamics, hydrogen bond forms, and complex free energies is made easier with the use of this information (Zhang et al., 2016). Example: selected asparagine bioactive compound to interact with targeted receptor JAK (Janus kinase), PDB code: 4C61, for screening of efficacy toward hair fall. It has been used for the treatment of antialopecia using in silico computation technique (Mustarichie et al., 2017).

The reverse docking method demonstrates the adaptability and effectiveness of molecular docking in locating putative pharmacological targets for tiny compounds, such as 4-hydroxycoumarin

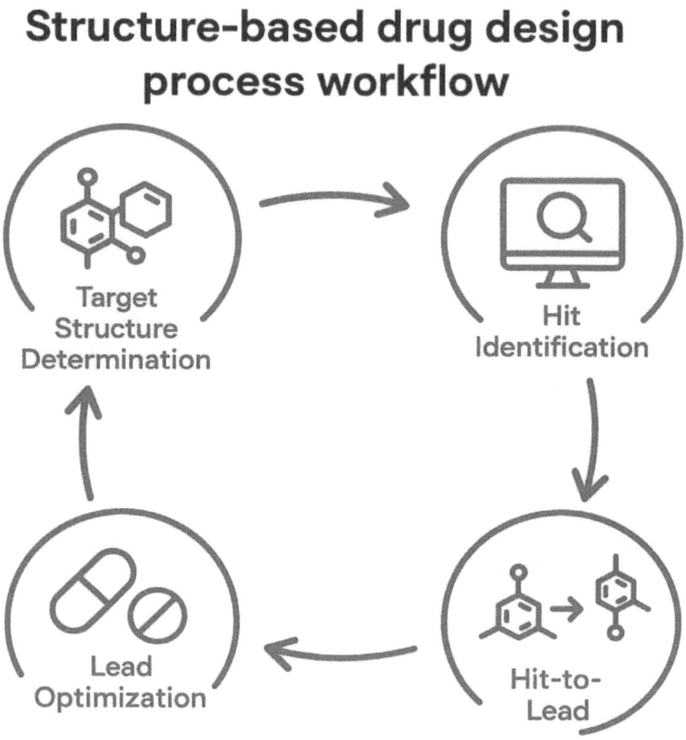

FIGURE 15.2 Structure-based drug design process workflow.

(4-HC). This methodology provides a solid framework for investigating the interactions of other bioactive chemicals with large protein databases, while also improving our understanding of the pharmacological properties of 4-HC. This reverse docking protocol's flexibility creates opportunities for further study, which may result in the identification of new therapeutic uses and enhanced medication safety profiles (Ruiz-Moreno et al., 2021).

These binding sites can be found using experimental methods, such as site-directed mutagenesis and X-ray crystallography (Pan et al., 2017). The abundant literature on crystallized proteins with known inhibitors or substrates provides valuable insights into these binding sites (Gomez-Gutierrez et al., 2017). When the binding site is unknown, various tools and databases, such as CASTp, NSiteMatch, metapocket, DEPTH, LISE, MSpocket, and DoGSite Scorer, are available for identification (Sahu et al., 2017; Sun et al., 2017; Volkamer et al., 2012). Computational tools like Epock, TRAPP, and POVME further evaluate the volume of the binding pocket. During lead identification, these tools facilitate the screening out of ligands that are too large to fit into the binding site.

Affinity selection–mass spectrometry (AS-MS) represents a transformative advancement in label-free binding assays. By leveraging UHPLC-MS for precise size-based separation, AS-MS efficiently distinguishes target compound complexes from unbound compounds, identifies binding sites, quantifies dissociation rate constants, and characterizes affinity-extracted ligands. As AS-MS continues to gain traction in HTS laboratories, it holds the promise of becoming a pivotal tool in the discovery and development of novel drugs, marking a significant step forward in pharmaceutical research (Ramatapa et al., 2022).

15.2.1.3 Structure-Based Virtual Screening

Computational techniques are used to screen libraries of small molecules (virtual libraries) to identify potential candidates that could bind to the identified binding sites on the protein. Techniques such as molecular docking are employed to predict the binding mode and affinity of these molecules. Various sources, such as natural ligands, public repositories, and commercial databases, provide the foundation for constructing a compound library. Compounds with known ligands or drug structure similarity are frequently used as selection criteria. These ligands need to adhere to Lipinski's rule of five in order to improve precision; they should pay particular attention to factors like molecular weight, logP values, and hydrogen bond donors and acceptors (Nogara et al., 2015; Huang et al., 2017; Lipinski, 2004; Raymond, 2002).

Prior to docking studies, it is crucial to evaluate ligands for risk factors, such as dimensions, concentration, solubility in water, volume of distribution, acute rat toxicity, carcinogenicity, increase of blood glutamic oxaloacetic transaminase, hepatotoxicity, and inhibition of Midazolam's 3A4 oxidation, and their ADMETrisk scores (Wang et al., 2009). These considerations aid in identifying bioactive compounds. Additionally, assessing the synthetic accessibility of these molecules is essential. Tools like SYLVIAXT can visually gauge the feasibility of synthesis (Boda et al., 2007). In cases where quantitative structure–activity relationship (QSAR) studies are lacking, options for selecting appropriate compounds may be more limited.

A developing subject that is critical to comprehending off-target effects and improving the logical design of selective medicines is computational target prediction for bioactive chemicals. Targets and binding modes can be correctly predicted with great potential using structure-based approaches, especially inverse screening techniques, like iRAISE (inverse rapid index-based screening engine). iRAISE provides customized statistical cutoffs for protein pockets and tackles issues like interprotein scoring noise by leveraging triangle descriptors and a strong scoring cascade. This approach represents a major breakthrough in computer target prediction, providing large-scale applications with both speed and accuracy (Schomburg et al., 2014; Laurent et al., 2015).

15.2.1.4 Experimental Validation

Promising candidates are identified by virtual screening, generated, and tested experimentally to verify that they are capable of binding to the desired protein. This stage aids in confirming the computational predictions and evaluating the efficacy of the compounds.

This work effectively created, validated experimentally, and implemented a virtual screening strategy to find new hERG channel inhibitors. The inhibitory action of 20 substances on hERG channels expressed in HEK 293 cells was confirmed by applying these models to both internal and subsequent experimental validation using patch clamp techniques. These results validate the efficacy of our virtual screening approach and highlight its potential as a tool in predicting cardiotoxicity during preclinical drug development. This makes it an invaluable resource in reducing the risk of cardiotoxicity associated with novel therapeutic candidates (Kratz et al., 2014).

15.2.1.5 *De Novo* Drug Design

In certain situations, the design of drugs based on structure (DDBS) entails creating brand-new molecules (de novo drug design) based on anticipated interactions with the target protein, as opposed to screening already-existing compounds. This strategy often makes use of molecular subunits and fragments that are anticipated or known to bind effectively to the target.

The findings of this study provide strong evidence supporting the use of generative AI in the design of novel molecules. The computational methods employed successfully identified new agonists for crucial nuclear receptors, showcasing notable potency and diverse selectivity profiles, particularly for RXRs and PPARs. The ease of synthesis from commercially available building blocks highlights the model's capability to inherently consider chemical synthesizability. Furthermore, the study underscores the significance of selecting appropriate compound libraries for model fine-tuning through transfer learning, paving the way for customized AI-driven de novo design. This approach may also extend to multitarget drug design, demonstrating generative AI's potential to revolutionize medicinal chemistry and chemical biology by providing swift, knowledge-based access to new small molecules.

15.2.1.6 Ligand and Protein Preparation

Docking and simulations cannot occur unless ligands and the Protein Data Bank (PDB) are ready. This will ensure accuracy in computational predictions. As part of this preparation, water molecules are eliminated, hydrogen atoms are added, charges are assigned, and structures are optimized.

The first step in the design of drugs based on structure (DDBS) process is the preparation of the target macromolecule structure. The Protein Data Bank (PDB) contains an increasing number of protein–ligand complex structures found by X-ray crystallography and nuclear magnetic resonance (NMR) spectroscopy. Even with recent advancements, many target protein structures remain difficult to reach using experimental approaches. In order to get beyond these experimental constraints, researchers employ computational methods, including as homology modeling, threading, and de novo folding, to predict structures from sequences (Gulerez et al., 2014; Laskowski et al., 2018; Svensson et al., 2018; Jo et al., 2015; Wang et al., 2014; Adhikari et al., 2017; Fischer et al., 2017; Karakaş et al., 2012).

15.2.1.7 Simulations

The dynamics of the protein–ligand interaction is understood by MD simulations. These simulations provide insight into the drug candidate's stability, the kinetics of the binding process, and its long-term interactions with the protein.

15.2.1.8 Scoring and Optimization

To assess and prioritize the binding affinity and additional characteristics of possible therapeutic options, computational scoring functions are employed. The potency, selectivity, and other pharmacological characteristics of the compounds are then enhanced by the application of optimization strategies. Design of drugs based on structure (DDBS) is frequently an iterative procedure, where computational forecasts and insights from experimental data are mutually reinforcing. Until medication candidates with the required qualities (efficacy, safety, and pharmacokinetics) are found, this iterative cycle is continued.

15.2.1.9 Docking, Scoring, and Predictions

Molecular docking is a critical component of the design of drugs based on structure (DDBS) used to predict how a ligand will bind to a protein receptor. It involves several key steps:

Target protein preparation. Obtaining and refining the 3D structure of the protein from databases like the Protein Data Bank (PDB).

Ligand preparation. Preparing the 3D structure of the ligand, often sourced from medicinal plant extracts or databases of natural compounds.

Docking simulation. This entails utilizing algorithms to predict the optimal binding configurations between the protein and the ligand.

Scoring and ranking. Assessing the binding affinities of different ligand configurations to rank them for further investigation. Ligands typically adopt low-energy conformations when binding to the protein, often requiring the generation of multiple initial conformations for each ligand. Various tools, such as Corina/Rotate, ConfGen, LigPrep, and MacroModel, facilitate the generation of 3D structures and different conformations of ligands for docking simulations (Feher et al., 2009; Corbeil et al., 2009; Vogt et al., 2014; Bharatham et al., 2014; Cappel et al., 2015; Chandra et al., 2015; Albadry et al., 2013).

Molecular docking faces a central challenge in developing an energy scoring function that can accurately and efficiently evaluate the fitness of protein–ligand interactions. Scoring functions are essential for post-docking analyses and have been extensively researched since the early 1990s (Liu, J. et al., 2015).

Three primary types of scoring functions are utilized: knowledge-based, empirical, and force field–based. Force field–based analysis, a significant approach, involves customizing van der Waals, electrostatic, and hydrophobic interactions specific to force fields, such as OPLS, CHARMM, Amber, GoldScore, and CHARMM. These methods aim to compute noncovalent interactions and provide essential attributes, like computational speed and accuracy, crucial to effectively representing the binding affinity between a ligand and a macromolecular target (Barradas et al., 2015; Debroise et al., 2017).

Predicting the binding free energy of protein–ligand complexes remains challenging, hindering technological advancement. Scientists are integrating modern force fields, quantum physics methods, and solvation models to improve predictions. Empirical scoring functions, notably the fastest, optimize terms like van der Waals forces, hydrogen bonds, electrostatic interactions, hydrophobicity, and desolvation energies to assess interactions within binding molecules and changes in solvent-accessible surface area (Eldridge et al., 1997). TS-Chemscore shows a robust correlation with virtual screening outcomes. Knowledge-based scoring algorithms leverage structural data from databases such as PDB (Lizunov et al., 2015) to model protein–ligand interactions, contrasting with empirical and force field–based approaches. These algorithms utilize distance-dependent pairwise potentials derived from structural insights of protein–ligand complexes. Evaluations of various docking systems and their scoring functions gauge their reliability in ranking ligands bound to protein targets. Figure 15.3 is a compilation of relevant databases and their websites.

There are many scoring functions available, but none of them achieves 100% generalizability or accuracy. Every scoring system has benefits and drawbacks. Consensus scoring, which combines scores from several scoring methods, significantly improves compound ranking accuracy. Consensus scoring makes the assumption that every scoring function adds predictive value to compound ranking, regardless of how precise they may be. By combining scores from several functions, they are used to emphasize more precise ligand–receptor binding affinities (Alexander et al., 2015).

Tools such as MultiScore (Terp et al., 2001, X-Cscore Wang et al., 2002; Betzi et al., 2006; Bar-Haim et al., 2009), and CONSENSUS-DOCK (Okamoto et al., 2010) exemplify consensus scoring approaches. In the design of drugs based on structure (DDBS), a number of medicines have successfully navigated clinical trials and regulatory approval. The development of novel drugs, such as

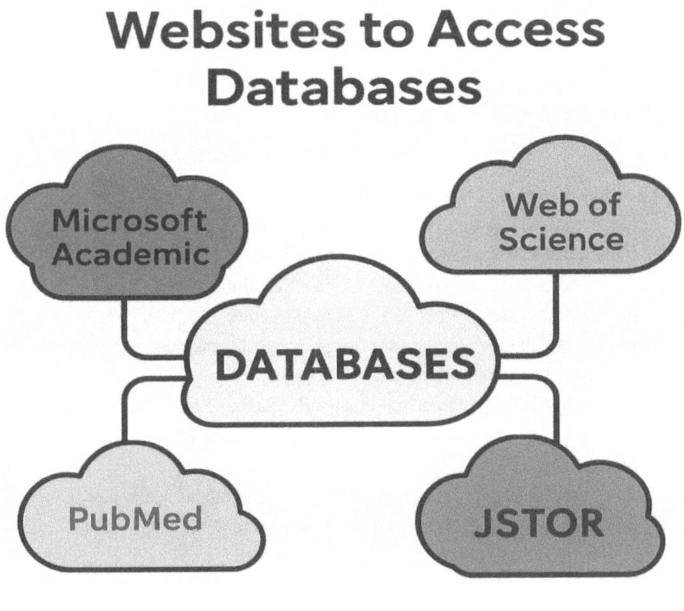

FIGURE 15.3 Websites to access databases.

inhibitors that target the NS5B RNA-dependent RNA polymerase for hepatitis viruses, is greatly aided by DDBS (Barreca et al., 2013).

Furthermore, dual-targeting antibiotics that specifically target the ATP-binding subunits of DNA gyrase and topoisomerase have been discovered by the design of drugs based on structure (DDBS). Innovative pyrimidoindole inhibitors created via design of drugs based on structure (DDBS) have broad-spectrum effectiveness against pathogens that are important for therapeutic settings (Tari et al., 2013). A list of the software used for docking molecules, with the version and their applications, is shown in Table 15.1.

15.2.1.10 Challenge and Prospection

DDBS methods are long-standing tools in designing and identifying bioactive compounds. Initially, during molecular docking modeling, it is essential to consider the flexibility of the target protein. Despite proteins being typically represented as static structures through techniques like X-ray crystallography and DDBS, they exhibit a dynamic behavior. Ligand binding can induce significant conformational changes in proteins, ranging from minor adjustments in side chains to complete reorganization of backbone structures (Salsbury et al., 2010; Ollikainen et al., 2015). Neglecting protein flexibility can lead to inaccurate results and diminish DDBS accuracy (Tian et al., 2014; Zhao, 2010). Flexible docking methods, such as CABS-dock, BP-Dock, Flexible CDOCKER, GalaxyDock, and MedusaDock, address these limitations effectively (Kurcinski et al., 2015; Bolia et al., 2014; Gagnon et al., 2016; Shin et al., 2012; Ding et al., 2013; Gasteiger, 2020).

Furthermore, structured water molecules are present in many protein–ligand complexes, which affect hydrogen binding networks and binding free energy (Kumar et al., 2013). Considering these water molecules improves docking accuracy, although predicting their specific relevance remains challenging (Lemmon et al., 2013). The enhanced Rosetta Ligand protocol incorporates water molecule effects to refine docking precision. Additionally, solvation effects are crucial in docking studies, particularly for highly solvated ligands, although their explicit contribution is complex to integrate into rating systems for interaction with proteins and ligands (Kitchen et al., 2004). Scoring functions are ultimately responsible for the precise prediction of binding affinity in molecular docking. These functions provide ranks and scores to projected protein–ligand complexes. Examples of these functions are knowledge-based potential functions, like drug score and PMF (potential of mean force) (Yan, X et al., 2015; Mitchell et al., 1999; Lima et al., 2016; Yan et al., 2015).

TABLE 15.1
Commonly Used Software for Docking Molecules

Software	Key Applications
DOCK Version	The identification of potential bioactive compounds through molecular docking. In predicting the binding affinity of small bioactive compounds to a protein or receptor of interest, it aids in the identification of possible therapeutic candidates.
Auto Dock	For structure-based drug design; to predict how bioactive compound will attach to a receptor or protein with a known three-dimensional structure.
Glide	Precise binding mode prediction; comprehensive databases of materials available for purchase from MolPort, MilliporeSigma, and Enamine.
FlexX Version	Identifying binding patterns and practically obtaining high-throughput screening.
Affinity	Estimate the targeted protein and bioactive compound binding models; predict the flexibility of the selected protein.
Fred Version 2013	Examine each recognized protein–ligand position methodically and provide a detailed scoring analysis. It is used for the rapid virtual screening of bioactive compounds.
SLIDE Version 3.4	It is used to obtain a list of the best possible bioactive compounds compatible with the receptor binding site. This software is used to screen several active compounds in a matter of days.

15.2.2 Designing Drugs Based on Ligands (DDBL)/ Ligand Based Drug Design (LBDD)

"Knowledge-based" drug design, or ligand-based drug design, creates models that forecast pharmacological characteristics by utilizing important chemical features seen in pharmaceuticals. Ligand-based drug design strategies were employed where pharmacophore modeling and QSAR connections were chosen, in the absence of receptor structure information. The predictive techniques employed are essential in determining the binding mechanism of a specific bioactive molecule to a given biological target, receptor, or protein, as well as the therapeutic efficacy in curing disorders. They generate predictive models essential for lead identification and optimization (Acharya et al., 2011).

It is predicated on the idea that ligand structures provide enough information to deduce their mode of action. Two chemical descriptors that are frequently used to predict the pharmacological and pharmacokinetic features of ligands are molecular weight and lipophilicity. One well-known example is Lipinski's rule of five, which describes standards based on molecular characteristics to separate chemicals that are not drugs from molecules that have drug-like characteristics (Lipinski, 2016). Chemical similarity concepts, which suggest that molecules with structural similarity probably share comparable biological features, are commonly used in similarity-based medication design (Yan, X et al., 2015). By utilizing these concepts, current chemical database search techniques seek to find molecules with increased bioactivity.

15.2.2.1 Outline of Medicinal Products Designed Using Ligands

Chemical similarity searching is a vital technique in ligand-based drug discovery. By comparing the structures of chemicals known to have biological activity, drug designers can identify similar molecules that may exhibit improved properties. These comparisons are facilitated by representing chemical structures as graphs, where atoms are vertices and bonds are edges (Le et al., 2004). A common method for determining how a compound's structure and physical/chemical characteristics relate to its biological activity is called ligand-based drug design, or LBDD. The prediction of molecules with better biological properties may then be based on the ensuing structure–activity relationship (SAR).

Chemoinformatics algorithms analyze these graphs to extract connectivity, vertex and edge counts, and other relevant properties, which are then translated into chemical features for machine learning–based similarity assessment (Yan & Wang, 2015). A common approach for chemical searching is the nearest neighbor method, using chemical fingerprints to find compounds in a database that are most similar to the query compound. Chemical fingerprints, such as path-based (e.g., Daylight fingerprints, Obabel FP2) or substructure-based (e.g., MACCS keys), encode structural information into mathematical representations. Path-based fingerprints emphasize pathway similarities across different bond lengths in a molecule's graph, offering high specificity in search results due to their path-dependent nature. In contrast, substructure-based fingerprints use binary arrays to indicate the presence or absence of predefined substructures, allowing for the identification of scaffold-hopping ligands by ignoring specific bond lengths (Lo et al., 2015; Panteleev et al., 2018; Luo, 2016). Figure 15.4 depicts the features of the ligand-based drug design (LBDD) process.

15.2.2.2 Selective Sites of Bioactive Compound Prediction

Ligand-based methods utilize the chemical similarity of annotated ligands to predict molecular targets (Davies et al., 2015). Limitations in ligand-based prediction, such as the absence of a clear bioactivity cliff, are addressed by techniques like the similarity ensemble approach (SEA) (Keiser et al., 2007), which computes similarity against a random background similar to BLAST. Ligand-based drug discovery employs various methods, including pharmacophore modeling, which identifies molecules with similar structures and binding characteristics (Klopmand et al., 1992).

Pharmacophore models, like Catalyst, DISCO, and GASP, utilize features such as hydrogen bond donors/acceptors and acidic/basic residues. QSAR (quantitative structure–activity relationship) analysis

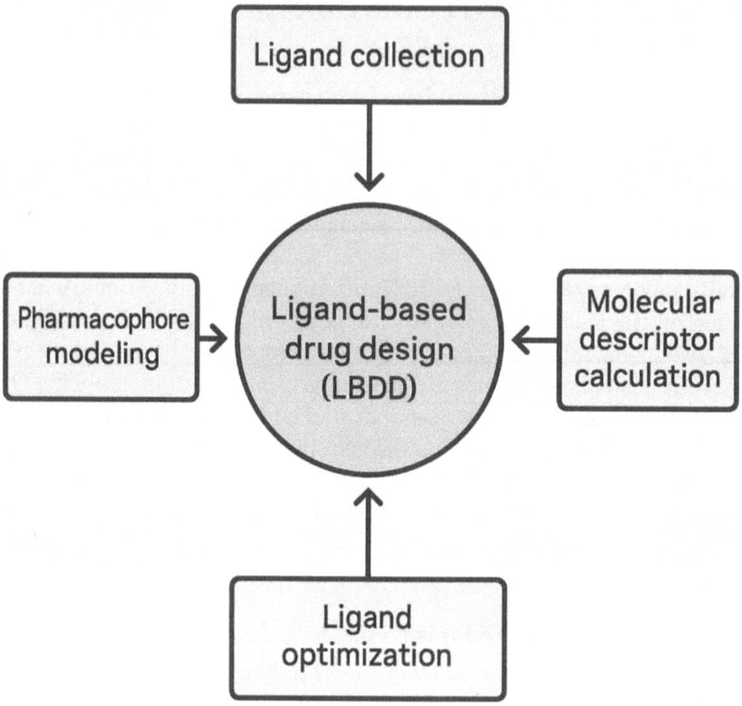

FIGURE 15.4 Features of the ligand-based drug design (LBDD) process.

correlates biological activity with structural characteristics of small ligands, enabling predictions for novel drug analogs. Machine learning algorithms are also utilized to identify patterns in experimental datasets for predictive modeling (Tarca et al., 2007; Sommer et al., 2013; Lavecchia et al., 2015).

Structure-based approaches focus on receptor binding site structures to predict targets. Techniques such as panel docking evaluate docking scores to identify likely targets. Binding site similarity algorithms compare the receptor environment of target ligands with databases of receptor pockets to predict interactions effectively. Network pharmacology offers a holistic approach by studying drug–target interactions beyond traditional single drug–single target paradigms (Westerhoff et al., 2015; Lo et al., 2006).

15.2.2.3 Side Effects Prediction of Drugs

Side effect prediction will need to be considered while creating new pharmaceuticals, because major clinical side effects have led to drug failure in the late stages of clinical trials. A multitude of statistical techniques, such as canonical component analysis (CCA), which determines a set of parameters that maximizes the link between drug binding qualities and side effect features, can be used to anticipate side effects based on a medication's binding affinity to different proteins. Predicting side effects has also been done using chemoinformatics analysis of the molecular structure. Numerous techniques, utilized since the initial stages of clinical drug research, precisely forecast the adverse reactions of pharmaceuticals (Liu, T et al., 2015).

15.2.2.4 QSAR

Quantitative structure–activity relationship (QSAR) models are employed to correlate the chemical structure of bioactive compounds with their biological activities. These models predict the activity of novel chemicals, facilitating the development of more potent medications. The QSAR process involves several key steps (Figure 15.5).

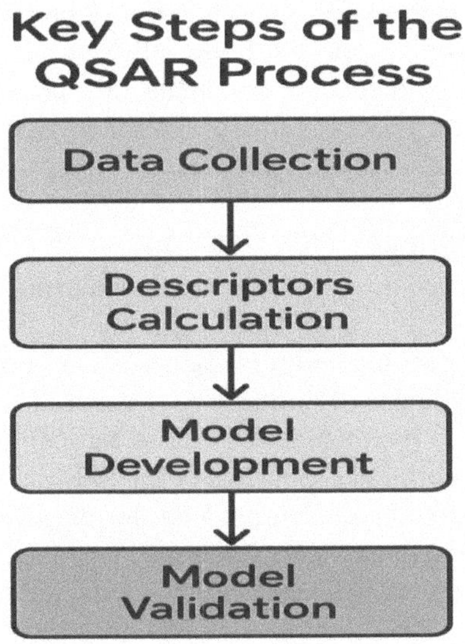

FIGURE 15.5 Key steps of the QSAR process.

Numerous plant species have been utilized for various medicinal purposes, leading to the discovery of economically significant pharmaceuticals. Notable examples include diosgenin from *Dioscorea nipponica* and aspirin derived from the willow bark (*Salix* spp.). Traditional ethnobotanical knowledge has played a crucial role in identifying and purifying important compounds and medicinal formulations.

Researchers have also leveraged traditional knowledge–based bio-screening techniques to identify potential new drug sources from plants. Computational methods can enhance this process by identifying new plant sources and their active components in a more cost-effective manner. These methods can also detect patterns in phytochemical data, potentially leading to the development of new medicinal products. Combining phytochemical research with phylogenetic analyses is one such approach. This strategy is based on the hypothesis that closely related plant species may share similar biological properties. Studies have demonstrated that the distribution of therapeutic properties aligns with the evolutionary history of plant species (Kang et al., 2011; Miner et al., 2007; Fabricant et al., 2001; Clarkson et al., 2004; Prinzing et al., 2001; Paton et al., 2004).

15.2.2.4.1 Phylogenetic-Based Technique

This approach leverages chemical similarities to identify potential candidates when searching for new plant sources. Phylogenetic-based methods have shown promise in discovering potential medications, as evidenced by successful applications in the *Pterocarpus* and *Narcissus* genera. Phylogenetic analyses indicate that therapeutic efficacy is heavily influenced by evolutionary history, guiding the selection of target species for further investigation (Ronsted et al., 2008; Saslis-Lagoudakis et al., 2011). Pharmaceutical development and testing typically make use of the QSAR-based technique. QSAR-based models have been used for the identification and prediction of the bioactive secondary metabolites from medicinal plants (Zhou et al., 2010; Petersen et al., 2011).

15.2.2.4.2 QSAR Models (A, B, and C)

This method utilizes the immunomodulatory activities of cleomiscosin compounds, with their logLD50 values predicted using QSAR models (A, B, and C). These prediction models demonstrate comparability to experimentally reported logLD50 values in in vivo toxicity assessments. QSAR research on medicinal plants aims to establish a connection between chemical structure and biological activity, providing insights into the mechanisms of bioactive compounds. By combining QSAR models, new active ingredients can be identified through structural comparisons with known compounds. Additionally, virtual screening, a computational technique, can aid in the development of plant-based medications by mining large datasets of plant-derived chemical compounds for novel pharmaceuticals (Sakkiah et al., 2012; Yadav et al., 2010).

15.2.2.4.3 QSAR Methods to Enhance the Usage and Treatment of Natural Products

In this study, a virtual pharmacophore model was developed using gamma partial agonists of peroxisome proliferator–activated receptors (PPARs). This model was subsequently employed in a digital screening of phytochemicals sourced from plants cataloged in the Chinese Natural Product Database (CNPD). Virtual screening holds promise but faces challenges due to the vast diversity of phytochemicals, the variability in receptor conformations caused by rotatable bonds, and the complexities in determining binding affinities (Shoichet et al., 2004).

Bioinformatics techniques offer significant potential for discovering novel drug candidates by analyzing trends in phytochemical data. Machine learning techniques, applied for computer modeling and phytochemical classification, help narrow down potential sources. Additionally, bioinformatics integrated with network systems biology supports drug development strategies by analyzing pharmacological components from a network perspective (Arrell et al., 2010). This approach aims to optimize research, reduce costs, and minimize reliance on animal testing.

The current study focuses on identifying drug-like compounds from natural extracts using computer modeling. It aims to assess these compounds' potential therapeutic actions and their ADME&T (adsorption, distribution, metabolism, excretion, and toxicity) properties. This computational approach seeks to facilitate in an efficient selection of lead molecules for subsequent wet laboratory testing, aiming ultimately to improve therapeutic outcomes (Abhigna et al., 2020).

15.2.3 Network Pharmacology

The goal of the developing area of network pharmacology (NP) is to understand the interactions between medications and different targets. NP carefully catalogs the chemical interactions of drug compounds inside of cells by utilizing computational skills. This method has proven essential in clarifying the complex interactions between herbal remedies and the human body. Regarding safety, efficacy, and sustainability, the one-drug/one-target/one-disease approach to drug discovery is today confronted with numerous obstacles. Recently, methods such as network biology and polypharmacology have gained popularity for developing multitarget drugs and integrating omics data. A novel paradigm known as network pharmacology (NP), which examines how medications affect both the interactome and the diseasome level, was created by combining these two techniques. The complex multicomponent formulations utilized in the ancient Indian medical system known as Ayurveda have no scientific basis, and its mechanisms are still mostly unknown (Hopkins et al., 2007; Zhang et al., 2013; Kibble et al., 2015; Ellingson et al., 2014). Figure 15.6 depicts various elements of network pharmacology involving data integration, network construction, and network analysis, leading to the improved safety and efficacy of current therapies.

15.2.3.1 Limitations and Solutions of Network Pharmacology

Network pharmacology (NP) enhances drug research by leveraging comprehensive data mining and literature searches across diverse datasets. However, several challenges persist in applying NP to conventional medicine, which may be addressed in the future:

1. *Data accuracy and chemical complexity.* Despite careful curation, NP depends on large datasets that are prone to errors, particularly when it comes to botanical components that have been altered by preparation techniques like boiling or chemical reactions. To properly identify bioactives in formulations, solutions entail using sophisticated chemical

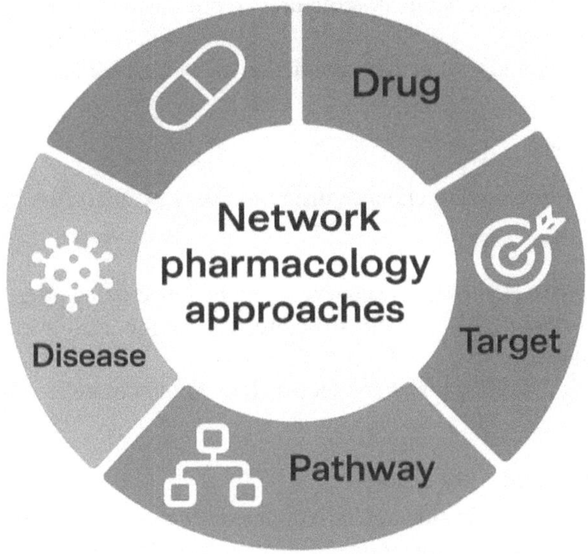

FIGURE 15.6 Network pharmacology approaches.

identification techniques, like ultra-performance liquid chromatography–electrospray ionization–tandem mass spectrometry (UPLC-ESI-MS/MS).

2. *ADMET considerations.* In vitro and in vivo models must be bridged in order to predict the absorption, distribution, metabolism, excretion, and toxicity (ADMET) of bioactives in therapeutic formulations. Formulation solvents have an impact on ADMET characteristics; hence, experimental validation is required to improve predictions.
3. *Target identification.* Limited access to comprehensive databases for target identification hampers NP's efficacy. Addressing this involves integrating multiple databases and validating target molecules through experimental methods, like protein–protein interaction studies and gene expression analyses.
4. *Synergistic effects and complex interactions.* Traditional medicines often involve synergistic interactions among multiple bioactives targeting various proteins and physiological pathways. Current in vitro and *in silico* methods inadequately capture these interactions, necessitating preliminary NP insights to guide subsequent in vivo and clinical investigations.
5. *Formulation complexity and multitarget effects.* Traditional formulations often modulate multiple targets simultaneously, necessitating comprehensive omics studies to understand their synergistic actions. High-throughput omics investigations based on network data offer insights into these complex interactions. In summary, while NP offers valuable insights into traditional medicine, addressing these challenges through advanced technologies and integrated approaches will enhance its application in drug discovery and development (Chandran et al., 2017).

15.2.4 Cheminformatics

Cheminformatics is the study of managing and analyzing chemical data via the use of databases and computer tools. Important instruments consist of:

1. *Chemical databases.* To obtain information about compounds, use resources like ZINC, ChEMBL, and PubChem.

2. *Molecular fingerprints.* Chemical structures represented for the purpose of finding similarities.
3. *Machine learning.* Biological activity prediction using algorithms.

15.2.5 Bioinformatics

Tools for bioinformatics are used to enhance drug discovery by analyzing biological data. Crucial instruments consist of:

1. *Sequence alignment.* Homologous sequences can be found using programs like BLAST.
2. *Protein structure prediction.* Protein structures can be modeled using programs such as the SWISS-MODEL.
3. *Pathway analysis.* Utilizing resources such as KEGG to comprehend signaling and metabolic pathways.

15.2.5.1 Applications and Case Studies

1. *Artemisinin from* Artemisia annua. One example of the promise of computational tools is the discovery of artemisinin, an antimalarial chemical.
2. *Curcumin from* Curcuma longa. Computational research has made it easier to comprehend how curcumin and its derivatives work on several targets.
3. *Resveratrol from* Vitis vinifera. The pathways underlying resveratrol's anti-inflammatory and anticancer properties have been clarified using network pharmacology

15.3 SCREENING OF POTENTIAL CANDIDATES OF BIOACTIVE COMPOUNDS

15.3.1 Pretreatment

Pretreatment is the process of analyzing the compounds found in plants. Several different bioactive compounds are present in medicinal plants. It can be obtained from therapeutic plants, albeit most of them do not have good pharmacological effects. The first step in improving screening efficiency is to remove these non-potential molecules and to refine the molecules that are left.

15.3.2 Drug Ability Prediction

In drug design, "drug-like qualities" refer to qualitative descriptors indicating a chemical's suitability based on parameters like bioavailability, determined by its molecular structure (Kerns et al., 2008). The potential medicinal value of a molecule is predicted by key structural properties. Pharmacologists place a strong emphasis on the early stages of drug discovery and pay close attention to characteristics including bonds of hydrogen, polarized area of surface, the lipophilicity shape, molecular mass, and pKa. Transport, protein binding, and metabolism are all governed by biochemical characteristics, whereas solubility, pH, permeability, or stability is covered by physicochemical characteristics. Half-life, bioavailability, medication interactions, and LD50 are all taken into account in pharmacokinetic and toxicological analyses. According to Lipinski's rule of five (LRO5), small compounds that are appropriate for oral treatment should have molecular weights of at least 500 Da (dalton), at least ten hydrogen bond acceptors, at least five hydrogen bond donors, and a logP of at least five in order to maximize bioavailability (Lipinski et al., 2001). Furthermore, rotatable bond lengths of at least 10 or 140 polar surface units, or hydrogen bonds of at least 12, are requirements for oral drug administration obtained from rat experiments (Veber et al., 2002). Computational tools are increasingly used to predict drug likeness (Eldehna et al., 2016).

15.3.3 ADME/T Analysis

Therapeutic candidates often fail in drug discovery due to various issues, like toxicity, poor pharmacokinetics, and incorrect target selection. An essential component of assessing potential drugs is predicting their ADME/T (absorption, distribution, metabolism, excretion, and toxicity) characteristics (Lyubimov et al., 2012). Algorithms for hit-to-lead and lead optimization rely on these predictions to select optimal candidates. *In silico* ADME/T models forecast how drugs move through the body, beginning with absorption across mucosal surfaces, distribution to tissues, metabolism into metabolites mainly by the liver, and eventual excretion (Patil et al., 2016).

Accumulation of foreign substances can hinder metabolism, necessitating effective excretion processes (Pichai et al., 2019). Support vector machines, deep neural networks, and artificial neural networks are examples of machine learning approaches that are used to create robust ADME/T prediction models (Lo et al., 2018). These models leverage extensive datasets from sources like ChEMBL and DrugBank, employing algorithms such as PCA, random forest, and SVM to interpret and predict drug behavior (Maltarollo et al., 2015; Guan et al., 2019; Vamathevan et al., 2019).

Supervised methods, like MLR, PLS, DT, RF, KNN, MLP, and SVM, as well as unsupervised techniques, like PCA and SOM, aid in analyzing data variability and ensuring model quality (Cano et al., 2017; Ferreira et al., 2019). These approaches are pivotal in enhancing drug discovery efficiency by predicting ADME/T properties accurately across diverse datasets. Several software tools are available for constructing prediction models in pharmaceutical research, such as ADMET predictor, Discovery Studio, ADMET lab, and Swiss ADME. ADMET SAR models in ADMET lab utilize quantitative structure property relationship research (QSPR) to predict ADME/T endpoints. This database includes five accurate quantitative regression and qualitative categorization methods based on a comprehensive dataset of 288,967 chemical compounds. QSPR models include descriptors such as MACCS keys or topological descriptors, which are examined using machine learning methods like support vector machines (SVM) for classification or regression. Integral to Swiss drug design, Swiss ADME uses multiple linear regression (MLR) and support vector machine (SVM) to predict 11 physicochemical properties, three models of water solubility (log S), six models of lipophilicity (log P), and nine models of pharmacokinetics (PK), covering skin permeation (log Kp), P-gp binding, permeability of the blood–brain barrier (BBB), gastrointestinal absorption, and CYP450 inhibition for five isoforms. Early ADME/T characteristic prediction in pharmaceutical development lowers costs and increases therapeutic candidate success rates (Yang et al., 2018; Cheng et al., 2012; Dong et al., 2018; Daina et al., 2017; Srivastava et al., 2023; Das et al., 2022; Pradhan et al., 2020). Various computational drug discovery methodologies have been utilized throughout the drug discovery and development process, encompassing target identification and verification, lead generation and enhancement, and preclinical experiments, as depicted in Figure 15.7.

15.3.4 False Positive Exclusion

The utilization of multiple receptor conformations (MRCs) in structure-based virtual screening (SBVS) poses a challenge due to the increased complexity in managing false positives (Totrov et al., 2008). The abundance of false positives in chemical libraries complicates the identification of genuine ligands or potential drug candidates, especially when considering receptor flexibility. This challenge becomes more pronounced when screening large compound libraries for drug discovery, where the majority of hits identified through docking may not exhibit actual inhibitory effects (Sink et al., 2010).

In this study, we hypothesized that true inhibitors may bind favorably to multiple conformations of the binding site, aligning with the concept of the binding energy landscape (Mobley et al., 2009). To address this, we developed a ligand selection method that aims to streamline the identification of genuine ligands by accounting for receptor flexibility in SBVS, thereby reducing the inclusion of false positives in the selection process.

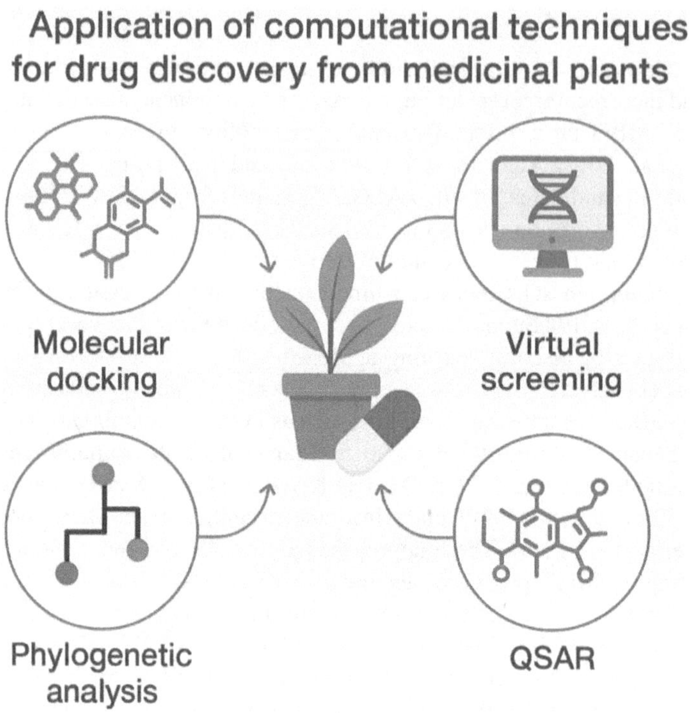

FIGURE 15.7 Application of computational techniques for drug discovery from medicinal plants.

This research underscores the significant issue of non-specific false positive hits in early drug discovery stages, which demand substantial time and resources to optimize. It is crucial for medicinal chemists to promptly exclude such compounds from further research and to recognize non-lead-like behaviors. Despite the use of computational filters to predict and filter out hazardous compounds, commercial screening collections may still contain undetected substances with undesirable structural characteristics, such as reactive false positives and warhead agents.

The common approaches of herbal bioactive compound screening, ADMET analysis, molecular docking with selected receptor or target, lead compound and pharmacophore modeling, are shown in a diagrammatic presentation in Figure 15.8.

Some compounds, such as aggregate-based inhibitors, pose challenges in predicting false positive outcomes solely from their structure. When experimental results deviate or appear unexpected, additional experiments are essential. These should include the rigorous selection of biochemical assays, meticulous data interpretation, and validation through secondary assays. The growing adoption of high-throughput screening (HTS) in academic biology, where quality standards for medicinal chemistry are less stringent compared to pharmaceutical industry standards, contributes to the issue of false positives and potentially diminishes publication quality. Baell et al. caution editors and reviewers about submissions describing specific structural types due to the proliferation of erroneous scientific literature (Sink et al., 2010).

Structure-based virtual screening encounters challenges in eliminating false positives from chemical libraries, particularly when receptor flexibility is considered. This study proposes that genuine binders can interact favorably with different conformations of binding sites, based on the theory of binding energy landscapes. Addressing receptor plasticity is crucial in reducing false positives in virtual screening. The study focuses on receptors such as influenza virus and bacterial actin MreB. The GOLD software targets chemical libraries against MreB and nucleoprotein binding sites, distinguishing between control compounds with varying affinities. Crystallographic validation of MreB with selected molecules supports the effectiveness of this

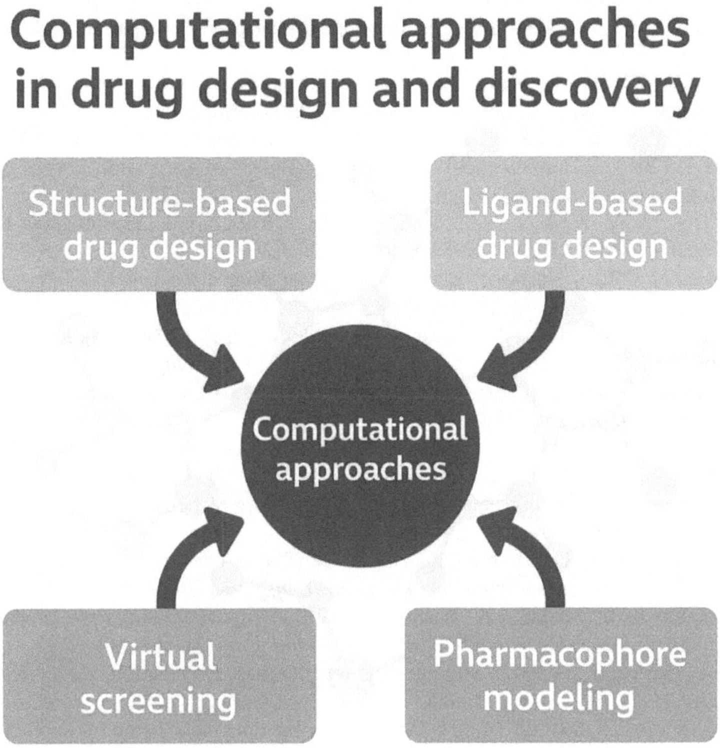

FIGURE 15.8 Computational approaches in drug design and discovery.

approach, offering a method to identify true ligands and mitigate false positives in chemical libraries (Sink et al., 2010).

15.4 CONCLUSION

The availability of biological data and the development of computer techniques have greatly improved the efficacy and accuracy of *in silico* drug target and treatment discovery in the last several years. Computational techniques such as computer-aided drug design (CADD) utilize structural data of therapeutic targets or bioactive ligands to identify potential drug candidates. This approach accelerates drug development by leveraging insights into ligand–receptor interactions, structural optimization, and synthesis. Successful drug development hinges on understanding key pharmacological factors, ADMET. Alongside the expansion of biological data, machine learning models have been increasingly employed to augment these efforts. Several drugs discovered via CADD methods have successfully reached the market, underscoring their efficacy and accessibility. However, ongoing advancements are necessary, particularly in refining molecular docking scoring functions, targeting receptors lacking structural knowledge, integrating solvent effects and molecular flexibility, developing force fields for molecular dynamics simulations, and enhancing computational efficiency. Addressing these challenges through CADD promises to fully unlock its potential in advancing drug discovery and development from medicinal plants.

REFERENCES

Abhigna, N., Kudagi, B., Subash, K.R., Sriharsha, R. (2020). In-Silico molecular properties prediction of compounds from carica papaya leaf extract. *Asian J. Pharm. Clin. Res.*, 13(11): 91–95.

Acharya, C., Coop, A., Polli, J.E., Mackerell, A.D. (2011). Recent advances in ligand-based drug design: Relevance and utility of the conformationally sampled pharmacophore approach. *Curr. Comput. Aided Drug Des.*, 7(1): 10–22.
Adhikari, B., Cheng, J. (2017). Improved protein structure reconstruction using secondary structures, contacts at higher distance thresholds, and non-contacts. *BMC Bioinform.*, 18(1): 380.
Albadry, M.A., Elokely, K.M., Wang, B., Bowling, J.J., Abdelwahab, M.F., Hossein, M.H., Doerksen, R.J., Hamann, M.T. (2013). Computationally assisted assignment of kahalalide Y configuration using an NMR-constrained conformational search. *J. Nat. Prod.*, 76(2): 178–185.
Alexander, D.L., Tropsha, A., Winkler, D.A. (2015). Beware of R(2), simple, unambiguous assessment of the prediction accuracy of QSAR and QSPR models. *J. Chem. Inf. Model.*, 55(7): 1316–1322.
Arrell, D.K., Terzic, A. (2010). Network systems biology for drug discovery. *Clin. Pharmacol. Ther.*, 88: 120–125.
Bar-Haim, S., Aharon, A., Ben-Moshe, T., Marantz, Y., Senderowitz, H. (2009). SeleX-CS: A new consensus scoring algorithm for hit discovery and lead optimization. *J. Chem. Inf. Model.*, 49(3): 623–633.
Barradas, D., Fernandez-Recio, J. (2015). A comprehensive analysis of scoring functions for protein-protein docking. *Proteins Sci.*, 24: 250–251.
Barreca, M.L., Manfroni, G., Leyssen, P., Winquist, J., Kaushik Basu, N., Paeshuyse, J., Krishnan, R., Iraci, N., Sabatini, S., Tabarrini, O., Basu, A., Danielson, U.H., Neyts, J., Cecchetti, V. (2013). Structure-based discovery of pyrazolo benzothiazine derivatives as inhibitors of hepatitis C virus replication. *J. Med. Chem.*, 56(6): 2270–2282.
Betzi, S., Suhre, K., Chétrit, B., Guerlesquin, F., Morelli, X. (2006). GFscore: A general nonlinear consensus scoring function for high-throughput docking. *J. Chem. Inf. Model.*, 46(4): 1704–1712.
Bharatham, N., Bharatham, K., Shelat, A.A., Bashford, D. (2014). Ligand binding mode prediction by docking: Mdm2/mdmx inhibitors as a case study. *J. Chem. Inf. Model.*, 54(2): 648–659.
Boda, K., Seidel, T., Gasteiger, J. (2007). Structure and reaction based evaluation of synthetic accessibility. *J. Comput. Aided Mol. Des.*, 21(6): 311–325.
Bolia, A., Gerek, Z.N., Ozkan, S.B. (2014). BP-Dock: A flexible docking scheme for exploring protein-ligand interactions based on unbound structures. *J. Chem. Inf. Model.*, 54(3): 913–925.
Cano, G., Garcia-Rodriguez, J., Garcia-Garcia, A., Perez-Sanchez, H., Benediktsson, J.A., Thapa, A., Barr, A. (2017). Automatic selection of molecular descriptors using random forest: Application to drug discovery. *Expert Syst. Appl.*, 72: 151–159.
Cappel, D., Dixon, S.L., Sherman, W., Duan, J. (2015). Exploring conformational search protocols for ligand-based virtual screening and 3-D QSAR modeling. *J. Comput. Aided Mol. Des.*, 29(2): 165–182.
Chandra, De U., Debnath, T., Sen, D., Debnath, S. (2015). Three-dimensional quantitative structure-activity relationships and docking studies of some structurally diverse flavonoids and design of new aldose reductase inhibitors. *J. Adv. Pharm. Technol. Res.*, 6(1): 13–18.
Chandran, U., Mehendale, N., Patil, S., Chaguturu, R., Patwardhan, B. (2017). *Innovative Approaches in Drug Discovery*. 127–164.
Chen, D.Q., Chen, H., Chen, L., Tang, D.D., Miao, H., Zhao, Y.Y. (2016). Metabolomic application in toxicity evaluation and toxicological biomarker identification of natural product. *Chem. Biol. Interact.*, 252: 114–130.
Cheng, F., Li, W., Zhou, Y., Shen, J., Wu, Z., Liu, G., Lee, P.W., Tang, Y. (2012). ADMET SAR: A comprehensive source and free tool for assessment of chemical ADMET properties. *J. Chem. Inf. Model.*, 52: 3099–3105.
Clark, D.E. (2006). What has computer-aided molecular design ever done for drug discovery. *Expert Opin. Drug Discov.*, 1: 103–110.
Clarkson, C., Maharaj, V.J., Crouch, N.R., et al. (2004). In vitro antiplasmodial activity of medicinal plants native to or naturalised in South Africa. *J. Ethnopharmacol.*, 92: 177–191.
Corbeil, C.R., Moitessier, N. (2009). Docking ligands into flexible and solvated macromolecules. 3. Impact of input ligand conformation, protein flexibility, and water molecules on the accuracy of docking programs. *J. Chem. Inf. Model.*, 49(4): 997–1009.
Daina, A., Michielin, O., Zoete, V. (2017). Swiss ADME: A free web tool to evaluate pharmacokinetics, drug-likeness and medicinal chemistry friendliness of small molecules. *Sci. Rep.*, 7: 42717.
Das, V., Bhatnagar, A., Pradhan, J. (2022). Herbal remedies for prophylaxis of COVID-19: A critical review on chemistry and pharmacology of eleven domestic herbs. *Int. J. Pharm. Clin. Res.*, 4(1): 7–18.
Davies, M., Nowotka, M., Papadatos, G., Dedman, N., Gaulton, A., Atkinson, F., et al. (2015). ChEMBL web services: Streamlining access to drug discovery data and utilities. *Nucleic Acids Res.*, 43(W1): W612–W620.
Debroise, T., Shakhnovich, E.I., Chéron, N. (2017). A hybrid knowledgebased and empirical scoring function for protein-ligand interaction: SmoG 2016. *J. Chem. Inf. Model.*, 57(3): 584–593.

Ding, F., Dokholyan, N.V. (2013). Incorporating backbone flexibility in MedusaDock improves ligand-binding pose prediction in the CSAR2011 docking benchmark. *J. Chem. Inf. Model.*, 53(8): 1871–1879.

Dong, J., Wang, N.N., Yao, Z.J., Zhang, L., Cheng, Y., Ouyang, D., Lu, A.P., Cao, D.S. (2018). ADMET lab: A platform for systematic ADMET evaluation based on a comprehensively collected ADMET database. *J. Cheminform.*, 10: 1–11.

Eldehna, W.M., et al. (2016). Synthesis and cytotoxic activity of biphenylurea derivatives containing indolin-2-one moieties. *Molecules.*, 21(6): 762.

Eldridge, M.D., Murray, C.W., Auton, T.R., Paolini, G.V., Mee, R.P. (1997). Empirical scoring functions: I. The development of a fast empirical scoring function to estimate the binding affinity of ligands in receptor complexes. *J. Comput. Aided Mol. Des.*, 11(5): 425–445.

Ellingson, S.R., Smith, J.C., Baudry, J. (2014). Polypharmacology and supercomputer-based docking: Opportunities and challenges. *Mol. Simul.*, 40: 848–854.

Fabricant, D.S., Farnsworth, N.R. (2001). The value of plants used in traditional medicine for drug discovery. *Environ. Health Perspect.*, 109(1): 69–75.

Feher, M., Williams, C.I. (2009). Effect of input differences on the results of docking calculations. *J. Chem. Inf. Model.*, 49(7): 1704–1714.

Feixas, F., Lindert, S., Sinko, W., McCammon, J.A. (2014). Exploring the role of receptor flexibility in structure-based drug discovery. *Biophys. Chem.*, 186: 31–45.

Ferreira, L.L.G., Andricopulo, A.D. (2019). ADMET modeling approaches in drug discovery. *Drug Discov. Today*, 24: 1157–1165.

Fischer, A.W., Anderson, D.M., Tessmer, M.H., Frank, D.W., Feix, J.B., Meiler, J. (2017). Structure and dynamics of type III secretion effector protein ExoU as determined by SDSL-EPR spectroscopy in conjunction with de novo protein folding. *ACS Omega.*, 2(6): 2977–2984.

Gagnon, J.K., Law, S.M., Brooks, C.L. (2016). Flexible CDOCKER: Development and application of a pseudo-explicit structure-based docking method within CHARMM. *J. Comput. Chem.*, 37(8): 753–762.

Gasteiger, J. (2020). Chemistry in times of artificial intelligence. *Chem Phys Chem.*, 21: 2233–2242.

Gomez-Gutierrez, P., Rubio-Martinez, J., Perez, J.J. (2017). Identification of potential small molecule binding pockets in p38α MAP kinase. *J. Chem. Inf. Model.*, 57(10): 2566–2574.

Guan, L., Yang, H., Cai, Y., Sun, L., Di, P., Liu, G., Tang, Y. (2019). ADMET-score-a comprehensive scoring function for evaluation of chemical drug-likeness. *Med. Chem. Comm.*, 10: 148–157.

Gulerez, I.E., Gehring, K. (2014). X-ray crystallography and NMR as tools for the study of protein tyrosine phosphatases. *Methods*, 65(2): 175–183.

Hopkins, A.L. (2007). Network pharmacology. *Nat. Biotechnol.*, 25: 1110–1111.

Huang, B. (2009). MetaPocket: A meta approach to improve protein ligand binding site prediction. *OMICS*, 13(4): 325–330.

Huang, T., Sun, J., Zhou, S., Gao, J., Liu, Y. (2017). Identification of direct activator of adenosine monophosphate-activated protein kinase (AMPK) by structure-based virtual screening and molecular docking approach. *Int. J. Mol. Sci.*, 18(7): 1408.

Jo, T., Cheng, J. (2014). Improving protein fold recognition by random forest. *BMC Bioinform.*, 15(11): 14.

Jo, T., Hou, J., Eickholt, J., Cheng, J. (2015). Improving protein fold recognition by deep learning networks. *Sci. Rep.*, 5: 17573.

Kang, T.H., Moon, E., Hong, B.N., et al. (2011). Diosgenin from dioscoreanipponica ameliorates diabetic neuropathy by inducing nerve growth factor. *Biol. Pharm. Bull.*, 34: 1493–1498.

Kapoor, K., Das, V., Bhatnagar, A., Panchawat, S., Pradhan, J. (2022). An in-silico multitargeted approach in search of potential drug candidate(s) against SARS-CoV-2 lung infection. *Coronaviruses*, 3(6): 39–52. Bentham Science Publisher. ISSN (Print): 2666–7967; ISSN (Online): 2666–7975. https://doi.org/10.2174/2666796704666221202143702.

Karakaş, M., Woetzel, N., Staritzbichler, R., Alexander, N., Weiner, B.E., Meiler, J. (2012). BCL: Fold-de novo prediction of complex and large protein topologies by assembly of secondary structure elements. *PLoS One*, 7(11): e49240.

Keiser, M.J., Roth, B.L., Armbruster, B.N., Ernsberger, P., Irwin, J.J., Shoichet, B.K. (2007). Relating protein pharmacology by ligand chemistry. *Nat. Biotechnol.*, 25(2): 197–206.

Kerns, E.H., Li, D. (2008). *Drug-Like Properties: Concepts, Structure Design and Methods.* Oxford: Elsevier LTD. 125–126.

Kibble, M., Saarinen, N., Tang, J., Wennerberg, K., Mäkelä, S., Aittokallio, T. (2015). Network pharmacology applications to map the unexplored target space and therapeutic potential of natural products. *Nat. Prod. Rep.*, 32: 1249–1266.

Kitchen, D.B., Decornez, H., Furr, J.R., Bajorath, J. (2004). Docking and scoring in virtual screening for drug discovery: Methods and applications. *Nat. Rev. Drug Discov.*, 3(11): 935–949.

Kratz, J.M., Schuster, D., Edtbauer, M., Saxena, P., Mair, C.E., Kirchebner, J., Matuszczak, B., Baburin, I., Hering, S., Rollinger, J.M. (2014). Experimentally validated HERG pharmacophore models as cardiotoxicity prediction tools. *J. Chem. Inf. Model.*, 54(10): 2887–2901. https://doi.org/10.1021/ci5001955.

Kumar, A., Zhang, K.Y. (2013). Investigation on the effect of key water molecules on docking performance in CSARdock exercise. *J. Chem. Inf. Model.*, 53(8): 1880–1892.

Kurcinski, M., Jamroz, M., Blaszczyk, M., Kolinski, A., Kmiecik, S. (2015). CABS-dock web server for the flexible docking of peptides to proteins without prior knowledge of the binding site. *Nucleic Acids Res.*, 43: 419–424.

Laskowski, R.A., Jabońska, J., Pravda, L., Vařeková, R.S., Thornton, J.M. (2018). PDBsum: Structural summaries of PDB entries. *Proteins Sci.*, 27(1): 129–134.

Laurent, B., Chavent, M., Cragnolini, T., Dahl, A.C., Pasquali, S., Derreumaux, P., Sansom, M.S., Baaden, M. (2015). Epock: Rapid analysis of protein pocket dynamics. *Bioinformatics*, 31(9): 1478–1480.

Lavecchia, A. (2015). Machine-learning approaches in drug discovery: Methods and applications. *Drug Discov. Today*, 20: 318–331.

Le, S.Q., Ho, T.B., Phan, T.T. (2004). A novel graph-based similarity measure for 2D chemical structures. *Genome Inform.*, 15(2): 82–91.

Lemmon, G., Meiler, J. (2013). Towards ligand docking including explicit interface water molecules. *PLoS One*, 8(6): e67536.

Lima, A.N., Philot, E.A., Trossini, G.H.G., Scott, L.P.B., Maltarollo, V.G., Honorio, K.M. (2016). *Expert Opin. Drug Discov.*, 11: 225–239.

Lipinski, C.A. (2004). Lead and drug-like compounds: The rule-of-five revolution. *Drug Discov. Today Technol.*, 1(4): 337–341.

Lipinski, C.A. (2016). Rule of five in 2015 and beyond: Target and ligand structural limitations, ligand chemistry structure and drug discovery project decisions. *Adv. Drug Deliv. Rev.*, 101: 34–41.

Lipinski, C.A., et al. (2001). Experimental and computational approaches to estimate solubility and permeability in drug discovery and development settings1. *Adv. Drug Deliv. Rev.*, 46(1–3): 3–26.

Liu, J., Wang, R. (2015). Classification of current scoring functions. *Chem. Inf. Model.*, 55(3): 475–482.

Liu, T., Altman, R.B. (2015). Relating essential proteins to drug side Õeffects using canonical component analysis: A structure-based approach. *J. Chem. Inf. Model.*, 55(7): 1483–1494.

Lizunov, A.Y., Gonchar, A.L., Zaitseva, N.I., Zosimov, V.V. (2015). Accounting for intraligand interactions in flexible ligand docking with a pmf-based scoring function. *J. Chem. Inf. Model.*, 55(10): 2121–2137.

Lo, Y.C., Rensi, S.E., Torng, W., Altman, R.B. (2018). Machine learning in chemoinformatics and drug discovery. *Drug Discov. Today*, 23: 1538–1546.

Lo, Y.C., Senese, S., Damoiseaux, R., Torres, J.Z. (2006). 3D chemical similarity networks for structure-based target prediction and scaffold hopping. *ACS Chem. Biol.*, 11(8): 2244–2253.

Lo, Y.C., Senese, S., Li, C.M., Hu, Q., Huang, Y., Damoiseaux, R., et al. (2015). Large-scale chemical similarity networks for target profiling of compounds identified in cell-based chemical screens. *PLoS Comput. Biol.*, 11(3): e1004153.

Luo, G. (2016). PredicT-ML: A tool for automating machine learning model building with big clinical data, Heal. *Inf. Sci. Syst.*, 4: 1–16.

Lyubimov, A.V., Hop, C.E.C.A. (2012). Role of ADME studies in selecting drug candidates: Dependence of ADME parameters on physicochemical properties, encycl. *Drug Metab. Interact.*, 3: 2–29.

Madariaga-Mazón, A., Naveja, J.J., Medina-Franco, J.L., Noriega-Colima, K.O., Martinez Mayorga, K. (2021). DiaNat-DB: A molecular database of antidiabetic compounds from medicinal plants. *RSC Adv.*, 11: 5172.

Maltarollo, V.G., Gertrudes, J.C., Oliveira, P.R., Honorio, K.M. (2015). Applying machine learning techniques for ADME-Tox prediction: A review. *Expert Opin. Drug Metab. Toxicol.*, 11: 259–271.

Miner, J., Hoffhines, A. (2007). The discovery of aspirin's antithrombotic effects. *Tex. Heart Inst.*, 34: 179–186.

Mitchell, J.B.O., Laskowski, R.A., Alex, A., Thornton, J.M. (1999). BLEEP—Potential of mean force describing protein–ligand interactions: I. Generating potential. *J. Comput. Chem.*, 20: 1165–1176.

Mobley, D.L., Dill, K.A. (2009). Binding of small-molecule ligands to proteins: "What you see" is not always "what you get". *Structure*, 17: 489–498.

Mustarichie, R., Megantara, S., Saptarini, N.M. (2017). In-silico study of bioactive compounds of natural materials as a JAK-signal transducer and activator of transcription inhibitor for anti-alopecia. *Asian J. Pharm. Clin. Res.*, 10(11): 331–336.

Nogara, P.A., Saraiva, Rde A., Caeran Bueno, D., Lissner, L.J., Lenz Dalla Corte, C., Braga, M.M., Rosemberg, D.B., Rocha, J.B. (2015). Virtual screening of acetylcholinesterase inhibitors using the Lipinski's rule of five and ZINC databank. *Biomed. Res. Int.*, 870389.

Okamoto, M., Masuda, Y., Muroya, A., Yasuno, K., Takahashi, O., Furuya, T. (2010). Evaluation of docking calculations on X-ray structures using CONSENSUS-DOCK. *Chem. Pharm. Bull (Tokyo).*, 58(12): 1655–1657.

Ollikainen, N., de Jong, R.M., Kortemme, T. (2015). Coupling protein side-chain and backbone flexibility improves the re-design of protein-ligand specificity. *PLoS Comput. Biol.*, 11(9): e1004335.

Pan, L., Gardner, C.L., Pagliai, F.A., Gonzalez, C.F., Lorca, G.L. (2017). Identification of the tolfenamic acid binding pocket in PrbP from liberibacter asiaticus. *Front. Microbiol.*, 8: 1591.

Panteleev, J., Gao, H., Jia, L. (2018). Recent applications of machine learning in medicinal chemistry. *Bioorg. Med. Chem. Lett.*, 28: 2807–2815.

Patil, P. (2016). Drug discovery and ADMET process: A review. *Int. J. Adv. Res. Biol. Sci.*, 3: 181–192.

Paton, A.J., Springate, D., Suddee, S., et al. (2004). Phylogeny and evolution of basils and allies (Ocimeae, Labiatae) based on three plastid DNA regions. *Mol. Phylogenet. Evol.*, 31: 277–299.

Petersen, R.K., Christensen, K.B., Assimopoulou, A.N., et al. (2011). Pharmacophore-driven identification of PPAR gamma agonists from natural sources. *J. Comp. Aided Mol. Design*, 25: 107–116.

Pichai, E., Lakshmanan, M. (2019). Drug elimination. In: *Introduction to Basics of Pharmacology and Toxicology General and Molecular Pharmacology: Principles of Drug Action*. Singapore: Springer. 1: 117–129.

Pradhan, J., Panchawat, S. (2020). Molecular docking studies and pharmacophore modeling of some insulin mimetic agents from herbal sources: A rational approach towards designing of orally active insulin mimetic agents. *Curr. Tradit. Med.*, 6(2): 121–133. https://doi.org/10.2174/221508380566619100120342.

Prinzing, A., Durka, W., Klotz, S., et al. (2001). The niche of higher plants: Evidence for phylogenetic conservatism. *Proc. Biol. Sci. Roy. Soc.*, 268: 2383–2389.

Rahman, M.M., Islam, M.R., Rahman, F., Rahaman, M.S., Khan, M.S., Abrar, S., Ray, T.K., Uddin, M.B., Kali, M.S.K., Dua, K., et al. (2022). Emerging promise of computational techniques in anti-cancer research: At a glance. *Bioengineering*, 9: 335.

Ramatapa, T., Msobo, A., Maphari, P.W., Ncube, E.N., Nogemane, N., Mhlongo, M.I. (2022). Identification of Plant-Derived Bioactive Compounds Using Affinity Mass Spectrometry and Molecular Networking. *Metabolites*, *12*, 863. https://doi.org/10.3390/metabo12090863

Raymond, J.W. (2002). Effectiveness of graph-based and fingerprint based similarity measures for virtual screening of 2D chemical structure databases. *J. Comput. Aided Mol. Des.*, 16(1): 59–71.

Ronsted, N., Savolainen, V., Molgaard, P., Jager, A.K. (2008). Phylogenetic selection of Narcissus species for drug discovery. *Biochem. Sys. Ecol.*, 36: 417–422.

Ruiz-Moreno, A.J., Dömling, A., Velasco-Velázquez, M.A. (2021). Reverse docking for the identification of molecular targets of anticancer compounds. *Methods Mol. Biol.*, 2174: 31–43.

Sahu, A., Patra, P.K., Yadav, M.K., Varma, M. (2017). Identification and characterization of ErbB4 kinase inhibitors for effective breast cancer therapy. *J. Recept. Signal Transduct. Res.*, 37(5): 470–480.

Sakkiah, S., Thangapandian, S., Lee, K.W. (2012). Ligand based virtual screening and molecular docking studies to identify the critical chemical features of potent cathepsin D inhibitors. *Chem. Biol. Drug Des.*, 80: 64–79.

Salsbury, F.R. (2010). Molecular dynamics simulations of protein dynamics and their relevance to drug discovery. *Curr. Opin. Pharmacol.*, 210(6): 738–744.

Saslis-Lagoudakis, C.H., Klitgaard, B.B., Forest, F., et al. (2011). The use of phylogeny to interpret cross-cultural patterns in plant use and guide medicinal plant discovery: An example from pterocarpus (leguminosae). *PLoS One*, 6: 22275.

Schomburg, K.T., Bietz, S., Briem, H., Henzler, A.M., Urbaczek, S., Rarey, M. (2014). Facing the challenges of structure-based target prediction by inverse virtual screening. *J. Chem. Inf. Model.*, 54(6): 1676–1686. https://doi.org/10.1021/ci500130e.

Shin, W.H., Seok, C. (2012). GalaxyDock: Protein-ligand docking with flexible protein side-chains. *J. Chem. Inf. Model.*, 52(12): 3225–3232.

Shoichet, B.K. (2004). Virtual screening of chemical libraries. *Nature*, 432: 862–865.

Sink, R., Gobec, S., Pecar, S., Zega, A. (2010). False positives in the early stages of drug discovery. *Curr. Med. Chem.*, 17: 4231–4255.

Sommer, C., Gerlich, D.W. (2013). Machine learning in cell biology – Teaching computers to recognize phenotypes. *J. Cell. Sci.*, 126: 5529–5539.

Srivastava, R., Patel, P.S., Dev, S.K., Pradhan, J., Panchawat, S. (2023). Insight to combat post COVID mortality: Complications and their biomarkers. *Curr. Mol. Med.*, 23(8). ISSN (Online): 1875–5666 2022 Aug. PMID: 35959906. https://doi.org/10.2174/1566524022666220812111830.

Sun, J., Chen, K. (2017). NSiteMatch: Prediction of binding sites of nucleotides by identifying the structure similarity of local surface patches. *Comput. Math. Methods Med.*, 5471607.

Svensson, F., Bender, A., Bailey, D. (2018). Fragment-based drug discovery of phosphodiesterase inhibitors. *J. Med. Chem.*, 61(4): 1415–1424.

Talele, T.T., Khedkar, S.A., Rigby, A.C. (2010). Successful applications of computer aided drug discovery: Moving drugs from concept to the clinic. *Curr. Top. Med. Chem.*, 10: 127–141.

Tan, K.P., Nguyen, T.B., Patel, S., Varadarajan, R., Madhusudhan, M.S. (2013). Depth: A web server to compute depth, cavity sizes, detect potential small-molecule ligand-binding cavities and predict the pKa of ionizable residues in proteins. *Nucleic Acids Res.*, 41: W314–W321.

Tarca, A.L., Carey, V.J., Chen, X.W., Romero, R., Drăghici, S. (2007). Machine learning and its applications to biology. *PLoS Comput. Biol.*, 3: 116.

Tari, L.W., Li, X., Trzoss, M., Bensen, D.C., Chen, Z., Lam, T., Zhang, J., Lee, S.J., Hough, G., Phillipson, D., Akers-Rodriguez, S., Cunningham, M.L., Kwan, B.P., Nelson, K.J., Castellano, A., Locke, J.B., Brown-Driver, V., Murphy, T.M., Ong, V.S., Pillar, C.M., Shinabarger, D.L., Nix, J., Lightstone, F.C., Wong, S.E., Nguyen, T.B., Shaw, K.J., Finn, J. (2013). Tricyclic GyrB/ParE (TriBE) inhibitors: A new class of broad-spectrum dual-targeting antibacterial agents. *PLoS One*, 8(12): e84409.

Terp, G.E., Johansen, B.N., Christensen, I.T., Jørgensen, F.S. (2001). A new concept for multidimensional selection of ligand conformations (Multi Select) and multidimensional scoring (MultiScore) of protein-ligand binding affinities. *J. Med. Chem.*, 44(14): 2333–2343.

Tian, S., Sun, H., Pan, P., Li, D., Zhen, X., Li, Y., Hou, T. (2014). Assessing an ensemble docking-based virtual screening strategy for kinase targets by considering protein flexibility. *J. Chem. Inf. Model.*, 54(10): 2664–2679.

Totrov, M., Abagyan, R. (2008). Flexible ligand docking to multiple receptor conformations: A practical alternative. *Curr. Opin. Struct. Biol.* 18: 178–184.

Vamathevan, J., Clark, D., Czodrowski, P., Dunham, I., Ferran, E., Lee, G., Li, B., Madabhushi, A., Shah, P., Spitzer, M. (2019). *Appl. Mach. Learn. Healthcare*, 18: 463–477.

Veber, D.F., et al. (2002). Molecular properties that influence the oral bioavailability of drug candidates. *J. Med. Chem.*, 45(12): 2615–2623.

Vogt, A.D., Pozzi, N., Chen, Z., Di Cera, E. (2014). Essential role of conformational selection in ligand binding. *Biophys. Chem.*, 186: 13–21.

Volkamer, A., Kuhn, D., Rippmann, F., Rarey, M. (2012). DoGSite scorer: A web server for automatic binding site prediction, analysis and druggability assessment. *Bioinformatics*, 28(15): 2074–2075.

Wang, J., Skolnik, S. (2009). Recent advances in physicochemical and ADMET profiling in drug discovery. *Chem. Biodivers.*, 6(11): 1887–1899.

Wang, Q., Yan, J., Li, X. (2014). Protein fold recognition based on functional domain composition. *Comput. Biol. Chem.*, 48: 71–76.

Wang, R., Lai, L., Wang, S. (2002). Further development and validation of empirical scoring functions for structure-based binding affinity prediction. *J. Comput. Aided Mol. Des.*, 16(1): 11–26.

Westerhoff, H.V. (2015). Network-based pharmacology through systems biology. *Drug Discov. Today Technol.*, 15: 15–16.

Xie, Z.R., Liu, C.K., Hsiao, F.C., Yao, A., Hwang, M.J. (2013). LISE: A server using ligand-interacting and site-enriched protein triangles for prediction of ligand-binding sites. *Nucleic Acids Res.*, 41: 292–296.

Yadav, D.K., Meena, A., Srivastava, A., et al. (2010). Development of QSAR model for immunomodulatory activity of natural coumarinolignoids. *Drug Des. Devel. Ther.*, 4: 173–186.

Yan, X., Liao, C., Liu, Z., Hagler, A.T., Gu, Q., Xu, J. (2015). Chemical structure similarity search for ligand-based virtual screening: Methods and computational resources. *Curr. Drug Targets*, 17(14): 1580–1585.

Yan, Z., Wang, J. (2015). Optimizing the affinity and specificity of ligand binding with the inclusion of solvation effect. *Proteins*, 283(9): 1632–1642.

Yang, M., Chen, J., Xu, L., Shi, X., Zhou, X., Xi, Z., An, R., Wang, X. (2018). A novel adaptive ensemble classification framework for ADME prediction. *RSC Adv.*, 8: 11661–11683.

Zhang, B., Wang, X., Li, S. (2013). An integrative platform of TCM network pharmacology and its application on a herbal formula, Qing-Luo-Yin. *Evid. Based Complement. Alternat. Med.*, 2013.

Zhang, Y., Hang, D., Tian, H., Jiao, Y., Shi, Z., Ran, T., Liu, H., Lu, S., Xu, A., Qiao, X., Pan, J., Yin, L., Zhou, W., Lu, T., Chen, Y. (2016). Identification of covalent binding sites targeting cysteines based on computational approaches. *Mol. Pharm.*, 13(9): 3106–3118.

Zhao, Q. (2010). Protein flexibility as a biosignal. *Crit. Rev. Eukaryot. Gene. Expr.*, 20(2): 157–170.

Zhou, X., Li, Y., Chen, X. (2010). Computational identification of bioactive natural products by structure activity relationship. *J. Mol. Graph. Model.*, 29: 38–45.

Zhu, H., Pisabarro, M.T. (2011). MSPocket: An orientation-independent algorithm for the detection of ligand binding pockets. *Bioinformatics*, 27(3): 351–358.

16 Biotechnological Approaches for the Microbial Production of Plant-Based Compounds

*Yuvaraj Sivamani, Hayagriba Swain, and Sumitha Elayaperumal**
* Corresponding author: sumithae@jssuni.edu.in

16.1 INTRODUCTION

The biotechnological manufacture of plant-based chemicals using microbial platforms has become a paradigm shift in a number of businesses in recent years. This novel method uses the metabolic abilities of bacteria, yeast, and fungi to create useful chemicals that are typically derived from plants (Ray et al. 2020). This change is being driven by the possibility of providing improved sustainability, scalability, and efficiency in the production of a wide range of compounds obtained from plants, thereby mitigating the drawbacks of conventional extraction techniques. This thorough investigation attempts to explore the complexities of biotechnological methods, looking at the fundamental ideas, technological instruments, and practical uses that characterize this emerging discipline.

Bioactive substances found in plants have historically been extracted using technologies that have molded industries, including pharmaceuticals, tastes, perfumes, and cosmetics. However, this strategy has its own set of problems, such as scarce resources, climate sensitivity, and frequently ineffective extraction techniques (Dincheva, Badjakov, and Galunska 2023). The search for alternative approaches has been spurred by the growing demand for natural products as well as the necessity for scalable and sustainable manufacturing techniques. Genetic engineering and synthetic biology–based biotechnological methods offer a convincing answer to these problems and have the potential to completely transform the chemical production industry (Burgos-Morales et al. 2021). The field of metabolic engineering is crucial to the microbial synthesis of plant-based chemicals. Using a methodical approach, this tactic modifies microbial metabolic pathways to increase the synthesis of particular chemicals (Mohd Kamal et al. 2022). Through the introduction or enhancement of gene expression linked to the manufacture of target chemicals, genetic engineering techniques are utilized to alter the genetic composition of microorganisms. The ability to manipulate microbial metabolism to such an extent enables the optimization of microorganisms as effective biofactories involved in the production of a wide range of chemicals obtained from plants (Cho et al. 2022b).

Utilizing synthetic biology as a foundation allows for the creation of completely new biological systems, making it an effective complement to metabolic engineering (Garner 2021). Synthetic biology offers researchers a level of customization and precision that was previously unattainable in the setting of microbial production of plant-based chemicals by enabling the creation of synthetic gene clusters or pathways. When designing microbial strains, this modular approach offers unmatched versatility and adaptability by enabling the development of microorganisms specifically suited for the synthesis of particular chemicals (Cho et al. 2022a). In the process of producing biotechnological products, choosing the right microbial strain is essential. It may be necessary to genetically modify some microbes to confer the ability to manufacture specific plant chemicals.

DOI: 10.1201/9781003326939-20

Microorganisms' biosynthetic pathways must be optimized to produce large yields of the desired chemicals (Francois, Alkim, and Morin 2020). This entails adjusting the metabolic flux, detecting and removing bottlenecks that could obstruct effective production, and adjusting the expression levels of important enzymes. Pathway optimization seeks to enhance the biosynthesis process's efficiency, which will ultimately result in higher yields and productivity (Su et al. 2023). It is informed by ideas from systems biology and computational modeling. The development of CRISPR/Cas9 technology has transformed genome editing and gives scientists a precise and effective tool for changing microorganisms' genetic composition (Wei and Li 2023). When it comes to biotechnological production, CRISPR/Cas9 makes it possible to introduce or remove particular genetic components from a microbial genome with precision. The combination of omics technologies, including proteomics, metabolomics, transcriptomics, and genomes, provides a thorough understanding of the biological mechanisms controlling the synthesis of compounds (Dai and Shen 2022).

Utilizing the data produced by these omics technologies, systems biology techniques simulate and optimize microbial metabolic networks. The entire effectiveness of the biotechnological production process is increased when researchers can discover viable targets for genetic alteration and route engineering to this comprehensive understanding (Amer and Baidoo 2021). The successful transition of microbial production from laboratory to industrial scales is contingent upon the implementation of efficient fermentation techniques. Controlling variables like temperature, pH, nutrient availability, and oxygen levels to foster microbial development and chemical synthesis is the process of optimizing fermentation conditions. Plant-based chemicals are consistently produced at high quality through the iterative improvement of fermentation methods, which address issues related to scale-up (Melini et al. 2023).

There are inherent scale-up issues when moving from laboratory-scale investigations to industrial manufacturing. Large-scale fermentation system engineering necessitates giving significant thought to elements like automation, process monitoring, and reactor design. Achieving high product purity and yield requires the use of downstream processing procedures, such as extraction, purification, and recovery of target components from the fermentation broth. Technological advancements in downstream processing make biotechnology techniques economically viable (Boodhoo et al. 2022). In the enzymatic processes that comprise the pathways leading to biosynthesis, cofactors are essential.

Cofactor engineering is the practice of altering an enzyme's cofactor specificity to increase the production process's overall yield and efficiency. Researchers can enhance the catalytic efficiency of important enzymes in the biosynthetic process and contribute to higher productivity and decreased by-product generation by optimizing cofactor use (Boodhoo et al. 2022). The process of creating unique DNA sequences for microbial engineering has become more efficient due to the integration of gene synthesis and bioinformatics. Technological developments in gene synthesis have made it possible to assemble synthetic gene clusters more effectively, which in turn has made it easier to create microbes that are more capable of generating chemicals derived from plants (Hoose et al. 2023). To create and optimize genetic constructs and gain insight into how well synthetic pathways work with microbial hosts, bioinformatics tools are essential.

16.2 METABOLIC ENGINEERING IN THE MODIFICATION OF CELLULAR PATHWAYS

One important biotechnological strategy used in the microbial synthesis of plant-based chemicals is metabolic engineering. This method entails altering and improving microorganisms' cellular metabolic pathways to increase the synthesis of particular target chemicals. By adding, modifying, or rerouting metabolic pathways in microbial hosts, metabolic engineering seeks to replicate or enhance the biosynthetic processes present in plants in the setting of plant-based chemicals (Hoose et al. 2023). The primary enzymatic processes in the biosynthesis of plant chemicals are the focus of metabolic engineers, who work to identify and optimize these steps. This could entail boosting the expression levels of enzymes that catalyze important processes or enhancing the catalytic efficiency of such enzymes. One basic tactic in metabolic engineering is the modification of gene expression.

The synthesis of target substances can be increased by overexpressing genes that encode enzymes in the desired biosynthetic pathway. On the other hand, suppressing regulatory components or rival pathways can increase the flow toward the intended products (Cravens, Payne, and Smolke 2019). Heterologous pathways are frequently inserted into microbial hosts by metabolic engineers. These pathways could originate from plants or other naturally occurring organisms that can make the desired chemicals (Markina, Kotlobay, and Tsarkova 2020). The process entails the introduction of foreign genes that encode the necessary enzymes for the production of the targeted plant-based chemicals. It is necessary to attain optimal metabolic flow via the designed routes. To ensure effective synthesis of the target chemicals without endangering the general health and growth of the cell, a balance must be struck between the supply of precursors, energy, and cofactors. For certain biosynthesis processes to perform as best as they can, certain cofactors are needed (Markina, Kotlobay, and Tsarkova 2020).

In metabolic engineering, altering cofactor availability can help increase the engineered pathways' efficiency. This can be accomplished by optimizing cofactor regeneration mechanisms or introducing substitute enzymes (Ku, Chen, and Lan 2020). The optimal production of plant-based chemicals can be achieved by fine-tuning microbial hosts through the systematic and combinatorial combination of numerous techniques, including pathway optimization, gene overexpression, and cofactor engineering. Temporal control of gene expression can be achieved by putting dynamic control systems, like feedback mechanisms or inducible promoters, into practice. This can be especially helpful in controlling metabolic flux and avoiding the buildup of intermediary metabolites that could be harmful to the host. Biofuels, nutraceuticals, and medicines have all been produced successfully using metabolic engineering in the microbial production of plant-based chemicals (Glasscock et al. 2021). Sustainable and profitable bioproduction procedures for a variety of chemicals obtained from plants are largely dependent on the iterative tuning of microbial strains using metabolic engineering techniques.

16.3 SYNTHETIC BIOLOGY FOR THE PRODUCTION OF PLANT-BASED COMPOUNDS

The development of biotechnological strategies for the microbial synthesis of plant-based chemicals is greatly aided by synthetic biology. It offers strong tools and processes for planning, building, and refining biological systems, making it possible to create specially designed microbes with improved capacities for the synthesis of particular substances derived from plants (Martins-Santana et al. 2018). Researchers can create whole new metabolic pathways inside microbes. This entails assembling genes from different sources to build artificial pathways that can generate particular chemicals derived from plants. Scientists can engineer microbes to biosynthesize chemicals that the selected host may not normally generate by creating these pathways. Microbial genomes can be precisely and specifically modified to methods like CRISPR-Cas9 (Zhao, Fang, and Zhang 2020) (Figure 16.1).

Utilizing genome editing, synthetic biology modifies, adds, or removes particular genes to optimize the host organism for increased synthesis of target plant chemicals. This accuracy makes it possible to fine-tune regulatory components and metabolic pathways (Zhao, Fang, and Zhang 2020). Genes from several creatures can be expressed heterologously in synthetic biology. Genes encoding enzymes involved in the manufacture of plant-based chemicals can be introduced into microbial hosts selectively by researchers. With this method, chemicals that are typically derived from intricate plant systems can be produced using microbes that are genetically manipulable (Rafeeq et al. 2023). The modular construction of genetic elements, including terminators, coding sequences, and promoters, into standardized components is encouraged by synthetic biology. Building genetic circuits and pathways with predictable activity is possible with this modular technique.

Through the precise regulation of gene expression levels made possible by synthetic biology, researchers can maximize the production of target chemicals. To ensure the best possible metabolic

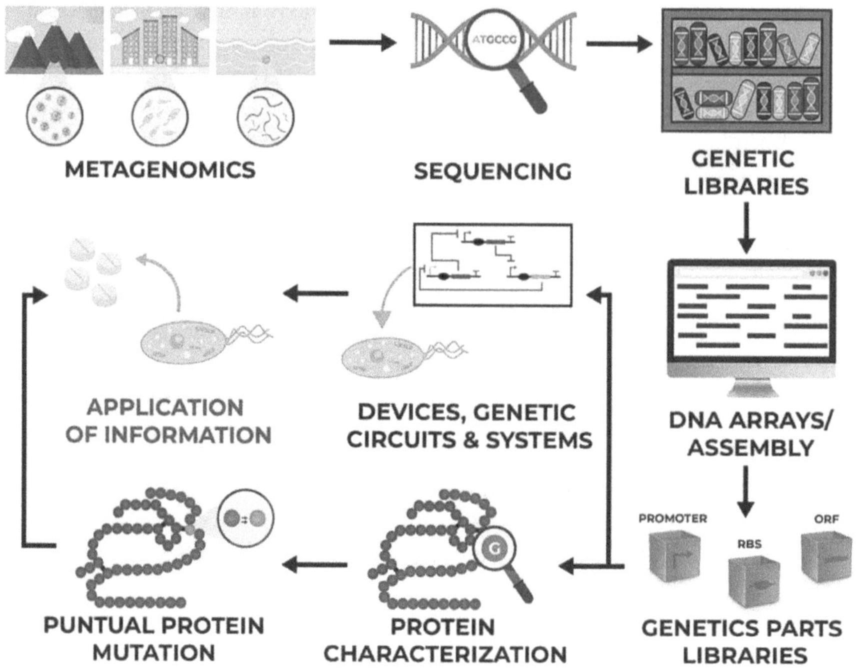

FIGURE 16.1 Different stages for the development of synthetic biology and metabolic engineering.

flux, this involves adjusting the expression of important genes in the designed pathways through the use of synthetic promoters and regulatory elements (Han et al. 2020). Developments in synthetic biology aid in the development of biosensors that track metabolite levels and cellular functions. By using these biosensors to dynamically regulate gene expression in response to particular metabolic or environmental circumstances, microbial production systems can be made even more efficient. Synthetic amino acids and other non-natural building blocks can be incorporated into proteins through the use of orthogonal and non-native systems made possible by synthetic biology. This adds to the arsenal of tools available for engineering microbes and presents fresh possibilities for enhancing enzyme activity in biosynthetic pathways (Fan, Evans, and Ling 2017). Synthetic biology plays a revolutionary role in biotechnological methods for microbial production, providing previously unheard-of control and precision in the design of microbial hosts for the efficient and sustainable synthesis of chemicals obtained from plants. The development of sustainable and scalable bioproduction systems can be facilitated by the creation of customized microbes that can produce a wide range of important plant-based chemicals as synthetic biology techniques continue to progress (Carruthers and Lee 2022). The biotechnological synthesis of plant-based chemicals on microbial platforms is largely dependent on genetic alteration and strain selection. This crucial stage of the procedure is selecting the right microbial host and then modifying its genetic composition to maximize the synthesis of particular chemicals taken from plants (Zhu et al. 2021). This investigation explores the role of genetic alteration and strain selection as essential elements of biotechnological approaches for the microbial synthesis of plant-based chemicals.

16.4 STRAIN SELECTION IN MICROBIAL PRODUCTION

The meticulous selection of a suitable host strain is the first step toward the success of microbial production. A wide range of viable possibilities, including bacteria, yeast, and fungi, each with specific metabolic properties, are available due to microbial diversity (Ali et al. 2023). Factors including growth characteristics, genetic modification ease, and the innate ability to manufacture or ingest target substances all play a role in selecting the right strain. Certain microbes are naturally able to synthesize particular chemicals derived from plants. By locating and using these naturally occurring producers, the biotechnological process can be streamlined, and the selected strain's biochemical pathways can be fully utilized. By utilizing the innate skills of microbes, this method is consistent with the concepts of environmentally benign and sustainable production. Non-native hosts can be investigated when native producers are unreliable or ineffective. Through genetic modification, researchers can introduce heterologous genes into a selected host, giving it the ability to generate plant-based substances that the organism is not normally capable of producing. Important factors to take into account include the host's suitability for the new pathway and its susceptibility to genetic modification (Health 2004). Genetic modification is key to improving a strain's biosynthetic capabilities once a suitable strain has been chosen. Key enzymes in the biosynthesis pathways or genes in charge of producing plant-based chemicals are either overexpressed or introduced. By increasing the expression of pertinent genes, this genetic modification seeks to increase the synthesis of particular target chemicals (Yang et al. 2018).

With the development of CRISPR/Cas9 technology, genetic modification has undergone a revolutionary change as a highly accurate and effective genome editing tool became available. With the use of this technique, scientists can precisely insert, remove, or replace genes by focusing on particular genomic regions. A key component of the arsenal for altering microbial genomes, CRISPR/Cas9 enables the quick creation of modified strains (Arroyo-Olarte, Bravo Rodríguez, and Morales-Ríos 2021). Biosynthetic pathways need to be carefully balanced to reach optimal production levels. Variations in substrate supply or enzyme activity can lead to imbalances. The main goal of genetic modification techniques is to optimize the expression levels of important enzymes in the pathway so that precursor molecules are effectively transformed into the intended plant-based products (Wu et al. 2022). Competing metabolic routes found in microbial hosts can direct precursor molecules away from the intended biosynthesis biosynthetic pathway.

To reroute metabolic flow toward the synthesis of desired chemicals, genetic manipulation may entail the downregulation or deletion of these rival pathways. Overall production process efficiency is increased by using this methodical technique (Ingram et al. 2010). Microbial cells may experience stress as a result of genetic alterations, which may lower their rates of growth or perhaps cause cell death. A crucial factor in ensuring the robustness and viability of the modified strains is controlling cellular stress. To lessen these difficulties, techniques including expression level tuning and promoter engineering are used. Regulatory oversight is applied to the use of genetically modified organisms (GMOs) in industrial operations. Narrowing the legal and ethical issues surrounding the use of genetically modified microorganisms requires adherence to national and international rules. Developing open lines of communication and following rules are essential to receiving regulatory approval. Maintaining the stability and productivity of modified strains presents hurdles in the effective transfer from laboratory-scale studies to industrial production (Ingram et al. 2010). Large-scale manufacturing requires the optimization of elements, including fermentation conditions, substrate supply, and downstream processing, while maintaining reliable and repeatable outcomes.

16.5 PATHWAY OPTIMIZATION FOR INCREASED YIELDS

Plant-based chemical synthesis by microbiological means is a potentially efficient and sustainable method of synthesis. Pathway optimization, the deliberate alteration of microbial metabolic pathways to increase the yield of target chemicals, lies at the heart of this biotechnological strategy (X. Sun et al. 2015). Complex metabolic pathways are used by microbial hosts to synthesize chemicals derived from plants. These routes transform precursor molecules into the intended end products through a sequence of enzyme processes. To enhance the efficiency of chemical production, pathway optimization aims to optimize and streamline various metabolic routes. One important factor influencing chemical formation is the genetic foundation of microbial metabolism. Genetic engineering is used in route optimization when particular metabolic pathway genes are altered to improve their expression or functionality.

Researchers may accurately target and edit the microbial genome with techniques like CRISPR/Cas9, optimizing it for greater synthesis of plant-based chemicals. A thorough study is the first stage in pathway optimization, which identifies the biosynthetic pathway's rate-limiting steps (X. Sun et al. 2015). These steps indicate the stages where metabolite flux is constrained, hence obstructing the overall efficiency of the process. Researchers can strategically act to improve compound yields and flow by identifying these bottlenecks. For the best chemical formation, the flow of metabolites through a pathway, or metabolic flux, must be precisely balanced. The goal of pathway optimization is to reallocate metabolic flux by modifying the enzymes' levels of expression. This equilibrium makes sure that precursor molecules are directed effectively toward the target chemical, preventing needless intermediate buildup.

Enzymes function as metabolic pathway catalysts, and effective chemical synthesis depends on their activity. Pathway optimization involves cofactor engineering or genetic alterations to fine-tune the activity of important enzymes (Khanijou et al. 2022). This guarantees that enzymatic reactions happen at the fastest possible pace, which raises the yields of chemicals derived from plants. By-products in microbial synthesis have the potential to divert metabolic flux away from the target chemical and compete for precursors. By modifying the expression of enzymes participating in competing processes or introducing genetic alterations to improve selectivity toward the target product, pathway optimization seeks to reduce the creation of by-products (Figure 16.2). Pathway optimization is greatly aided by omics technologies, such as transcriptomics, proteomics, metabolomics, and genomes (Wiltschi et al. 2020). These resources help identify important targets and offer a thorough grasp of cellular functions.

Optimizing the entire biosynthetic route at the systems level is made possible by the integration of multiomics data. Pathway optimization is frequently an iterative procedure that follows systems biology guidelines. To get the desired results, researchers iteratively fine-tune genetic alterations and route tweaks using computational modeling and experimental data analysis. This comprehensive method optimizes the overall system for higher yields while taking into account how linked cellular functions are. Pathway optimization methodologies are increasingly using principles of synthetic biology (Jamil et al. 2020). To accurately manage and improve metabolic pathways, this

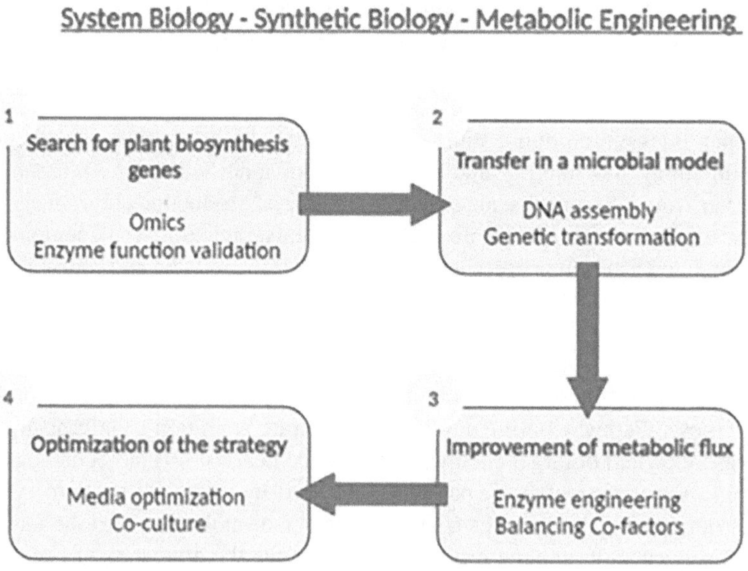

FIGURE 16.2 Steps in developing microbial synthetic biology.

includes building synthetic gene circuits, feedback control systems, and engineering regulatory elements. Tools from synthetic biology provide creative ways to maximize microbial hosts for effective chemical synthesis. Pathway optimization takes into account process and environmental factors in addition to genetic alterations.

Microbial growth and chemical synthesis are influenced by various factors, including temperature, pH, availability of nutrients, and oxygen levels. By fine-tuning these variables, one may make sure that the selected microbial strain thrives in environments that promote higher yields. Pathway optimization continues to present issues despite recent developments. Considerations such as substrate constraints, cellular toxicity, and the possibility of unforeseen repercussions from genetic alterations are important to take into account. One of the main challenges is continuing to strike a balance between the demand for higher yields and the stability and viability of the modified microbial strain (Jamil et al. 2020).

16.6 MICROBIAL PRODUCTION OF PLANT-BASED COMPOUNDS: CRISPR/CAS9 TECHNOLOGY

The introduction of CRISPR/Cas9 technology has resulted in a significant advance in the fast-changing field of microbial synthesis of plant-based chemicals. With the use of this precision genome editing technology, researchers can now accurately alter the genetic composition of microbial hosts, creating new opportunities for pathway optimization and improving the production of important compounds derived from plants (Wei and Li 2023). The bacterial immune system gave rise to the flexible and incredibly effective genome editing technology known as CRISPR/Cas9 (Figure 16.3). The Cas9 enzyme is directed to a particular DNA sequence by a guide RNA (gRNA), which makes targeted alterations in the DNA. In the realm of microbiological production, CRISPR/Cas9 has

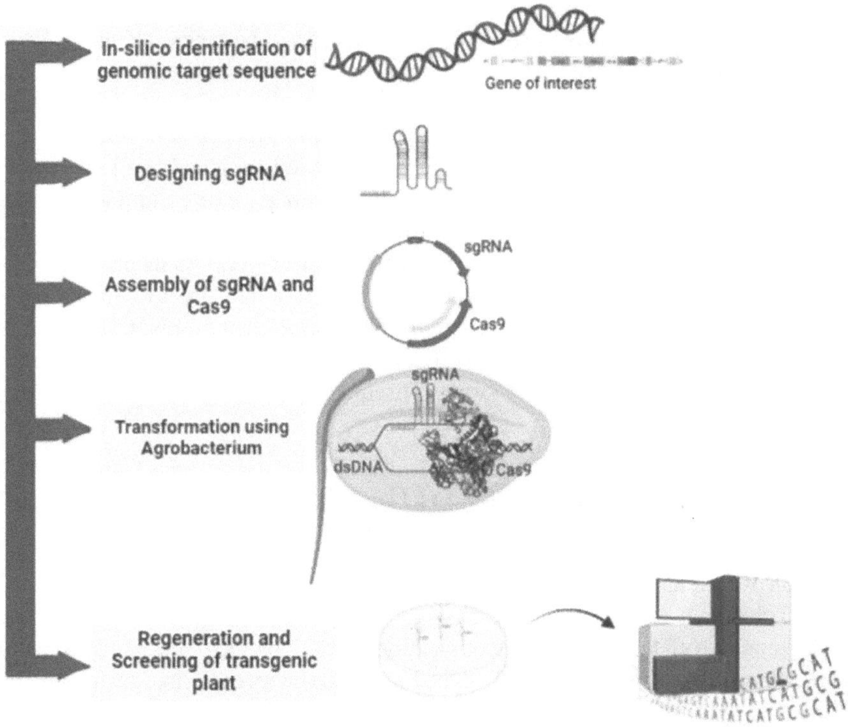

FIGURE 16.3 Gene manipulation by CRISPR/Cas9 gene editing system.

emerged as a vital instrument for accurate and regulated genetic modification. Researchers may create precise genetic alterations within the microbial genome to CRISPR/Cas9. Gene insertions, deletions, and replacements are examples of this that allow strains to be engineered to have improved capacities for the synthesis of particular plant-based chemicals. Targeting particular regions within the genome enables a level of control that was previously difficult to attain. Optimizing biosynthetic pathways is one of the main uses of CRISPR/Cas9 in microbial production. Researchers can increase the flow toward the manufacture of specific plant-based chemicals by strategically editing genes within the pathway to raise the expression of critical enzymes or to remove bottlenecks (Wei and Li 2023).

The entire biosynthetic process is made more efficient by this precision engineering. The technology of CRISPR/Cas9 enables the regulation of enzyme activity involved in biosynthetic pathways. Researchers can maximize an enzyme's catalytic performance, substrate selectivity, or cofactor preferences by carefully altering its coding sequences. This fine-tuning makes sure the enzymes work at their best during the production of chemicals derived from plants. To produce compounds at their best, microbial hosts must have their metabolic flux balanced (Chen et al. 2019). Through the use of CRISPR/Cas9, scientists can modify genes that control metabolic flow and maintain a balanced distribution of precursor molecules throughout the biosynthetic process. The pathway's efficiency is improved by this flux regulation, which raises the yields of chemicals derived from plants. With CRISPR/Cas9 technology, the production of by-products can be reduced by carefully altering the genes that compete with the pathways.

By minimizing the amount of metabolic flux diverted from the intended chemical, this tailored method helps avoid the accumulation of undesirable by-products (Soma et al. 2014). As a result, the biosynthetic process is more simplified and effective. Pathway optimization is not the only use of CRISPR/Cas9 for genetic alteration. Researchers can improve microbial hosts' aptitude for chemical production by the introduction or modification of genes associated with growth characteristics, stress tolerance, and cellular fitness. This all-encompassing strategy guarantees that the modified strains are resilient and suitably tailored to the manufacturing setting (Parmar et al. 2017). Despite the great precision of CRISPR/Cas9, off-target effects must still be taken into account. The goal of ongoing research is to improve the technology to reduce off-target alterations and boost specificity. Improving the accuracy of genome editing involves creating new Cas proteins and refining gRNA structures. In microbial manufacturing, the strain formation process is accelerated by CRISPR/Cas9 technology. Conventional techniques for improving strain frequently call for labor- and time-intensive procedures. Researchers can quickly modify microbial strains with CRISPR/Cas9, which reduces development times and makes it easier to put improved strains into large-scale manufacturing. CRISPR/Cas9 can be transformative, yet there are drawbacks to using it in microbial production (Y. Mu et al. 2022). Careful regulatory control is required due to ethical concerns about the release of genetically modified organisms into the environment and worries about unforeseen repercussions. Other continuous hurdles in the field include mitigating the possibility of horizontal gene transfer and managing potential off-target consequences.

16.7 OMICS TECHNOLOGIES AND SYSTEMS BIOLOGY: MICROBIAL PRODUCTION OF PLANT-BASED COMPOUNDS

The application of systems biology and omics technology has revolutionized the microbial synthesis of plant-based chemicals. Omics techniques, which include transcriptomics, proteomics, metabolomics, and genomes, offer a thorough understanding of the biological mechanisms controlling the synthesis of compounds (Y. Mu et al. 2022). When integrated with the principles of systems biology, these technologies provide a comprehensive framework for optimizing microbial hosts and routes, hence leading to increased yields of valuable compounds obtained from plants. A key component of biotechnological methods is genomics, the study of an organism's whole gene pool. Genomic analysis enables the identification of important genes involved in the manufacture of plant-based chemicals in microbial production. The process of choosing appropriate microbial strains is aided by comparative genomics, while genome sequencing makes it easier to construct specific genetic alterations that optimize pathways for higher output (Lasken and McLean 2014). The study of all RNA transcripts in a cell is known as transcriptomics because it offers a dynamic snapshot of gene expression. Transcriptomics provides light on the regulation of genes involved in chemical biosynthesis in the context of microbial production. Identifying rate-limiting processes, comprehending pathway dynamics, and creating interventions to improve the expression of important genes all depend on this knowledge (Lowe et al. 2017). The study of all the proteins in a cell is the main emphasis of proteomics. Proteomic analyses provide information about the precise protein levels involved in biosynthetic pathways in microbial production (Lowe et al. 2017). With the help of this data, transcriptome data can be validated, post-translational changes can be found, and enzyme activity can be optimized for the increased synthesis efficiency of plant-based drugs.

The thorough examination of tiny molecules, or metabolites, within a biological system is known as metabolomics. Metabolomics offers a comprehensive comprehension of the metabolic landscape in microbial production, facilitating the identification of intermediates and end products of biosynthetic pathways (Pang and Hu 2023). This data directs the optimization of pathways, making it easier to reroute metabolic flow in the direction of higher target compound yields. When data from proteomics, metabolomics, transcriptomics, and genomes are combined, the full potential of omics technologies becomes apparent. A knowledge of microbial metabolism at the systems level is made possible by this multiomics approach. Researchers can rationally create strategies for better compound production by gaining a full understanding of the interactions within the cell through the correlation of genetic information with gene expression, protein levels, and metabolite concentrations.

The multidisciplinary discipline of systems biology uses mathematical and computer models to study and comprehend intricate biological processes. Systems biology integrates data from omics technology to build predictive models of cellular behavior in the microbial synthesis of plant-based chemicals (Yahya et al. 2021). These models make it easier to identify important regulatory nodes, forecast bottlenecks in pathways, and plan treatments logically to maximize output. Researchers can simulate and forecast the behavior of modified microbial strains through the use of computational models that are informed by omics data. By forecasting the effects of route modifications, finding possible targets for genetic change, and directing the creation of strains with improved capacities for the synthesis of plant-based chemicals, this modeling approach helps optimize pathways (Ahmad 2023).

Optimization driven by systems biology is an iterative process. The design of genetic alterations or route adjustments is guided by computational predictions and tested experimentally. Continuous refinement is made possible by the feedback loop between experimental validation and computational modeling, which guarantees that the modified strains show the intended increases in chemical production (Pfau, Christian, and Ebenhöh 2011). Systems biology methods and omics technology have limitations even with their revolutionary potential. Ongoing issues include the necessity for robust computer models, the intricacy of biological systems, and data integration. Furthermore, proficiency in bioinformatics and computational biology is necessary for the analysis of omics data.

16.8 SCALE-UP AND DOWNSTREAM PROCESSING

The synthesis of plant-based chemicals using microbial production has shown great promise for an efficient and sustainable synthesis. Scale-up procedures and downstream processing must be carefully considered when moving from laboratory-scale research to industrial output. This investigation explores the difficulties, approaches, and inventions related to increasing microbial production and streamlining downstream procedures for the purification and extraction of plant-based chemicals (Cho et al. 2022c). There are inherent difficulties with transitioning from small-scale laboratory research to large-scale industrial manufacturing. At bigger sizes, variables like sustaining consistent conditions, effectively dissipating heat, and distributing nutrients become more complicated. These issues must be resolved to achieve a reliable and profitable microbial production. The essential component of industrial microbial production is the bioreactor (Abdur Razzak et al. 2023).

Vessel geometry, agitation methods, aeration techniques, and temperature control are all taken into account during design. By maximizing nutrient availability, ensuring consistent growing conditions, and facilitating effective gas exchange, these characteristics can be optimized to increase the yields of chemicals derived from plants (Cho et al. 2022c). Several scale-up techniques are designed to lessen the difficulties brought on by higher production quantities. The three types of similarity scaling are dynamic, constant power per unit volume, and geometric. Through the application of these tactics, scientists hope to preserve comparable conditions at various scales, enabling a seamless shift from lab to industrial production. Optimizing fermentation procedures is essential to producing high quantities of chemicals derived from plants. Precise control is required over parameters like pH, temperature, dissolved oxygen, and nutrient concentrations. Throughout the fermentation process, constant observation and feedback mechanisms help maintain ideal conditions, boosting growth and productivity of microbial strains.

Target chemicals are extracted, purified, and recovered from the broth of microbial fermentation as part of downstream processing. To achieve high product purity and yield, this phase is essential. The downstream processing techniques are customized to the unique properties of the microbial host and the plant-based chemicals (Gupta et al., 2002). Microbial cells function as factories that produce chemicals derived from plants. Intracellular chemicals are released into the fermentation broth by the use of cell disruption techniques. The properties of the microbial cells and the stability of the target substances are taken into consideration when selecting techniques like mechanical disruption, enzymatic lysis, or chemical treatments. Techniques for separation and filtering are used to extract the target components from the fermentation broth after cell destruction (Nemer et al. 2021).

Common techniques for eliminating cell fragments and other contaminants and leaving behind a cleared solution enhanced with the required plant-based chemicals include centrifugation, filtration, and sedimentation. Plant-based compound purification heavily relies on chromatography, namely, preparative chromatography and high-performance liquid chromatography (HPLC) (Nemer et al. 2021). To separate and isolate the target molecules with high purity, these techniques make use of variations in molecular characteristics, like size, charge, or affinity. Techniques for precipitation and crystallization are used to further purify chemicals derived from plants. Impurities can be eliminated, and the target chemicals can be obtained in a highly pure crystalline form by adjusting the precipitation conditions and solubility. Drying the refined chemicals is the last stage in downstream processing that produces a stable, usable product.

Common techniques include lyophilization, spray drying, and freeze-drying. To improve stability and usability, plant-based compounds are also formulated into certain products, including powders or liquid formulations (Adali et al. 2020). To achieve cost-effective production, both upstream and downstream processes must be continuously optimized. This entails limiting waste production, maximizing extraction and purification techniques, and fine-tuning fermentation conditions. Technological developments in automation and process control lead to lower manufacturing costs and higher productivity. Sustainability is an important factor in downstream processing as well as scale-up. The environmental impact of the microbial production processes is reduced by employing techniques, such as the development of environmentally friendly extraction techniques, the recycling of water and solvents, and the use of renewable energy sources.

16.9 COFACTOR ENGINEERING

Cofactors, also known as *co-substrates*, or *vital coenzymes*, are crucial to the enzymatic processes that control the production of molecules derived from plants. Cofactor engineering is a biotechnological tactic that aims to increase the overall yield and efficiency of the biosynthetic process by altering the cofactor specificity of enzymes (Sun et al. 2023). This investigation explores the tenets, tactics, and uses of cofactor engineering in the microbial synthesis of chemicals derived from plants. During biosynthesis, cofactors like NAD(P)H and FAD play a critical role as electron carriers or as participants in redox processes. For their catalytic activity, many of the enzymes involved in the production of chemicals derived from plants require particular cofactors. The specificity and availability of cofactors can be changed to greatly affect how efficiently biosynthetic pathways work (Barra, Awakawa, and Abe 2022).

The term "cofactor engineering" refers to a variety of techniques used to maximize cofactor use in microbial hosts. This involves introducing different pathways that produce or renew particular cofactors, modifying the intracellular cofactor pools, and changing the cofactor selectivity of enzymes. All the strategies are designed to increase the biosynthesis process's overall efficiency (Barra, Awakawa, and Abe 2022). The cofactor selectivity of enzymes involved in biosynthetic pathways can be changed through the use of genetic modification techniques. The enzyme's binding pocket can be changed to favor a certain cofactor, or mutations can be introduced into the active site to do this. Improving the compatibility between the enzyme and the cofactor needed for the best catalytic activity is the objective.

The availability of cofactors in microbial cells is impacted by both metabolic flux and cellular redox equilibrium. *Cofactor engineering* is the process of maximizing intracellular cofactor pools through the control of gene expression levels related to their manufacture or regeneration. This guarantees that the enzymes that catalyze biosynthetic reactions have an adequate supply of cofactors (Sun et al. 2023). Alternative mechanisms for the generation or regeneration of particular cofactors can be implemented to get around restrictions on cofactor availability. This could entail the introduction of heterologous genes that code for enzymes that can synthesize cofactors from various sources. Microbial strains grow more robust and adaptive to the requirements of plant-based chemical biosynthesis by broadening their pathways for cofactor production. Cofactor availability and microbial cell redox balance are intimately related.

Cofactor engineering refers to techniques for regulating the intracellular redox state to guarantee the presence of the necessary cofactors in their active forms. Genes encoding redox-active proteins may be introduced, or metabolic pathways may be optimized to improve electron flow (Su et al. 2017).

The influence of cofactor engineering on metabolic flux in biosynthetic pathways is significant. Researchers can control the direction and rate of biosynthesis by adjusting the cofactor specificity of important enzymes. Better yields of plant-based chemicals follow as a result of improved cofactor utilization and better biosynthetic process efficiency (Kuska and O'Reilly 2020). Applications for cofactor engineering can be found in particular biosynthetic pathways that are important for the synthesis of chemicals derived from plants

For example, maximizing an enzyme's cofactor specificity can greatly increase the overall efficiency and yield of substances when redox reactions are common, such as the production of alkaloids or flavonoids. Cofactor engineering is an advanced tactic that necessitates a thorough comprehension of both the particular cofactor requirements of important enzymes and the target biosynthetic pathway. Potential negative effects on cellular metabolism, inadvertent effects on other pathways, and the requirement for meticulous optimization to produce the intended results are some of the challenges.

16.10 GENE SYNTHESIS AND BIOINFORMATICS

Bioinformatics and gene synthesis are essential in improving the microbial production of plant-based chemicals. These biotechnological methods include the synthesis of synthetic genes and the modeling, analysis, and optimization of biosynthetic pathways through the use of computational tools. This thorough investigation explores the concepts, methods, and uses of bioinformatics and gene synthesis concerning the microbial synthesis of plant-based chemicals (Carbonell et al. 2016). The synthetic assembling of nucleotides to create genes is known as gene synthesis. This refers to the process of generating synthetic DNA sequences that code for the enzymes or pathways involved in the biosynthesis of chemicals derived from plants in the context of microbial production. Through gene synthesis, genetic sequences can be tailored, codon usage can be optimized, and specific changes can be added to improve expression and function (Mauro 2018).

Genetic sequences can be tailored to the specific traits of the selected microbial host through the process of gene synthesis. This entails eliminating secondary structures, adjusting regulatory elements to guarantee effective gene expression, and optimizing codon usage to fit the host's preferred codon usage bias (Mauro 2018). Customizing artificial genes to the host organism improves overall performance and compatibility. A crucial step in the synthesis of genes is codon optimization, particularly when transferring genes between animals that have differing codon usage preferences. Enzymes involved in the production of plant-based chemicals are expressed at higher levels when the synthetic gene is modified to match the codon preferences of the microbial host. This improves translation efficiency.

Gene expression can be precisely controlled by introducing synthetic regulatory elements and promoters into synthetic genes. This customization raises the overall yield and efficiency of producing compounds from plants by ensuring that biosynthetic enzymes are expressed at optimal amounts at various phases of microbial growth. An essential tool for analyzing biosynthetic routes is bioinformatics. To forecast, simulate, and examine the metabolic processes involved in the synthesis of chemicals derived from plants, computational techniques are employed (Liu et al. 2023). This includes identifying important enzymes, charting pathways, and forecasting any bottlenecks or limiting steps. One bioinformatics tool that measures the flow of metabolites across biological pathways is metabolic flux analysis. It sheds light on the distribution of metabolic flux within the plant-based chemical biosynthesis pathway in the context of microbial production. The selection of possible targets for genetic change to improve route efficiency is guided by this knowledge.

Biosynthetic pathways are optimized and simulated using computational models. Researchers can forecast how genetic mutations, changes in enzyme kinetics, and variations in food availability would affect the pathway's overall performance with the use of bioinformatics tools. The logical design of microbial hosts for enhanced plant-based chemical synthesis is guided by these models. Protein structures can be predicted using bioinformatics methods, which helps with the engineering of enzymes for improved functionality (Mu, Wang, and Yuan 2022). Computational modeling–guided rational design permits the introduction of mutations that enhance the stability, substrate selectivity, and catalytic efficiency of enzymes. This strategy of protein engineering aids in the improvement of biosynthetic pathways.

16.11 CONCLUSION

In conclusion, the biotechnological synthesis of plant-based chemicals using microbial platforms is a revolutionary strategy that will have a significant impact on several businesses. To fully utilize microorganisms as bio-factories, this multidisciplinary discipline integrates concepts from fermentation technology, synthetic biology, genetics, and bioinformatics. As this sector develops, it presents the chance to address issues with conventional extraction techniques while offering scalable, sustainable, and financially feasible substitutes. The integration of metabolic engineering, synthetic biology, and sophisticated bioprocessing methods establishes microbial production as a fundamental element for the sustainable synthesis of compounds in the future. The biotechnological synthesis of plant-based chemicals has the potential to make a significant contribution toward the creation of more environmentally friendly and efficient industrial processes. This can be achieved via continuous research, innovation, and collaboration across scientific fields.

REFERENCES

Abdur Razzak, Shaikh, Khairul Bahar, K.M. Oajedul Islam, Abdul Khaleel Haniffa, Mohammed Omar Faruque, S.M. Zakir Hossain, and Mohammad M. Hossain. 2023. "Microalgae Cultivation in Photobioreactors: Sustainable Solutions for a Greener Future." *Green Chemical Engineering* (October). Elsevier. https://doi.org/10.1016/J.GCE.2023.10.004.

Adali, Merve B., Antonello A. Barresi, Gianluca Boccardo, and Roberto Pisano. 2020. "Spray Freeze-Drying as a Solution to Continuous Manufacturing of Pharmaceutical Products in Bulk." *Processes 2020* 8 (6). Multidisciplinary Digital Publishing Institute: 709. https://doi.org/10.3390/PR8060709.

Ahmad, M. 2023. "Plant Breeding Advancements with 'CRISPR-Cas' Genome Editing Technologies Will Assist Future Food Security." *Frontiers in Plant Science* 14. Frontiers Media SA. https://doi.org/10.3389/FPLS.2023.1133036.

Ali, Sajid, Sumera Afzal Khan, Muhammad Hamayun, and In Jung Lee. 2023. "The Recent Advances in the Utility of Microbial Lipases: A Review." *Microorganisms* 11 (2). Multidisciplinary Digital Publishing Institute (MDPI). https://doi.org/10.3390/MICROORGANISMS11020510.

Amer, Bashar, and Edward E.K. Baidoo. 2021. "Omics-Driven Biotechnology for Industrial Applications." *Frontiers in Bioengineering and Biotechnology* 9 (February). Frontiers Media SA: 613307. https://doi.org/10.3389/FBIOE.2021.613307.

Arroyo-Olarte, Ruben D., Ricardo Bravo Rodríguez, and Edgar Morales-Ríos. 2021. "Genome Editing in Bacteria: CRISPR-Cas and Beyond." *Microorganisms* 9 (4). Multidisciplinary Digital Publishing Institute (MDPI). https://doi.org/10.3390/MICROORGANISMS9040844.

Barra, Lena, Takayoshi Awakawa, and Ikuro Abe. 2022. "Noncanonical Functions of Enzyme Cofactors as Building Blocks in Natural Product Biosynthesis." *JACS Au* 2 (9). American Chemical Society: 1950–1963. https://doi.org/10.1021/JACSAU.2C00391/ASSET/IMAGES/LARGE/AU2C00391_0015.JPEG.

Boodhoo, K.V.K., M.C. Flickinger, J.M. Woodley, and E.A.C. Emanuelsson. 2022. "Bioprocess Intensification: A Route to Efficient and Sustainable Biocatalytic Transformations for the Future." *Chemical Engineering and Processing—Process Intensification* 172 (February). Elsevier: 108793. https://doi.org/10.1016/J.CEP.2022.108793.

Burgos-Morales, O., M. Gueye, L. Lacombe, C. Nowak, R. Schmachtenberg, M. Hörner, C. Jerez-Longres, H. Mohsenin, H.J. Wagner, and W. Weber. 2021. "Synthetic Biology as Driver for the Biologization of Materials Sciences." *Materials Today Bio* 11 (June). Elsevier: 100115. https://doi.org/10.1016/J.MTBIO.2021.100115.

Carbonell, Pablo, Andrew Currin, Adrian J. Jervis, Nicholas J.W. Rattray, Neil Swainston, Cunyu Yan, Eriko Takano, and Rainer Breitling. 2016. "Bioinformatics for the Synthetic Biology of Natural Products: Integrating across the Design–Build–Test Cycle." *Natural Product Reports* 33 (8). Royal Society of Chemistry: 925. https://doi.org/10.1039/C6NP00018E.

Carruthers, David N., and Taek Soon Lee. 2022. "Translating Advances in Microbial Bioproduction to Sustainable Biotechnology." *Frontiers in Bioengineering and Biotechnology* 10 (August). Frontiers Media SA. https://doi.org/10.3389/FBIOE.2022.968437.

Chen, Kunling, Yanpeng Wang, Rui Zhang, Huawei Zhang, and Caixia Gao. 2019. "CRISPR/Cas Genome Editing and Precision Plant Breeding in Agriculture." *https://Doi.Org/10.1146/Annurev-Arplant-050718-100049* 70 (April). Annual Reviews: 667–697. https://doi.org/10.1146/ANNUREV-ARPLANT-050718-100049.

Cho, Jae Sung, Gi Bae Kim, Hyunmin Eun, Cheon Woo Moon, and Sang Yup Lee. 2022a. "Designing Microbial Factories for the Production of Chemicals." *JACS Au* 2 (8). American Chemical Society: 1781. https://doi.org/10.1021/JACSAU.2C00344.

Cho, Jae Sung, Gi Bae Kim, Hyunmin Eun, Cheon Woo Moon, and Sang Yup Lee. 2022b. "Designing Microbial Cell Factories for the Production of Chemicals." *JACS Au* 2 (8). American Chemical Society: 1781–1799. https://doi.org/10.1021/JACSAU.2C00344/ASSET/IMAGES/LARGE/AU2C00344_0008.JPEG.

Cho, Jae Sung, Gi Bae Kim, Hyunmin Eun, Cheon Woo Moon, and Sang Yup Lee. 2022c. "Designing Microbial Cell Factories for the Production of Chemicals." *JACS Au* 2 (8). American Chemical Society: 1781–1799. https://doi.org/10.1021/JACSAU.2C00344/ASSET/IMAGES/LARGE/AU2C00344_0008.JPEG.

Cravens, Aaron, James Payne, and Christina D. Smolke. 2019. "Synthetic Biology Strategies for Microbial Biosynthesis of Plant Natural Products." *Nature Communications* 10 (1). Nature Publishing Group. https://doi.org/10.1038/S41467-019-09848-W.

Dai, Xiaofeng, and Li Shen. 2022. "Advances and Trends in Omics Technology Development." *Frontiers in Medicine* 9 (July). Frontiers Media SA: 911861. https://doi.org/10.3389/FMED.2022.911861.

Dincheva, Ivayla, Ilian Badjakov, and Bistra Galunska. 2023. "New Insights into the Research of Bioactive Compounds from Plant Origins with Nutraceutical and Pharmaceutical Potential." *Plants* 12 (2). Multidisciplinary Digital Publishing Institute (MDPI): 258. https://doi.org/10.3390/PLANTS12020258.

Fan, Yongqiang, Christopher R. Evans, and Jiqiang Ling. 2017. "Rewiring Protein Synthesis: From Natural to Synthetic Amino Acids." *Biochimica et Biophysica Acta* 1861 (11 Pt B). NIH Public Access: 3024. https://doi.org/10.1016/J.BBAGEN.2017.01.014.

Francois, Jean Marie, Ceren Alkim, and Nicolas Morin. 2020. "Engineering Microbial Pathways for Production of Bio-Based Chemicals from Lignocellulosic Sugars: Current Status and Perspectives." *Biotechnology for Biofuels* 13 (1). BioMed Central: 1–23. https://doi.org/10.1186/S13068-020-01744-6/FIGURES/5.

Garner, Kathryn L. 2021. "Principles of Synthetic Biology." *Essays in Biochemistry* 65 (5). Portland Press Ltd: 791. https://doi.org/10.1042/EBC20200059.

Glasscock, Cameron J., Bradley W. Biggs, John T. Lazar, Jack H. Arnold, Lisa A. Burdette, Aliki Valdes, Min Kyoung Kang, Danielle Tullman-Ercek, Keith E.J. Tyo, and Julius B. Lucks. 2021. "Dynamic Control of Gene Expression with Riboregulated Switchable Feedback Promoters." *ACS Synthetic Biology* 10 (5). NIH Public Access: 1199. https://doi.org/10.1021/ACSSYNBIO.1C00015.

Gupta, R., Beg, Q.K., Khan, S., Chauhan, B. (2002). An overview on fermentation, downstream processing and properties of microbial alkaline proteases. *Appl Microbiol Biotechnol. 60*, 381–95. https://doi.org/10.1007/S00253-002-1142-1

Han, Laichuang, Qiaoqing Chen, Qiao Lin, Jintao Cheng, Li Zhou, Zhongmei Liu, Junling Guo, Linpei Zhang, Wenjing Cui, and Zhemin Zhou. 2020. "Realization of Robust and Precise Regulation of Gene Expression by Multiple Sigma Recognizable Artificial Promoters." *Frontiers in Bioengineering and Biotechnology* 8 (February). Frontiers Media SA. https://doi.org/10.3389/FBIOE.2020.00092/FULL.

Health, National Research Council (US) Committee on Identifying and Assessing Unintended Effects of Genetically Engineered Foods on Human. 2004. "Methods and Mechanisms for Genetic Manipulation of Plants, Animals, and Microorganisms." National Academies Press (US). www.ncbi.nlm.nih.gov/books/NBK215771/.

Hoose, Alex, Richard Vellacott, Marko Storch, Paul S. Freemont, and Maxim G. Ryadnov. 2023. "DNA Synthesis Technologies to Close the Gene Writing Gap." *Nature Reviews Chemistry* 7 (3). Nature Publishing Group: 144–161. https://doi.org/10.1038/s41570-022-00456-9.

Ingram, Lonnie O., Laura R. Jarboe, Xueli Zhang, Xuan Wang, Jonathan C. Moore, and K.T. Shanmugam. 2010. "Metabolic Engineering for Production of Biorenewable Fuels and Chemicals: Contributions of Synthetic Biology." *Journal of Biomedicine and Biotechnology* 2010. https://doi.org/10.1155/2010/761042.

Jamil, Ili Nadhirah, Juwairiah Remali, Kamalrul Azlan Azizan, Nor Azlan Nor Muhammad, Masanori Arita, Hoe Han Goh, and Wan Mohd Aizat. 2020. "Systematic Multi-Omics Integration (MOI) Approach in Plant Systems Biology." *Frontiers in Plant Science* 11 (June). Frontiers Media SA: 540561. https://doi.org/10.3389/FPLS.2020.00944/BIBTEX.

Khanijou, Jasmeet Kaur, Hanna Kulyk, Cécilia Bergès, Leng Wei Khoo, Pnelope Ng, Hock Chuan Yeo, Mohamed Helmy, Floriant Bellvert, Wee Chew, and Kumar Selvarajoo. 2022. "Metabolomics and Modelling Approaches for Systems Metabolic Engineering." *Metabolic Engineering Communications* 15 (December). Elsevier: e00209. https://doi.org/10.1016/J.MEC.2022.E00209.

Ku, Jason T., Arvin Y. Chen, and Ethan I. Lan. 2020. "Metabolic Engineering Design Strategies for Increasing Acetyl-CoA Flux." *Metabolites* 10 (4). Multidisciplinary Digital Publishing Institute (MDPI). https://doi.org/10.3390/METABO10040166.

Kuska, Justyna, and Elaine O'Reilly. 2020. "Engineered Biosynthetic Pathways and Biocatalytic Cascades for Sustainable Synthesis." *Current Opinion in Chemical Biology* 58 (October). Elsevier Current Trends: 146–154. https://doi.org/10.1016/J.CBPA.2020.08.006.

Lasken, Roger S., and Jeffrey S. McLean. 2014. "Recent Advances in Genomic DNA Sequencing of Microbial Species from Single Cells." *Nature Reviews: Genetics* 15 (9). NIH Public Access: 577. https://doi.org/10.1038/NRG3785.

Liu, Xinyu, Peijun Zhang, Qiao Zhao, and Ancheng C. Huang. 2023. "Making Small Molecules in Plants: A Chassis for Synthetic Biology-Based Production of Plant Natural Products." *Journal of Integrative Plant Biology* 65 (2). John Wiley & Sons, Ltd: 417–443. https://doi.org/10.1111/JIPB.13330.

Lowe, Rohan, Neil Shirley, Mark Bleackley, Stephen Dolan, and Thomas Shafee. 2017. "Transcriptomics Technologies." *PLoS Computational Biology* 13 (5). PLOS. https://doi.org/10.1371/JOURNAL.PCBI.1005457.

Markina, N.M., A.A. Kotlobay, and A.S. Tsarkova. 2020. "Heterologous Metabolic Pathways: Strategies for Optimal Expression in Eukaryotic Hosts." *Acta Naturae* 12 (2). National Research University Higher School of Economics: 28. https://doi.org/10.32607/ACTANATURAE.10966.

Martins-Santana, Leonardo, Luisa C. Nora, Ananda Sanches-Medeiros, Gabriel L. Lovate, Murilo H. Murilo, and Rafael Silva-Rocha. 2018. "Systems and Synthetic Biology Approaches to Engineer Fungi for Fine Chemical Production." *Frontiers in Bioengineering and Biotechnology* 6 (October). Frontiers Media SA: 384079. https://doi.org/10.3389/FBIOE.2018.00117/BIBTEX.

Mauro, Vincent P. 2018. "Codon Optimization in the Production of Recombinant Biotherapeutics: Potential Risks and Considerations." *BioDrugs* 32 (1). Springer: 69–81. https://doi.org/10.1007/S40259-018-0261-X.

Melini, Francesca, Francesca Luziatelli, Paolo Bonini, Anna Grazia Ficca, Valentina Melini, and Maurizio Ruzzi. 2023. "Optimization of the Growth Conditions through Response Surface Methodology and Metabolomics for Maximizing the Auxin Production by Pantoea Agglomerans C1." *Frontiers in Microbiology* 14 (March). Frontiers Media SA: 1022248. https://doi.org/10.3389/FMICB.2023.1022248/BIBTEX.

Mohd Kamal, Khairunnisa, Mohd Hafidz Mahamad Maifiah, Nusaibah Abdul Rahim, Yumi Zuhanis Has Yun Hashim, Muhamad Shirwan Abdullah Sani, and Kamalrul Azlan Azizan. 2022. "Bacterial Metabolomics: Sample Preparation Methods." *Biochemistry Research International* 2022. Hindawi Limited. https://doi.org/10.1155/2022/9186536.

Mu, Huiying, Baoshan Wang, and Fang Yuan. 2022. "Bioinformatics in Plant Breeding and Research on Disease Resistance." *Plants* 11 (22). Multidisciplinary Digital Publishing Institute (MDPI): 3118. https://doi.org/10.3390/PLANTS11223118.

Mu, Yulin, Chengxiao Zhang, Taihua Li, Feng Jie Jin, Yun Ju Sung, Hee Mock Oh, Hyung Gwan Lee, and Long Jin. 2022. "Development and Applications of CRISPR/Cas9-Based Genome Editing in Lactobacillus." *International Journal of Molecular Sciences* 23 (21). Multidisciplinary Digital Publishing Institute (MDPI). https://doi.org/10.3390/IJMS232112852.

Nemer, Georgio, Nicolas Louka, Eugène Vorobiev, Dominique Salameh, Jean Marc Nicaud, Richard G. Maroun, and Mohamed Koubaa. 2021. "Mechanical Cell Disruption Technologies for the Extraction of Dyes and Pigments from Microorganisms: A Review." *Fermentation* 7 (1). Multidisciplinary Digital Publishing Institute: 36. https://doi.org/10.3390/FERMENTATION7010036.

Pang, Huanhuan, and Zeping Hu. 2023. "Metabolomics in Drug Research and Development: The Recent Advances in Technologies and Applications." *Acta Pharmaceutica Sinica B* 13 (8). Elsevier: 3238–3251. https://doi.org/10.1016/J.APSB.2023.05.021.

Parmar, Nehanjali, Kunwar Harendra Singh, Deepika Sharma, Lal Singh, Pankaj Kumar, J. Nanjundan, Yasin Jeshima Khan, Devendra Kumar Chauhan, and Ajay Kumar Thakur. 2017. "Genetic Engineering Strategies for Biotic and Abiotic Stress Tolerance and Quality Enhancement in Horticultural Crops: A Comprehensive Review." *3 Biotech* 7 (4). Springer. https://doi.org/10.1007/S13205-017-0870-Y.

Pfau, Thomas, Nils Christian, and Oliver Ebenhöh. 2011. "Systems Approaches to Modelling Pathways and Networks." *Briefings in Functional Genomics* 10 (5). Oxford Academic: 266–279. https://doi.org/10.1093/BFGP/ELR022.

Rafeeq, Hamza, Nadia Afsheen, Sadia Rafique, Arooj Arshad, Maham Intisar, Asim Hussain, Muhammad Bilal, and Hafiz M.N. Iqbal. 2023. "Genetically Engineered Microorganisms for Environmental Remediation." *Chemosphere* 310 (January). Pergamon: 136751. https://doi.org/10.1016/J.CHEMOSPHERE.2022.136751.

Ray, Prasun, Venkatachalam Lakshmanan, Jessy L. Labbé, and Kelly D. Craven. 2020. "Microbe to Microbiome: A Paradigm Shift in the Application of Microorganisms for Sustainable Agriculture." *Frontiers in Microbiology* 11 (December). Frontiers Media SA: 622926. https://doi.org/10.3389/FMICB.2020.622926/BIBTEX.

Soma, Yuki, Keigo Tsuruno, Masaru Wada, Atsushi Yokota, and Taizo Hanai. 2014. "Metabolic Flux Redirection from a Central Metabolic Pathway toward a Synthetic Pathway Using a Metabolic Toggle Switch." *Metabolic Engineering* 23. Academic Press Inc.: 175–184. https://doi.org/10.1016/J.YMBEN.2014.02.008.

Su, Hongfei, Xiaolin Chen, Shijing Chen, Mingzhang Guo, and Huilin Liu. 2023. "Applications of the Whole-Cell System in the Efficient Biosynthesis of Heme." *International Journal of Molecular Sciences* 24 (9). Multidisciplinary Digital Publishing Institute (MDPI). https://doi.org/10.3390/IJMS24098384.

Su, Liqiu, Yanbing Shen, Wenkai Zhang, Tian Gao, Zhihua Shang, and Min Wang. 2017. "Cofactor Engineering to Regulate NAD+/NADH Ratio with Its Application to Phytosterols Biotransformation." *Microbial Cell Factories* 16 (1). BMC: 182. https://doi.org/10.1186/S12934-017-0796-4.

Sun, Xinxiao, Xiaolin Shen, Rachit Jain, Yuheng Lin, Jian Wang, Jing Sun, Jia Wang, Yajun Yan, and Qipeng Yuan. 2015. "Synthesis of Chemicals by Metabolic Engineering of Microbes." *Chemical Society Reviews* 44 (11). The Royal Society of Chemistry: 3760–3785. https://doi.org/10.1039/C5CS00159E.

Sun, Yang, Ting Zhang, Bingqian Lu, Xiangfei Li, and Ling Jiang. 2023. "Application of Cofactors in the Regulation of Microbial Metabolism: A State of the Art Review." *Frontiers in Microbiology* 14. Frontiers Media SA. https://doi.org/10.3389/FMICB.2023.1145784.

Wei, Junwei, and Yingjun Li. 2023. "CRISPR-Based Gene Editing Technology and Its Application in Microbial Engineering." *Engineering Microbiology* 3 (4). Elsevier: 100101. https://doi.org/10.1016/J.ENGMIC.2023.100101.

Wiltschi, Birgit, Tomislav Cernava, Alexander Dennig, Meritxell Galindo Casas, Martina Geier, Steffen Gruber, Marianne Haberbauer, et al. 2020. "Enzymes Revolutionize the Bioproduction of Value-Added Compounds: From Enzyme Discovery to Special Applications." *Biotechnology Advances* 40 (May). Elsevier: 107520. https://doi.org/10.1016/J.BIOTECHADV.2020.107520.

Wu, Sijia, Wenjuan Chen, Sujuan Lu, Hailing Zhang, and Lianghong Yin. 2022. "Metabolic Engineering of Shikimic Acid Biosynthesis Pathway for the Production of Shikimic Acid and Its Branched Products in Microorganisms: Advances and Prospects." *Molecules* 27 (15). Multidisciplinary Digital Publishing Institute (MDPI). https://doi.org/10.3390/MOLECULES27154779.

Yahya, Farizatul Akmawati, Nur Fariesha Md Hashim, Daud Ahmad Israf Ali, Tham Chau Ling, and Manraj Singh Cheema. 2021. "A Brief Overview to Systems Biology in Toxicology: The Journey from in to Vivo, in-Vitro and—Omics." *Journal of King Saud University—Science* 33 (1). Elsevier: 101254. https://doi.org/10.1016/J.JKSUS.2020.101254.

Yang, Miaomiao, Junhua Yun, Huanhuan Zhang, Tinashe A. Magocha, Hossain Zabed, Yanbo Xue, Ernest Fokum, Wenjing Sun, and Xianghui Qi. 2018. "Genetically Engineered Strains: Application and Advances for 1,3-Propanediol Production from Glycerol." *Food Technology and Biotechnology* 56 (1). Faculty of Food Technology and Biotechnology University of Zagreb: 3. https://doi.org/10.17113/FTB.56.01.18.5444.

Zhao, Jing, Huan Fang, and Dawei Zhang. 2020. "Expanding Application of CRISPR-Cas9 System in Microorganisms." *Synthetic and Systems Biotechnology* 5 (4). KeAi Publishing: 269. https://doi.org/10.1016/J.SYNBIO.2020.08.001.

Zhu, Xiaoxi, Xiaonan Liu, Tian Liu, Yina Wang, Nida Ahmed, Zhichao Li, and Huifeng Jiang. 2021. "Synthetic Biology of Plant Natural Products: From Pathway Elucidation to Engineered Biosynthesis in Plant Cells." *Plant Communications* 2 (5). Elsevier: 100229. https://doi.org/10.1016/J.XPLC.2021.100229.

17 DNA-Based Approaches for the Discrimination of Medicinal Plants from Their Adulterants

*Yuvaraj Sivamani, Afrah Kounin P., and Sumitha Elayaperumal**
* Corresponding author: sumithae@jssuni.edu.in

17.1 INTRODUCTION

DNA-based techniques have become essential sources in the field of medicinal plant authentication, providing a sophisticated way to distinguish true species from substitutes or adulterants. The integrity of medicinal plant products is paramount, as any substitution or adulteration can not only compromise their therapeutic efficacy but also pose potential health risks to consumers. By leveraging DNA-based approaches, researchers and industry stakeholders can delve deep into the genetic signatures of medicinal plants, allowing for precise identification and authentication (Sucher and Carles 2008). These methods rely on the unique genetic sequences inherent to each species, enabling the discrimination of closely related taxa and the detection of adulterants, even in complex mixtures. The significance of DNA-based techniques lies in their accuracy, reliability, and objectivity. Unlike traditional methods that may rely on subjective morphological or chemical characteristics, DNA analysis provides an objective means of authentication, immune to external factors, such as environmental variation or processing methods). Furthermore, DNA-based methods offer scalability and adaptability to diverse matrices, accommodating the varied forms in which medicinal plants are traded, whether as raw materials, powders, or processed products. This versatility makes DNA analysis suitable for integration into quality control measures along the entire supply chain, from cultivation and harvesting to manufacturing and distribution. Due to the wide range of therapeutic benefits that medicinal plants offer as well as their potential uses in traditional medicine, pharmaceuticals, and nutraceuticals, the demand for these products has skyrocketed worldwide. But as demand has grown, so too have adulteration techniques—in which inferior or unrelated species are used in place of or contaminated with real therapeutic herbs (Nyakudya et al. 2020). Robust authentication techniques are essential, since adulteration not only jeopardizes the effectiveness of herbal treatments but also poses major health hazards.

Conventional approaches to medical plant authentication, such as morphological and biochemical investigation, frequently fall short when it comes to identifying intricate adulteration techniques. As a result, the use of sophisticated molecular techniques is increasing, and DNA-based methods are shown to be effective instruments for precise species identification and discrimination (Noviana, Indrayanto, and Rohman 2022). This introduction gives a general overview of the importance of medicinal plants, the difficulties caused by adulteration, and the function of DNA-based methods in guaranteeing the genuineness and caliber of herbal goods. Throughout history, medicinal plants have had a significant impact on human health. An in-depth examination of the historical and cultural significance of medicinal plants, the scientific underpinnings of their therapeutic effects, the business and economics of the herbal industry, and the opportunities and problems related to their sustainable use are all part

DOI: 10.1201/9781003326939-21

of this thorough review (Fitzgerald, Heinrich, and Booker 2019). The versatility of medicinal plants affects a wide range of industries, such as pharmaceuticals, nutraceuticals, and cosmetics, in addition to their conventional applications in medicine. Ever since the beginning of human civilization, medicinal plants—also known as nature's pharmacy—have played a crucial role. Cultural customs, conventional medical systems, and Indigenous knowledge all have a strong foundation in the usage of plants for health and wellness. The use of medicinal plants has a long history dating back to ancient societies, when the use of herbal treatments was essential to medical procedures.

17.2 SCIENTIFIC RATIONALE FOR MEDICINAL PLANTS' THERAPEUTIC BENEFITS

Chemical substances, each with specific characteristics and possible therapeutic applications, can be found in medicinal plants. The pharmacological properties of these bioactive substances are what give the plants their traditional therapeutic usage (Fitzgerald, Heinrich, and Booker 2019). Alkaloids, flavonoids, terpenoids, and polyphenols are a few of the main types of bioactive substances that are present in medicinal plants. Alkaloids are nitrogen-containing substances that are present in a wide range of plant species and frequently have significant physiological effects on people (Roy et al. 2022). Quinine from the cinchona tree (*Cinchona officinalis*), morphine from the opium poppy (*Papaver somniferum*), and caffeine from coffee plants (*Coffea* spp.) are a few examples. These substances have stimulant, antimalarial, and analgesic effects, in that order. A broad class of polyphenolic chemicals called flavonoids is well-known for having anti-inflammatory and antioxidant characteristics. Green tea, berries, and citrus fruits are excellent providers of flavonoids. Two flavonoids with a reputation for improving health include quercetin (found in onions and apples) and epigallocatechin gallate (EGCG), which is present in large amounts in green tea. Another family of bioactive molecules commonly found in medicinal plants are terpenoids, sometimes known as terpenes (Roy et al. 2022). Terpenoids found in essential oils are derived from plants such as peppermint, eucalyptus, and lavender. These substances are antibacterial, anti-inflammatory, and anti-anxious. For example, peppermint's calming properties are partly attributed to its terpenoid menthol content. Polyphenols are recognized for their anti-inflammatory and antioxidant properties. Examples of these chemicals are resveratrol, found in grapes; curcumin, found in turmeric; and catechins, found in green tea. The potential of these chemicals to prevent chronic diseases, such as cardiovascular disorders and some types of cancer, has been thoroughly investigated.

17.2.1 Mechanism of Action

Deciphering the mechanisms underlying the therapeutic qualities of medicinal plants requires an understanding of the interactions between bioactive chemicals and biological systems. Medicinal plants include a variety of bioactive substances that act as antioxidants, lowering oxidative stress and neutralizing reactive oxygen species (ROS) (Cory et al. 2018). Numerous illnesses, such as cardiovascular ailments and neurological problems, are linked to oxidative stress. Antioxidants support cellular health by scavenging free radicals, which may reduce the chance of oxidative damage. An intricate biological reaction, inflammation is linked to several illnesses, such as inflammatory bowel disorders, arthritis, and some types of cancer. By altering inflammatory pathways and lowering the synthesis of pro-inflammatory mediators, bioactive substances like flavonoids and polyphenols frequently have anti-inflammatory properties (Yahfoufi et al. 2018). The antibacterial qualities of many therapeutic plants have long been employed in traditional medicine. The growth of bacteria, fungi, and viruses can be inhibited by bioactive substances like terpenoids and alkaloids. These substances may function by rupturing the membranes surrounding microbial cells, interfering with biological functions, or altering the immunological response of the host. Bioactive substances can influence gene expression and cellular responses through interactions with cellular signaling networks (Yahfoufi et al. 2018). For instance, it has been demonstrated that polyphenols, such as resveratrol, activate the sirtuin protein family, which is linked to longevity and cellular stress response. These interactions may have consequences for age-related disorders and help regulate cellular functioning. Neurotransmitter activity can be modulated by certain medicinal plants, especially those utilized in traditional herbal therapy for cognitive health. For example, flavonoids and terpenoids found in ginkgo biloba may boost blood flow to the brain and alter neurotransmitter levels, which may lead to better cognitive function (Yahfoufi et al. 2018).

Clinical applications and ongoing research endeavors have been made possible by the scientific understanding of the medicinal characteristics of medicinal plants. A wide range of chemicals originating from plants have found their way into supplements, prescription drugs, and alternative therapies. Examples include the discovery of aspirin, which was initially extracted from willow bark, and the use of artemisinin derivatives from *Artemisia annua* for the treatment of malaria. The discovery of chemicals in medicinal plants has led to the creation of medications used in contemporary medicine. For example, the painkiller codeine is produced from opium poppies, whereas the anticancer medication vincristine is obtained from the Madagascar periwinkle (*Catharanthus roseus*). The combination of several chemicals found in medicinal plants frequently enhances the therapeutic efficiency of these medications. Complementary and alternative medicine (CAM) approaches continue to use a variety of therapeutic plants. Plant-based interventions are used in herbal treatments, dietary supplements, and traditional medical systems to address a variety of health issues. Utilizing the holistic advantages of medicinal plants, complementary and alternative medicine often aims to improve overall health. Thanks to a recent study, we are becoming more knowledgeable about the therapeutic potential of medicinal plants. Currently, studies are being carried out in many areas, such as the discovery of new bioactive compounds, screenings of the possible synergistic effects of plant combinations, and the use of state-of-the-art technologies, like metabolomics and systems biology, to comprehend the intricate relationships that exist between humans and plants. For centuries, people have used medicinal plant species to treat and cure a wide range of illnesses in both humans and animals. The toxicity and adverse effects of synthetic pharmaceuticals and antibiotics have brought herbal medicine back to the forefront of the health-care system, even though its use and practice dropped with the advent of antibiotics and sulfa drugs. Over 2 billion people may rely heavily on medical plants; according to WHO and other sources, about 80% of the world's population still uses botanical medications. The strategy for preventing and treating severe acute respiratory syndrome heavily relies on medicinal plants. This system plays a significant part in the approach to prevent and cure severe acute respiratory syndrome. In essence, DNA-based approaches serve as a powerful ally in the fight against adulteration and fraud within the medicinal plant industry, ensuring product integrity, consumer safety, and the preservation of traditional knowledge and practices (Techen et al. 2014; Gurib-Fakim 2006; Karunamoorthi et al. 2012).

17.3 MOLECULAR MARKERS FOR DISCRIMINATION OF ADULTERANTS

Adulteration is the practice of using inferior or unrelated species to replace or contaminate real therapeutic plants. It is a pervasive problem with far-reaching effects. Molecular markers, which are distinct genetic variants or DNA sequences, offer a strong and trustworthy way to identify adulterants and authenticate medicinal plants (Ichim 2019). There are significant physical similarities between a lot of therapeutic plants and their adulterants, especially when they are made into herbal remedies. It is possible that using morphological traits alone is not enough to differentiate between closely related species in traditional ways. Molecular markers probe the genetic composition, where minute changes become visible, to provide a more objective and accurate method of identification. Plant material that has been processed or pulverized is sold in the herbal market as extracts, powders, or capsules. With such treated materials, traditional morphological identification becomes all but impossible. Even with processed or deteriorated plant material, molecular markers—specifically, DNA barcoding—remain effective and guarantee precise identification in a range of product compositions (Abubakar and Haque 2020). The development of herbal goods requires the procurement of medicinal plants from all over the world, which adds complexity to the supply chain. It is difficult to detect and regulate adulteration because it can happen at any point, from production to growing. Molecular identifiers make it possible to confirm a plant's botanical origin at every stage of the supply chain, increasing transparency and avoiding the unintentional spread of contaminated goods (Abubakar and Haque 2020).

Adulteration is a common practice used by dishonest people to make money, whereby they replace rare or pricey medicinal plant species with less-expensive ones. This affects the genuine manufacturers' capacity to make a living, as well as the effectiveness and quality of herbal goods. Molecular markers facilitate trustworthy identification and encourage ethical trade practices, which function as a deterrent. Due to their distinct chemical makeup, medicinal plants are prized for their particular therapeutic qualities. The effectiveness of herbal therapies may be impacted by adulteration, which can result in the lack or the modification of essential bioactive ingredients (Wallace et al. 2020). Molecular markers play a role in maintaining the predicted therapeutic advantages and consumer safety by helping verify the authenticity of medicinal plants. Global regulatory agencies are realizing how critical it is for herbal goods to have precise botanical authentication. Plant identification techniques must be consistent and dependable to comply with strict requirements. Molecular markers offer a method that is based on strong science, complies with legal requirements, and encourages industry adherence to quality control procedures. Advances in molecular biology technology, including digital PCR, high-throughput sequencing, and SNP analysis, have greatly improved the accuracy, speed, and affordability of DNA-based methods (Meena et al. 2023). Thanks to these developments, molecular markers are now more widely available for routine usage in authentication and quality control processes. The loss of biodiversity can result from adulteration techniques, overharvesting, and the extinction of some plant species. Molecular markers can help ensure that valuable species are not overexploited, by helping to monitor and manage the sustainable use of therapeutic plants.

17.4 AUTHENTICATION OF MEDICINAL PLANTS BY MOLECULAR MARKERS

DNA analysis based on markers uses a unique genetic structure, which is a better alternative than other marker systems since they are resistant to age, environmental impacts, and physiological circumstances (Ganie et al., 2015a, 2015b). Furthermore, these markers are not limited to specific tissues, allowing for identification at any stage of plant development (Amiteye 2021). In contrast to phenotypic and chemical markers, DNA-based technology provides a more efficient, precise, and cost-effective technique for certifying several samples at the same time, thanks to its automation capabilities. DNA-based authentication for medicinal plants is an important method for verifying the quality and safety of herbal medicines and nutraceuticals, which improves their medical efficacy and economic feasibility. A genetic marker, whether a gene or a nucleotide sequence on a chromosome, has the potential to distinguish across cells, people, and species. Given the very precise character of DNA sequences, molecular markers may accurately identify individual DNA sequences within a collection of unknown samples (Alberts et al. 2002).

17.5 TYPES OF MOLECULAR MARKERS

Restriction fragment length polymorphisms (RFLPs), amplified fragment length polymorphisms (AFLPs), randomly amplified polymorphic DNA (RAPD), simple sequence repeats (SSRs), inter simple sequence repeats (ISSR), sequence-characterized amplified regions (SCAR), loop-mediated isothermal amplification (LAMP), single-nucleotide polymorphisms (SNPs), and others are among the many molecular markers. DNA barcoding, microarray-based markers, and next-generation sequencing (NGS-based markers) are relatively new developments. Each methodology, whether targeted at a specific component of the genome or wholly random, faces distinct methodological, technological, and material obstacles; therefore, no DNA marker can be called optimal. The researcher's aim determines the usage of a certain marker (Kumari, Jaiswal, and Tripathy 2021).

17.5.1 Restriction Fragment Length Polymorphisms (RFLPs)

To differentiate therapeutic plants from adulterants, restriction fragment length polymorphisms, or RFLPs, are an effective molecular technique. RFLP analysis takes advantage of inherent differences to identify distinct restriction sites and fragment patterns in DNA sequences among various plant species. Within the context of medicinal plants, we explore here the guiding concepts, methods, and benefits of employing RFLPs for adulterant discrimination. The discovery of DNA sequence variants that lead to variations in restriction fragment lengths is the foundation of RFLP analysis (Li et al. 2021). The RFLP analysis process involves several crucial components. Plant samples that may include both genuine medical plants and possible adulterants are used to isolate genomic DNA. Restriction enzymes are then used to carry out an enzymatic digestion of the isolated DNA. These enzymes cleave DNA at certain restriction sites, or specific DNA sequences, that they have identified. Different genomic DNA sequences in different plant species lead to different restriction sites. Using gel electrophoresis, the digested DNA fragments are sorted according to size. The distinct DNA sequence of the plant species is reflected in the fragment pattern that results. To move the separated fragments from the gel to a membrane for hybridization with a tagged probe, Southern blotting may be used in certain circumstances (Li et al. 2021; Rafalski et al. 1996; Kumari, Jaiswal, and Tripathy 2021).

The RFLP analysis's sensitivity and specificity are improved by this procedure. Staining or autoradiography is used to see the separated fragments, and the resulting banding patterns are studied. Variations in the lengths of the fragments reveal genetic differences between possible adulterants and the real therapeutic plant. Because RFLP analysis may identify differences in the DNA sequence at particular loci for discrimination, it has a high degree of discriminating power. Because of this, it is useful in differentiating between closely related plant species. RFLP analysis is a flexible method for identifying possible adulterants and authenticating a variety of medicinal plants, because it may be used on a broad range of plant species. RFLP analysis is historically significant as one of the first approaches to DNA-based genetic fingerprinting, even if more advanced methods have since been developed. Its credibility and widespread adoption are aided by its heritage. RFLP analysis is a cost-effective solution for some applications since it can be done with basic laboratory equipment, unlike some newer approaches that would require complex equipment. RFLP analysis is renowned for being reliable and consistent. The trustworthiness of outcomes is enhanced by standard operating procedures and well-established protocols. Polymerase chain reaction (PCR) and DNA sequencing are more sophisticated methods that have fully replaced restriction fragment length polymorphism (RFLP) analysis as a useful tool for genetic fingerprinting in recent years. Nonetheless, in several investigations, RFLP analysis has been used to identify and separate adulterants from medicinal plants. Here are a few instances: (1) *Differentiating between genuine* Panax ginseng *and possible adulterant,* Panax quinquefolius *(American ginseng).* RFLP analysis is aimed at the ribosomal DNA's internal transcribed spacer (ITS) region. Different RFLP patterns were found in the study, which made it possible to distinguish between *Panax ginseng* and *Panax quinquefolius* (Li et al. 2021). (2) *Identifying* Glycyrrhiza glabra *(European licorice) as a possible adulterant and* Glycyrrhiza uralensis *(genuine licorice) as the second example.* RFLP investigation aimed at the trnL–trnF intergenic spacer region of chloroplast DNA. Research showed that *Glycyrrhiza glabra* and *Glycyrrhiza uralensis* could be distinguished from one another using RFLP analysis, which helped prevent adulteration. (3) *Identifying genuine* Astragalus membranaceus *from any impurities.* Nuclear ribosomal DNA's ITS region is the focus of an RFLP investigation. Utilizing RFLP analysis, the study was able to distinguish *Astragalus membranaceus* from other species that are frequently employed as adulterants by identifying distinctive fragment patterns (Raghavendra, 2020). (4) *Identifying possible adulterants and verifying the authenticity of* Hypericum perforatum. RFLP analysis aimed at the trnL intron of chloroplast DNA. Saint-John's-wort items were guaranteed to be authentic since *Hypericum perforatum* could be distinguished from other *Hypericum* species using RFLP analysis. (5) *Differentiating between real* Ginkgo biloba *and looking for any adulterants.* RFLP analysis aimed at the trnL intron of chloroplast DNA. By exposing distinctive fragment patterns by RFLP analysis, *Ginkgo biloba* could be identified, and the use of adulterants in herbal products may be avoided (Liu et al. 2023).

17.5.2 Randomly Amplified Polymorphic DNA (RAPD)

Molecular biology techniques such as randomly amplified polymorphic DNA (RAPD) analysis have been applied in the identification of adulterants from medicinal plants. To identify polymorphisms in the genomic DNA, RAPD analysis uses the polymerase chain reaction (PCR) to amplify random DNA segments. Here are a few instances of how RAPD analysis has been used to identify adulterants in therapeutic plants: (1) *Distinction between* Panax quinquefolius *(American ginseng, possible adulterant) and* Panax ginseng *(genuine)* (El-Haggar et al. 2023). RAPD analysis with primers chosen at random. A dependable technique for authentication and adulteration detection was made possible by the identification of particular DNA markers by RAPD analysis that set *Panax ginseng* apart from *Panax quinquefolius.* (2) *Differentiation between* Rhodiola *species to avoid adulteration.* Using random primers, RAPD analysis was performed before the method. By identifying species-specific banding patterns, RAPD made it possible to distinguish genuine *Rhodiola rosea* from other species that might be utilized as adulterants. (3) Actaea racemosa, *or black cohosh, authentication and possible adulterant detection.* Employing random primers for RAPD analysis. By effectively distinguishing genuine black cohosh from closely similar species, RAPD fingerprints offer a means of preventing adulteration in herbal goods (Terletskaya et al. 2023). (4) *Determination of the identity of genuine turmeric,* Curcuma longa, *and identification of any possible adulterants.* Random primers were used for the RAPD study. RAPD markers were utilized to differentiate *Curcuma longa* from other *Curcuma* species and avoid the use of less-expensive substitutes as adulterants. (5) *Discrimination between several species of* Echinacea *to avoid adulteration.* Employing random primers for RAPD analysis. To ensure the correct botanical identity and high quality of *Echinacea* products, RAPD profiles were employed to distinguish between *Echinacea purpurea* and *Echinacea angustifolia* (Ganie et al. 2015c). With only a few random primers and little prior sequence knowledge needed, RAPD analysis is a rather easy and affordable method. RAPD analysis is appropriate for routine quality control in the herbal industries since it enables the quick screening of a large number of samples. Since RAPD may be used on a wide variety of plant species, it is flexible in achieving the goal of authenticity across numerous medicinal plants. With RAPD, one does not need to know the genome sequence beforehand, in contrast to other molecular approaches that might. Because RAPD is susceptible to experimental circumstances, results must be carefully optimized to ensure reproducibility. Without additional sequencing or analysis, RAPD can only identify polymorphisms and not the precise identity of the amplified fragments. RAPD analysis is a flexible and affordable method for botanical authentication in the herbal sector. It has been effectively used to separate adulterants from medicinal plants. To get accurate results, nevertheless, like with any molecular approach, thorough optimization and awareness of its limitations are essential (Ganie et al. 2015c).

17.5.3 Simple Sequence Repeats (SSRs)

SSRs are repetitive DNA sequences made up of one to six base pair motifs repeated in tandem. They are sometimes referred to as microsatellites or short tandem repeats. It has become common practice to separate genuine therapeutic plants from adulterants using these molecular markers. There is a discussion of instances where SSR analysis has been used in the framework of botanical authentication to detect and stop adulteration: (1) *Differentiating between German chamomile (*Matricaria recutita*), which may be an adulterant, and true chamomile (*Matricaria chamomilla*)* (Guzelmeric et al. 2017). SSR analysis focused on microsatellite areas with particular primers. SSR markers allowed for the precise differentiation of *Matricaria chamomilla* and *Matricaria recutita* by revealing variations in the quantity and arrangement of microsatellite repeats. (2) *Identifying possible adulterants and verifying the authenticity of* Withania somnifera, *or genuine ashwagandha.* SSR research aimed at particular microsatellite loci. The study used SSR markers to separate Withania somnifera from closely related species and avoid the use of adulterants in ashwagandha products. (3) *Distinguishing between* Commiphora africana *(possibly an adulterant) and* Commiphora wightii *(genuine guggul).* SSR analysis using microsatellite region-specific primers. Commiphora wightii and Commiphora africana were successfully distinguished by SSR markers, offering a dependable technique for guggul authenticity. (4) *Recognizing genuine* Aloe vera *and spotting possible adulterants.* Utilizing microsatellite markers for SSR analysis (Guzelmeric et al. 2017). To separate *Aloe vera* from other *Aloe* species and stop adulteration of *Aloe vera* products, SSR markers were used.

Benefits of SSR Analysis for Adulterant Discrimination

1. Because SSRs are highly polymorphic, they offer a high degree of discrimination between species of closely related plants.
2. SSRs show codominant inheritance, which makes it possible to distinguish between heterozygotes and provide more precise details regarding genetic diversity.
3. SSRs can be applied to a wide variety of therapeutic plants since they can be transferred between related species.
4. SSR analysis produces quantitative data that makes population structure and genetic diversity evaluation possible.
5. SSRs can be utilized to create distinct genetic profiles for individual plant species using DNA fingerprinting (Guzelmeric et al. 2017).

17.5.4 Inter Simple Sequence Repeats (ISSR)

PCR amplification of the areas between microsatellite sequences, also known as simple sequence repeats (SSRs), or microsatellites, is used in the molecular approach known as inter simple sequence repeats (ISSR) study. This technique has been used to separate adulterants from therapeutic herbs. Simpleness, economy, and the capacity to produce polymorphic markers without previous sequence information are among the benefits of ISSR analysis. Illustrative cases of adulterant discrimination using ISSR analysis: (1) *Differentiating between possible adulterants and genuine* Melaleuca alternifolia *(Tea Tree).* Targeting microsatellite areas, certain primers were used for ISSR analysis. By revealing distinctive banding patterns, ISSR markers made it possible to identify genuine tea tree oil and set it apart from other *Melaleuca* species. (2) *Identifying authentic oregano,* Origanum vulgare, *and finding possible adulterants* (Gemmill and Grierson 2021). ISSR analysis using primers that target microsatellites. By making it easier to distinguish *Origanum vulgare* from closely related species, ISSR fingerprints helped avoid adulteration in oregano-based goods. (3) *Identifying potential adulterants and verifying the authenticity of authentic* Aloe vera. ISSR research aimed at microsatellite areas. In order to differentiate *Aloe vera* from related *Aloe* species and guarantee the validity of *Aloe vera* gel products, ISSR markers were used. (4) *Differentiating between possible adulterants and genuine lavender,* Lavandula angustifolia. Using SSR-specific primers for ISSR analysis. By creating distinct profiles using ISSR markers, it was possible to identify *Lavandula angustifolia* and avoid the use of adulterants in lavender oil products (Gemmill and Grierson 2021). (5) *Verifying the authenticity of* Zingiber officinale, *or ginger, and looking for possible adulterants.* ISSR research aimed at microsatellite areas. Distinctive banding patterns offered by ISSR markers made it easier to distinguish *Zingiber officinale* from other species that resemble ginger.

Benefits of ISSR Analysis for Adulterant Discrimination

1. *No prior sequence information.* ISSR analysis can be used for a variety of plant species because it does not necessitate prior knowledge of the genomic sequence.
2. *High polymorphism.* ISSR markers frequently show high polymorphism, which improves the analysis's ability to discriminate.

3. *Cost-effectiveness.* Considering how inexpensive ISSR analysis is in comparison to other molecular methods, it is a good choice for regular quality control applications.
4. *Fast screening.* ISSR analysis enables huge sample sets to be screened quickly, which makes it appropriate for quality control in the herbal product industry (Gemmill and Grierson 2021).

17.5.5 Sequence-Characterized Amplified Regions (SCAR)

A molecular biology approach called sequence-characterized amplified regions (SCAR) uses particular DNA sequences to create primers for target region selective amplification. SCAR markers are frequently used to separate real therapeutic plants from adulterants. The main goal is to locate distinct DNA sequences or polymorphisms that are exclusive to the real medicinal plant and can be amplified selectively. A plant's authentic DNA is used to choose DNA sequences, which are commonly obtained using amplified fragment length polymorphism (AFLP), random amplified polymorphic DNA (RAPD), or other DNA fingerprinting procedures, creating primers that will amplify the DNA sequences that have been found (Sairkar, Sharma, and Shukla 2016). A crucial stage in the process is primer design, which entails creating oligonucleotide primers that correspond to the DNA sequences that have been found. These primers should not amplify the DNA of any adulterants; instead, they should be unique to the real medicinal plant: (1) The targeted DNA sequences are amplified selectively using the SCAR primers that have been created. DNA that has been isolated from plant materials is used in PCR. Target DNA sequences are specifically amplified by the SCAR primers, resulting in distinctive banding patterns. (2) The amplified DNA fragments' visualization and analysis. After gel electrophoresis is used to separate the PCR products, the resulting banding patterns are seen. Specific bands show whether something is legitimate or if it may be adulterated. (3) Identifying genuine medicinal plants from possible adulterants by looking for particular SCAR marker bands and excluding others. Adulterants that do not possess the unique DNA sequences linked to the genuine plant will not generate the distinctive SCAR marker bands, which facilitates differentiation.

SCAR Markers' Benefits for Discrimination

1. To ensure precise discrimination, SCAR markers are engineered to be highly unique to the target DNA sequences.
2. Once established, SCAR markers can be used for routine quality control because of their excellent repeatability.
3. Because SCAR markers are so sensitive, they can find even minute levels of possible adulterants.
4. SCAR markers offer adaptability in medicinal plant authentication, as they can be employed for a variety of plant species.

Uses for Authenticating Medicinal Plants

1. Ginseng authentication
    - Goal: Differentiating *Panax ginseng* from possible adulterants.
    - Technique: To specifically identify *Panax ginseng*, SCAR markers were created based on distinct DNA sequences linked to the plant.
2. *Echinacea* species discrimination
    - Goal: To stop adulteration, and to discriminate between several *Echinacea* species.
    - Method: To accurately identify *Echinacea purpurea* and distinguish it from other species, SCAR markers were created to target particular DNA sequences (Sairkar, Sharma, and Shukla 2016).

3. Authentication of *Aloe Vera*
   - Goal: Differentiating between real *Aloe vera* and possible adulterants
   - Method: To verify the authenticity of *Aloe vera* products, particular DNA sections linked to the plant were amplified using SCAR markers.

17.5.6 Loop-Mediated Isothermal Amplification (LAMP)

DNA is amplified under isothermal conditions using a molecular biology technique called loop-mediated isothermal amplification (LAMP), which offers a quick and accurate way to identify and differentiate between DNA sequences. LAMP has been used extensively in many domains, such as in the identification of medicinal plants by identifying adulterants: (1) *Creating LAMP primers with a focus on DNA sequences connected to the real medicinal plant.* Usually, LAMP uses four to six primers that each identifies different sections of the target DNA. Based on particular DNA sequences that are exclusive to the real plant, these primers were created. (2) Targeting DNA sequences amplified isothermally. The temperature of LAMP processes is always between 60 and 65°C; hence, a thermal cycler is not necessary.

LAMP's Discrimination Benefits

1. *Quick reaction.* LAMP reactions can be finished quickly, usually in 30 to 60 minutes, which enables quick on-site detection.
2. *Sensitivity.* LAMP is ideal for trace-level adulteration because of its exceptional sensitivity and ability to identify small amounts of target DNA.
3. *Specificity.* To minimize the possibility of false positive results, LAMP primers are made to be extremely specific to the target DNA sequences.
4. *Ease of use.* The approach is suitable for both laboratory and field applications because LAMP reactions can be carried out under straightforward settings (Soroka, Wasowicz, and Rymaszewska 2021).

Uses for Authenticating Medicinal Plants

1. Ginseng authentication
   - Goal: Differentiating *Panax ginseng* from possible adulterants.
   - Technique: *Panax ginseng* may be specifically detected thanks to the invention of LAMP primers, which target specific DNA sequences linked to the plant (Yang et al. 2023).
2. *Curcuma* authentication
   - Goal: Differentiating between possible adulterants and genuine *Curcuma longa*.
   - Approach: To guarantee the validity of turmeric items, LAMP was utilized to target particular DNA sequences linked to *Curcuma longa* (Núñez et al. 2020).
3. *Aloe Vera* authentication
   - Goal: Differentiating between real *Aloe vera* and possible imposters.
   - Technique: To detect and distinguish adulterants, certain DNA areas linked to *Aloe vera* were amplified using LAMP primers.

17.5.7 Single-Nucleotide Polymorphisms (SNPs)

Single-nucleotide polymorphisms, or SNPs, are differences in DNA sequences at the single nucleotide level. They are essential for distinguishing real therapeutic plants from adulterants. Because of their great frequency in the genome and generational stability, SNPs are useful molecular markers: (1) *Determination of which SNPs are unique to the real medicinal plant and can serve as discriminatory markers.* SNP arrays, focused SNP discovery techniques, or high-throughput sequencing are used to find single-nucleotide differences in the genomic DNA of the real plant. (2) *Primer design for SNP detection.* For selective amplification or detection, create primers or probes that are unique to the SNP sites that have been discovered. For SNP detection to be specific, primer design is essential. The purpose of allele-specific primers and probes is to amplify or bind to the SNP-containing DNA segment only. To genotype and differentiate between various SNP alleles, PCR-based techniques, such as TaqMan tests, high-resolution melting (HRM) studies, and allele-specific PCR are frequently utilized. (3) *Adulterant discrimination.* Using SNP profiling, differentiate real therapeutic plants from possible adulterants. Examine the genotyping data to pinpoint particular SNP alleles that are exclusive to the real plant. Adulterants are present when SNP profiles are absent or vary (Morgil et al. 2020).

Uses for Authenticating Medicinal Plants

1. Authenticity of tea plant
   - Goal: Distinguishing genuine tea plant, *Camellia sinensis*, from other kinds of Camellia.
   - Approach: SNP analysis provided a method for authenticating tea plants by focusing on particular genomic areas linked to *Camellia sinensis* (Lee et al. 2017).
2. Veracity of olive oil
   - Goal: Identifying genuine olive oil and spotting any adulterants.
   - Approach: SNP analysis allowed for product certification by focusing on particular areas linked to different types of olive oil.
3. Herbal product quality control
   - Goal: Including identifying adulteration and ensuring the authenticity of herbal products.
   - Approach: SNP analysis is used to confirm whether the herbal mixture contains particular SNP markers linked to the therapeutic plant.

17.5.8 Next-Generation Sequencing (NGS)

Next-generation sequencing (NGS) has revolutionized genomic research, enabling high-throughput sequencing of entire genomes and providing valuable insights into genetic variations. NGS-based markers, such as single-nucleotide polymorphisms (SNPs), have been instrumental in discriminating adulterants from authentic medicinal plants (Satam et al. 2023). SNPs are exclusive to genuine medicinal plant species. The whole genome or particular target sections of the real plant and any adulterants are sequenced using NGS technologies. After that, computational analysis is performed to compare the sequences and find SNPs. Based on the found SNPs, markers are created for ensuing genotyping. The detected SNPs are the exclusive target of primers or probes. In further research, these markers will be utilized to capture or amplify the areas harboring the SNPs. Genuine therapeutic plants, as well as possible adulterants, have their genomic DNA extracted. Targeted sequencing, microarrays, and PCR-based tests are some of the methods used in SNP genotyping to find the precise SNPs linked to the real plant.

1. *Ginseng authenticity.* Distinguish between genuine *Panax ginseng* and possible adulterant, *Panax quinquefolius* (American ginseng). Both *Panax ginseng* and *Panax quinquefolius* have their entire genomes sequenced using next-generation sequencing (NGS), which made it possible to identify a large number of single-nucleotide polymorphisms (SNPs) and other genomic variants. Species-specific differences were identified by the thorough genomic study, which made it possible to create SNP markers that accurately distinguish between the two species of ginseng (Ichim and de Boer 2020).
2. *Authenticity of licorice.* Differentiating between European licorice (*Glycyrrhiza glabra*, perhaps adulterant) and true licorice, *Glycyrrhiza uralensis*. Using NGS, particular

genomic regions of both licorice species were sequenced, with an emphasis on locations with known genetic variability. SNPs and other genetic markers unique to each species were found throughout the study. Afterwards, a genetic assay for licorice authentication was created using these markers (Ichim and de Boer 2020).

3. *Identification of turmeric adulterants.* This involves determining which turmeric is real, or *Curcuma longa*, and identifying any possible adulterants. The genomes of several *Curcuma* species were sequenced using NGS, with an emphasis on areas linked to the synthesis of bioactive substances. Through the comparison of genomic data, *Curcuma longa*–specific SNPs and genetic variations were found, paving the way for the creation of molecular markers that can be used to identify adulteration in turmeric-derived products (Parvathy et al. 2015).
4. Rhodiola *species discrimination.* To stop adulteration, *Rhodiola* species should be distinguished from one another. The genomes of several *Rhodiola* species were sequenced using NGS, with an emphasis on areas that might contain genetic markers.
5. Astragalus *authentication.* Making a distinction between probable adulterants and authentic *Astragalus membranaceus*. The genomes of *Astragalus* species were sequenced using NGS, with an emphasis on areas recognized for their genetic diversity. The identification of *Astragalus membranaceus*–specific SNPs and other variants in the genomic data made it possible to create molecular markers for authentication (Zheng et al. 2014).

17.5.9 DNA Barcoding

DNA barcoding refers to the utilization of short sequences from nuclear or organelle DNA to identify organisms. Unlike determining patterns of relationships, the primary aim of DNA barcoding is to identify unknown samples using existing classification systems. This technique has been in existence for over a decade since its proposal in 2003 by Hebert et al., who suggested the use of the mitochondrial enzyme CO1 (cytochrome oxidase 1) coding DNA as a DNA barcode for animals (Zheng et al. 2014; Hebert et al. 2003). However, mitochondrial genes in plant systems have slow evolutionary rates and low substitution rates, making them unsuitable for barcoding. Alternatively, chloroplast or nuclear genomes with higher substitution rates are used. Several chloroplast genomic regions, including rbcL, matK, trnH-psbA, trnL-F, rpl36-rps8, ITS, and 5S rRNA, have been evaluated in plants by the Consortium for the Barcode of Life Plant Working Group (CBOL) (Wong et al., 2013). Among these, rbcL is considered universal but has low species resolution, while matK has the opposite characteristic. Combining these two markers may yield better results according to (Letsiou et al., 2024). Li et al. nuclear ITS to the combination of matK + rbcL to improve discriminating ability, particularly for closely related species (Li et al., 2011). In DNA barcoding, the focus is on finding a universal DNA sequence that balances conserved sequences with enough diversity to differentiate organisms. Primers are designed to complement conserved sequences and flank variable barcode sites. However, designing primers becomes challenging when primer binding sequences accumulate sufficient sequence divergence over evolutionary time. To address this challenge, barcode primers should accommodate sequence variation or degeneracy at one or several nucleotide positions. Since there are numerous chloroplast genomes in a single plant cell, amplification of barcode regions of rbcL and COI is higher than that of nuclear loci, enabling higher quantities of product from minute material and simultaneously reducing contaminant concentration that could affect PCR. Reliable DNA barcodes require bidirectional sequencing of chosen DNA amplicons (e.g., rbcL, COI, or ITS) in both orientations.

Several important research have been conducted to authenticate medicinal plants using DNA barcoding. *Crocus sativus*, a highly valued medicinal spice, is vulnerable to adulteration with a variety of materials, including *Carthamus tinctorius* L. and *Calendula officinalis* L. flowers, *Hemerocallis* L. petals, *Daucus carota* L. fleshy root, *Curcuma longa* L. rhizomes, and *Zea mays* L. and *Nelumbo nucifera* Gaertn. stigmas (Jiang et al., 2014) created an accurate detection approach for these adulterants in traded saffron, called barcoding melting curve analysis (Bar-MCA). Another instance is the use of the internal transcribed spacer 2 (ITS2) region to discriminate between *Schisandra chinensis* species and populations (Zhu et al., 2022). A transition from C → A substitution at position 86-bp in wild populations against cultivated ones, and the ITS2 region efficiently differentiated *S. chinensis* from *S. sphenanthera* (Li et al., 2013). El-Sherif and Ibrahim (2020) investigated the identification of *Astragalus* species, which frequently have physical characteristics. They investigated four coding (trnH-psbA, rpoC1, rbcL, matK) and two non-coding areas (ITS, ITS2) among 319 species and discovered that ITS2 and ITS barcodes were more successful at species differentiation. In addition, El-Sherif and Ibrahim (2020) used DNA barcoding to distinguish *Peucedanum praeruptorum* from common replacements and adulterants.

17.5.9.1 Limitations of DNA Barcoding

1. DNA barcoding methods may encounter challenges in distinguishing recently diverged species.
2. It should be noted that the effectiveness of barcoding relies heavily on the amplification of degraded DNA. This aspect presents a significant challenge, particularly because the sequences targeted are typically longer than 500 base pairs, making amplification problematic (Zhou et al., 2014). Consequently, DNA barcoding cannot be regarded as universally applicable compared to other conventional markers.

17.5.9.2 Using DNA-Based Techniques to Identify Numerous Medicinal Plants within Intricate Blends

With the escalating global demand, there is a pressing need for a streamlined approach to authenticate medicinal plant products. This research endeavors to create a DNA-oriented technique for verifying and validating plant species, suitable for integration into industrial processes, to facilitate regulatory compliance. The study focuses *on Hypericum perforatum* L. (commonly known as Saint-John's-wort, or Guan ye Lian Qiao), which holds significant economic importance (Seethapathy et al. 2019). DNA-based approaches are favored due to their efficiency, cost-effectiveness, reduced time requirements, lower plant material demands, and reliable species-level discrimination. Nevertheless, existing identification techniques, whether DNA-based or chemical, are unable to simultaneously resolve mixtures containing multiple plant species.

The need for identifying species both independently and simultaneously within complex mixtures has gained significance for regulators due to two primary reasons. Firstly, instances of adulteration and contamination are prevalent, especially concerning rare and costly medicinal plant species substituted with low-grade alternatives. These substitutes may lack biological activity or pose potential health risks. Hence, it becomes crucial to identify both the target plant and any hazardous adulterants to evaluate product safety. Secondly, synergistic polyherbal formulations play a vital role in complementary medicine (CM) practices, where specific combinations of medicinal plants lead to enhanced outcomes, surpassing the individual effects of each component. To authenticate such preparations, each included plant must be identified. Investigation aimed at developing DNA-based identification assays adhering to industry standards. *Hypericum perforatum* L. (commonly known as Saint-John's-wort, or Guan ye Lian Qiao) was chosen as the focal point. This herb is renowned in complementary medicine (CM) for its anti-inflammatory and antimicrobial characteristics and is commonly used in Europe to treat mild to moderate depression.

The study focused on utilizing the nuclear ribosomal internal transcribed spacer (nrITS) regions for assay design. It involved analyzing sequences from 15 *Hypericum* species to pinpoint the most distinct regions. PCR primers were then specifically designed to target these regions across different species. These primers were selected to generate species-specific PCR products with varying sizes, enabling differentiation using capillary electrophoresis with fluorescently labelled primers. This method effectively identifies closely related species, such as *Hypericum androsaemum*, *Hypericum athoum*, *Hypericum ascyron*, and *Hypericum perforatum*, both individually and in any combination within a DNA mixture containing seven *Hypericum* species. Notably, it confirms the presence of the desired plant material and detects any adulterants in a single reaction. Furthermore, this design approach can be replicated for other medicinal plants and their potential adulterants.

17.5.9.3 Methods

1. The initial assessment involved aligning and analyzing nrITS sequences from seven *Hypericum* species to identify distinct regions.
2. Specific PCR primers were designed to target regions where sequences of four target species differed from others.
3. Nineteen primer pairs were created, six for *H. androsaemum*, four each for *H. ascyron* and *H. athoum*, and five for *H. perforatum*.
4. The developed PlantID assay can simultaneously detect these closely related species within mixtures, crucial for authenticating polyherbal preparations.

The process began with analyzing nrITS sequences from seven *Hypericum* species to pinpoint unique regions. PCR primers were then crafted to specifically bind to areas where sequences of four target species differed from others. A total of 19 primer pairs were developed and distributed among the different *Hypericum* species. The resulting PlantID assay can concurrently detect these closely related species within mixtures, offering significant value in the authentication of polyherbal preparations. The nrITS sequences extracted from seven *Hypericum* species were meticulously aligned and scrutinized to serve as the foundation for designing species-specific primers. Primer design focused on regions where the sequences of the four target species (*H. androsaemum*, *H. ascyron*, *H. athoum*, and *H. perforatum*) exhibited discrepancies from all other species. Each primer pair's sequence was tailored to be species-specific, aligning exclusively with one sequence from the input data. A total of 19 primer pairs were crafted, all of which successfully yielded amplicons of the anticipated size in conventional PCR when tested with the respective target DNA templates: six pairs for *H. androsaemum*, four pairs for *H. ascyron*, four pairs for *H. athoum*, and five pairs for *H. perforatum*. The outcomes for the six *H. androsaemum* pairs are depicted in Figure 17.1.

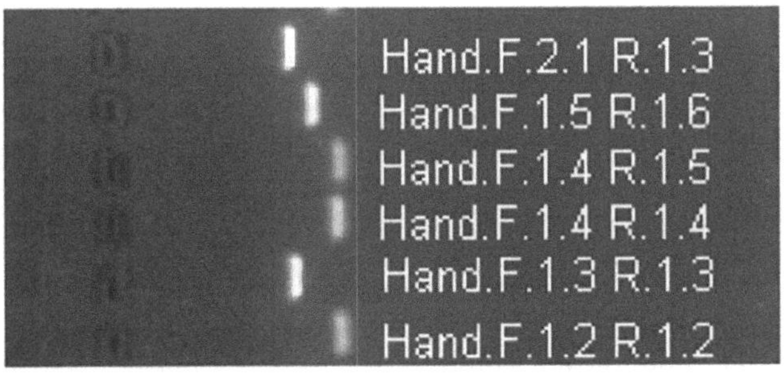

FIGURE 17.1 Image of gel containing PCR products from *H. androsaemum* primers. The product sizes, from top to bottom, are 137, 98, 63, 67, 135, and 65 bp. The primer pair H and F.1.4 and R.1.4 was selected for the PlantID assay.

17.6 CONCLUSION

To conclude, the issues presented by the extensive adulteration in the herbal sector have been addressed by the development of nucleic acid (NA)–based techniques as effective instruments in differentiating therapeutic plants from their adulterants. The precision, accuracy, and dependability of authentication procedures have been greatly improved by the incorporation of molecular approaches that take advantage of the distinct genetic information encoded in the DNA of plant species. When separating real medicinal plants from possible adulterants, nucleic acid–based methods, like DNA barcoding, single-nucleotide polymorphism (SNP) analysis, restriction fragment length polymorphism (RFLP), sequence-characterized amplified regions (SCAR), and next-generation sequencing (NGS), provide excellent accuracy and precision. These techniques take advantage of molecular genetic differences to provide a resolution not possible with conventional morphological techniques. Because of their increased sensitivity, NA-based methods may identify minute quantities of adulterants in intricate combinations. These techniques' specificity guarantees that even closely related species can be distinguished with accuracy, which is essential for preserving the integrity of herbal products. These methods work with a wide variety of therapeutic plants, which makes them useful instruments for plant species authentication. Natural remedies based on plants, such as ginseng and turmeric, or more specialized plants, provide a general remedy. In the herbal sector, DNA-based verification techniques are being included more and more in standard quality control procedures. Molecular markers, such as SNP markers or DNA barcodes, have made it possible to efficiently screen and verify raw materials, processed herbal products, and polyherbal formulations. While conventional morphological and chemical methods are limited in their ability to discriminate between closely related species or processed plant material, NA-based approaches serve as a valuable supplement, offering an extra degree of certainty and addressing the drawbacks of conventional methods. The increasing acceptance of NA-based strategies is partly due to the ongoing developments in molecular biology technologies, such as the availability of high-throughput sequencing platforms and reasonably priced genotyping tests. The viability and the scalability of applying DNA-based authentication techniques in academic and industrial contexts are improved by these technological developments. The regulatory authorities' growing appreciation of the value of precise botanical authentication emphasizes the importance of NA-based methods even more. Standardized and scientifically proven DNA-based techniques make it easier to comply with strict rules. It is critical to use NA-based methods to verify the legitimacy of medicinal plants to protect consumer health. The medicinal effectiveness of herbal treatments is compromised by adulteration, which also presents health risks since potentially dangerous or unknown chemicals are introduced.

REFERENCES

Abubakar, Abdullahi R., and Mainul Haque. 2020. "Preparation of Medicinal Plants: Basic Extraction and Fractionation Procedures for Experimental Purposes." *Journal of Pharmacy & Bioallied Sciences* 12 (1). Wolters Kluwer—Medknow Publications: 1. https://doi.org/10.4103/JPBS.JPBS_175_19.

Alberts, Bruce, Alexander Johnson, Julian Lewis, Martin Raff, Keith Roberts, and Peter Walter. 2002. "Studying Gene Expression and Function." *Garland Science*. www.ncbi.nlm.nih.gov/books/NBK26818/.

Amiteye, Samuel. 2021. "Basic Concepts and Methodologies of DNA Marker Systems in Plant Molecular Breeding." *Heliyon* 7 (10). Elsevier. https://doi.org/10.1016/J.HELIYON.2021.E08093.

Cory, Hannah, Simone Passarelli, John Szeto, Martha Tamez, and Josiemer Mattei. 2018. "The Role of Polyphenols in Human Health and Food Systems: A Mini-Review." *Frontiers in Nutrition* 5 (September). Frontiers Media SA: 87. https://doi.org/10.3389/FNUT.2018.00087.

El-Haggar, Miran A., Yasmin A. Mahgoub, Hesham M. Aly, Nabila M. Ghazy, Fathy K. El-Fiky, and Amr M. El-Hawiet. 2023. "DNA Barcodes, ISSR, RAPD and SCAR Markers as Potential Quality Control Tools for Molecular Authentication of Black and White Mulberry." *Food Control* 152 (October). Elsevier: 109821. https://doi.org/10.1016/J.FOODCONT.2023.109821.

El-Sherif, N., and M. Ibrahim. 2020. "Implications of rbcL and rpoC1 DNA Barcoding in Phylogenetic Relationships of Some Egyptian *Medicago sativa* L. Cultivars." *Egyptian Journal of Botany* 60: 451–460. https://doi.org/10.21608/EJBO.2020.20028.1399.

Fitzgerald, Martin, Michael Heinrich, and Anthony Booker. 2019. "Medicinal Plant Analysis: A Historical and Regional Discussion of Emergent Complex Techniques." *Frontiers in Pharmacology* 10. Frontiers Media SA. https://doi.org/10.3389/FPHAR.2019.01480.

Ganie, Showkat Hussain, Priti Upadhyay, Sandip Das, and Maheshwer Prasad Sharma. 2015a. "Authentication of Medicinal Plants by DNA Markers." *Plant Gene* 4 (December). Elsevier: 83. https://doi.org/10.1016/J.PLGENE.2015.10.002.

Ganie, Showkat Hussain, Priti Upadhyay, Sandip Das, and Maheshwer Prasad Sharma. 2015b. "Authentication of Medicinal Plants by DNA Markers." *Plant Gene* 4 (December). Elsevier BV: 83–99. https://doi.org/10.1016/J.PLGENE.2015.10.002.

Ganie, Showkat Hussain, Priti Upadhyay, Sandip Das, and Maheshwer Prasad Sharma. 2015c. "Authentication of Medicinal Plants by DNA Markers." *Plant Gene* 4 (December). Elsevier: 83. https://doi.org/10.1016/J.PLGENE.2015.10.002.

Gemmill, Chrissen E.C., and Ella R.P. Grierson. 2021. "Inter-Simple Sequence Repeats (ISSR), Microsatellite-Primed Genomic Profiling Using Universal Primers." *Methods in Molecular Biology (Clifton, N.J.)* 2222 (January). Humana Press Inc.: 249–262. https://doi.org/10.1007/978-1-0716-0997-2_14.

Gurib-Fakim, Ameenah. 2006. "Medicinal Plants: Traditions of Yesterday and Drugs of Tomorrow." *Molecular Aspects of Medicine* 27 (1): 1–93. https://doi.org/10.1016/J.MAM.2005.07.008.

Guzelmeric, Etil, Petar Ristivojević, Irena Vovk, Dušanka Milojković-Opsenica, and Erdem Yesilada. 2017. "Quality Assessment of Marketed Chamomile Tea Products by a Validated HPTLC Method Combined with Multivariate Analysis." *Journal of Pharmaceutical and Biomedical Analysis* 132 (January). Elsevier BV: 35–45. https://doi.org/10.1016/J.JPBA.2016.09.030.

Hebert, Paul D.N., Alina Cywinska, Shelley L. Ball, and Jeremy R. De Waard. 2003. "Biological Identifications through DNA Barcodes." *Proceedings. Biological Sciences* 270 (1512). Proc Biol Sci: 313–321. https://doi.org/10.1098/RSPB.2002.2218.

Ichim, Mihael Cristin. 2019. "The DNA-Based Authentication of Commercial Herbal Products Reveals Their Globally Widespread Adulteration." *Frontiers in Pharmacology* 10 (October). Frontiers Media SA: 479249. https://doi.org/10.3389/FPHAR.2019.01227/BIBTEX.

Ichim, Mihael Cristin, and Hugo J. de Boer. 2020. "A Review of Authenticity and Authentication of Commercial Ginseng Herbal Medicines and Food Supplements." *Frontiers in Pharmacology* 11 (January). Frontiers Media SA. https://doi.org/10.3389/FPHAR.2020.612071.

Jiang, C., L. Cao, Y. Yuan, M. Chen, Y. Jin, and L. Huang. 2014. "Barcoding Melting Curve Analysis for Rapid, Sensitive, and Discriminating Authentication of Saffron (*Crocus sativus* L.) from Its Adulterants." *BioMed Research International* 2014: 809037. https://doi.org/10.1155/2014/809037.

Karunamoorthi, Kaliyaperumal, Kaliyaperumal Jegajeevanram, Jegajeevanram Vijayalakshmi, and Embialle Mengistie. 2012. "Traditional Medicinal Plants." 18 (1). SAGE Publications Sage CA: Los Angeles, CA: 67–74. https://doi.org/10.1177/2156587212460241.

Kumari, Leena, Pranita Jaiswal, and S. Swarupa Tripathy. 2021. "Various Techniques Useful for Determination of Adulterants in Valuable Saffron: A Review." *Trends in Food Science and Technology* 111 (May). Elsevier Ltd: 301–321. https://doi.org/10.1016/j.tifs.2021.02.061.

Lee, Shih Chieh, Chia Hsiang Wang, Cheng En Yen, and Chieh Chang. 2017. "DNA Barcode and Identification of the Varieties and Provenances of Taiwan's Domestic and Imported Made Teas Using Ribosomal Internal Transcribed Spacer 2 Sequences." *Journal of Food and Drug Analysis* 25 (2). No longer published by Elsevier: 260–274. https://doi.org/10.1016/J.JFDA.2016.06.008.

Letsiou, S., P. Madesis, E. Vasdekis, C. Montemurro, M. E. Grigoriou, G. Skavdis, et al. 2024. "DNA Barcoding as a Plant Identification Method." *Applied Sciences* 14: 1415. https://doi.org/10.3390/APP14041415.

Li, D. Z., L. M. Gao, H. T. Li, H. Wang, X. J. Ge, J. Q. Liu, et al. 2011. "Comparative Analysis of a Large Dataset Indicates that Internal Transcribed Spacer (ITS) should be Incorporated into the Core Barcode for Seed Plants." *Proceedings of the National Academy of Sciences of the United States of America* 108: 19641–19646. https://doi.org/10.1073/PNAS.1104551108.

Li, X. K., B. Wang, R. C. Han, Y. C. Zheng, H. B. Y. Yin, L. Xu, et al. 2013. "Identification of Medicinal Plant *Schisandra chinensis* Using a Potential DNA Barcode ITS2." *Acta Societatis Botanicorum Poloniae* 82: 283–288. https://doi.org/10.5586/asbp.2013.032.

Li, Jiao Sheng, Luo Yu Wu, Hui Zhang, Xiu Shi Song, Jian Xin Wang, Ming Guo Zhou, and Yi Ping Hou. 2021. "PCR-RFLP for Detection of Fusarium Graminearum Genotypes with Resistance to Phenamacril." *Plant Disease* 105 (4). American Phytopathological Society: 889–895. https://doi.org/10.1094/PDIS-06-20-1156-RE/ASSET/IMAGES/LARGE/PDIS-06-20-1156-REF3.JPEG.

Liu, Xinyu, Yuran Bai, Yachao Wang, Yifeng Chen, Wenpan Dong, and Zhixiang Zhang. 2023. "Complete Chloroplast Genome of Hypericum Perforatum and Dynamic Evolution in Hypericum (Hypericaceae)." *International Journal of Molecular Sciences* 24 (22). Multidisciplinary Digital Publishing Institute (MDPI): 16130. https://doi.org/10.3390/IJMS242216130/S1.

Meena, Vijay Kamal, Harsh Vardhan Singh Shekhawat, Subhash Chand, Kapil Choudhary, Jitendra Kumar Sharma, Lekha Lekha, Vijay Kamal Meena, et al. 2023. "Advances in Molecular Marker Technology and Their Significance in Plant Improvement Strategies." *Updates in Plant Breeding [Working Title]* (October). IntechOpen. https://doi.org/10.5772/INTECHOPEN.1002773.

Morgil, Hande, Yusuf Can Gercek, Isil Tulum, Hande Morgil, Yusuf Can Gercek, and Isil Tulum. 2020. "Single Nucleotide Polymorphisms (SNPs) in Plant Genetics and Breeding." *The Recent Topics in Genetic Polymorphisms* (March). IntechOpen. https://doi.org/10.5772/INTECHOPEN.91886.

Noviana, Eka, Gunawan Indrayanto, and Abdul Rohman. 2022. "Advances in Fingerprint Analysis for Standardization and Quality Control of Herbal Medicines." *Frontiers in Pharmacology* 13 (June). Frontiers Media SA. https://doi.org/10.3389/FPHAR.2022.853023.

Núñez, Nerea, Oscar Vidal-Casanella, Sonia Sentellas, Javier Saurina, and Oscar Núñez. 2020. "Characterization, Classification and Authentication of Turmeric and Curry Samples by Targeted LC-HRMS Polyphenolic and Curcuminoid Profiling and Chemometrics." *Molecules* 25 (12). Multidisciplinary Digital Publishing Institute (MDPI). https://doi.org/10.3390/MOLECULES25122942.

Nyakudya, Trevor T., Thulani Tshabalala, Rachael Dangarembizi, Kennedy H. Erlwanger, and Ashwell R. Ndhlala. 2020. "The Potential Therapeutic Value of Medicinal Plants in the Management of Metabolic Disorders." *Molecules* 25 (11). Multidisciplinary Digital Publishing Institute (MDPI). https://doi.org/10.3390/MOLECULES25112669.

Parvathy, V. A., V. P. Swetha, T. E. Sheeja, and B. Sasikumar. 2015. "Detection of Plant-Based Adulterants in Turmeric Powder Using DNA Barcoding." *Pharmaceutical Biology* 53 (12). Pharm Biol: 1774–1779. https://doi.org/10.3109/13880209.2015.1005756.

Rafalski, J. Antoni, Julie M. Vogel, Michele Morgante, Wayne Powell, Chaz Andre, and Scott V. Tingey. 1996. "Generating and Using DNA Markers in Plants." *Nonmammalian Genomic Analysis*. Elsevier: 75–134. https://doi.org/10.1016/B978-012101285-4/50005-9.

Raghavendra, P. 2020. "ITS-2 Marker for Identification of Medicinal Plants in Herbal Industry: A Supporting Tool to Differentiate Adulterants." *~ 26 ~ The Pharma Innovation Journal* 9 (7): 26–29. www.keygene.com.

Roy, Arpita, Ariba Khan, Irfan Ahmad, Saad Alghamdi, Bodour S. Rajab, Ahmad O. Babalghith, Mohammad Y. Alshahrani, Saiful Islam, and Md Rabiul Islam. 2022. "Flavonoids a Bioactive Compound from Medicinal Plants and Its Therapeutic Applications." *BioMed Research International* 2022. Hindawi Limited. https://doi.org/10.1155/2022/5445291.

Sairkar, Pramod Kumar, Anjana Sharma, and N. P. Shukla. 2016. "SCAR Marker for Identification and Discrimination of Commiphora Wightii and C. Myrrha." *Molecular Biology International* 2016 (March). Hindawi Limited: 1–10. https://doi.org/10.1155/2016/1482796.

Satam, Heena, Kandarp Joshi, Upasana Mangrolia, Sanober Waghoo, Gulnaz Zaidi, Shravani Rawool, Ritesh P. Thakare, et al. 2023. "Next-Generation Sequencing Technology: Current Trends and Advancements." *Biology* 12 (7). Multidisciplinary Digital Publishing Institute (MDPI). https://doi.org/10.3390/BIOLOGY12070997.

Seethapathy, Gopalakrishnan Saroja, Ancuta Cristina Raclariu-Manolica, Jarl Andreas Anmarkrud, Helle Wangensteen, and Hugo J. de Boer. 2019. "DNA Metabarcoding Authentication of Ayurvedic Herbal Products on the European Market Raises Concerns of Quality and Fidelity." *Frontiers in Plant Science* 10 (February). Frontiers Media SA: 430518. https://doi.org/10.3389/FPLS.2019.00068/BIBTEX.

Soroka, Marianna, Barbara Wasowicz, and Anna Rymaszewska. 2021. "Loop-Mediated Isothermal Amplification (LAMP): The Better Sibling of PCR?" *Cells* 10 (8). Multidisciplinary Digital Publishing Institute (MDPI). https://doi.org/10.3390/CELLS10081931.

Sucher, Nikolaus J., and Maria C. Carles. 2008. "Genome-Based Approaches to the Authentication of Medicinal Plants." *Planta Medica* 74 (6): 603–623. https://doi.org/10.1055/S-2008-1074517.

Techen, Natascha, Iffat Parveen, Zhiqiang Pan, and Ikhlas A. Khan. 2014. "DNA Barcoding of Medicinal Plant Material for Identification." *Current Opinion in Biotechnology* 25 (February). Elsevier Current Trends: 103–110. https://doi.org/10.1016/J.COPBIO.2013.09.010.

Terletskaya, Nina V., Ainur S. Turzhanova, Oxana N. Khapilina, Moldir Z. Zhumagul, Nataliya D. Meduntseva, Nataliya O. Kudrina, Nazym K. Korbozova, Serik A. Kubentayev, and Ruslan Kalendar. 2023. "Genetic Diversity in Natural Populations of Rhodiola Species of Different Adaptation Strategies." *Genes* 14 (4). MDPI: 794. https://doi.org/10.3390/GENES14040794/S1.

Wallace, E. Diane, Daniel A. Todd, James M. Harnly, Nadja B. Cech, and Joshua J. Kellogg. 2020. "Identification of Adulteration in Botanical Samples with Untargeted Metabolomics." *Analytical and Bioanalytical Chemistry* 412 (18). NIH Public Access: 4273. https://doi.org/10.1007/S00216-020-02678-6.

Wong, K. L., P. P. H. But, and P. C. Shaw. 2013. "Evaluation of Seven DNA Barcodes for Differentiating Closely Related Medicinal Gentiana Species and their Adulterants." *Chinese Medicine* (United Kingdom) 8: 1–12. https://doi.org/10.1186/1749-8546-8-16/FIGURES/7.

Yahfoufi, Nour, Nawal Alsadi, Majed Jambi, and Chantal Matar. 2018. "The Immunomodulatory and Anti-Inflammatory Role of Polyphenols." *Nutrients* 10 (11). Multidisciplinary Digital Publishing Institute (MDPI). https://doi.org/10.3390/NU10111618.

Yang, Zhengxiu, Yat Tung Lo, Zheng Quan, Junchen He, Yanjun Chen, Adam Faller, Tiffany Chua, et al. 2023. "Application of a Modified Tetra-Primer ARMS–PCR Assay for Rapid Panax Species Identity Authentication in Ginseng Products." *Scientific Reports* 13 (1). Nature Publishing Group: 14396. https://doi.org/10.1038/S41598-023-39940-7.

Zheng, Sihao, Dewang Liu, Weiguang Ren, Juan Fu, Linfang Huang, and Shilin Chen. 2014. "Integrated Analysis for Identifying Radix Astragali and Its Adulterants Based on DNA Barcoding." *Evidence-Based Complementary and Alternative Medicine: ECAM* 2014. Hindawi Limited. https://doi.org/10.1155/2014/843923.

Zhou, J., W. Wang, M. Liu, and Z. Liu. 2014. "Molecular Authentication of the Traditional Medicinal Plant *Peucedanum praeruptorum* and Its Substitutes and Adulterants by DNA – Barcoding Technique." *Pharmacognosy Magazine* 10: 385. https://doi.org/10.4103/0973-1296.141754.

Zhu, S., Q. Liu, S. Qiu, J. Dai, and X. Gao. 2022. "DNA Barcoding: An Efficient Technology to Authenticate Plant Species of Traditional Chinese Medicine and Recent Advances." *Chinese Medicine* 17: 1–17. https://doi.org/10.1186/S13020-022-00655-Y.

18 Medicinal Plants and Their Chemical Compounds in Ruminant Nutrition and Antimethanogenic Effects

The Biotechnological Approach

*Edwin Rafael Alvarado-Ramírez, Mohamed Z. M. Salem, Ahmed Abdel-Megeed, Pedro Enrique Hernández-Ruiz, and Abdelfattah Z. M. Salem**

* Corresponding author: salem@uaemex.mx

18.1 INTRODUCTION: BACKGROUND FOR THE BIOTECHNOLOGICAL APPROACH

Many plants produce secondary metabolites, which the plants utilize for growth and reproduction, but which may also be useful for animal nutrition due to their chemical features. According to Kamra et al. (2006), thousands of plant secondary metabolites have been identified. It has been demonstrated that phenolic compounds, flavonoid compounds, essential oils (i.e., EO), and sarsaponins exhibited antimicrobial activities (Bakkali et al. 2008; Elbanoby et al. 2022; Shakam et al. 2022; El-Hefny et al. 2023; Taha et al. 2024).

Animal performance was increased using alternative feed sources, such as woody plant foliage, especially during the dry season (Elghandour et al. 2018). Secondary metabolites are also being researched to substitute antibiotics in animal feed (Reddy et al. 2020; Salem et al. 2020). Once identified, these compounds may be used to modify ruminal fermentation by increasing feed energy efficiency while lowering methane emissions. By *in vitro* screening 450 plants, a species that can reduce methane production by more than 25% without sacrificing digestibility was found (Bodas et al., 2008). Similar results were reported by Jayanegara et al. (2009), who discovered a substantial inverse association between plant total phenol, total tannins, or tannin activity and *in vitro* methane formation.

Two of the most important objectives for animal nutritionists include altering the rumen microbial ecology to improve the digestion of fibrous diets and ruminant performance by lowering methane emission and nitrogen excretion. Methane is a much more potent greenhouse gas than CO_2 (Dijkstra et al. 2011; Patra and Saxena 2011; Cobellis et al. 2016). Strategies to reduce urinary nitrogen excretion may result in greater methane emissions, depending on the degree of fiber present (Gerber et al. 2013). Understanding the trade-off between enteric methane generation at the animal scale is important for trustworthy data to be used at the level of the entire farm (Dijkstra et al. 2011).

Most of the studied essential oils (EO) were discovered as antimethanogenic agents with no negative effects on ruminal fermentation or NDF degradability in an *in vitro* screening trial, namely, *Origanum vulgare* leaf material EO (Tekippe et al. 2011). With higher doses, the EOs of clove, eucalyptus, garlic, oregano, and peppermint all significantly decreased methane production

DOI: 10.1201/9781003326939-22

(Patra Amlan and Yu 2012). In four rumen-cannulated cows (*Bos taurus* × *Bos indicus*), canola oil supplementation (1.2 g/kg SBW) significantly reduced the total ruminal protozoa population, the acetate–propionate ratio, and enteric methane output (g/kg DMI) by 9, 5.3, and 17.5%, respectively (Avilés-Nieto et al. 2023).

The lowest net methane generation was determined (0.12 mL/mL) of net gas, and 0.03 mL/mg on a genuinely degraded substrate basis for *Carduus*, when rumen fluid was obtained from an animal that had been fed hay. These values for *Carduus* were found in the rumen liquor of an animal that had been fed concentrate: 0.116 mL/mL of net gas and 0.025 mL/mg of fully decomposed substrate (Goel et al. 2008). Saponin-rich fenugreek and sesbania did not show any methane inhibition on a volume basis, even though there was a roughly 50% decrease in protozoal populations in the nutrients in hay or concentrate hay (68:32)–based diets (Goel et al. 2008).

Tropical plants that contain saponin limit or eliminate the rumen protozoa also reduce methane and ammonia production (Kamra et al. 2000; Śliwiński et al. 2002b). Similar to this, some plant extracts rich in flavonoids reduce methane generation and cause widespread stimulation of microbial metabolism, increasing the efficiency and productivity of microbial biomass production, as well as the degradability of crude protein and cell wall elements (Bodas et al. 2012; Ku-Vera et al. 2020). Tannins are also known for reducing the formation of methane (Goel and Makkar 2012). However, the efficacy of plants or plant extracts with high amounts of saponins, flavonoids, and tannins was influenced by the type, source, and concentration of secondary metabolites (Patra and Saxena 2010; Reddy et al. 2020). Potential exists for the antimethanogenic characteristics of *Rheum undulatum*, *Vaccinium vitis-idaea*, *Bergenia crassifolia*, *Rhus typhina*, and *P. peltatum* leaves and roots to be related to tannin concentration (Jayanegara et al. 2009).

Reducing methane emissions and nitrogen excretion from the ruminant cattle industry is imperative, as ruminants account for 16–25% of global greenhouse gas emissions. It has been demonstrated that adding EO to feed can help lower emissions of ammonia and methane. Regarding feed degradability and volatile fatty acids (VFA) production, research on the effects and activities of EO on the rumen microbiota has produced contradictory findings. The outcomes of *in vitro* experiments cannot, however, be directly applied to in vivo settings. *In vitro,* assays help screen bioactive compounds for antimicrobial action, but real animal studies are needed to assess the genuine value of these plant extracts in changing rumen microbial fermentation *in vitro.*

18.2 SOURCES OF THE NATURAL PRODUCTS FROM MEDICINAL PLANTS

Leaves, flowers, wood, bark, seeds, fruits, branches, and roots of medicinal plants can be used for the extraction of EO and other extracts with several methods (Mansour et al. 2015; Salem et al. 2018; Abdelsalam et al. 2019; Hamad et al. 2019; Mohamed et al. 2019a; Elbanoby et al. 2022; Shakam et al. 2022; El-Hefny et al. 2023; Alkharpotly et al. 2024; Awad et al. 2024).

Since extracts, EO, and medicinal herbs are relatively new feed additives in ruminant nutrition, studies have demonstrated their impact on ruminal fermentation and general nutritional advantages. Because of this, scientists have mostly depended on brief laboratory experiments (i.e., *in vitro*) to assess their ability to regulate ruminal fermentation and predict *in vivo* results. Numerous *in vitro* experiments using batch or continuous culture systems have been conducted on a wide variety of medicinal plants, as well as their extracts and EO.

18.3 EFFECTS OF ESSENTIAL OILS ON METHANE PRODUCTION

EOs are secondary metabolites and volatile substances that are found in nature and have the potential to be antimicrobial agents against a wide range of pathogens, including bacteria, protozoa, fungi, and viruses (Dehsheikh et al. 2020). However, EOs and their chemical compounds (Figure 18.1) have been shown to increase the productivity and performance of ruminant animals (Cobellis et al. 2016) and have a greater effect in reducing ammonia N concentration. Because of differences in dose and EO type, in vivo trials frequently produce contradictory results. Additionally, it is challenging to estimate the dosage of EO needed to obtain an effective EO concentration in the rumen, due to variations in the chemical makeup of various EO, diet, and feeding (Bodas et al. 2012; Kholif and Olafadehan 2021; Dorantes-Iturbide et al. 2022).

Thyme applied at 140 and 280 µL/L, pistachio applied at 280 µL/L, and cinnamon and thyme applied at 280 L/L all reduced DM disappearance in comparison to the control (Jahani-Azizabadi et al. 2014). When employing goat rumen liquor with *Salvia officinalis* EO in levels greater than 10 g/mL, the production of methane decreased in a dose-dependent manner (Saber et al. 2022). When applied to buffalo rumen liquor, *Allium sativum* EO had the highest level of methane inhibition (38.5%), followed by cinnamon bark EO (34.9%), *Curcuma aromatica* EO (15.5%), *Syzygium aromaticum* bud EO (CBO; 12.9%), *S. aromaticum* leaf EO (8.0%), *Cymbopogon citratus* EO (7.3%), and *Curcuma longa* EO (no effect) (Pawar et al. 2014). Low-dose (33.33 µL/L of rumen fluid) supplementation with garlic EO has a good effect on decreasing enteric methane generation production from buffaloes (*Bubalus bubalis*) by *in vitro* rumen fermentation (Dey et al. 2021). According to the finding of the studied EO from coriander, oregano, pistachio hull, caraway, cinnamon, cumin, and thyme, they may allow manipulation of rumen microbial fermentation (Jahani-Azizabadi et al. 2014). Using 0.33 mL/L of peppermint EO resulted in a 20% decrease in methane production and an increase in methanogen abundance, with the production of VFA remaining unaltered (Agarwal et al., 2009).

Compounds having phenolic structures, such as thymol and carvacrol, are believed to have a greater antimicrobial impact than other nonphenolic secondary plant metabolites because of the presence of a hydroxyl group in their structure (Helander et al. 1998; Załuski et al. 2015; Marinelli et al. 2018; Pisoschi et al. 2018). The primary active component in oregano is carvacrol (Sivropoulou et al. 1996). Carvacrol has counterintuitive effects on ruminal microbial fermentation that vary with dosage. Both carvacrol and oregano EO had similar effects on pH, individual and total VFA and ammonia N concentrations; however, carvacrol was more inhibitory than oregano EO (Busquet et al. 2006). At 300 mg/L, carvacrol increased the proportion of acetate, decreased the proportion of propionate, and increased the proportion of butyrate when compared to the control. At 3,000 mg/L, however, it did not affect acetate, raised the proportion of propionate, and lowered the fraction of butyrate (Busquet et al. 2006).

Carvacrol's hydroxyl proton (from the phenolic group) was thought to be swapped for a monovalent cation like potassium, serving as a transmembrane carrier. These occurrences would result in the reduction of ATP synthesis, the lack of a proton motive force, and ultimately, cell death. Carvacrol may not be effective in a complex microbial environment such as the rumen because all

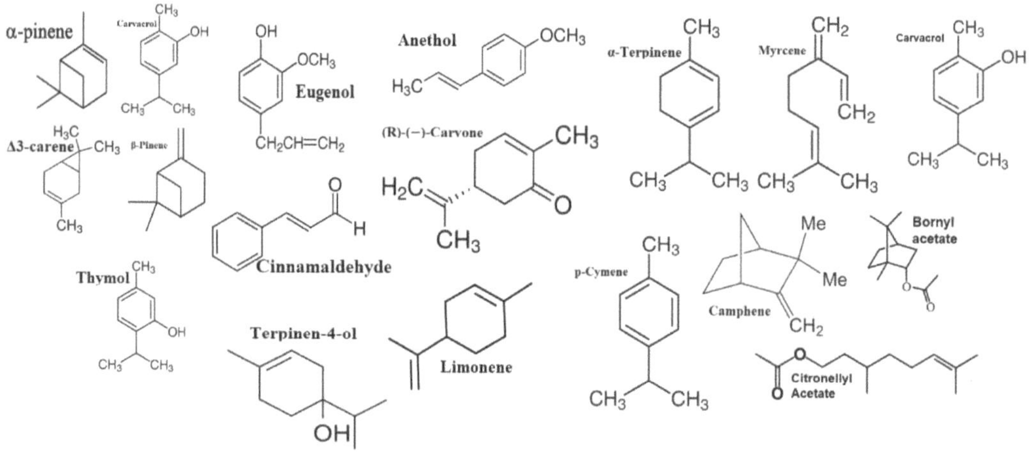

FIGURE 18.1 Chemical composition of the primary essential oil chemical components that have received the most attention due to their effects on rumen fermentation.

the effects indicate that it has a low antimicrobial specificity and may inhibit gram-positive (most acetate- and butyrate-producing bacteria) or gram-negative (usually propionate-producing bacteria) bacteria in a dose-dependent manner (Nazzaro et al. 2013; Ultee et al. 2002; Castañeda-Correa et al. 2019; Kiprotich 2019).

When compared to the control, thymol at 400 mg/L lowered the amount of acetate and propionate (Evans and Martin 2000). In batch cultures, thymol decreased ammonia N concentration by 1,000 mg/L, which seems to be in line with the observation that ammonia N concentration decreased when carvacrol (3,000 mg/L) was introduced (Borchers 1965; Broderick and Balthrop 1979). At doses of 5 and 50 mg/L, thymol did not affect ruminal N metabolism, but at doses of 500 and 5,000 mg/L, it decreased the amount of ammonia in 24-hour batch cultures (Castillejos et al. 2008).

Eugenol is a phenolic molecule as well, and the presence of a hydroxyl group is thought to play a role in how it works as an antimicrobial (Gallucci et al. 2009; Ulanowska and Olas 2021; Di Consiglio et al. 2023). It has been demonstrated that clove bud EO's main antimicrobial component is eugenol (Lin et al. 2013; Marchese et al. 2017; Hamad et al. 2019). In an *in vitro* study using a continuous culture system, low dosages of clove bud EO (2.2 mg/L) resulted in reduced molar proportions of acetate and BCVFA and a higher molar percentage of propionate, but the same amount of eugenol had no impact *in vitro* (Busquet et al. 2005).

Different effects were produced by cinnamon EO and cinnamaldehyde, the compound's primary active ingredient. Cinnamaldehyde demonstrated larger inhibitory effects than cinnamon EO in lowering the amounts of total VFA, BCVFA, and ammonia N (Busquet et al. 2006). However, cinnamaldehyde alone increased the pH and proportion of propionate, whereas cinnamon EO alone increased the fraction of acetate (Busquet et al. 2006).

Despite being the main and most active component of cinnamon EO, cinnamaldehyde may interact with other compounds in the EO. Cinnamaldehyde at low concentrations (2.2 mg/L) quantitatively increased the butyrate proportion and decreased the acetate proportion but did not affect the propionate proportion (Busquet et al. 2005). The observed periplasmic and deeper cell penetration of cinnamaldehyde can be attributed to the porin proteins that traverse the outer membrane, as suggested by the absence of outer membrane disintegrating activity and the decrease in intracellular ATP (Nikaido 1994; Helander et al. 1998).

Similar effects were observed on the rumen microbial fermentation profile by the anise EO and its main ingredient, anethole, which reduced total VFA, increased butyrate, and decreased the amounts of acetate and propionate. However, anethole had larger effects than anise EO (Busquet et al. 2006). Ammonia N content was unaffected by anise EO or anethole. Numerous investigations have shown that the primary antibacterial active component in anise EO is anethol (Davidson and Naidu 2000; El-Essawy et al. 2021; AlBalawi et al. 2023). The main building blocks for the synthesis of fat and glucose in ruminants are acetate and propionate, respectively, which were lowered by anethole despite their significant capacity to alter rumen microbial fermentation. This raises the possibility that anethole may not be nutritionally advantageous to dairy cattle (Calsamiglia et al. 2007; Ansari et al. 2022; Patra et al. 2010; Janini et al. 2023).

Numerous research has looked in-depth at how EO affects ruminal N metabolism. Even though they only decreased the ammonia N level at 3,000 mg/L, carvone and tea tree EO had effects that were comparable to those of anethole and anise EO (Busquet et al. 2006). Dill EO contains 47% carvone, which is thought to be the substance responsible for its antibacterial properties (Davidson and Naidu 2000; Kaur et al. 2021; Lu-Lu et al. 2022), because it has one double bond and a ketonic group on the molecule, is highly soluble (low saturation) in water, and is active against both gram-positive and gram-negative bacteria (Dorman and Deans 2000; Guliani et al. 2021).

Chaves et al. (2008) detected no difference in the amount of ruminal ammonia in lambs treated with carvacrol or cinnamaldehyde (200 mg/kg of dry matter intake [DMI]). Cattle in feedlots given 400, 800, or 1,600 mg/d of either eugenol or cinnamaldehyde had no change in ruminal ammonia content (Yang et al. 2010a; Yang et al. 2010b). The ruminal fermentation of lactation dairy cows was altered by a mixture of cinnamaldehyde and garlic EO, which decreased methane generation, decreased acetate molar proportions *in vitro*, and increased overall rumen volatile concentrations (Blanch et al. 2016). Feeding dairy cows cinnamon EO (50 mg/kg DM intake), cinnamaldehyde (43 and 50 mg/kg DM intake), or eugenol (25, 50, and 75 mg/kg DM intake) did not affect ruminal protein breakdown and ammonia levels (Benchaar et al. 2008a).

Tea tree EO's primary active ingredient, terpinen-4-ol (4-carvomenthenol), is thought to account for up to 40% of its activity (Iacovelli et al. 2023), whereas the remaining (about 50%) constituents of this EO are nonoxygenated terpenoid hydrocarbons, such as terpinene and p-cymene (Cox et al. 2001). The 30-day incidence of mastitis in one cattle over the first year was considerably lower in the terpinen-4-ol group. Terpinen-4-ol post-milking teat disinfection demonstrated comparable efficacy in reducing bacterial mastitis occurrence (Dore et al. 2019).

Douglas fir needle monoterpene hydrocarbons, such as α/β-pinene, limonene, myrcene, camphene, and Δ3-carene, promoted activity in sheep rumen microbes, but only slightly or not at all in deer rumen microbes. Terpinolene and Δ3-carene, however, inhibited activity (Oh Hi et al. 1967).

There were no effects of feed additives (cinnamaldehyde, garlic, and juniper berry EO) at the rate of 200 mg/kg of DM on ruminal pH and concentrations of ammonia and total VFA (Chaves et al. 2008). In dairy cows, cinnamon EO, cinnamaldehyde, or monensins are not practical CH_4 mitigation measures (Benchaar 2016). Since most research has found that EO decreases ruminal ammonia concentration, the increase in ammonia N concentration seen in this experiment is somewhat unexpected (Benchaar et al. 2009). Ammonia content in the rumen is controlled by processes such as deamination of feed AA, microbial lysis, absorption, and uptake (Hristov et al. 2005). Additionally, it was observed that oregano EO's carvacrol content may be to blame for the elevated ammonia levels seen with *Origanum vulgare* L. leaves (Tekippe et al. 2011). *In vitro* studies by Busquet et al. (2006) showed that carvacrol raised ammonia N concentrations and lowered big peptide concentrations 2 hours after feeding.

Citronellyl acetate and citronellal suppressed rumen microbial activity in sheep but encouraged activity in both deer and deer. Bornyl acetate exhibited modest inhibition for both sheep and deer bacteria and other rumen (Oh Hi et al. 1967; Foggi et al. 2022; Daning et al. 2022). Another study showed that the quantities of acetate were lowered and raised by benzyl salicylate (300 and 3,000 mg/L) and garlic EO (300 and 3,000 mg/L), indicating that methane generation was inhibited (Busquet et al. 2006). Incorporating a supplement containing micro-encapsulated carvacrol, thymol, and cinnamaldehyde into lamb feed improved antioxidant response without changing the serum levels of free radicals and lipid peroxidation, increased weight gain, and had anti-inflammatory effects without endangering lamb health (Favaretto et al. 2020). The amount of acetate and propionate was reduced by various doses of anethol, anise EO, carvone, and tea tree EO, indicating that these substances would not be advantageous to dairy cattle diets (Busquet et al. 2006).

18.4 EFFECTS OF NATURAL EXTRACTS AND MAIN CHEMICAL GROUPS ON RUMEN FEED

Natural products from medicinal plants, whether as pure compounds or as standardized extracts, provide countless opportunities for new therapeutic discoveries due to the hitherto unheard-of abundance of chemical variability. Food plants in particular are becoming more and more popular as a result of the need for chemical variety in screening processes to meet the growing need for medical pharmaceuticals from natural sources. Medicinal botanicals and herbal medicines contain a variety of bioactive chemicals (Sasidharan et al. 2011). These chemical compounds are related to various chemical groups, like saponins, tannins, polyphenolic compounds, flavonoids, and alkaloids (Salem et al. 2014a; Salem et al. 2014b; Ashmawy et al. 2018; Abdelkhalek et al. 2020; Ashmawy et al. 2020b; Hayat et al. 2020; Salem et al. 2021).

Addressed was how to analyze the bioactive chemicals found in plant extracts using conventional phytochemical screening assays, and the effective method for the analysis of herbal products is the coupling of column liquid chromatography (LC) or gas chromatography (GC) with a UV-Vis or mass spectrometer (HPLC-DAD, CE-DAD, GC-MS, and LC-MS, respectively) (Kamboj 2012; Elufioye et al. 2020) (Figure 18.2). However, a variety of hyphenated techniques, such as LC-IR, LC-MS, LC-NMR, GC-MS, HPLC-DAD, HPLC-MS, hyphenation of EC, and many others, can be used, depending on the objectives of the study and the availability of the necessary tools (Kamboj 2012; Mohamed et al. 2019b; Arora et al. 2023).

These extracts or chemical groups or pure compounds have shown potential effects on ruminant fermentation. Altering rumen microbial fermentation with culture fluid of various plant extracts (garlic, cinnamon, yucca, anise, oregano, or pepper) and secondary plant metabolites at dosages ranging from 0.22 to 2.2 mg/L may not have been sufficient (Cardozo et al. 2004).

The main chemical factor affecting feed digestibility is lignin, which is found in large quantities in many tropical fodder plants, along with saponins and tannins (Goel and Makkar 2012). These compounds may be harmful to animals' health and the rumen microbial population when supplied at high levels (Anee et al. 2021). They show some ability to enhance rumen fermentation at low doses. For instance, these substances might block microbial urease to decrease ammonia output in the rumen (Maxiselly et al. 2022). Tannins have also been found to reduce methane production (Aboagye and Beauchemin 2019).

In the study on the effects of natural plant extracts on ruminal protein degradation and fermentation profiles, the average ammonia N concentration was 31% higher in anise extract and 25.5% lower in garlic extract when compared with the control (treatment with no extract) (5.5 mg/100 mL) in a continuous culture of fermenters fed a diet with a 52:48 forage–concentrate ratio (DM basis) (Cardozo et al. 2004).

High doses of garlic oil resulted in a decrease in total volatile fat and the proportion of acetate, while the quantities of propionate and butyrate increased (Busquet et al. 2006). This was concurred upon in the presence of 2-bromoethanosulfonic acid, or amicloral (Martin and Macy 1985), which directly inhibits methanogens. With increasing doses of *Terminalia chebula* meal supplementation in the diet, NH_3_N concentrations tended to rise. Additionally, supplementation with *T. chebula* at 8, 12, and 16 mg enhanced total volatile fatty acid (VFA) and propionate concentrations while decreasing acetate concentration, acetate-to-propionate ratio, CH_4 generation, and protozoal populations (Anantasook et al. 2016).

Fenugreek extract, unlike other essential oils and their secondary components, did not alter the concentration of total VFA even at large dosages, indicating that it did not affect the availability of energy and the fermentability of the diet (Benchaar et al. 2008a). Fenugreek raised the proportion of propionate,

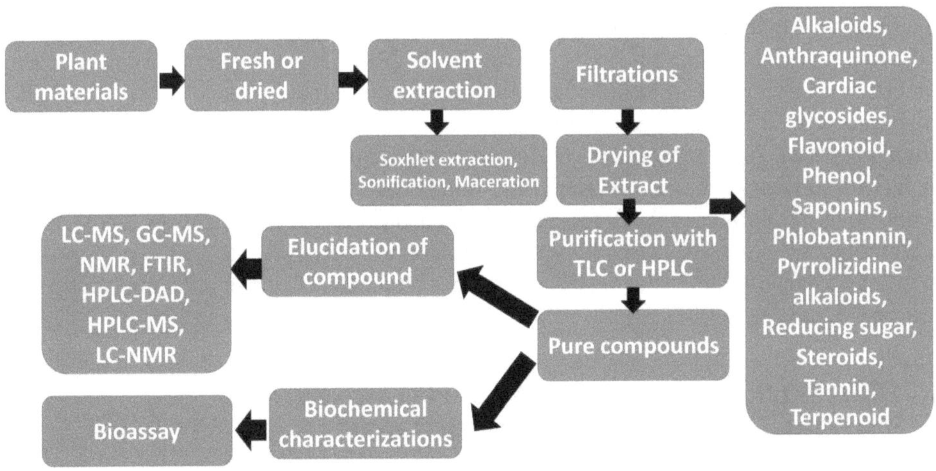

FIGURE 18.2 General methods for isolating, characterizing, and extracting bioactive compounds from plant extracts.

lowered the concentration of ammonia N, and decreased the fraction of BCVFA (Cao and Yang 2011). The elevated concentration of steroidal saponins is responsible for these effects. The saponin-containing yucca extract had no impact on the rumen's microbial fermentation. The high pH is probably to blame for the yucca extract's lack of impact on rumen microbial fermentation (Benchaar et al. 2008b). For male crossbred Holstein calves, ruminal NH_3_N content showed a quadratic decline with increasing amounts of *Corymbia citriodora* leaf extract supplementation (Hassan et al. 2020).

The pH of the media had a significant impact on how yucca saponins affected rumen microbial fermentation (Cardozo et al. 2005). A consistent observation is a decrease in ruminal ammonia N content when saponins are given to ruminants (Patra and Saxena 2009b; Śliwiński et al. 2002a). Typically, these results have been ascribed to saponins' potent antiprotozoal properties (Patel 2012; Maxiselly et al. 2022), where protozoa are the main generators of rumen ammonia (Khejornsart et al. 2021).

The basis for saponins' antiprotozoal activity is their capacity to create long-lasting complexes with the cholesterol in protozoal cell membranes, which cause membrane rupture, cell lysis, and death (Ramos-Morales et al. 2017). Decreased ruminal pH was associated with saponin supplementation (Ramírez-Restrepo et al. 2016). The saponin–nitrate combination reduced the abundance of protozoa and methanogens, which could have further decreased the number of protozoa. The reduction of protozoa is also mediated by the pH lowering caused by saponins (Patra and Yu 2013). Although the reduction of protozoa is the main action of saponins on ruminal microorganisms, inhibitory effects on ruminal bacteria have also been noted (Hui-Ling et al. 2010; Patra and Yu 2014).

Adding *Terminalia chebula*, which contains tannins and saponins, increased the concentrations of propionate and decreased the concentrations of acetate (Anantasook et al. 2016). A decrease in gram-positive bacteria has been linked to an increase in propionate content in the rumen (Marques and Cooke 2021), which are normally acetate, but not propionate, producers (Wallace et al. 1994). The ability of saponins to inhibit bacteria and protozoa not involved in the formation of propionate (Hui-Ling et al. 2010) increased the accumulation of propionate in the rumen by allowing gram-negative, propionate-producing organisms to fill the niche (Christophersen 2007; O'Hara 2019). Tannin extracts from *A. luederitzii* may be used as a dietary substitute to minimize methane generation, but it is important to establish the ideal degree of inclusion that will not impair the effectiveness of rumen fermentation and the diet's overall digestibility (Hassen et al. 2016).

Nitrate has the added advantage of reducing intestinal methane production and increasing the average daily gain of lambs when utilized as a source of non-protein nitrogen (Adejoro et al. 2020), while the tannin extract failed to exhibit any beneficial effects to support its usage in dietary contexts.

Research has shown that condensed tannin extracts from *Acacia luederitzii* and several warm-season perennial legumes that sheep and goats consume frequently have antiparasitic and enteric methane-reducing qualities in addition to the enhanced bypass protein that the animal has access to in the intestine (Naumann et al. 2014; Hassen et al. 2016). However, providing ruminants with a high dietary concentration of tannin may cause them to consume less feed (Eckard et al. 2010).

To reduce enteric methane generation, plant extracts have been added. *Emblica officinalis*, *Terminalia chebula*, *Terminalia bellirica*, *Acacia concinna* pods, and *Azadirachta indica* seed kernel in various solvents (ethanol, methanol, and water) have been used (Patra et al. 2006). Among them, only *T. chebula* (methanol extract) demonstrated antimethanogenic activity among three seed pulps with high antiprotozoal activity; methane emission is not always linked to protozoa activity. There may be a purpose for the *T. chebula* seed pulp's methanol extract in reducing enteric methane production (Patra et al. 2006).

18.5 EFFECTS OF MEDICINAL PLANT SUPPLEMENTATION ON METHANE OR AMMONIA MITIGATION

Owing to the substantial variations in methane emission estimates both within and across animals, the significant reduction in rumen methane generation with the addition of *Origanum vulgare* leaves occurred within 8 hours after feeding (Tekippe et al. 2011). *Rheum nobile* and *Carduus pycnocephalus* consistently reduced methane generation across the 450 samples evaluated while having no negative effects on other rumen fermentation metrics (Bodas et al. 2008). *Salix babylonica* extract, administered at a rate of 40 mL per lamb per day, improved daily weight growth in growing lambs and increased asymptotic *in vitro* ruminal gas production (Cedillo et al. 2014). By supplementing lactation dairy cows with *Origanum vulgare* L. leaves, the milk fat content of the animals increased, and 3.5% fat-corrected milk output tended to increase when compared to the control (3.29 vs. 3.12% and 42.4 vs. 41.0 kg/d, respectively) (Tekippe et al., 2011).

The leaves of *Acacia luederitzii* and *Marrubium incanum* reduced methane generation by more than 50%, but they also reduced digestibility and rumen fermentation indices, including the content of volatile fatty acids (Hassen et al. 2016). When calves were given 20 mL of *Corymbia citriodora* leaf extract as a supplement, intakes of milk, calf starter, hay, and total dry matter (DM) were shown to be significantly reduced (Hassan et al. 2020).

Increased amounts of *Moringa oleifera* leaf incorporation were seen to result in a decrease in CH_4 (%) and CH_4 (mL/100 mg dDM). All three of the nutrition planes that *M. oleifera* leaves used showed no change in the ratios of ammonia, acetate, and propionate (Leitanthem et al. 2022). *Leucaena leucocephala*, *Piscidia piscipula*, *Neomillspaughia emarginata*, and *Tabernaemontana amygdalifolia* treatments reduced methane production in ruminant diets based on *Pennisetum purpureum* (25.8, 29.5, 30.6, and 31.8 L/kg of digested dry matter, respectively) (Albores-Moreno et al. 2019). The growth, health, and blood metabolites of Holstein dairy calves were improved by adding fennel seed powder or oregano leaf powder to the liquid diet, particularly fennel seed powder (Ansari et al. 2022). The concentration of ruminal ammonia was enhanced in dairy cows when 500 g of oregano leaves (90.8% carvacrol of the total EO) were pulse-dosed daily into the rumen (Tekippe et al. 2011). In contrast, it was reported that calves fed 250, 500, or 750 g/day of oregano leaves (78.5% carvacrol of total EO) saw a drop in ruminal ammonia content (Hristov et al. 2013).

Due to the fact that a variety of factors might influence the methanogenesis process and that the genes involved in the methanogenesis pathway could be expressed differently in various methanogen species, the decreased methane generation was not always accompanied by a decrease in archaeal abundance (Liechty et al. 2020).

18.6 EFFECTS OF PLANT SECONDARY METABOLITES ON RUMINAL MICROBES AND PROTOZOA

The antimicrobial properties of plant secondary compounds, such as phenolic compounds, which act against bacteria, protozoa, and fungi, are well-known (Salem et al. 2013; Adegbeye et al. 2020; Ashmawy et al. 2020a; Ashmawy et al. 2020b; Reddy et al. 2020). The plant species they ingest and the chemical makeup of the plant both influence how they affect the activity of ruminal bacteria (Waghorn and McNabb 2003; Frutos et al. 2004; Li 2021). The rumen pH, the feed they are consumed with, and the preparation and extraction techniques all affect how these chemicals behave (Patra and Saxena 2009a).

Several essential oils, tannins, and saponins exhibit pronounced antiprotozoal action (Kuralkar and Kuralkar 2021; Patra and Saxena 2010). The mechanism by which saponins cause cell membrane degradation, cell lysis, and death in protozoa appears to be their ability to create irreversible compounds with cholesterol in the cell membrane (Fuchs et al. 2009; Bodas et al. 2012; Ramos-Morales et al. 2017). The antiprotozoal properties of *Yucca schidigera* saponins have been observed *in vitro* (Wina et al. 2005; Fuchs et al. 2009; Zúñiga-Serrano et al. 2022) and alfalfa (Kozłowska et al. 2020). However, even though saponins can inhibit rumen ciliate growth (Hui-Ling et al. 2010; Jia-Ku et al. 2012), it is unclear at this time if all protozoa species respond to this PSM equally (Patra and Saxena 2009a). When it comes to tannins, the antiprotozoal activity of the combination of hydrolyzable and condensed tannins is greater than that of hydrolyzable tannins alone (Yoshida et al. 2005; Batiha et al. 2022).

In vitro extract and compounds from the stem bark of *Combretum molle* were found to have strong to intermediate antiprotozoal activity against *Leishmania donovani*, *Trypanosoma cruzi*, *Plasmodium falciparum*, and *Trypanosoma brucei rhodesiense* (Asres et al. 2001). The antiprotozoal activity of dietary supplements containing tannin, saponin, and linoleic acid was examined in four ruminally andx duodenally cannulated heifers. The supplements included safflower oil (a source of linoleic acid), quillaja (a source of saponin), quebracho (a source of tannin), and control (base diet). Protozoa populations in the rumen of heifers given quabracho and quillaja were only slightly lower than those of the control, whereas safflower oil induced a log drop in protozoa counts compared to the control (Baah et al. 2007).

The EO has varying effects on protozoal populations; certain studies show that they have no effect (Newbold et al. 2004), whilst others have discovered that using EO stimulates protozoa (Patra and Saxena 2009a). Some research suggested that dome additives have an antiprotozoal impact that lowers the concentrations of total, entodiniomorph, and holotrich protozoa, that is, anise and capsicum (Fandiño et al. 2008), or from the extracts of garlic, cinnamon, yucca, anise, oregano, and capsicum and pure cinnamaldehyde, anethole, and eugenol (Cardozo et al. 2005). The mechanism of action might be connected to the lipophilic properties of substances like anethole, which make it easier for essential oils to pass through protozoal membranes (Cobellis et al. 2016; Andrade-Ochoa et al. 2021; Kholif and Olafadehan 2021).

When EO compounds were given to dairy cows at a dose of 750 mg/d, Benchaar et al. (2007) found no change in the populations of protozoa. α-Terpineol, terpinen-4-ol, linalool, citronellol, and fenchyl alcohol strongly inhibited the rumen microbial activity of both sheep and deer (Oh Hi et al. 1967). When cinnamon EO, cinnamaldehyde, or monensin was ingested by dairy cows, the feed additions did not change the characteristics of ruminal fermentation or the number of protozoa that were present (Benchaar 2016).

The addition of an EO preparation (containing a blend of natural and nature-identical EO) did not affect the overall counts of viable and cellulolytic bacteria and protozoa in dairy ewes of the Chios breed, but counts of hyper-ammonia-producing bacteria were reduced at the two highest supplementation levels in comparison to the control group (Giannenas et al. 2011). Clove, eucalyptus, garlic, oregano, and peppermint essential oils lowered the quantities of the main cellulolytic bacteria linearly with increasing EO doses: *Fibrobacter succinogenes*, *Ruminococcus flavefaciens*, and *R. albus* (Patra and Yu 2012). With the addition of a mixture of equal ratios of oil from clove, oregano, cinnamon, and lemon, or eugenol, carvacrol, citral, and cinnamaldehyde, the population of fungus and *R. flavefaciens* did not change in four Hu sheep (Lin et al. 2013).

Improved rumen function without a corresponding drop in animal performance might be possible, with a better understanding of the interactions between methanogens and other rumen microbial communities (bacteria, protozoa, and fungus).

18.7 CONCLUSIONS

The primary chemical components of, as well as the essential oils and unrefined extracts from, medicinal plants may be efficient natural remedies that improve rumen fermentations, lower rumen methane and ammonia output, and reduce the environmental impact of ruminant production. Several problems, including their effective doses, mechanism of action, effect on the organoleptic aspects of animal products, and cost-benefit ratio, need to be resolved before their use as feed additives in managing ruminant nutrition at the farm level.

REFERENCES

Abdelkhalek A., Salem MZM, Hafez E, Behiry SI, Qari SH. 2020. The phytochemical, antifungal, and first report of the antiviral properties of Egyptian *Haplophyllum tuberculatum* extract. Biology 9(9):248. https://doi.org/10.3390/biology9090248

Abdelsalam NR, Salem, MZM, Ali HM, Mackled MI, El-Hefny M, Elshikh MS, Hatamleh AA. 2019. Morphological, biochemical, molecular, and oil toxicity properties of *Taxodium* trees from different locations. Industrial Crops and Products 139:111515. https://doi.org/10.1016/j.indcrop.2019.111515

Aboagye IA, Beauchemin KA. 2019. Potential of molecular weight and structure of tannins to reduce methane emissions from ruminants: A review. Animals 9(11):856. https://doi.org/10.3390/ani9110856

Adegbeye MJ, Elghandour MMMY, Faniyi TO, Rivero-Perez N, Barbabosa-Pilego A, Zaragoza-Bastida A, Salem AZM. 2020. Antimicrobial and antihelminthic impacts of black cumin, pawpaw and mustard seeds in livestock production and health. Agroforestry Systems 94:1255–1268. https://doi.org/10.1007/s10457-018-0337-0

Adejoro FA, Hassen A, Akanmu AM, Morgavi DP. 2020. Replacing urea with nitrate as a non-protein nitrogen source increases lambs' growth and reduces methane production, whereas acacia tannin has no effect. Animal Feed Science and Technology 259:114360. https://doi.org/10.1016/j.anifeedsci.2019.114360

Agarwal N, Shekhar C, Kumar R, Chaudhary LC, Kamra DN. 2009. Effect of peppermint (*Mentha piperita*) oil on *in vitro* methanogenesis and fermentation of feed with buffalo rumen liquor. Animal Feed Science and Technology 148(2–4):321–327. https://doi.org/10.1016/j.anifeedsci.2008.04.004

AlBalawi AN, Elmetwalli A, Baraka DM, Alnagar HA, Alamri ES, Hassan MG. 2023. Chemical Constituents, Antioxidant Potential, and Antimicrobial Efficacy of Pimpinella anisum Extracts against Multidrug-Resistant Bacteria. Microorganisms 11:1024. https://doi.org/10.1016/j.anifeedsci.2008.04.004

Albores-Moreno S, Alayón-Gamboa JA, Miranda-Romero LA, Alarcón-Zúñiga B, Jiménez-Ferrer G, Ku-Vera JC, Piñeiro-Vázquez AT. 2019. Effect of tree foliage supplementation of tropical grass diet on *in vitro* digestibility and fermentation, microbial biomass synthesis and enteric methane production in ruminants. Tropical Animal Health and Production 51:893–904. https://doi.org/10.1007/s11250-018-1772-7

Alkharpotly AA, Abd-Elkader DY, Salem MZM, Hassan HS. 2024. Growth, productivity and phytochemicals of Coriander in responses to foliar application of Acacia saligna fruit extract as a biostimulant under field conditions. Scientific Reports 14(1):2921. https://doi.org/10.1038/s41598-024-53378-5

Anantasook N, Wanapat M, Gunun P, Cherdthong A. 2016. Reducing methane production by supplementation of *Terminalia chebula* RETZ containing tannins and saponins. Animal Science Journal 87(6):783–790. https://doi.org/10.1111/asj.12494C

Andrade-Ochoa S, Chacón-Vargas KF, Sánchez-Torres LE, Rivera-Chavira BE, Nogueda-Torres B, Nevárez-Moorillón GV. 2021. Differential antimicrobial effect of essential oils and their main components: Insights based on the cell membrane and external structure. Membranes 11(6):405. https://doi.org/10.3390/membranes11060405

Anee IJ, Alam S, Begum RA, Shahjahan RM, Khandaker AM. 2021. The role of probiotics on animal health and nutrition. The Journal of Basic and Applied Zoology 82:52. https://doi.org/10.1186/s41936-021-00250-x

Ansari M, Kargar S, Eslami MA, Falahati R, Albenzio M, Caroprese M, Zamiri MJ, Kanani M. 2022. Potential benefits of early-life supplementation of liquid feed with fennel (*Foeniculum vulgare*) seeds or oregano (*Origanum vulgare*) leaves on growth, health, and blood metabolites in Holstein dairy calves. Journal of Dairy Science 105(8):6639–6653. https://doi.org/10.3168/jds.2022-21776

Arora C, Bharti D, Tiwari BK, Kumar A, Verma DK, Singh B. 2023. Characterization techniques used for the analysis of phytochemical constituents. Phytochemicals in Medicinal Plants: Biodiversity, Bioactivity and Drug Discovery, 131. https://doi.org/10.1515/9783110791891-006

Ashmawy NA, Behiry SI, Al-Huqail AA, Ali HM, Salem MZM. 2020a. Bioactivity of selected phenolic acids and hexane extracts from *Bougainvilla spectabilis* and *Citharexylum spinosum* on the growth of Pectobacterium carotovorum and *Dickeya solani* bacteria: An opportunity to save the environment. Processes 8(4):482. https://doi.org/10.3390/pr8040482

Ashmawy NA, Salem MZM, El-Hefny M, Abd El-Kareem MSM, El-Shanhorey NA, Mohamed AA, Salem AZM. 2018. Antibacterial activity of the bioactive compounds identified in three woody plants against some pathogenic bacteria. Microbial Pathogenesis 121:331–340. https://doi.org/10.1016/j.micpath.2018.05.032

Ashmawy NA, Salem MZM, El Shanhorey N, Al-Huqail A, Ali HM, Behiry SI. 2020b. Eco-friendly wood-biofungicidal and antibacterial activities of various *Coccoloba uvifera* L. leaf extracts: HPLC analysis of phenolic and flavonoid compounds. BioResources 15(2):4165–4187.

Asres K, Bucar F, Knauder E, Yardley V, Kendrick H, Croft SL. 2001. *In vitro* antiprotozoal activity of extract and compounds from the stem bark of *Combretum molle*. Phytotherapy Research 15(7):613–617. https://doi.org/10.1002/ptr.897

Avilés-Nieto JN, Márquez-Mota CC, Hernández-Medrano JH, Ramírez-Bribiesca JE, Castillo-Gallegos E, Plascencia A, Castrejón-Pineda FA, Corona L. 2023. Effect of canola oil supplementation level on total tract digestion, ruminal fermentation, and methane emissions of cows grazing Urochloa sp. supplemented with a fixed amount of concentrate. Tropical Animal Health and Production 55(2):77. https://doi.org/10.1007/s11250-02303485-8

Awad M, Hassan NN, Alfuhaid NA, Amer A, Salem MZM, Fónagy A, Moustafa MA. 2024. Insecticidal and biochemical impacts with molecular docking analysis of three essential oils against *Spodoptera littoralis* (Lepidoptera: Noctuidae). Crop Protection 180:106659. https://doi.org/10.1016/j.cropro.2024.106659

Baah J, Ivan M, Hristov AN, Koenig KM, Rode LM, McAllister TA. 2007. Effects of potential dietary antiprotozoal supplements on rumen fermentation and digestibility in heifers. Animal Feed Science and Technology 137(1–2):126–137. https://doi.org/10.1016/j.anifeedsci.2006.11.004

Bakkali F, Averbeck S, Averbeck D, Idaomar M. 2008. Biological effects of essential oils—A review. Food and Chemical Toxicology 46(2):446–475. https://doi.org/10.1016/j.fct.2007.09.106

Batiha GES, Akhtar N, Alsayegh AA, Abusudah WF, Almohmadi NH, Shaheen HM, Singh TG, De Waard M. 2022. Bioactive compounds, pharmacological actions, and pharmacokinetics of genus Acacia. Molecules 27(21):7340. https://doi.org/10.3390/molecules27217340

Benchaar C. 2016. Diet supplementation with cinnamon oil, cinnamaldehyde, or monensin does not reduce enteric methane production of dairy cows. Animal 10(3):418–425. https://doi.org/10.1017/S175173111500230X

Benchaar C, Calsamiglia S, Chaves AV, Fraser GR, Colombatto D, McAllister TA, Beauchemin KA. 2008a. A review of plant-derived essential oils in ruminant nutrition and production. Animal Feed Science and Technology 145(1–4):209–228. https://doi.org/10.1016/j.anifeedsci.2007.04.014

Benchaar C, Hristov AN, Greathead H. 2009. Essential Oils as Feed Additives in Ruminant Nutrition. Nuttingham University Press: Nottingham, UK, pp. 111–146.

Benchaar C, McAllister TA, Chouinard PY. 2008b. Digestion, ruminal fermentation, ciliate protozoal populations, and milk production from dairy cows fed cinnamaldehyde, quebracho condensed tannin, or *Yucca schidigera* Saponin extracts1. Journal of Dairy Science 91(12):4765–4777. https://doi.org/10.3168/jds.2008-1338

Benchaar C, Petit HV, Berthiaume R, Ouellet DR, Chiquette J, Chouinard PY. 2007. Effects of essential oils on digestion, ruminal fermentation, rumen microbial populations, milk production, and milk composition in dairy cows fed alfalfa silage or corn silage. Journal of Dairy Science 90(2):886–897. https://doi.org/10.3168/jds.S0022-0302(07)71572-2

Blanch M, Carro MD, Ranilla MJ, Viso A, Vázquez-Añón M, Bach A. 2016. Influence of a mixture of cinnamaldehyde and garlic oil on rumen fermentation, feeding behavior and performance of lactating dairy cows. Animal Feed Science and Technology 219:313–323. https://doi.org/10.1016/j.anifeedsci.2016.07.002

Bodas R, López S, Fernández M, García-González R, Rodríguez AB, Wallace RJ, González JS. 2008. *In vitro* screening of the potential of numerous plant species as antimethanogenic feed additives for ruminants. Animal Feed Science and Technology 145(1–4):245–258. https://doi.org/10.1016/j.anifeedsci.2007.04.015

Bodas R, Prieto N, García-González R, Andrés S, Giráldez FJ, López S. 2012. Manipulation of rumen fermentation and methane production with plant secondary metabolites. Animal Feed Science and Technology 176(1–4):78–93. https://doi.org/10.1016/j.anifeedsci.2012.07.010

Borchers R. 1965. Proteolytic activity of rumen fluid *In vitro*. Journal of Animal Science 24(4):1033–1038. https://doi.org/10.2527/jas1965.2441033x

Broderick GA, Balthrop JE Jr. 1979. Chemical inhibition of amino acid deamination by ruminal microbes *in vitro*. Journal of Animal Science 49(4):1101–1111. https://doi.org/10.2527/jas1979.4941101x

Busquet M, Calsamiglia S, Ferret A, Kamel C. 2005. Screening for the effects of natural plant extracts and secondary plant metabolites on rumen microbial fermentation in continuous culture. Animal Feed Science and Technology 123:597–613. https://doi.org/10.1016/j.anifeedsci.2005.03.008

Busquet M, Calsamiglia S, Ferret A, Kamel C. 2006. Plant extracts affect *In vitro* rumen microbial fermentation. Journal of Dairy Science 89:761–771. https://doi.org/10.1016/j.anifeedsci.2005.03.008

Calsamiglia S, Busquet M, Cardozo PW, Castillejos L, Ferret A. 2007. Invited review: Essential oils as modifiers of rumen microbial fermentation. Journal of Dairy Science 90(60):2580–2595. https://doi.org/10.3168/jds.2006-644

Cao YC, Yang HJ. 2011. Ruminal digestibility and fermentation characteristics *in vitro* of fenugreek and alfalfa hay combination with or without the inoculation of Neocallimastix sp. YAK11. Animal Feed Science and Technology 169(1–2):53–60. https://doi.org/0.1016/j.anifeedsci.2011.05.010

Cardozo PW, Calsamiglia S, Ferret A, Kamel C. 2004. Effects of natural plant extracts on ruminal protein degradation and fermentation profiles in continuous culture. Journal of Animal Science 82(11):3230–3236. https://doi.org/10.2527/2004.82113230x

Cardozo PW, Calsamiglia S, Ferret A, Kamel C. 2005. Screening for the effects of natural plant extracts at different pH on *in vitro* rumen microbial fermentation of a high-concentrate diet for beef cattle1. Journal of Animal Science 83(11):2572–2579. https://doi.org/10.2527/2005.83112572x

Castañeda-Correa A, Corral-Luna A, Hume ME, Anderson RC, Ruiz-Barrera O, Castillo-Castillo, Y, Rodriguez-Almeida F, Salinas-Chavira J, Arzola-Alvarez C. 2019. Effects of thymol and carvacrol, alone or in combination, on fermentation and microbial diversity during *in vitro* culture of bovine rumen microbes. Journal of Environmental Science and Health, Part B 54(3):170–175. https://doi.org/10.1080/03601234.2018.1536580

Castillejos L, Calsamiglia S, Martín-Tereso J, Ter Wijlen H. 2008. *In vitro* evaluation of effects of ten essential oils at three doses on ruminal fermentation of high concentrate feedlot-type diets. Animal Feed Science and Technology 145(1–4):259–270. https://doi.org/10.1016/j.anifeedsci.2007.05.037

Cedillo J, Vázquez-Armijo JF, González-Reyna A, Salem AZM, Kholif AE, Hernández-Meléndez J, Martínez-González JC, Jiménez RMdO, Rivero N, López D. 2014. Effects of different doses of *Salix babylonica* extract on growth performance and diet *in vitro* gas production in pelibuey growing lambs. Italian Journal of Animal Science 13(3):3165. https://doi.org/10.4081/ijas.2014.3165

Chaves AV, Stanford K, Dugan MER, Gibson LL, McAllister TA, Van Herk F, Benchaar C. 2008. Effects of cinnamaldehyde, garlic and juniper berry essential oils on rumen fermentation, blood metabolites, growth performance, and carcass characteristics of growing lambs. Livestock Science 117(2–3):215–224. https://doi.org/10.1016/j.livsci.2007.12.013

Christophersen C. 2007. Grain and Artificial Stimulation of the Rumen Change the Abundance and Diversity of Methanogens and Their Association with Ciliates. University of Western Australia.

Cobellis G, Trabalza-Marinucci M, Yu Z. 2016. Critical evaluation of essential oils as rumen modifiers in ruminant nutrition: A review. Science of The Total Environment 545:556–568. https://doi.org/10.1016/j.scitotenv.2015.12.103

Cox SD, Mann CM, Markham JL. 2001. Interactions between components of the essential oil of *Melaleuca alternifolia*. Journal of Applied Microbiology 91(3):492–497. https://doi.org/10.1046/j.1365-2672.2001.01406.x

Daning D, Yusiati L, Hanim C, Widyobroto B. 2022. Dietary supplementation of galangal (*Alpinia galangal*) essential oil affects rumen fermentation pattern. Advances in Animal and Veterinary Sciences 10(8):323. https://doi.org/10.17582/journal.aavs/2022/10.2.323.334

Davidson P, Naidu A. 2000. Phyto-phenols. Natural Food Antimicrobial Systems 226:284–287.

Dehsheikh AB, Sourestani MM, Dehsheikh PB, Mottaghipisheh J, Vitalini S, Iriti M. 2020. Monoterpenes: Essential oil components with valuable features. Mini Reviews in Medicinal Chemistry 20(11):958–974. https://doi.org/10.2174/1389557520666200122144703

Dey A, Paul SS, Lailer PC, Dahiya SS. 2021. Reducing enteric methane production from buffalo (*Bubalus bubalis*) by garlic oil supplementation in *in vitro* rumen fermentation system. SN Applied Sciences 3(187). https://doi.org/10.1007/s42452-021-04264-6

Di Consiglio M, Sturabotti E, Brugnoli B, Piozzi A, Migneco LM, Francolini I. 2023. Synthesis of sustainable eugenol/hydroxyethylmethacrylate-based polymers with antioxidant and antimicrobial properties. Polymer Chemistry 14(4):432–442.

Dijkstra J, Oenema O, Bannink A. 2011. Dietary strategies to reducing N excretion from cattle: Implications for methane emissions. Current Opinion in Environmental Sustainability 3(5):414–422. https://doi.org/10.1016/j.cosust.2011.07.008

Dorantes-Iturbide G, Orzuna-Orzuna JF, Lara-Bueno A, Mendoza-Martínez GD, Miranda-Romero LA, Lee-Rangel HA. 2022. Essential oils as a dietary additive for small ruminants: A meta-analysis on performance, rumen parameters, serum metabolites, and product quality. Veterinary Sciences 9(9):475. https://doi.org/10.3390/vetsci9090475

Dore S, Ferrini AM, Appicciafuoco B, Massaro MR, Sotgiu G, Liciardi M, Cannas EA. 2019. Efficacy of a terpinen-4-ol based dipping for post-milking teat disinfection in the prevention of mastitis in dairy sheep. Journal of Essential Oil Research 31(1):19–26. https://doi.org/10.1080/10412905.2018.1523069

Dorman HJD, Deans SG. 2000. Antimicrobial agents from plants: Antibacterial activity of plant volatile oils. Journal of Applied Microbiology 88(2):308–316. https://doi.org/10.1046/j.1365-2672.2000.00969

Eckard RJ, Grainger C, de Klein CAM. 2010. Options for the abatement of methane and nitrous oxide from ruminant production: A review. Livestock Science 130(1–3):47–56. https://doi.org/10.1016/j.livsci.2010.02.010

El-Essawy AM, Anele UY, Abdel-Wahed AM, Abdou AR, Khattab IM. 2021. Effects of anise, clove and thyme essential oils supplementation on rumen fermentation, blood metabolites, milk yield and milk composition in lactating goats. Animal Feed Science and Technology 271:114760. https://doi.org/10.1016/j.anifeedsci.2020.114760

El-Hefny M, Mohamed AA, Abdelkhalek A, Salem MZM. 2023. Productivity and phytochemicals of *Asclepias curassavica* in response to compost and silver nanoparticles application: HPLC analysis and antibacterial activity of extracts. Plants 12(12):2274. https://doi.org/10.3390/plants12122274

Elbanoby NE, El-Settawy AAA, Mohamed AA, Salem MZM. 2022. Phytochemicals derived from *Leucaena leucocephala* (Lam.) de Wit (Fabaceae) biomass and their antimicrobial and antioxidant activities: HPLC analysis of extracts. Biomass Conversion and Biorefinery. https://doi.org/10.1007/s13399-022-03420-1

Elghandour MMMY, Kanth Reddy PR, Salem AZM, Ranga Reddy PP, Hyder I, Barbabosa-Pliego A, Yasaswini D. 2018. Plant bioactives and extracts as feed additives in horse nutrition. Journal of Equine Veterinary Science 69:66–77. https://doi.org/10.1016/j.jevs.2018.06.004

Elufioye TO, Habtemariam S, Adejare A. 2020. Chemistry and pharmacology of alkylamides from natural origin. Revista Brasileira de Farmacognosia 30:622–640. https://doi.org/10.1007/s43450-020-00095-5

Evans JD, Martin SA. 2000. Effects of thymol on ruminal microorganisms. Current Microbiology 41:336–340. https://doi.org/10.1007/s002840010145

Fandiño I, Calsamiglia S, Ferret A, Blanch M. 2008. Anise and capsicum as alternatives to monensin to modify rumen fermentation in beef heifers fed a high concentrate diet. Animal Feed Science and Technology 145(1–4):409–417. https://doi.org/10.1016/j.anifeedsci.2007.04.018

Favaretto JA, Alba DF, Marchiori MS, Marcon HJ, Souza CF, Baldissera MD, Bianchi AE, Zanluchi M, Klein B, Wagner R, Vedovatto M, Da Silva AS. 2020. Supplementation with a blend based on micro-encapsulated carvacrol, thymol, and cinnamaldehyde in lambs feed inhibits immune cells and improves growth performance. Livestock Science 240:104144. https://doi.org/10.1016/j.livsci.2020.104144

Foggi G, Terranova M, Conte G, Mantino A, Amelchanka SL, Kreuzer M, Mele M. 2022. *In vitro* screening of the ruminal methane and ammonia mitigating potential of mixtures of either chestnut or quebracho tannins with blends of essential oils as feed additives. Italian Journal of Animal Science 21(1):1520–1532. https://doi.org/10.1080/1828051X.2022.2130832

Frutos P, Hervás G, Giráldez FJ, Mantecón AR. 2004. Review: Tannins and ruminant nutrition. Spanish Journal of Agricultural Research 2(2):191–202. https://doi.org/10.5424/sjar/2004022-73

Fuchs H, Bachran D, Panjideh H, Schellmann N, Weng A, Melzig MF, Sutherland M, Bachran C. 2009. Saponins as tool for improved targeted tumor therapies. Current Drug Targets 10(2):140–151. https://doi.org/10.2174/138945009787354584

Gallucci MN, Oliva M, Casero C, Dambolena J, Luna A, Zygadlo J, Demo M. 2009. Antimicrobial combined action of terpenes against the food-borne microorganisms *Escherichia coli*, *Staphylococcus aureus* and *Bacillus cereus*. Flavour and Fragrance Journal 24(6):348–354. https://doi.org/10.1002/ffj.1948

Gerber PJ, Hristov AN, Henderson B, Makkar H, Oh J, Lee C, Meinen R, Montes F, Ott T, Firkins J, Rotz A, Dell C, Adesogan AT, Yang WZ, Tricarico JM, Kebreab E, Waghorn G, Dijkstra J, Oosting S. 2013. Technical options for the mitigation of direct methane and nitrous oxide emissions from livestock: A review. Animal 7(s2):220–234. https://doi.org/10.1017/S1751731113000876

Giannenas I, Skoufos J, Giannakopoulos C, Wiemann M, Gortzi O, Lalas S, Kyriazakis I. 2011. Effects of essential oils on milk production, milk composition, and rumen microbiota in Chios dairy ewes. Journal of Dairy Science 94(11):5569–5577. https://doi.org/10.3168/jds.2010-4096

Goel G, Makkar HPS. 2012. Methane mitigation from ruminants using tannins and saponins. Tropical Animal Health and Production 44:729–739. https://doi.org/10.1007/s11250-011-9966-2

Goel G, Makkar HPS, Becker K. 2008. Effects of *Sesbania sesban* and *Carduus pycnocephalus* leaves and Fenugreek (*Trigonella foenum-graecum* L.) seeds and their extracts on partitioning of nutrients from roughage- and concentrate-based feeds to methane. Animal Feed Science and Technology 147(1–3):72–89. https://doi.org/10.1016/j.anifeedsci.2007.09.010

Guliani A, Pooja Verma M, Kumari A, Acharya A. 2021. Retaining the 'essence' of essential oil: Nanoemulsions of citral and carvone reduced oil loss and enhanced antibacterial efficacy via bacterial membrane perturbation. Journal of Drug Delivery Science and Technology 61:102243. https://doi.org/10.1016/j.jddst.2020.102243

Hamad YK, Abobakr Y, Salem MZM, Ali HM, Al-Sarar AS, Al-Zabib AA. 2019. Activity of plant extracts/essential oils against three plant pathogenic fungi and mosquito larvae: GC/MS analysis of bioactive compounds. BioResources 14(2):4489–4511.

Hassan A, Abu Hafsa SH, Elghandour MMY, Kanth Reddy PR, Salem MZM, Anele UY, Ranga Reddy PP, Salem AZM. 2020. Influence of *Corymbia citriodora* leaf extract on growth performance, ruminal fermentation, nutrient digestibility, plasma antioxidant activity and faecal bacteria in young calves. Animal Feed Science and Technology 261:114394. https://doi.org/10.1016/j.anifeedsci.2020.114394

Hassen A, Theart JJF, Adriaan van Niekerk W, Adejoro FA, Gemeda BS. 2016. *In vitro* methane and gas production characteristics of *Eragrostis trichopophora* substrate supplemented with different browse foliage. Animal Production Science 56(3):634–640. https://doi.org/10.1071/AN15612

Hayat J, Akodad M, Moumen A, Baghour M, Skalli A, Ezrari S, Belmalha S. 2020. Phytochemical screening, polyphenols, flavonoids and tannin content, antioxidant activities and FTIR characterization of *Marrubium vulgare* L. from 2 different localities of Northeast of Morocco. Heliyon 6:e05609. https://doi.org/10.1016/j.heliyon.2020.e05609

Helander IM, Alakomi HL, Latva-Kala K, Mattila-Sandholm T, Pol I, Smid EJ, Gorris LGM, von Wright A. 1998. Characterization of the action of selected essential oil components on gram-negative bacteria. Journal of Agricultural and Food Chemistry 46(9):3590–3595. https://doi.org/10.1021/jf980154m

Hristov AN, Lee C, Cassidy T, Heyler K, Tekippe JA, Varga GA, Corl B, Brandt RC. 2013. Effect of *Origanum vulgare* L. leaves on rumen fermentation, production, and milk fatty acid composition in lactating dairy cows. Journal of Dairy Science 96(2):1189–1202. https://doi.org/10.3168/jds.2012-5975

Hristov AN, McAllister TA, Ouellet DR, Broderick GA, 2005. Comparison of purines and nitrogen-15 as microbial flow markers in beef heifers fed barley- or corn-based diets. Canadian Journal of Animal Science 85(2):211–222. https://doi.org/10.4141/A04-05

Hui-Ling M, Jia-Kun W, Yi-Yi Z, Jian-Xin L. 2010. Effects of addition of tea saponins and soybean oil on methane production, fermentation and microbial population in the rumen of growing lambs. Livestock Science 129(1–3):56–62. https://doi.org/10.1016/j.livsci.2009.12.011

Iacovelli F, Romeo A, Lattanzio P, Ammendola S, Battistoni A, La Frazia S, Vindigni G, Unida V, Biocca S, Gaziano R. 2023. Deciphering the broad antimicrobial activity of *Melaleuca alternifolia*tea tree oil by combining experimental and computational investigations. International Journal of Molecular Sciences 24(15):12432. https://doi.org/10.3390/ijms241512432

Jahani-Azizabadi H, Danesh Mesgaran M, Vakili A, Rezayazdi K. 2014. Effect of some plant essential oils on *in vitro* ruminal methane production and on fermentation characteristics of a mid-forage diet. Journal of Agricultural Science and Technology 16(7):1543–1554.

Janini LCZ, Dellaqua TT, Membrive CMB, Oba E, Nichi M, Rizzoto G, da Cruz Landim-Alvarenga F. 2023. Effects of anethole supplementation on bovine embryo production and quality. Livestock Science 274:105262. https://doi.org/10.1016/j.livsci.2023.105262

Jayanegara A, Togtokhbayar N, Makkar HPS, Becker K. 2009. Tannins determined by various methods as predictors of methane production reduction potential of plants by an *in vitro* rumen fermentation system. Animal Feed Science and Technology 150(3–4):230–237. https://doi.org/10.1016/j.anifeedsci.2008.10.011

Jia-Ku W, Jun-An Y, Jian-Xin L. 2012. Effects of tea saponins on rumen microbiota, rumen fermentation, methane production and growth performance—A review. Tropical Animal Health and Production 44:697–706. https://doi.org/10.1007/s11250-011-9960-8

Kamboj A. 2012. Analytical evaluation of herbal drugs. Drug Discovery Research in Pharmacognosy 3:23–55.

Kamra DN, Agarwal N, Chaudhary LC. 2006. Inhibition of ruminal methanogenesis by tropical plants containing secondary compounds. International Congress Series 1293:156–163. https://doi.org/10.1016/j.ics.2006.02.002

Kamra DN, Singh R, Chaudhary L, Neeta A, Pathak N. 2000. Soapnut as natural defaunating agent: Its effect on rumen fermentation and in sacco degradability of jowar hay in buffaloes. Buffalo Journal 16(1):99–104.

Kaur V, Kaur R, Bhardwaj U. 2021. A review on dill essential oil and its chief compounds as natural biocide. Flavour and Fragrance Journal 36:412–431. https://doi.org/10.1002/ffj.3633

Khejornsart P, Cherdthong A, Wanapat M. 2021. *In vitro* screening of plant materials to reduce ruminal protozoal population and mitigate ammonia and methane emissions. Fermentation 7(3):166. https://doi.org/10.3390/fermentation7030166

Kholif AE, Olafadehan OA. 2021. Essential oils and phytogenic feed additives in ruminant diet: Chemistry, ruminal microbiota and fermentation, feed utilization and productive performance. Phytochemistry Reviews 20:1087–1108. https://doi.org/10.1007/s11101-021-09739-3

Kiprotich SS. 2019. Application of Thyme Oil or Cinnamaldehyde for the Inactivation of Enteric Pathogens in Marinated Raw Chicken and Apple Juice. United States—Iowa: Iowa State University, p. 107.

Kozłowska M, Cieślak A, Jóźwik A, El-Sherbiny M, Stochmal A, Oleszek W, Kowalczyk M, Filipiak W, Szumacher-Strabel M. 2020. The effect of total and individual alfalfa saponins on rumen methane production. Journal of the Science of Food and Agriculture 100(5):1835–2313.https://doi.org/10.1002/jsfa.10204

Kuralkar P, Kuralkar SV. 2021. Role of herbal products in animal production—An updated review. Journal of Ethnopharmacology 278:114246. https://doi.org/10.1016/j.jep.2021.114246

Ku-Vera JC, Jiménez-Ocampo R, Valencia-Salazar SS, Montoya-Flores MD, Molina-Botero IC, Arango J, Gómez-Bravo CA, Aguilar-Pérez CF, Solorio-Sánchez FJ. 2020. Role of secondary plant metabolites on enteric methane mitigation in ruminants. Frontiers in Veterinary Science 7:584. https://doi.org/10.3389/fvets.2020.00584

Leitanthem VK, Chaudhary P, Bhakat M, Mohini M, Mondal G. 2022. Impact of *Moringa oleifera* on rumen fermentation and methane emission under *in vitro* condition. AMB Express 12:141. https://doi.org/10.1186/s13568-022-01480-0

Li X. 2021. Plant cell wall chemistry: Implications for ruminant utilisation. Journal of Applied Animal Nutrition 9(1):31–56. https://doi.org/10.3920/JAAN2020.0017

Liechty Z, Santos-Medellín C, Edwards J, Nguyen B, Mikhail D, Eason S, Phillips G, Sundaresan V. 2020. Comparative analysis of root microbiomes of rice cultivars with high and low methane emissions reveals differences in abundance of methanogenic archaea and putative upstream fermenters. mSystems 5(1):e00897–19. https://doi.org/10.1128/msystems.00897-00819

Lin B, Lu Y, Salem AZM, Wang JH, Liang Q, Liu JX. 2013. Effects of essential oil combinations on sheep ruminal fermentation and digestibility of a diet with fumarate included. Animal Feed Science and Technology 184(1–4):24–32. https://doi.org/10.1016/j.anifeedsci.2013.05.011

Lu-Lu W, Chen Y, Zhi-Jian L, Li X., Fan G. 2022. Bioactive properties of the aromatic molecules of spearmint (*Mentha spicata* L.) essential oil: A review. Food & Function 13:3110–3132.

Mansour MM, Abdel-Megeed A, Nasser RA, Salem, MZM. 2015. Comparative evaluation of some woody tree methanolic extracts and Paraloid B-72 against phytopathogenic mold fungi *Alternaria tenuissima* and *Fusarium culmorum*. BioResources 10(2):2570–2584.

Marchese A, Barbieri R, Coppo E, Orhan IE, Daglia M, Nabavi SF, Izadi M, Abdollahi M, Nabavi SM, Ajami M. 2017. Antimicrobial activity of eugenol and essential oils containing eugenol: A mechanistic viewpoint. Critical Reviews in Microbiology 43(6):668–689. https://doi.org/10.1080/1040841X.2017.1295225

Marinelli L, Di Stefano A, Cacciatore I. 2018. Carvacrol and its derivatives as antibacterial agents. Phytochemistry Reviews 17:903–921. https://doi.org/10.1007/s11101-018-9569-x

Marques RdS, Cooke RF. 2021. Effects of ionophores on ruminal function of beef cattle. Animals 11(10):2871. https://doi.org/10.3390/ani11102871

Martin SA, Macy JM. 1985. Effects of monensin, pyromellitic diimide and 2-bromoethanesulfonic acid on rumen fermentation *in vitro*. Journal of Animal Science 60(2):544–550. https://doi.org/10.2527/jas1985.602544x

Maxiselly Y, Chiarawipa R, Somnuk K, Hamchara P, Cherdthong A, Suntara C, Prachumchai R, Chanjula P. 2022. Digestibility, blood parameters, rumen fermentation, hematology, and nitrogen balance of goats after receiving supplemental coffee cherry pulp as a source of phytochemical nutrients. Veterinary Sciences 9(10):532. https://doi.org/10.3390/vetsci9100532

Mohamed AA, Behiry SI, Younes HA, Ashmawy NA, Salem MZM, Márquez-Molina O, Barbabosa-Pilego A. 2019a. Antibacterial activity of three essential oils and some monoterpenes against *Ralstonia solanacearum* phylotype II isolated from potato. Microbial Pathogenesis 135:103604. https://doi.org/10.1016/j.micpath.2019.103604

Mohamed WA, Mansour MMA, Salem MZM. 2019b. *Lemna gibba* and *Eichhornia crassipes* extracts: Clean alternatives for deacidification, antioxidation and fungicidal treatment of historical paper. Journal of Cleaner Production 219:846–855. https://doi.org/10.1016/j.jclepro.2019.02.097

Naumann HD, Armstrong SA, Lambert BD, Muir JP, Tedeschi LO, Kothmann MM. 2014. Effect of molecular weight and concentration of legume condensed tannins on *in vitro* larval migration inhibition of *Haemonchus contortus*. Veterinary Parasitology 199(1–2):93–98. https://doi.org/10.1016/j.vetpar.2013.09.025

Nazzaro F, Fratianni F, De Martino L, Coppola R, De Feo V. 2013. Effect of essential oils on pathogenic bacteria. Pharmaceuticals 6(12):1451–1474. https://doi.org/10.3390/ph6121451

Newbold CJ, McIntosh FM, Williams P, Losa R, Wallace RJ. 2004. Effects of a specific blend of essential oil compounds on rumen fermentation. Animal Feed Science and Technology 114(1–4):105–112. https://doi.org/10.1016/j.anifeedsci.2003.12.006

Nikaido H. 1994. Prevention of drug access to bacterial targets: Permeability barriers and active efflux. Science 264(5157):382–388. https://doi.org/10.1126/ciencia.81536

O'Hara E. 2019. Investigating Early Life Microbial and Host Transcriptomic Dynamics in the Bovine Gastrointestinal Tract. University of Alberta.

Oh Hi K, Sakai T, Jones MB, Longhurst WM. 1967. Effect of various essential oils isolated from Douglas fir needles upon sheep and deer rumen microbial activity. Applied Microbiology 15(4):777–784. https://doi.org/10.1128/am.15.4.777-784.1967

Patel S. 2012. Yucca: A medicinally significant genus with manifold therapeutic attributes. Natural Products and Bioprospecting 2:231–234. https://doi.org/10.1007/s13659-012-0090-4

Patra AK, Kamra DN, Agarwal N. 2006. Effect of plant extracts on *in vitro* methanogenesis, enzyme activities and fermentation of feed in rumen liquor of buffalo. Animal Feed Science and Technology 128(3–4): 276–291. https://doi.org/10.1016/j.anifeedsci.2005.11.001

Patra AK, Kamra DN, Agarwal N. 2010. Effects of extracts of spices on rumen methanogenesis, enzyme activities and fermentation of feeds *in vitro*. Journal of the Science of Food and Agriculture 90:511–520. https://doi.org/10.1002/jsfa.3849

Patra AK, Saxena J. 2009a. Dietary phytochemicals as rumen modifiers: A review of the effects on microbial populations. Antonie van Leeuwenhoek 96(3):363–375. https://doi.org/10.1007/s10482-009-9364-1

Patra AK, Saxena J. 2009b. The effect and mode of action of saponins on the microbial populations and fermentation in the rumen and ruminant production. Nutrition Research Reviews 22(2):204–219. https://doi.org/10.1017/S0954422409990163

Patra AK, Saxena J. 2010. A new perspective on the use of plant secondary metabolites to inhibit methanogenesis in the rumen. Phytochemistry 71(11–12):1198–1222. https://doi.org/10.1016/j.phytochem.2010.05.010

Patra AK, Saxena J. 2011. Exploitation of dietary tannins to improve rumen metabolism and ruminant nutrition. Journal of the Science of Food and Agriculture 91(1):24–37. https://doi.org/10.1002/jsfa.4152

Patra AK, Yu Z. 2012. Effects of essential oils on methane production and fermentation by, and abundance and diversity of, rumen microbial populations. Applied and Environmental Microbiology 78(12):4271–4280. https://doi.org/10.1128/AEM.00309-12

Patra AK, Yu Z. 2013. Effective reduction of enteric methane production by a combination of nitrate and saponin without adverse effect on feed degradability, fermentation, or bacterial and archaeal communities of the rumen. Bioresource Technology 148:352–360. https://doi.org/10.1016/j.biortech.2013.08.140

Patra AK, Yu Z. 2014. Effects of vanillin, quillaja saponin, and essential oils on *in vitro* fermentation and protein-degrading microorganisms of the rumen. Applied Microbiology and Biotechnology 98:897–905. https://doi.org/10.1007/s00253-013-4930-x

Pawar MM, Kamra DN, Agarwal N, Chaudhary LC. 2014. Effects of essential oils on *in vitro* methanogenesis and feed fermentation with buffalo rumen liquor. Agricultural Research 3:67–74. https://doi.org/10.1007/s40003-014-0092-z

Pisoschi AM, Pop A, Georgescu C, Turcuş V, Olah NK, Mathe E. 2018. An overview of natural antimicrobials role in food. European Journal of Medicinal Chemistry 143:922–935. https://doi.org/10.1016/j.ejmech.2017.11.095

Ramírez-Restrepo CA, Tan C, ONeill CJ, López-Villalobos N, Padmanabha J, Wang J, McSweeney CS. 2016. Methane production, fermentation characteristics, and microbial profiles in the rumen of tropical cattle fed tea seed saponin supplementation. Animal Feed Science and Technology 216:58–67. https://doi.org/10.1016/j.anifeedsci.2016.03.005

Ramos-Morales E, De La Fuente G, Duval S, Wehrli C, Bouillon M, Lahmann M, Preskett D, Braganca R, Newbold CJ. 2017. Antiprotozoal effect of saponins in the rumen can be enhanced by chemical modifications in their structure. Frontiers in Microbiology 8:399. https://doi.org/10.3389/fmicb.2017.00399

Reddy PRK, Elghandour MMMY, Salem AZM, Yasaswini D, Reddy PPR, Reddy AN, Hyder I, 2020. Plant secondary metabolites as feed additives in calves for antimicrobial stewardship. Animal Feed Science and Technology 264:114469. https://doi.org/10.1016/j.anifeedsci.2020.114469

Saber J, Houcine S, Foued A, Ameni D, Houcem S, Chaabane A, Hichem S. 2022. Antioxidant and antibacterial properties of *Salvia officinalis* essential oils and their effects on *in vitro* feed fermentation with goat rumen liquor. Animal Nutrition and Feed Technology 22(2):297–310. https://doi.org/10.1016/j.anifeedsci.2020.114469

Salem AZM, Kunst CR, Jose S. 2020. Alternative animal feeds from agroforestry plants. Agroforestry Systems 94:1133–1138. https://doi.org/10.1007/s10457-020-00525-2

Salem MZM, Ali HM, El-Shanhorey NA, Abdel-Megeed A. 2013. Evaluation of extracts and essential oil from *Callistemon viminalis* leaves: Antibacterial and antioxidant activities, total phenolic and flavonoid contents. Asian Pacific Journal of Tropical Medicine 6(10):785–791. https://doi.org/10.1016/S1995-7645(13)60139-X

Salem MZM, Ali HM, Mansour MM. 2014b. Fatty acid methyl esters from air-dried wood, bark, and leaves of *Brachychiton diversifolius* R. Br: Antibacterial, antifungal, and antioxidant activities. BioResources 9(3):3835–3845.

Salem MZM, El-Hefny M, Ali HM, Abdel-Megeed A, El-Settawy, AAA, Böhm M, Mansour MMA, Salem AZM. 2021. Plants-derived bioactives: Novel utilization as antimicrobial, antioxidant and phytoreducing agents for the biosynthesis of metallic nanoparticles. Microbial Pathogenesis 158:105107. https://doi.org/10.1016/j.micpath.2021.105107

Salem MZM, El-Hefny M, Ali HM, Elansary HO, Nasser RA, El-Settawy AAA, El Shanhorey N, Ashmawy NA, Salem AZM. 2018. Antibacterial activity of extracted bioactive molecules of *Schinus terebinthifolius* ripened fruits against some pathogenic bacteria. Microbial Pathogenesis 120:119–127. https://doi.org/10.1016/j.micpath.2018.04.040

Salem MZM, Olivares-Pérez J, Salem AZM. 2014a. Studies on biological activities and phytochemicals composition of *Hibiscus* species-A review. Life Science Journal 11(5):1–8.

Sasidharan S, Chen Y, Saravanan D, Sundram KM, Yoga Latha L. 2011. Extraction, isolation and characterization of bioactive compounds from plants' extracts. African Journal of Traditional, Complementary and Alternative Medicines 8(1):1–10. https://doi.org/10.4314/ajtcam.v8i1.60483

Shakam HM, Mohamed AA, Salem MZM. 2022. Down-regulatory effect of essential oils on fungal growth and Tri4 gene expression for some *Fusarium oxysporum* strains: GC-MS analysis of essential oils. Archives of Phytopathology and Plant Protection 55(8):951–972. https://doi.org/10.1080/03235408.2022.2064081

Sivropoulou A, Papanikolaou E, Nikolaou C, Kokkini S, Lanaras T, Arsenakis M. 1996. Antimicrobial and cytotoxic activities of origanum essential oils. Journal of Agricultural and Food Chemistry 44(5):1202–1205. https://doi.org/10.1021/jf950540t

Śliwiński BJ, Kreuzer M, Wettstein HR, Machmüller A. 2002a. Rumen fermentation and nitrogen balance of lambs fed diets containing plant extracts rich in tannins and saponins, and associated emissions of nitrogen and methane. Archiv für Tierernaehrung 56(6):379–392. https://doi.org/10.1080/00039420215633

Śliwiński BJ, Soliva CR, Machmüller A, Kreuzer M. 2002b. Efficacy of plant extracts rich in secondary constituents to modify rumen fermentation. Animal Feed Science and Technology 101(1–4):101–114. https://doi.org/10.1016/S0377-8401(02)00139-6

Taha AS, Abo-Elgat WAA, Fares YGD, Salem MZM. 2024. Isolated essential oils as antifungal compounds for organic materials. Biomass Conversion and Biorefinery 14(3):3853–3873. https://doi.org/10.1007/s13399-022-02815-4

Tekippe JA, Hristov AN, Heyler KS, Cassidy TW, Zheljazkov VD, Ferreira JFS, Karnati SK, Varga, GA. 2011. Rumen fermentation and production effects of *Origanum vulgare* L. leaves in lactating dairy cows. Journal of Dairy Science 94(10):5065–5079. https://doi.org/10.3168/jds.2010-4095

Ulanowska M, Olas B. 2021. Biological properties and prospects for the application of eugenol—A review. International Journal of Molecular Sciences 22(7):3671. https://doi.org/10.3390/ijms22073671

Ultee A, Bennik MHJ, Moezelaar R. 2002. The phenolic hydroxyl group of carvacrol is essential for action against the food-borne pathogen *Bacillus cereus*. Applied and Environmental Microbiology 68(4):1561–1568. https://doi.org/10.1128/AEM.68.4.1561-1568.2002

Waghorn GC, McNabb WC. 2003. Consequences of plant phenolic compounds for productivity and health of ruminants. Proceedings of the Nutrition Society 62(2):383–392. https://doi.org/10.1079/PNS2003245

Wallace RJ, Arthaud L, Newbold CJ. 1994. Influence of *Yucca shidigera* extract on ruminal ammonia concentrations and ruminal microorganisms. Applied and Environmental Microbiology 60(6):1762–1767. https://doi.org/10.1128/aem.60.6.1762-1767.1994

Wina E, Muetzel S, Becker K. 2005. The impact of saponins or saponin-containing plant materials on ruminant production—A Review. Journal of Agricultural and Food Chemistry 53(21):8093–8105. https://doi.org/10.1021/jf048053d

Yang WZ, Ametaj BN, Benchaar C, Beauchemin KA. 2010a. Dose response to cinnamaldehyde supplementation in growing beef heifers: Ruminal and intestinal digestion. Journal of Animal Science 88(2):680–688. https://doi.org/10.2527/jas.2008-1652

Yang WZ, Ametaj BN, Benchaar C, He ML, Beauchemin KA. 2010b. Cinnamaldehyde in feedlot cattle diets: Intake, growth performance, carcass characteristics, and blood metabolites. Journal of Animal Science 88(3):1082–1092. https://doi.org/10.2527/jas.2008-1608

Yoshida T, Hatano T, Ito H. 2005. High molecular weight plant polyphenols. Chemical Ecology and Phytochemistry of Forest Ecosystems: Proceedings of the Phytochemical Society of North America 39, 163.

Załuski D, Cieśla Ł, Janeczko Z. 2015. Chapter 7—The structure–Activity relationships of plant secondary metabolites with antimicrobial, free radical scavenging and inhibitory activity toward selected enzymes, in: Atta ur, R. (Ed.), Studies in Natural Products Chemistry. Elsevier, pp. 217–249.

Zúñiga-Serrano A, Barrios-García HB, Anderson RC, Hume ME, Ruiz-Albarrán, M, Bautista-Martínez Y, Sánchez-Guerra NA, Vázquez-Villanueva J, Infante-Rodríguez F, Salinas-Chavira J. 2022. Antimicrobial and digestive effects of *Yucca schidigera* extracts related to production and environment implications of ruminant and non-ruminant animals: A review. Agriculture 12(8):1198. https://doi.org/10.3390/agriculture12081198

19 Isolation Techniques of the Phytocompounds of Medicinal Plants

*Dipak D. Gadade, Vimal Kumar, Ekta Gurnany, Sourabh Jain, Aakanchha Jain, and Anuj Modi**
* Corresponding author: pharmaanuj@gmail.com

19.1 INTRODUCTION

The exploration of natural products, like extracts from plants, opens a new horizon for finding new therapeutic agents. The utilization of medicine and medicinal plants is prevalent in developing nations serving as standard practice to maintain good health. It is worth noting that 80% of the population relies on herbal remedies. Plants possess an array of phytoconstituents that hold potential for treating a variety of diseases (Sasidharan et al., 2011). *Phytoconstituents* are defined as chemical compounds that occur naturally in plants, where the word *phyto* means "plant" in Greek. Plant extracts are intricate mixtures that encompass hundreds or thousands of different constituents, which makes the isolation of single phytoconstituents more difficult (Liu, 2004). Due to the complex nature of chemical components within plants, it is essential to first extract and isolate pure phytochemicals before proceeding with tasks such as the identification of the chemical structures and the determination of the biological activity of these components. The initial phase involves extraction as a prerequisite for isolating the active constituents. This extraction process aims at the separation and isolation of the desired natural components from the plant material to the utmost extent, minimizing the presence of undesired elements in the solution (Feng et al., 2019). The extraction process for natural products involves several critical steps, including "(a) penetration of the solvent into the solid matrix; (b) dissolution of the solute in the solvent; (c) diffusion of the solute out of the solid matrix; (d) collecting of the extracted solutes" (Zhang et al., 2018).

According to the principle of extraction methods, these are classified as (1) solvent extraction method, (2) distillation method, (3) pressing method, and (4) sublimation method. Among them, the first method is preferred by researchers, and the selection of solvent is pivotal for the extraction. Consider factors like selectivity, solubility, cost, and safety. Similarity and miscibility principles guide solvent selection, favoring those with matching polarities for efficient dissolution. Factors improving diffusivity and solubility enhance extraction. Key considerations include solvent characteristics, particle size of material, solute–solvent ratio, temperature, and time extent, all impacting extraction efficacy (Table 19.1) (Li et al., 2008, 2014; Yi et al., 2012; Zhou et al., 2012; Zhang et al., 2018).

This chapter primarily emphasizes the diverse methodologies encompassing extraction techniques and the isolation of phytoconstituents from plant extracts, achieved through a range of chromatographic applications.

19.2 EXTRACTION OF PHYTOCHEMICALS

This section outlines the prevalent methods in the extraction of phytochemicals. An overview of the various extraction methods used for natural products is shown in Table 19.2.

DOI: 10.1201/9781003326939-23

TABLE 19.1
Factors Affecting Extraction Efficiency (EE)

Key Aspect	Information
Solvent selection	• Polar solvents (e.g., methanol, water) for polar compound extraction. • Nonpolar solvents (e.g., hexane, dichloromethane) for nonpolar compound extraction. • Alcohols (EtOH, MeOH) are versatile solvents for phytochemical extraction.
Particle size	• A finer particle size enhances solvent penetration, and solute diffusion yields better extraction results. • Excessively fine particles lead to solute absorption and filtration challenges.
Temperature	• Higher temperatures increase solubility and diffusion. • Excessive heat can lead to solvent loss, generation of undesired products/by-products, and degradation of sensitive compounds.
Extraction duration	• Enhances within a certain time range. • After solute equilibrium is established, increasing the extraction time does not alter the extraction process.
Solvent-to-solid ratio	• An increased ratio leads to greater extraction yield. • Excessive higher solvent-to-solid ratio requires longer concentration time and results in excess solvent usage.
Fractionation solvent sequence	• Sequence based on increasing polarity: n-C_6H_6(nonpolar) to H_2O (most polar). • Typical solvent approach: low-polarity (n-C_6H_{14}, $CHCl_3$), medium-polarity (CH_2Cl_2, n-C_4H_9OH), and high-polarity (H_2O)

Source: Zhang et al. (2018); Abubakar and Haque (2020).

19.2.1 Maceration Method

Maceration involves immersing a plant material (coarse powder or pieces)ins ide a container in an appropriate solvent and allowing it to stand at RT for at least 3 days. Following a three-day period, the mixture is either pressed or strained by filtration. Continuous stirring or intermittent shaking during maceration enhances the dissolution rate of the target compounds into the solvent. This is a simple but slow extraction method, relatively less efficient compared to more modern extraction methods, resulting in a lower extraction ratio, yet appropriate for thermolabile components. Additionally, for aqueous extracts, there is a risk of mold formation, necessitating the use of suitable preservatives to maintain the extract's quality (Feng et al., 2019; Abubakar & Haque, 2020; Tzanova et al., 2020).

19.2.2 Percolation Method

The process begins by moistening finely powdered, dried, and ground plant material with an extraction solvent in a clean container. Additional solvent is introduced, and the blend is allowed to stand for 4 hours. Subsequently, the mixture is placed in a percolator and lower-end-sealed, then further allowed to stand for 24 hrs. Solvent is added to the percolator until saturation, and then the lower end is opened, allowing the slow dripping of liquid. The continuous addition of solvent facilitates extraction via gravitational force, pushing it through the material to be extracted. The solvent addition is ceased when the solvent volume reaches 3/4 of the required quantity. Then the extract is separated through a process of filtration and decantation, while the remaining marc is expressed, and further solvent may be added as required to reach the desired final volume. This process typically proceeds at a moderate rate, often around six drops/min, till extraction is accomplished. Percolation is more efficient than maceration due to its continuous solvent replacement. This approach offers

TABLE 19.2
Overview of Diverse Techniques for Extracting Natural Products

Methods	Maceration	Percolation	Decoction	Reflux Extraction	Soxhlet Extraction	Supercritical Fluid (SF) Extraction	Ultrasound-Assisted Extraction	Microwave-Assisted Extraction
Solvent	H_2O, aqueous (aq.) and non-aq.	H_2O, aq. and non-aq.	H_2O	Aq. and non-aq.	Organic solvents	SF (usually Sc-CO_2), sometimes with a modifier	H_2O, aq. and non-aq.	H_2O, aq. and non-aq.
Temperature	Room temperature (RT)	RT	At elevated temperature	At elevated temperature	At elevated temperature	Near RT	RT	RT
Pressure	Atmospheric	Atmospheric	Atmospheric	Atmospheric	Atmospheric	High	Atmospheric	Atmospheric
Sample size	--	--	--	--	--	1–5 g	1–30 g	1–10 g
Extraction Time	2–7 days	1–3 days	30–60 mins	6–8 hrs	6–8 hrs	10–60 mins	10–90 mins	3–30 mins
Volume of solvent consumed	H (high)	H	None (N)	M (moderate)	M	H	M	N/S (small)
Cost	L (low)	L	L	L	M	H	L	M
Polarity of natural products extracted	Solvent-dependent (SD)	SD	Polar compounds	SD	SD	Nonpolar to moderately polar compounds	SD	SD
Advantage	Simple method; no thermal degradation up on extraction	Simple method suitable for heat-sensitive compounds	Simple method suitable for the extraction of water-soluble compounds	Simple method; no further filtration required; ability to recover solvent	Simple method; no filtration needed after extraction	Eco-friendly method and can be used with thermolabile substances	Easy-to-use, eco-friendly method suitable for thermolabile substances	Simple method, short extraction time, and low volume of solvent required
Disadvantage	Slow and time-consuming process; high volume of solvent used	High solvent consumption, long extraction time	Not suitable for extraction of heat-sensitive compounds	Slow and time-consuming process; high volume of solvent used; thermal degradation of compounds	Slow process; risk of thermal destruction of compounds; required large amounts of solvent increase cost	High pressure required, and many parameters to optimize	Additional filtration required; limited extraction volume and yield	Moderate cost of microwave setup and limited extraction volume

Source: Zhang et al. (2018); Suwal and Marciniak (2018); Mihaylova and Lante (2019); Dobgima et al. (2020).

controlled extraction and the ability to extract diverse compounds, yet it might not match the efficiency of modern techniques employing higher pressures and shorter times for extraction (Feng et al., 2019; Abubakar & Haque, 2020; Wang et al., 2020).

19.2.3 Decoction Method

This method involves extraction using water as a solvent. First, finely powdered dry plant material is introduced to the vessel, following which water is added, then the mixture is heated under specified conditions. Decoction is typically carried out at atmospheric pressure and involves moderate heat. The process, as typical, lasts for a brief duration, typically around 15 mins. The usual proportion of crude plant powder–solvent is either 1:4 or 1:16. After the extraction process, the resulting liquid, known as the decoction, is separated from the plant material using filtration or decantation. Decoction is not suitable for extracting thermolabile or volatile components (Feng et al., 2019; Abubakar & Haque, 2020).

19.2.4 Reflux Extraction Method

The reflux extraction (RE) method is based on the principle of repeated evaporation and condensation of the solvent over a specified period within a closed system. Heat is applied to the mixture of drug and solvent in a round bottom flask, causing the solvent to vaporize and rise through a reflux condenser. Upon reaching the condenser, the vapor is cooled and condensed back into liquid form, ensuring a continuous cycling process. The cyclic process continues for a specified duration, allowing the solvent to extract target compounds from the sample. This system finds extensive application in herbal industries due to its competence, convenient operation, and worthiness. RE includes its time-consuming nature, need for specialized equipment, relatively more solvent volume requirement, and increased energy consumption as some newer alternatives (Feng et al., 2019; Abubakar & Haque, 2020; Tzanova et al., 2020; Ma et al., 2022).

19.2.5 Soxhlet Extraction Method

In the extraction chamber of a Soxhlet, finely powdered, dried, and ground plant material is enclosed within a porous bag or thimble made from muslin cloth or robust filtration support. Then the extracting solvent is poured and heated in the round-bottom flask, causing it to evaporate. The vapor then ascends through the condenser, where it condenses back into liquid form. This condensed solvent flows down to the extraction chamber, coming into contact with the plant material in the thimble. As a result, the solvent extracts the desired compounds from the plant material. As the solvent reaches the siphon's top, the solvent and the extracted plant material return to the bottom of the flask due to gravity. This continuous cycling occurs until the extracted compounds have been thoroughly collected, as indicated by the absence of residue in the flowing solvent. This method is well-suited for plant material with partial solubility in the selected solvent or containing insoluble impurities but is unsuitable for thermolabile plant materials. Soxhlet extraction offers automated and continuous extraction, boasting heightened efficiency compared to maceration or percolation. It demands less time and solvent, eliminating the need for filtration. Nevertheless, prolonged extraction at high temperatures in the Soxhlet method can elevate the risk of thermal degradation (Słominska et al., 2012; Feng et al., 2019; Abubakar & Haque, 2020; Tzanova et al., 2020).

19.2.6 Supercritical Fluid Extraction (SFE) Method

SFE employs SF as a solvent for extraction. SFs are highly compressed gases that combine the properties of both liquids and gases in an intriguing manner. Their solvating properties can be fine-tuned with slight pressure and temperature adjustments near their critical points. Supercritical carbon dioxide (Sc-CO_2) is favored as a solvent due to its low critical temperature (31°C), selectivity, safety, ability to extract heat-sensitive compounds, and economic nature. Its low polarity makes it more suitable for the extraction of hydrophobic compounds, like lipids and volatile oils, but can also handle polar substances with the addition of a polar modifier to Sc-CO_2. SFE can seamlessly integrate with SF chromatography, allowing automated high-throughput extraction and separation processes and enabling the analysis of compounds in darkness and under anoxic conditions. In contrast to other methods, the extraction process leaves no residual solvent (Herrero et al., 2010; Feng et al., 2019; Abubakar & Haque, 2020; Tzanova et al., 2020).

19.2.7 Microwave-Assisted Extraction (MAE)

MAE uses microwave radiation to heat the sample, promoting solvent movement into the drug material for faster extraction. MAE relies on the mechanism of dipole rotation and ionic transfer, utilizes electromagnetic radiation from the 300 MHz to 300 GHz frequency range, and optimal wavelengths. It operates at about 2,450 Hz frequency and produce energy around 600–700 W. With polar solvent, it activates dipole rotation and ion migration, enhancing solvent penetration and assisting in the extraction process. However, in the case of nonpolar solvents, microwave radiation generates minimal heat, making this method less conducive to use of nonpolar solvents. MAE finds application in the preparation of medicinal plants, the extraction of specific compounds, such as phenolics and flavonoids, while its suitability for other substances may vary. MAE offers distinct benefits, including reduced solvent usage and shorter extraction time, resulting in higher yield and is environmentally friendly. Nevertheless, compounds like tannins and anthocyanins may undergo degradation due to the high temperatures involved (Eskilsson & Björklund, 2000; Kaufmann & Christen, 2002; Feng et al., 2019; Abubakar & Haque, 2020; Tzanova et al., 2020; Bagade & Patil, 2021).

19.2.8 Ultrasound-Assisted Extraction Method

Ultrasound-assisted extraction (UAE) is a technique that employs high-frequency sound waves (greater than 20 KHz) to disrupt the cell walls of plant material or samples loaded into an ultrasonic extractor. This disruption increases the surface area of the drug or compound, facilitating better penetration of the extraction solvent. Consequently, secondary metabolites are released more efficiently. In UAE, thoroughly dried plant material is finely ground and appropriately sieved. The sample is subsequently combined with a suitable extraction solvent and added to the extractor. The use of high-frequency sound energy speeds up the extraction process, reducing the necessity for heat. However, its selectivity and equipment requirements should be considered when choosing this extraction technique. UAE significantly reduces the extraction time, solvent requirement, risk of thermal degradation of sensitive compounds, and maximizes yield. Like other methods, UAE is not applicable to all types of samples or compounds, and excessive energy can generate free radicals, potentially degrading phytochemicals (Esclapez et al., 2011; Feng et al., 2019; Abubakar & Haque, 2020; Tzanova et al., 2020).

19.3 SEPARATION AND PURIFICATION OF PHYTOCHEMICALS

Components present within extract from plants are complex and encompass a diverse range of phytochemicals. To isolate the pure phytochemical from the extract, additional separation and purification steps are necessary. Phytochemical separation involves isolating individual phytoconstituents from extracts and isolating them into single components based on their physicochemical properties. Classical methods, like solvent extraction, precipitation, crystallization, distillation, and dialysis, are still in use. Meanwhile, modern chromatographic technologies are essential for phytochemical separation. This section outlines the prevalent methods in isolation of methods.

19.3.1 Classical Isolation Methods

19.3.1.1 Solvent Method

The solvent-based separation methods for plant extracts can be categorized into two main techniques.

19.3.1.1.1 Acid and Basic Solvent Method

This method relies on the acidity and alkalinity levels in the components of the mixture. It uses the principle that water does not dissolve neutral molecules but dissolves ionic salts. By controlling acidity, contact time, heating conditions, and duration, this method prevents irreversible changes in chemical compounds. Acid-based extraction is frequently used for the separation of chemicals and natural products like alkaloids from crude extracts, compounds having lactone or lactam substructures, etc. Water-insoluble alkaloids can form salts with inorganic acids, allowing for the separation of non-alkaline components that are soluble in water. Conversely, bases can convert acid components containing carboxyl or phenolic hydroxyl groups into water-soluble salts. Compounds with lactone or lactam substructures can also become water-soluble through saponification. The process involves extracting the mixture with lipophilic organic solvents and acidified or alkaline water, dividing it into acidic, alkaline, and neutral components (Feng et al., 2019).

19.3.1.1.2 Polarity Gradient Extraction Method

This technique separates plant extract components based on their varying polarities and partition coefficients in two-phase solvents. The choice of solvent system depends on component polarity. For instance, n-butanol and water are used for highly polar compounds, ethyl acetate and water for moderately polar ones, and chloroform (or ether) and water for less-polar compounds. The plant extract is first dissolved in water and then processed in a separating funnel with immiscible organic solvents. The layers obtained contain different types of molecules: low-polarity compounds in the petroleum ether layer, moderately polar substances in the ethyl acetate layer, strongly polar substances in the n-butanol layer, and highly polar chemicals in the water layer, including glycosides, carbohydrates, amino acids, proteins, and other water-soluble substances (Feng et al., 2019).

19.3.1.2 Precipitation Method

The precipitation method involves the formation of specific phytochemical precipitates by interacting with reagents or altering solution conditions to reduce solubility. For this method to work effectively, it should be reversible when the desired components need to be isolated. According to the addition of solvent or reagent, it is classified as:

19.3.1.2.1 Solvent Precipitation Method

In mixed solutions, changing solvent polarity can be induced when a specific solvent, mutually soluble with the solution, is introduced, leading to the precipitation of certain components. For example, in the extraction of phytochemicals using water as the solvent, adding ethanol increases the alcohol content to over 80%. This causes proteins, polysaccharides, starch, cellulose, and other chemicals to precipitate and be removed through filtration. Terpenoids, carbohydrates, and alkaloid derivatives are only a few of the phytochemicals purified by ethanol-based precipitation in traditional Chinese medicine (Tai et al., 2020). The alkaline ethanol precipitation procedure is explored as an example for removing tannins derived from *Salvia miltiorrhiza* and *Radix et Rhizoma* for the manufacturing of Danshen injection. It is possible to remove more than 90% of the tannins by the precipitation method. The recoveries of danshensu, rosmarinic acid, and salvianolic acid B, on the other hand, were less than 60% (Gong et al., 2014).

Elda et al. employed the solvent-induced precipitation technique to separate azadirachtin and other limonoids from neem oil. Hexane, chosen for its low polarity, was used as the solvent due to its ability to precipitate the more polar limonoids from the neem oil. The researchers investigated various factors, including temperature and the solvent-to-oil ratio, to determine the optimal precipitation conditions. Their findings indicated that the hexane-induced precipitation method, particularly during the pre-concentration stage, effectively separated azadirachtin and other limonoids from neem oils (Melwita & Ju, 2010).

19.3.1.2.2 Exclusive Reagent Precipitation Method

In this method, selective reagents can interact with specific chemical components, leading to reversible precipitation. The choice of precipitation reagent depends on the properties of the target constituents and impurities found in plants. For instance, alkaloid precipitation reagents like Reynolds ammonium salt can selectively precipitate alkaloids, enabling the separation of alkaloids from non-alkaloids or water-soluble alkaloids from other alkaloids. Additionally, gelatin can be employed to precipitate tannins, facilitating their separation or removal (Feng et al., 2019).

19.3.1.3 Dialysis Method

Dialysis method employs selectively permeable membranes to separate small molecules from larger ones, permitting only the smaller molecules to pass through. It is commonly used to adjust the salt (small molecule) concentration in a solution containing macromolecules (Wang et al., 2011). In order to screen and analyze the bioactive chemicals, Guo's team used dialysis in conjunction with HPLC-MS to study the interactions of multicomponents in herbal medicine with protein, DNA, and MCF-7 cells (Guo et al., 2006).

19.3.1.4 Fractional Distillation Method

To separate miscible liquids that are naturally volatile, fractional distillation is performed. These liquids' boiling points are rather near. To mimic the separation, a piece of fractionating column equipment is employed. Since the vapor is partially condensed and then returned as a liquid, it is sometimes referred to as rectification. Essentially, it is a process where a liquid combination is evaporated to produce a mixture of elements, which is then followed by the extraction of the desired component in its purest form (Ali et al., 2019). For instance, the two alkaloids coniine and conhydrine, which make up the total alkaloids of *Cicutavirosa*, have very different boiling points—166–167°C for the former and 226°C for the latter—and may be distinguished from one another using the fractional distillation method (Altemimi et al., 2017).

19.3.1.5 Crystallization Method

When a pure crystalline material with certain particle qualities needs to be extracted from a multicomponent process solution, crystallization is frequently employed as a purification, concentration, and separation procedure. Organic compound crystallization is often followed by a synthesis, extraction, or bioprocess. Controlling crystal shape, crystal size distribution, and polymorphism is the biggest hurdle to commercial crystallization (Yu et al., 2007).

The purifying process known as crystallization comes the closest to being optimal. It can be done on a range from a few milligrams up to hundreds of kilograms, is operationally straightforward, and is very inexpensive. The results are often exceedingly pure (unlike the mixes one may occasionally acquire with distillation).

19.3.2 Chromatographic Methods

Chromatography stands as the predominant method for separating chemical constituents in natural products, owing to its exceptional attributes: high separation efficiency, rapidity, and user-friendly nature. While it is not feasible to cover every aspect of chromatography in this chapter, there are specific areas that merit discussion, given as follows.

19.3.2.1 High-Performance Thin-Layer Chromatography (HPTLC)

Based on TLC, HPTLC is an analytical technique that has been improved to allow for the quantitative analysis of compounds and to improve the resolution of the compounds that need to be separated. Some of these improvements allow for increased resolution, such as the use of higher-quality TLC plates with smaller particle sizes in the stationary phase (Sudberg et al., 2010). The most popular methods for creating reference fingerprints of herbs, which may be used to assess completed products and analyze raw materials, are HPLC, HPTLC, and capillary electrophoresis. HPTLC, commonly referred to as planar chromatography, is a cutting-edge method that outperforms traditional TLC in terms of separation power and repeatability (Gujrati et al., 2021). In today's analytical environment, HPTLC is significant, not as a substitute for HLPC, but as a complementing technique. Because of its simplicity, adaptability, and dependability, the fingerprint analysis methodology employing HPTLC has emerged as the most effective method in ensuring the quality of herbal medicines. It may be used as a tool for the authenticity, quality assurance, and identification of herbal medications (Kamath & Salvi, 2023).

19.3.2.2 High-Performance Liquid Chromatography (HPLC)

HPLC may offer both qualitative and quantitative data in a single operation. HPLC is now the most widely used technology for phytochemical separation. The most popular analytical separation method for determining the composition of natural product extracts, fractions, and finished goods is HPLC (Govindarajan et al., 2019). HPLC is used by multiple researchers to analyze and measure secondary metabolites in plant extracts, primarily phenol chemicals, steroids, flavonoids, and alkaloids. Due to its numerous applications, reverse-phase chromatography is the

TABLE 19.3
Application of HPTLC in Isolation of Phytochemicals

S. No.	Plant Name	Isolate	Mobile Phase	Detector	References
1.	*Gymnosporia senegalensis*	β-carotene	*n*-hexane–ethyl acetate (6:4, *v/v*)	UV at 254 nm	(Jain et al., 2023)
2.	*Oroxylum indicum*	**Baicalin**	Acetone–ethyl acetate–water–formic acid (2:10:0.5:0.5, *v/v*)	UV at 318 nm	(Adin et al., 2022)
3.	*Mangifera indica*	Mangiferin	Acetone–chloroform–formic acid–water (2:10:0.5:1.0, *v/v*)	UV at 350 nm	(Adin et al., 2023)
4.	*Hypericum perforatum*	Hyperforin, hypericin, hyperoside	*n*-hexane-ethyl acetate (8:2) (for hyperforin); toluene–chloroform–ethyl acetate–formic acid (8:5:3.5:0.6) (for hypericin); ethyl acetate–formic acid–acetic acid–water (15:2:2:1) (for hyperoside)		(Saçıcı & Yesilada, 2022)
5.	*Trikatu* and *Sitopaladi* (Ayurvedic formulation)	Piperine	Toluene–ethyl acetate (7:3, *v/v*)	342 nm	(Hazra et al., 2019)
6.	*Hygrophila schulli*	Betulin	Toluene–methanol (8:1, *v/v*)	366 nm	(Takale et al., 2023)
7.	*Withania somnifera*	Withanolide	Dichloromethane–methanol–toluene–acetone (5:1:1:0.5 *v/v*)	365 nm	(Ahmad et al., 2023)
8.	*Guiera senegalensis*	Rutin and quercetin	Acetonitrile–water (4:6 *v/v*)	360 nm	(Alam et al., 2017)
9.	*Uraria picta*	β-sitosterol	Toluene–methanol–chloroform (8:1:1, *v/v/v*)	525nm	(Saxena et al., 2023)
10.	Jatyadi taila (polyherbal formulation)	Curcumin	Toluene–ethyl acetate–formic acid (6:2:1, *v/v/v*)	366 nm	(Patel et al., 2023)

HPLC (RP-HPLC) separation method that is most frequently utilized. According to Li et al., the RP-HPLC mode is used for more than 65% of all HPLC separations (Li et al., 2020). To isolate and identify 32 aconitum lipo-alkaloids from three plants of the genus *Aconitum*, Liu et al. devised an HPLC/ESI-MS/MS technique (Liu et al., 2010). Alali et al. (2008) investigated the alkaloid-rich fraction of *Colchicum brachyphyllum* using HPLC/MS and HPLC-PDA methods. In order to simultaneously detect and measure six predominant steroidal saponins in the rhizomes of *Paris polyphylla* var., Zhang et al. (2010) devised an HPLC/ESI-MS/MS technique. For the simultaneous quantitative measurement of strychnine and brucine in Ayurvedic tablets, Babu et al. devised an HPLC approach. The HPLC process parameters reported were Phenomenex C-18 ODS column with isocratic elution using a 15:85 *v/v* acetonitrile–phosphate buffer at a flow rate of 1 mL/min at 264 nm (Ganesan et al., 2010).

19.3.2.3 Ion Exchange Chromatography (IEC)

IEC is often employed in the domains of life sciences, biological sciences, and other associated fields. The separation of charged molecules such as proteins, peptides, and nucleic acids according to their net charge is the fundamental idea of IEC (Ngere et al., 2023). The method employs a stationary phase with charged groups, commonly in the form of an ion exchange resin, which may draw and hold charged molecules according to their net charge. The technique is frequently used to isolate, purify, and analyze biomolecules.

TABLE 19.4
Application of HPLC in Isolation of Phytochemicals

S. No.	Plant Name	Isolate	Colum	Mobile Phase	Flow Rate	Retention Time	Reference
1	*Croton blanchetianus*	Rutin	C18 column	Ultrapure water and methanol (acidified with trifluoroacetic acid)	0.8 mL/min	24.7 min	(Dos Santos Dantas et al., 2023)
2	*Anthocleista nobilis*	Brucine	Supelcosil C18 column	Methanol and 0.1% *v/v* glacial acetic acid (65:35 % *v/v*)	0.8 mL/min	2.37 min	(Bawa et al., 2023)
3	*Morus alba*	Quercetin and rutin	C18 column (Phenomenex Luna)	Acetonitrile and 0.1% *v/v* glacial acetic acid	1 mL/min	3.43 min	(Sarita et al., 2023)
4	*Saraca along*	Protocatechuic acid, catechin (C)	C18 column	0.1% acetic acid in water (solvent A) and acetonitrile (solvent B)	0.8 μL/min	7.82 (Protocatechuic acid), 13.21 (catechin)	(Sil et al., 2022)
5	Thatbunjob (Thai herbal formulation)	Gallic acid, eugenol, rutin	C18 reverse-phase column (Phenomene x)	Water (0.1% phosphoric acid) and acetonitrile	1.0 mL/min	7.3 min (gallic acid), 30.39 min (rutin) 49.37 (eugenol)	(Jaiboonya et al., 2023)

IEC may be divided into two primary categories: anion exchange chromatography (AEC) and cation exchange chromatography (CEC). The stationary phase in AEC contains positively charged groups, while the stationary phase in CEC contains groups that are negatively charged. The charge of the molecules being separated determines which kind should be used. The capacity of IEC to segregate molecules based on charge, a basic characteristic of many biomolecules, is one of its main benefits. The capability to analyze a diverse array of compounds in the pharmaceutical, biological sciences, natural, agricultural, and other industries has enhanced the appeal of IEC in recent years (Fritz, 2004).

Alkaloids and other protonated bases or deprotonated acids found in nature that may be separated and purified using ion exchange chromatography include fatty acids and derivatives of amino acids (Dragull & Beck, 2012).

19.3.2.4 Macroporous Adsorption Resin Chromatography (MDRC)

As organic polymer adsorbents, macroporous resins typically feature a macroporous network structure and a sizable specific surface area. Exchange groups are not present in the macroporous resin. Through the van der Waals forces, the resin attracts molecules, which are then separated and purified after being eluted by a particular eluent (Lou et al., 2012). In the separation and purification of active components such as total flavonoids, polyphenols, cyan pigments, carbohydrates, alkaloid compounds, and saponins, macroporous resins are commonly utilized.

Two different macroporous resin types, XAD-7HP and HP-20, were tested by Xian et al. for their ability to adsorb and desorb bioactive phenolics derived from *Phanerochaete chrysosporium*. The highest recovery of bioactive phenolics using XAD-7HP resins at 50% (*v/v*) ethanol was 94.3% (Idris et al., 2017).

Hou et al. developed a macroporous resin column chromatography method for salvianolic acids using near-infrared spectroscopy as a process analysis tool. A multivariate statistical process control model was also created to test the model's monitoring capability (Hou et al., 2016).

TABLE 19.5
Application of IEC in Isolation of Phytochemical

S. No.	Plant Name	Stationary Phase	Eluent	Detection	Isolate	References
1	*Nigella sativa*	XK50/30 column (5 × 15 cm) of DEAE sephadex A50	0.05 M phosphate buffer (pH 6.4) containing 0.01 M NaCl	UV at 280 nm	Numerous protein bands with molecular masses between 94 and 10 kDa	(Haq et al., 1999)
2	*Morus alba*	A DEAE-sephacel column (2 × 9 cm) with AEC	20 mM Tris-HCl eluted progressively using the buffer containing NaCl at a pH of 7.5 and a flow rate of 20 mL min^{-1}	UV at 280 nm	Lectins, MLL 1 and MLL 2	(Ratanapo et al., 1998)
3	*Coprinus comatus*	DEAE-sepharose CL-6B column (3.5 cm × 30 cm)	20 mM Tris-HCl, pH 7.5, flow rate 20 mL/min^{-1}, and then eluted progressively with NaCl-containing buffer	UV at 490 nm	α-L-fuco-α-Dgalactan	(Li et al., 2013a)
4	*Echium plantagineum*	Strata SCX; benzyl sulphonate, silica-based cartridges	Ammoniated MeOH	UV at 290 nm	Pyrrolizidine alkaloids	(Colegate et al., 2005)

In order to extract and isolate the four main isoflavonoids from *Dalbergia odorifera*, Li et al. (2016) created an eco-friendly and effective deep eutectic solvent–based negative pressure cavitation–assisted extraction, followed by macroporous resin column chromatography, namely, tectorigenin, prunetin, biochanin A, and genistein, where the highest extraction yields were 1.204, 1.057, 0.911, and 2.448 mg/g dry weight, respectively.

19.3.2.5 Optimal Performance Laminar Chromatography (OPLC)

A pressurized flow chromatography technique called OPLC, in contrast, uses a planar 2D column structure. It may boost loading capacity by using the whole column width for a single sample, increase throughput by allowing together analysis of many samples in a single run, and increase separation effectiveness by allowing two-dimensional (2D) chromatography directly on a square column (Fullas et al., 1989).

Research on plants, the development of pharmaceuticals, the metabolism of drugs, and forensic investigations into drug abuse, and other fields, have all made use of optimal-performance laminar chromatography technology. Online OPLC was used to separate physiologically active compounds from *Rhamnus frangula*, *Heracleum sphondylium*, and *Gentiana purpurea* using 2 mm thick layers with 50–500 mg loading (Nyiredy, 2001). Components of *Matricaria chamomilla* oil were isolated using an analytical silica gel layer, a 22.5 mg loading, and a fully online FEW-OPLC system with a one-channel configuration (Mincsovics, 2004). Furthermore, ginsenosides were isolated using the most recent generation of OPLC instruments and eluent optimized using the PRISMA model (Ludwiczuk et al., 2005).

19.3.2.6 High-Speed Counter-Current Chromatography (HSCCC)

Ma and Ito, 2014 first invented HSCCC, which is a solid supportless liquid–liquid partition chromatographic technique, in the beginning of the 1980s and further improved by other research teams. In this technique, two immiscible liquids are used, one of which is pumped through the column, while the other, which is in the mobile phase, is kept in the column by centrifugal partition power, that is, in the spinning coil (Zhang et al., 2013). In a coiled column, the unilateral phase distribution of the two solvents that are immiscible can be given a characteristic hydrodynamic phenomenon, HSCCC employs a particular arrangement of planetary movement and coil orientations. This hydrodynamic feature may be used to efficiently carry out a variety of counter-current chromatographies (Sun et al., 2017). In order to identify, separate, and purify a wide range of various compounds, including primary as well as secondary metabolites, HSCCC has been widely utilized. This technique enables the analysis of several substances, including hydrocarbons, from environmental samples, heavy metal ions, antibodies, and natural or man-made materials.

19.3.2.7 High-Performance Capillary Electrophoresis (HPCE)

One technique for separating, recognizing, and quantifying different chemical components in food items is called high-performance capillary electrophoresis (HPCE). It has gained popularity as a method for analyte separation in recent years (D'Orazio et al., 2016). In a narrow-bore capillary, separations via HPCE take place under the influence of an applied electrical field. On the basis of physicochemical characteristics, such as ionic charge, charge-to-mass ratio, lipid solubility, and spatial orientation, components of the sample matrix are separated and migrate at various speeds (Colombo & Papetti, 2020). The method offers a number of benefits, including short analytical times, automation, and minimal sample volume needs. The high concentration limit of the detection of HPCE is a considerable drawback, though. The electric field is what separates things in HPCE. The electro-osmotic flow (EOF) of the buffer and the influence of the electric field control the movement of the mixture's constituent parts. The potential difference is what causes the liquid to flow in the capillary. The charge, size, shape, and resistance of an electrically charged particle all affect its speed. The chemicals under analysis, which contain an electric charge, are distinguished

TABLE 19.6
Application of HSCCC in Isolation of Phytochemicals

S. No.	Plant Name	Isolate	Mobile Phase	Reference
1	*Vinca minor*	Vincine and vincamine	Ethanol–*n*-hexane–water (5:6:5)	(Zhou et al., 1990)
2	*Radix phytolaccae*	Esculentosides	Chloroform–methanol–water (4:4:2)	(Jie et al., 2010)
3	*Andrographis paniculata*	Andrographolide	Water–ethyl acetate–Methanol–*n*-hexane (2.5:4:2.5:1 *v/v/v/v*)	(Du et al., 2003)
4	*Berberis sibirica*	p-Coumaric acid	Methyl tert-butyl ether–water (1:1 *v/v*)	(Kukula-Koch et al., 2016)
5	*Carthamus tinctorius*	Phenyl propanoid	n-Butanol/HCl–0.1 mol/L (1:1 *v/v*)	(Li et al., 2013b)
6	*Platycodon grandiflorum*	Platycosides	Water–methanol–isopropanol–chloroform (3:2:2:3 *v/v*)	(Ha et al., 2011)
7	*Rheum officinale*	Hydroxyanthraquinones	80 mL of 0.2% NaOH, 55 mL of 0.7% $Na2CO_3$, and 35 mL of 4.0% $NaHCO_3$	(Yang et al., 1999)
8	*Vaccaria segetalis*	Segetalin B and segetalin A	Petroleum ether–ethyl acetate–methanol–water (0.5:3.5:1:5, *v/v/v/v*)	(Wang et al., 2011)

TABLE 19.7
Application of HPCE in Isolation of Phytochemicals

S. No.	Plant Name	Isolate	Applied Voltage	Detection	Reference
1.	*Cynanchum chinense*	7-O-α-l-rhamnopyranosyl-kaempferol-3-O-β-d-glucopyranoside	20 kV	UV 254 nm	(Liu et al., 2006)
2.	*Adhatoda vasica*	Vasicine and vasicinone	19 kV	—	(Avula et al., 2008)
3.	*Dracaena cochinchinensis*	Loureirin B	15 kV	UV 280 nm	(Yang et al., 2012)
4.	*Aconitum napellus*	Aesculin	30 kV	UV (194 and 206 nm)	(Bao et al., 2011)

by electric mobility, which leads to their segregation. The overall charge in the capillary walls and matrix affects CE resolution (Liu et al., 2006).

19.4 CONCLUSION

A major challenge in the study of natural products from plant extracts lies in the identification and characterization of the diverse mixtures of components, each with varying polarities. Extraction assumes a pivotal role in the separation and characterization of these natural products. In practice, the majority of these compounds necessitate purification through a multifaceted approach, encompassing both chromatographic and non-chromatographic techniques, alongside a variety of purification methods, in order to successfully isolate these valuable natural products.

REFERENCES

Abubakar, A. R., & Haque, M. (2020). Preparation of medicinal plants: Basic extraction and fractionation procedures for experimental purposes. Journal of Pharmacy & Bioallied Sciences, 12(1), 1. https://doi.org/10.4103/jpbs.JPBS_175_19

Adin, S. N., Gupta, I., Ahad, A., Aqil, M., & Mujeeb, M. (2022). A developed high-performance thin-layer chromatography method for the determination of baicalin in Oroxylum indicum L. and its antioxidant activity. JPC-Journal of Planar Chromatography, 35(4), 383–393. https://doi.org/10.1007/s00764-022-00182-4

Adin, S. N., Gupta, I., Aqil, M., Mujeeb, M., & Najmi, A. K. (2023). Development of a high-performance thin-layer chromatography method for the quantification of mangiferin in different extracts of Mangifera indica L. and its antioxidant activity. JPC–Journal of Planar Chromatography, 1–11. https://doi.org/10.1007/s00764-023-00241-4

Ahmad, S., Ganbase, M., Deshmukh, N., Patil, K., Barde, L., Deshpande, M., Gaikwad, A., & Tajane, P. (2023). Development of validated HPTLC and UHPLC methods for quantification of Withanolide in extract and formulation. International Journal of Experimental Research and Review, 30, 375–387. https://doi.org/10.52756/ijerr.2023.v30.035

Alali, F. Q., Tahboub, Y. R., Al-Daraysih, I. S., & El-Elimat, T. (2008). LC-MS and LC-PDA vs. phytochemical analysis of Colchicum brachyphyllum. Die Pharmazie-An International Journal of Pharmaceutical Sciences, 63(12), 860–865. https://doi.org/10.1691/ph.2008.8639

Alam, P., Parvez, M. K., Arbab, A. H., & Al-Dosari, M. S. (2017). Quantitative analysis of rutin, quercetin, naringenin, and gallic acid by validated RP-and NP-HPTLC methods for quality control of anti-HBV active extract of Guiera senegalensis. Pharmaceutical Biology, 55(1), 1317–1323. https://doi.org/10.1016/j.trac.2019.05.040

Ali, A., Chua, B. L., & Chow, Y. H. (2019). An insight into the extraction and fractionation technologies of the essential oils and bioactive compounds in Rosmarinus officinalis L.: Past, present and future. TrAC Trends in Analytical Chemistry, 118, 338–351. https://doi.org/10.1016/j.trac.2019.05.040

Altemimi, A., Lakhssassi, N., Baharlouei, A., Watson, D. G., & Lightfoot, D. A. (2017). Phytochemicals: Extraction, isolation, and identification of bioactive compounds from plant extracts. Plants, 6(4), 42. https://doi.org/10.3390/plants6040042

Avula, B., Begum, S., Ahmed, S., Choudhary, M. I., & Khan, I. A. (2008). Quantitative determination of vasicine and vasicinone in *Adhatodavasica* by high performance capillary electrophoresis. Die Pharmazie-An International Journal of Pharmaceutical Sciences, 63(1), 20–22. https://doi.org/10.1691/ph.2008.7175

Bagade, S. B., & Patil, M. (2021). Recent advances in microwave-assisted extraction of bioactive compounds from complex herbal samples: A review. Critical Reviews in Analytical Chemistry, 51(2), 138–149. https://doi.org/10.1080/10408347.2019.1686966

Bao, Y., Yang, F., & Yang, X. (2011). CE-electrochemiluminescence with ionic liquid for the facile separation and determination of diester-diterpenoid aconitum alkaloids in traditional Chinese herbal medicine. Electrophoresis, 32(12), 1515–1521. https://doi.org/10.1002/elps.201100040

Bawa, R. A., Addotey, J. N., Brobbey, A. A., Oppong-Kyekyeku, J., & Hlordzi, C. K. (2023). Bioassay and RP-HPLC method development for the analysis of extracts and herbal products containing anthocleista nobilis. International Journal of Analytical Chemistry, 2023. https://doi.org/10.1155/2023/6957519

Colegate, S. M., Edgar, J. A., Knill, A. M., & Lee, S. T. (2005). Solid-phase extraction and HPLC-MS profiling of pyrrolizidine alkaloids and their N-oxides: A case study of Echium plantagineum. Phytochemical Analysis: An International Journal of Plant Chemical and Biochemical Techniques, 16(2), 108–119. https://doi.org/10.1002/pca.828

Colombo, R., & Papetti, A. (2020). Pre-concentration and analysis of mycotoxins in food samples by capillary electrophoresis. Molecules, 25(15), 441. https://doi.org/10.3390/molecules25153441

Dobgima, J. F., EF, E. R. T., Bayaga, H., Teke, N. G., Ngono, M. R., Njinkio, N. B., . . . Fokunang, C. N. (2020). An overview of methods of extraction, isolation and characterization of natural medicinal plant products in improved traditional medicine research. Asian Journal of Research in Medical and Pharmaceutical Sciences, 61961, 2457–0745. https://doi.org/10.9734/ajrimps/2020/v9i230152

D' Orazio, G., Asensio-Ramos, M., Fanali, C., Hernández-Borges, J., & Fanali, S. (2016). Capillary electrochromatography in food analysis. TrAC Trends in Analytical Chemistry, 82, 250–267. https://doi.org/10.1016/j.trac.2016.06.012

Dos Santos Dantas, T., Machado, J. C. B., Dos Santos, E. C. F., de Oliveira, A. M., Raimundo E Silva, J. P., Tavares, J. F., Assunção Ferreira, M. R., & Soares, L. A. L. (2023). Phytochemical profile analysis by LC-ESI-MSn and LC-HR-ESI-MS and validated HPLC method for quantification of rutin in herbal drug and products from leaves of Croton blanchetianus. Biomedical Chromatography: BMC, 37(9), e5665. https://doi.org/10.1002/bmc.5665

Dragull, K., & Beck, J. J. (2012). Isolation of natural products by ion-exchange methods. Methods in Molecular Biology (Clifton, N.J.), 864, 189–219. https://doi.org/10.1007/978-1-61779-624-1_8

Du, Q., Jerz, G., & Winterhalter, P. (2003). Separation of andrographolide and neoandrographolide from the leaves of Andrographis paniculata using high-speed counter-current chromatography. Journal of Chromatography A, 984(1), 147–151. https://doi.org/10.1016/S0021-9673(02)01831-9

Esclapez, M. D., García-Pérez, J. V., Mulet, A., & Cárcel, J. A. (2011). Ultrasound-assisted extraction of natural products. Food Engineering Reviews, 3, 108–120. https://doi.org/10.1007/s12393-011-9036-6

Eskilsson, C. S., & Björklund, E. (2000). Analytical-scale microwave-assisted extraction. Journal of Chromatography A, 902(1), 227–250. https://doi.org/10.1016/s0021-9673(00)00921-3

Feng, W., Li, M., Hao, Z., & Zhang, J. (2019) Analytical Methods of Isolation and Identification. Intech Open. https://doi.org/10.5772/intechopen.88122

Fritz, J. S. (2004). Early milestones in the development of ion-exchange chromatography: A personal account. Journal of Chromatography A, 1039(1–2), 3–12. https://doi.org/10.1016/j.chroma.2003.12.068

Fullas, F., Kim, J., Compadre, C. M., & Kinghorn, A. D. (1989). Separation of natural product sweetening agents using overpressured layer chromatography. Journal of Chromatography, 464(1), 213–219. https://doi.org/10.1016/s0021-9673(00)94239-0

Ganesan, B., Perumal, P., Manickam, V., Srikakolapu, S., Gotteti, S., & Thirumurthy, L. (2010). Simultaneous determination of strychnine and brucine in herbal formulation by RP-HPLC. Asian Journal of Research in Chemistry, 3(3), 720–723. www.i-scholar.in/index.php/Ajrc/article/view/164972

Gong, X., Li, Y., & Qu, H. (2014). Removing tannins from medicinal plant extracts using an alkaline ethanol precipitation process: A case study of Danshen injection. Molecules, 19(11), 18705–18720. https://doi.org/10.3390/molecules191118705

Govindarajan, R., Tejas, V., & Pushpangadan, P. (2019). High-Performance Liquid Chromatography (HPLC) as a tool for standardization of complex herbal drugs. Journal of AOAC International, 102(4), 986–992. https://doi.org/10.5740/jaoacint.18-0378

Gujrati, A., Jain, S., Joshi, V., Shukla, S. S., Vyas, A., & Jain, V. (2021). Importance of chromatography techniques in phytomedicine research. Evidence-Based Validation of Traditional Medicines: A Comprehensive Approach, 279–297. https://doi.org/10.1007/978-981-15-8127-4_14

Guo, M., Su, X., Kong, L., Li, X., & Zou, H. (2006). Characterization of interaction property of multicomponents in Chinese Herb with protein by microdialysis combined with HPLC. Analytica Chimica Acta, 556(1), 183–188. https://doi.org/10.1016/j.aca.2005.07.064

Ha, I. J., Kang, M., Na, Y. C., Park, Y., & Kim, Y. S. (2011). Preparative separation of minor saponins from Platycodi Radix by high-speed counter-current chromatography. Journal of Separation Science, 34(19), 2559–2565. https://doi.org/10.1002/jssc.201100326

Haq, A., Lobo, P. I., Al-Tufail, M., Rama, N. R., & Al-Sedairy, S. T. (1999). Immunomodulatory effect of Nigella sativa proteins fractionated by ion exchange chromatography. International Journal of Immunopharmacology, 21(4), 283–295. https://doi.org/10.1016/S0192-0561(99)00010-7

Hazra, A. K., Chakraborty, B., Mitra, A., & Sur, T. K. (2019). A rapid HPTLC method to estimate piperine in Ayurvedic formulations. Journal of Ayurveda and Integrative Medicine, 10(4), 248–254. https://doi.org/10.1016/j.jaim.2017.07.006

Herrero, M., Mendiola, J. A., Cifuentes, A., & Ibáñez, E. (2010). Supercritical fluid extraction: Recent advances and applications. Journal of Chromatography A, 1217(16), 2495–2511. https://doi.org/10.1016/j.chroma.2009.12.019

Hou, X. M., Zhang, L., Yue, H. S., Ju, A. C., & Ye, Z. L. (2016). Monitoring method for macroporous resin column chromatography process of salvianolic acids based on near infrared spectroscopy. China Journal of Chinese Materia Medica, 41(13), 2435–2441. https://doi.org/10.4268/cjcmm20161311

Idris, Z.M., Dzahir, M.I.H.M., Jamal, P., Barka, A.A., Xian, R.L.W. (2017). *IOP Conference Series: Materials Science Engineering*, 206, 012070 doi:10.1088/1757-899X/206/1/012070.

Jaiboonya, J., Intouch, S., & Sumalee, P. (2023). Validation of HPLC method for quantitative determination of active compounds in Thai traditional herbal medicine to treat gastrointestinal disease. Talanta Open, 7, 100227. https://doi.org/10.1016/j.talo.2023.100227

Jain, D., Meena, M., Singh., D., & Janmeda, P. (2023). Isolation, development and validation of HPTLC method for the estimation of β-carotene from *Gymnosporia senegalensis* (Lam.) Loes. Plant Physiology and Biochemistry, 201, 107843. https://doi.org/10.1016/j.plaphy.2023.107843

Jie, M., Chen, Q., Lai, D., & Sun, W. (2010). Separation and purification of triterpene saponins from roots of Radix Phytolaccae by high-speed countercurrent chromatography coupled with evaporative light

scattering detection. Journal of Liquid Chromatography & Related Technologies, 33(4), 563–571. https://doi.org/10.1080/10826070903574659
Kamath, C.R., Salvi, S.A. (2023). Identification and quantification of berberine in a polyherbal formulation by HPTLC method. International Journal of Pharmaceutical Sciences and Research, 14(9), 4544–4547. https://doi.org/10.13040/IJPSR.0975-8232.
Kaufmann, B., & Christen, P. (2002). Recent extraction techniques for natural products: Microwave-assisted extraction and pressurised solvent extraction. Phytochemical Analysis: An International Journal of Plant Chemical and Biochemical Techniques, 13(2), 105–113. https://doi.org/10.1002/pca.631
Kukula-Koch, W., Koch, W., Angelis, A., Halabalaki, M., & Aligiannis, N. (2016). Application of pH-zone refining hydrostatic countercurrent chromatography (hCCC) for the recovery of antioxidant phenolics and the isolation of alkaloids from Siberian barberry herb. Food Chemistry, 203, 394–401. https://doi.org/10.1016/j.foodchem.2016.02.096
Li, B., Dobruchowska, J. M., Gerwig, G. J., Dijkhuizen, L., & Kamerling, J. P. (2013a). Structural investigation of water-soluble polysaccharides extracted from the fruit bodies of Coprinus comatus. Carbohydrate Polymers, 91(1), 314–321. https://doi.org/10.1016/j.carbpol.2012.08.045
Li, L., Liu, J. Z., Luo, M., Wang, W., Huang, Y. Y., Efferth, T., . . . Fu, Y. J. (2016). Efficient extraction and preparative separation of four main isoflavonoids from *Dalbergia odorifera* T. Chen leaves by deep eutectic solvents-based negative pressure cavitation extraction followed by macroporous resin column chromatography. Journal of Chromatography B, 1033, 40–48. https://doi.org/10.1016/j.jchromb.2016.08.005
Li, L. N., Yang, Y., Hou, X. L., Gu, D. Y., Ba, H., Abdulla, R., Wu, G. R., Xin, X. L., & Aisa, H. A. (2013b). Bioassay-guided separation and purification of water-soluble antioxidants from *Carthamus tinctorius* L. by combination of chromatographic techniques. Separation and Purification Technology, 104, 200–207. https://doi.org/10.1016/j.seppur.2012.11.027
Li, P., Xu, G., Li, S. P., Wang, Y. T., Fan, T. P., Zhao, Q. S., & Zhang, Q. W. (2008). Optimizing ultraperformance liquid chromatographic analysis of 10 diterpenoid compounds in Salvia miltiorrhiza using central composite design. Journal of Agricultural and Food Chemistry, 56(4), 1164–1171. https://doi.org/10.1021/jf073020u
Li, Y., Shen, Y., Yao, C. L., & Guo, D. A. (2020). Quality assessment of herbal medicines based on chemical fingerprints combined with chemometrics approach: A review. Journal of Pharmaceutical and Biomedical Analysis, 185, 113215. https://doi.org/10.1016/j.jpba.2020.113215
Liu, H., Gao, Y., Wang, K., & Hu, Z. (2006). Determination of active components in *Cynanchumchinense* R. Br. by capillary electrophoresis. Biomedical Chromatography, 20(5), 451–454. https://doi.org/10.1002/bmc.576
Liu, R. H. (2004). Potential synergy of phytochemicals in cancer prevention: Mechanism of action. The Journal of Nutrition, 134(12), 3479S–3485S. https://doi.org/10.1093/jn/134.12.3479S. https://doi.org/10.1080/10826076.2013.809545
Liu, W., Pi, Z., Wang, X., Song, F., & Liu, S. (2010). HPLC/ESI-MSn and ESI-MS studies on the Aconitum alkaloids in three Chinese medicinal herbs. Journal of Separation Science, 33(17–18), 2898–2906. https://doi.org/10.1002/jssc.201000285
Lou, S., Chen, Z., & Liu, Y. (2012). Synthesis of functional adsorption resin and its adsorption properties in purification of flavonoids from *Hippophaerhamnoides* L. leaves. Industrial & Engineering Chemistry Research, 51(6), 2682–2696. https://doi.org/10.1021/ie201494k
Ludwiczuk, A., Nyiredy, S., & Wolski, T. (2005). Separation of the ginsenosides fraction obtained from the roots of Panax quinquefolium L. cultivated in Poland. JPC–Journal of Planar Chromatography–Modern TLC, 18, 104–107. https://doi.org/10.1556/JPC.18.2005.2.3
Ma, Y., & Ito, Y. (2014). Chiral high-speed counter-current chromatography: Future strategies for chiral selector development. Current Chromatography, 1(1), 69–80. https://doi.org/10.2174/2213240611309999001
Ma, Y., Meng, A., Liu, P., Chen, Y., Yuan, A., Dai, Y., & Li, Z. (2022). Reflux extraction optimization and antioxidant activity of phenolic compounds from Pleioblastus amarus (Keng) Shell. Molecules, 27(2), 362. https://doi.org/10.3390/molecules27020362
Melwita, E., & Ju, Y. H. (2010). Separation of azadirachtin and other limonoids from crude neem oil via solvent precipitation. Separation and Purification Technology, 74, 219–224. https://doi.org/10.1016/j.seppur.2010.06.008
Mihaylova, D., & Lante, A. (2019). Water an eco-friendly crossroad in green extraction: An overview. The Open Biotechnology Journal, 13(1). https://doi.org/10.2174/1874070701913010155
Mincsovics, E. (2004). Flowing eluent wall processes in OPLC: Using segmentation of non-segmented adsorbent layer for single and parallel separations. Journal of Planar Chromatography, 17, 411–419. https://doi.org/10.1556/JPC.17.2004.6.3

Ngere, J. B., Ebrahimi, K. H., Williams, R., Pires, E., Walsby-Tickle, J., & McCullagh, J. S. O. (2023). Ion-exchange chromatography coupled to mass spectrometry in life science, environmental, and medical research. Analytical Chemistry, 95(1), 152–166. https://doi.org/10.1021/acs.analchem.2c04298

Nyiredy, S. (2001). The bridge between TLC and HPLC: Over pressured layer chromatography (OPLC). TrAC Trends in Analytical Chemistry, 20(2), 91–101. https://doi.org/10.1016/S0165-9936(00)00047-9

Patel, M. N., Nandpal, M. N., Patel, A. J., Raval, M. A., & Patel, S. G. (2023). Development and validation of a green stability-indicating HPTLC method for estimation of curcumin, gallic acid, and ursolic acid from polyherbal formulation Jatyadi Taila. Journal of AOAC International, 106(4), 979–991. https://doi.org/10.1093/jaoacint/qsac147

Ratanapo, S., Ngamjunyaporn, W., & Chulavatnatol, M. (1998). Sialic acid binding lectins from leaf of mulberry (Morus alba). Plant Science, 139(2), 141–148. https://doi.org/10.1016/S0168-9452(98)00185-X

Saçıcı, E., & Yesilada, E. (2022). Development of new and validated HPTLC methods for the qualitative and quantitative analysis of hyperforin, hypericin and hyperoside contents in Hypericum species. Phytochemical Analysis: PCA, 33(3), 355–364. https://doi.org/10.1002/pca.3093

Sarita, G., Rubal, C., Deepak, K., Rakesh, K., & Vineet, M. (2023). RP-HPLC method development and validation for simultaneous estimation of rutin and quercetin in *Morus alba* L. leaf extract. Research Journal of Pharmacy and Technology, 16(5), 2327–2325. https://doi.org/10.52711/0974-360X.2023.00383

Sasidharan, S. et al. (2011) Extraction, isolation and characterization of bioactive compounds from plants' extracts. African Journal of Traditional, Complementary, and Alternative Medicines, 8(1), 1–10. https://doi.org/10.4314/ajtcam.v8i1.60483

Saxena, H. O., Parihar, S., Pawar, G., & Rao, G. R. (2023). Simultaneous densitometric determination of β-sitosterol and lupeol through validated HPTLC method in different plant parts of Uraria picta (Jacq.) Desv. Ex DC.—A dashmool species. Acta Chromatographica, 35(1), 99–105. https://doi.org/10.1556/1326.2022.01018

Sil, S., De, K. K., & Ghosh, A. (2022). RP-HPLC-based phytochemical screening of different polyphenolic compounds from floral extract of four species of Saraca L. (Leguminosae). Future Journal of Pharmaceutical Sciences, 8(1), 41. https://doi.org/10.1186/s43094-022-00430-8

Słominska, M., Zygler, A., & Namieśnik, J. (2012). Soxhlet extraction and new developments such as Soxtec. In: Comprehansive Sampling and Sample Preparation. 65–82. Academic Press, Amsterdam. https://doi.org/10.1016/B978-0-12-381373-2.00037-5

Sudberg, S., Sudberg, E. M., Terrazas, J., Sudberg, S., Patel, K., Pineda, J., & Fine, B. (2010). Fingerprint analysis and the application of HPTLC to the determination of identity and quality of botanicals, from an industry perspective. Journal of AOAC International, 93(5), 1367–1375. https://doi.org/10.1093/jaoac/93.5.1367

Sun, Z. L., He, J. M., Lan, J. E., & Mu, Q. (2017). High-speed counter-current chromatography with an online storage technique for the preparative isolation and purification of dihydroflavonoids from Sophora alopecuroides L. Phytochemical Analysis, 28(6), 496–504. https://doi.org/10.1002/pca.2698

Suwal, S., & Marciniak, A. (2018). Technologies for the extraction, separation and purification of polyphenols—A review. Nepal Journal of Biotechnology, 6(1), 74–91. https://doi.org/10.3126/njb.v6i1.22341

Tai, Y., Shen, J., Luo, Y., Qu, H., & Gong, X. (2020). Research progress on the ethanol precipitation process of traditional Chinese medicine. Chinese Medicine, 15, 84. https://doi.org/10.1186/s13020-020-00366-2

Takale, N., Wadibhasme, S., & Ghule, B. (2023). Development of betulin as phytochemical reference standard for the analysis of *Hygrophilaschulli* plant by a validated high-performance thin-layer chromatography method. JPC-Journal of Planar Chromatography, 2023. https://doi.org/10.1007/s00764-023-00235-2

Tzanova, M., Atanasov, V., Yaneva, Z., Ivanova, D., & Dinev, T. (2020). Selectivity of current extraction techniques for flavonoids from plant materials. Processes, 8(10), 1222. https://doi.org/10.3390/pr8101222

Wang, W. Y., Qu, H. B., & Gong, X. C. (2020). Research progress on percolation extraction process of traditional Chinese medicines. China Journal of Chinese Materia Medica, 45(5), 1039–1046. https://doi.org/10.19540/j.cnki.cjcmm.20191221.305

Wang, X., Dong, H., Liu, Y., Yang, B., Wang, X., & Huang, L. (2011). Application of high-speed counter-current chromatography for preparative separation of cyclic peptides from Vaccariasegetalis. Journal of Chromatography B, 879(11–12), 811–814. https://doi.org/10.1016/j.jchromb.2011.02.001

Yang, F., Zhang, T., Tian, G., Cao, H., Liu, Q., & Ito, Y. (1999). Preparative isolation and purification of hydroxyanthraquinones from Rheum officinale Baill by high-speed counter-current chromatography using pH-modulated stepwise elution. Journal of Chromatography. A, 858(1), 103–107. https://doi.org/10.1016/s0021-9673(99)00827-4

Yang, X., Hu, X., Li, F., Wang, X., & Cao, Q. (2012). Determination of loureirin A and loureirin B in dragon's blood by capillary zone electrophoresis. Chinese Journal of Chromatography, 30(5), 501–506. https://doi.org/10.3724/sp.j.1123.2011.12006

Yi, Y., Zhang, Q. W., Li, S. L., Wang, Y., Ye, W. C., Zhao, J., & Wang, Y. T. (2012). Simultaneous quantification of major flavonoids in "Bawanghua", the edible flower of Hylocereus undatus using pressurised liquid extraction and high-performance liquid chromatography. Food Chemistry, 135(2), 528–533. https://doi.org/10.1016/j.foodchem.2012.05.010

Yu, Z. Q., Chew, J. W., Chow, P. S., & Tan, R. B. H. (2007). Recent advances in crystallization control: An industrial perspective. Chemical Engineering Research and Design, 85(7), 893–905. https://doi.org/10.1205/cherd06234

Zhang, Q. W., Lin, L. G., & Ye, W. C. (2018). Techniques for extraction and isolation of natural products: A comprehensive review. Chinese Medicine, 13, 1–26. https://doi.org/10.1186/s13020-018-0177-x

Zhang, T., Liu, H., Liu, X. T., Xu, D. R., Chen, X. Q., & Wang, Q. (2010). Qualitative and quantitative analysis of steroidal saponins in crude extracts from Paris polyphylla var. yunnanensis and P. polyphylla var. chinensis by high performance liquid chromatography coupled with mass spectrometry. Journal of Pharmaceutical and Biomedical Analysis, 51(1), 114–124. https://doi.org/10.1016/j.jpba.2009.08.020

Zhang, X., Ito, Y., Liang, J., Su, Q., Zhang, Y., Liu, J., & Sun, W. (2013). Preparative isolation and purification of five steroid saponins from *Dioscorea zingiberensis* CH Wright by counter-current chromatography coupled with evaporative light scattering detector. Journal of Pharmaceutical and Biomedical Analysis, 84, 117–123. https://doi.org/10.1016/j.jpba.2013.02.005

Zhou, J. Y., Fang, Q. C., & Lee, Y. W. (1990). The application of high-speed countercurrent chromatography to the semipreparative separation of vincamine and vincine. Phytochemical Analysis, 1(2), 74–76. https://doi.org/10.1002/pca.280001020

Zhou, Y. Q., Zhang, Q. W., Li, S. L., Yin, Z. Q., Zhang, X. Q., & Ye, W. C. (2012). Quality evaluation of semen oroxyli through simultaneous quantification of 13 components by high performance liquid chromatography. Current Pharmaceutical Analysis, 8(2), 206–213. https://doi.org/10.2174/1573412911208020206

Zhang, Q.W., Lin, L.G., and Ye, W.C. (2018). Techniques for extraction and isolation of natural products: a comprehensive review. *Chin Med.*, *13*, 20. https://doi.org/10.1186/s13020-018-0177-x

20 Potential Role of Metabolomics in Medicinal Plant Biotechnology Research

*Devyani Rajput, Neha Sisodiya, Sampat Singh Tanwar, Rahul Singh Thakur, G. N. Darwhekar, and Rakhi Khabiya**
* Corresponding author: rakhikhabiya@gmail.com

20.1 INTRODUCTION

Metabolomics is the study of minute molecules, which are known as metabolites, with molecular weight <1,500 Da, which are present within the cell, tissue, tissue fluid, lymph, and other biological fluids within the organism. In this process, mainly samples of blood, urine, and feces are analyzed for studying the metabolites from various metabolic pathways (Yang et al., 2019b).

Medicinal plants are a rich source of chemical constituents that are beneficial for organisms. Therefore, they produce huge clusters of chemicals, as compared to other organisms, which can be genetically modified using biotechnology to regulate the production of metabolites to improve the quality of plants. Hence, metabolomics plays vital role in medicinal plant biotechnology research (Saito & Matsuda, 2010).

20.2 FOUNDATIONS OF METABOLOMICS

The advancements in science and technology from the past few years have been tremendous, but still many scientists have found that "diagnosis, prevention, and cure" is not the only option for all biomedical problems, so the concept of omics from a holistic perspective came into existence. Professor Jeremy Nicholson was the first scientist to bring up the concept of metabolomics, in the year 1999 (Yang et al., 2019b). Since then, many researches have been going on this concept of metabolomics.

However, this concept of omics, metabolomics, is a difficult and time-consuming process. But mass spectroscopy came out as an emerging trend for the identification of endogenous substances, like sugar, amino acids, lipids, organic compounds, etc., in most living organisms.

Therefore, it was concluded that the study of metabolomics with mass spectroscopy is of great importance in determining and identifying metabolite biomarkers. This process generally involves the preparation and analysis of samples, analysis of data, and analysis of metabolic pathways. Other than mass spectroscopy, many analytical techniques, like GC-MS (gas chromatography–mass spectrometry), LC-MS (liquid chromatography–mass spectrometry), and NMR (nuclear magnetic resonance), are used for metabolomics research. But since metabolomics is the complete study of large data which is very complicated and complex, it cannot just be studied by conventional means, so it involves a special three-step data analysis procedure for a simple understanding of data (Haug et al., 2017). This step includes the extraction of data, pre-processing of data, and recognition of patterns. Through these steps we extract the required data from large raw data, through statistical screening, and elaborate on the biological significance, which is the motto of metabolomics, ultimately applicable to any disease (Figure 20.1).

 DOI: 10.1201/9781003326939-24

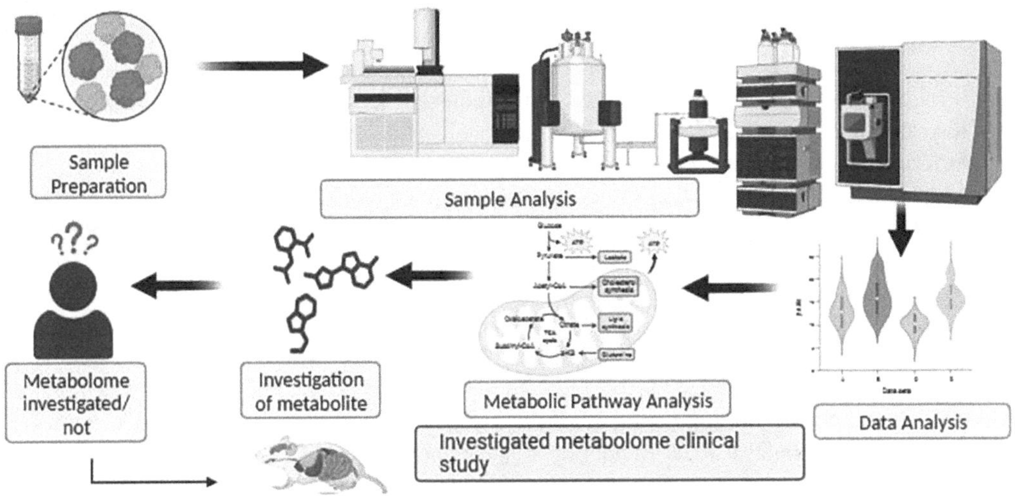

FIGURE 20.1 Identification, determination, and analysis of metabolite biomarkers through the following pathway that led to the foundation of metabolomics.

All this understanding of data, failures, and achievement in the field of omics led to the foundation of the importance of metabolomics in different fields, including biotechnology (Yang et al., 2019b). The study of metabolomics on phytomedicinal plants was done to cure diseases like cancer, and the many other cytotoxic effects involved, using a multi-analyses technique. (Aravindaram & Yang, 2013).

Medicinal plants are a great source of medicine that produces a large number of metabolites beneficial for humans. The ratio of these metabolites is influenced by many factors, like the site of synthesis, intermediates in the metabolic pathway, biosynthetic gene mimicking their pathway, genotype, etc. (Crispin & Wurtele, 2013).

A high quality of plant-based medicine can only be obtained from good-quality medicinal plant cultivation used to manufacture herbal medicines, which has given TCM (traditional Chinese plant) internationally. This led to the breeding and cultivation of TCM worldwide. This advancement has led to plant tissue culture and in vitro micropropagation techniques for the metabolic engineering of plants. Bioactive markers, such as microsatellites (SSR: simple sequence repeat) and single-nucleotide polymorphism, are used for the quality assessment of breeding. NGS (next-generation sequencing) technology has become the first choice for the characterization of genetic markers. NGS is used in transcriptomic, epigenetic, and metabolomics studies (Sucher et al., 2013).

20.3 METABOLOMICS APPLICATIONS IN MEDICINAL PLANT RESEARCH

Plant researchers have found that medicinal plants contain a large number of natural metabolites which have the high potency of pharmacological effect. Innumerable experiments have been carried out on medicinal plants for the identification of bioactive markers to understand the metabolic pathway of bioactive metabolites. In contrast to this experiment, it was concluded that single omics failed to explain the biological response, but multiomics studies on medicinal plants surprisingly gave astonishing results (Yang et al., 2023). Based on these studies, various applications of metabolomics in medicinal plant research have been found (Figure 20.2):

- **Genetic modification.** Biotechnology plays a vital role in genetically modified plant metabolites to improve the quality of plants, increase crop yield, and reduce environmental stress. By genetic engineering, plant cells are modified to mimic their metabolic pathway through metabolome. As an example, first studies were made on rice and *Arabidopsis* (Okazaki & Saito, 2012).
- **Safety assessment of medicinal plants.** The collaboration of advancements in technology with analytical instrumentation has led to the conjugation of biology and chemistry to deeply understand complex biological phenomena, such as cell homeostasis and protein–protein interaction. Metabolomics is used for the safety assessment of biotechnological products by enhancing metabolomics assessment to determine the consequences of biotechnology in medicinal plants. This is done through a deep understanding of statistical data and metabolite-level data processing algorithms (Oh et al., 2023).
- **Intestinal/colonic health.** DNA sequencing technology has greatly improved intestinal and colonic health through the metabolic activity of the microbiome with the determination of short-chain fatty acids. Metabolite dietary fiber intake has impacted on metabolome and improved intestinal and colonic health by working on intestinal microbiota, producing prebiotic effect. Agrarian diets, a rich source of fiber, have improved the microbial activity of the intestine and decreased bacteroides (Simpson & Campbell, 2015).
- **Prevention and diagnosis of disease.** Metabolomics aids in the investigation of a number of diseases by improving diagnosis and prevention and designing an appropriate therapeutic approach (Gomez-Casati et al., 2013). It helps in the detection and analysis of chiral molecules of metabolites that help in the analysis of health status. It identifies the disease biomarkers of any known abnormal enantiomers of diseases like cancer, brain disorders, and kidney diseases (Liu et al., 2023).
- **Pharmacological and pharmacodynamical studies.** Metabolomics helps in the study of pharmacological effects. Many studies have found that the processes of different treatments of drug can be differentiated by different metabolic methods. For example, *Forsythia* betulinic acid was used to determine anticancer drugs, but it was found that *Forsythia* oil extracts show an anti-inflammatory effect by acting on a different metabolic pathway. Twenty-seven different biomarkers were found with many other pharmacological effects, including gastrointestinal effect, effect on cerebral disorders, etc. (Yang et al., 2019a)
- **Plant breeding.** Metabolomics generated metabolome, which improves plant breeding. These studies were made through GC-MS and LC-MS by regulating the metabolic changes occurring in plant cells (Alseekh & Fernie, 2021). In a recent study, around 160 crops, including medicinal plants, were examined for plant breeding (Sakurai, 2022).
- **Toxicological studies.** Despite nanomaterials' having a lot of advantages in the fields of biomedicine, food, cosmetic, sensing, energy, etc., it has lot of disadvantages that are harmful for human beings as well as the environment when they are overly produced. Therefore, nanotoxicology is the field of study of nanomaterials. Metabolomics here plays an important role in the analysis of toxicity of different materials. This phenomenon works on the mechanism of reflex action of stimuli, where toxicity induced by nanomaterials is revealed by molecular information (Awashra & Młynarz, 2023).
- **Agriculture and food research.** Metabolomics is a phenotype that has played a vital role in improving agricultural and food science research. A basic model of *Arabidopsis thaliana* was used in agricultural research to investigate environmental factors, like heat, light, salinity, drought, flooding, etc. (Tian et al., 2016). Other than this, many studies have also been made on the soybean plant (*Glycine max*) under abiotic stress (Feng et al., 2020).
- **Chronic diseases.** Diseases like tuberculosis, cancer, kidney-related diseases, etc. can be diagnosed, treated, and moreover, prevented by metabolomics. In this, metabolites that are

closely related are identified as biological markers in medicinal plants used to treat diseases like TB, pancreatic cancer, breast cancer, ovarian cancer, colorectal cancer, metastasis, etc. Clinical samples of cancer are difficult to obtain. Therefore, further research has been hindered (Yang et al., 2019b). Other than this, orange juice also affects fatty acid β-oxidation through the mitochondria and the peroxidase pathway. Nine biomarkers were identified with potential effect.

- **Nutrition.** The study of nutrition with metabolomics is called *nutritional metabolomics*. There are a number of examples of nutritional metabolites found in many fruits and vegetables. These include metabolites found in cucumber, plum, ginger, cabbage, strawberry, kiwi, etc., prone to many abiotic conditions to find out secondary metabolites. Among these, kiwi showed an anti-inflammatory effect due to the biologically active metabolite Danshensu (Yang et al., 2019b).

20.4 BIOACTIVE MARKERS AND QUALITY CONTROL

20.4.1 Role of Metabolomics in Identifying Bioactive Markers

Metabolomics has emerged as one of the most rapidly evolving fields of "omics" technology in the last decade. Metabolomics involves the documentation, quantification, and detection of metabolites in various living systems using contemporary analytical technologies. Modern analytical technologies, such as gas chromatography–mass spectrometry gas, non-destructive nuclear magnetic resonance spectroscopy, and liquid chromatography–mass spectroscopy have aided in the reporting of metabolites in complicated biological matrices. Metabolomics, like transcriptomics, plays a very important part in uncovering links among metabolite phenotyping, genetic regulation, and biomarker identification (Marchev et al., 2021)

By doing metabolomics studies about various diseases, one can easily identify the various biomarkers involved in how the disease develops at the metabolic level. Additionally, a lot of metabolomics research has been carried out over the past ten years with the goal of determining how different herbal medications work. A recent in vitro and in vivo study involved a large number of samples for the metabolomic studies of medicinal plant mechanisms and their antidiabetic activities, reporting

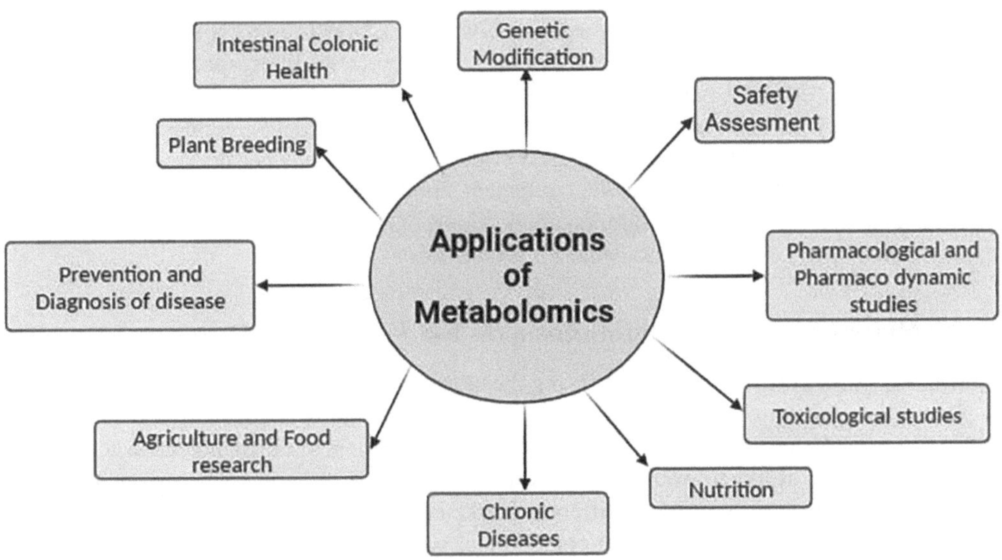

FIGURE 20.2 Application of metabolomics in medicinal plant research.

on the possible biomarker involved in the progression of diabetes and targeted by medicinal plants, on which treatment with *Ampelopsis grossdentata* of the diabetic model identified biomarkers like xanthine, proline, histidine, and threonine by using LC-MS (Hasanpour et al., 2020). Also, metabolomic studies on traditional Chinese medicine used in lowering blood pressure in animal models, reported as treatment with ginsenosides to hypertensive model, identified an elevated level of biomarkers like glycerol-2,3-diphosphate, glycerol-3-diphosphate, H-indole-3-propanoate, glycerate, and ornithine using GC-TOF/MS (Yang & Lao, 2019). Another metabolomic study revealed 25 biomarkers and linked 9 metabolic pathway networks of coronary heart disease models and found potential biomarkers and therapeutic mechanisms of JZGXP (Jing Zhi Guan XinPian) in the treatment of CHD using ESI-MS data (Sun et al., 2019). In metabolomic studies to identify biomarkers and to do multivariate statistical analyses of data obtained from LC-MS, GC-MS, and NMR methods, like partial least square discriminant analysis, principal component analysis and orthogonal partial least square discriminant analysis are the techniques that are most frequently utilized (Sun et al., 2014; Hasanpour et al., 2020; Sun et al., 2019).

20.4.2 Ensuring Quality and Consistency of Herbal Products

The safety, efficacy, and consistency of herbal medications, particularly those with numerous botanical origins, required metabolomics-assisted source identification (Pan et al., 2020). Recent research indicates that herbal remedies may be useful in the treatment of a number of ailments. The development of herbal medicines is now hampered by the difficulties inherent in their chemical compositions, which include ambiguous bioactive substances, modes of action, unidentified therapeutic targets, non-specific properties for drug metabolism, etc. Metabolomics is a powerful tool in understanding and improving herbal medicines at the level of small-molecule metabolism in order to get around these problems. From a metabolic perspective, metabolomics may be able to address scientific problems with natural remedies and to advance the search for new drugs. For the in vivo and in vitro investigation of herbal ingredients in recent years, mass spectrometry–based metabolomics has been frequently used (Wang et al., 2019). Several recent studies were carried out to check the quality of medicinal plants, such as metal content, inhibitory glucosidase action, and phytochemical profiling of Green Chiretta from various geographical origins based on UHPLC-Q-Orbitrap HRMS and FTIR metabolomics (Rafi et al., 2021). Chinese medicinal plants (CMP) have curative and preventative effects on illnesses. Over 10,000 different types of medicinal plants may be found in China, which accounts for about 87% of all CMM resources. Plant metabolomics has been used in the past year to analyze metabolites to recognize habitats, species, growth stages, harvesting seasons, different parts, and adulteration of CMM, having a tremendous potential in evaluating quality. Plant metabolomics is currently in its early phases of progress, despite its tremendous growth. Plant metabolomics still has considerable limitations because of the complexity and diversity of metabolites, as well as the equipment's poor performance. With the emergence of new technology and multifaceted integration applications, the use of plant metabolomics in the area of quality control will have greater potential for growth (Xiao et al., 2022).

20.4.3 Metabolomics-Assisted Authentication and Adulteration Detection

Metabolomics plays a crucial role in identifying the biomarkers as well as adulterants in the medicinal plants; therefore, proteomics-, genomics-, and metabolomics-based approaches are being established to check the authenticity of various food commodities. Massive molecular tools based on omics can be useful to get around the constraints of previous methods (Böhme et al., 2019). As a recent study carried out on five different marketed trades of saffron analyzed via UPLC-FT-MS/MS found out, several bioactive compounds, such as glycosylated picrocrocin, kaempferols, safranal, and crocin, with adulterants like Ellis, auramine-O (illegal synthetic dye), and geniposide, provided an adequate basis for monitoring possible contaminants, their effects on the acetylcholinesterase inhibitory (AChEI) activity of saffron, and their impact on saffron quality assessment (Younis et al., 2023). Also, Wen et al. used the RT metabolomics profiling approach to detect lampante virgin oil adulteration with four types of different oils by using rapid evaporative ionization mass spectrometry (Wen et al., 2023). Araceli et al. applied untargeted metabolomics with chemometrics and GC-Orbitrap-HRMS for the validation of *Piper nigrum* L. and the detection of processing markers; results from the study revealed the existence of monoterpenes, diterpenes, amides, and other miscellaneous compounds, with plenty of sesquiterpenes and alkenyl benzenes. Also, with the prospective markers for sample differentiation based on geographic origin, six metabolites were discovered (Rivera-Pérez et al., 2021). Rongbo et al. identified the markers of goat milk adulterated with cow's milk based on the application of metabolomics and proteomics (Fan et al., 2023). Petros V. et al. used metabolomics for the authentication and identification of adulteration in seafood (Kotsanopoulos et al., 2022). Thus, an omics-based approach serves as a potential tool for authentication and adulteration detection in a variety of commercial products, food, pharmaceuticals, and medicinal plants.

20.5 INTEGRATION WITH SYSTEM BIOLOGY

20.5.1 Interplay of Metabolomics with Genomics and Transcriptomics

Multiomics research, including proteomics, genomics, microbiomics, transcriptomics, metabolomics, radiomics, and pathomics, is currently a trendy topic. Scientists have given close consideration to the interaction between multiomics data, diseases and their medication. At the same time, multiomics can exactly predict disease prognosis, diagnosis, and treatment. In essence, the network can represent these study entities at several levels, including protein genes, microbes, RNAs, metabolite pathways, and pathological imaging data. In addition, nearly all computing and medical researchers have tried to employ computational tools to explore potential links among biological elements (Pan et al., 2022). Plant diseases, in addition to abiotic factors, such as dryness and saline soils, are a reason for the major share of loss of crop in the world. Prior to the creation of metabolomics, genomics, transcriptomics, and proteomics made substantial contributions to our understanding of plant diseases and the methods by which pathogens gain nutrients while avoiding plant immunity. Transcriptomic research showed that the host's genes are altered or reprogrammed by pathogens for a successful defense response (Castro-Moretti et al., 2020). Metabolomics has proved very effective in the natural product industry for the identification of new molecules that could be related to the bioactivity of plant extracts used for human health and illness treatment (Mumtaz et al., 2017). The same concept can be extended to understanding plant responses to infections, whether in a compatible interaction, where disease spreads, or in an incompatible connection, where the pathogen does not produce disease. In a current study of soybeans infected with the oomycete pathogen, for example, metabolomics studies revealed that many secondary metabolites and sugars were differentially collected in resistant plants versus susceptible plants, implying that these metabolites may play an important role in the defense (Zhu et al., 2018). It is now obvious that the metabolome influences cellular physiology through influencing other "omics" levels, such as the genome, epigenome, transcriptome, and proteome (Rinschen et al., 2019).

20.5.2 Understanding Metabolic Pathways and Regulatory Networks

A recent study used transcriptomics and metabolomics to detect distinct anthocyanin metabolites to analyze the different types of genes involved in anthocyanin production in white and purple-and-white flowers of *Salvi*. Insight into the biosynthetic pathways of medicinal products, a total of 100 unigenes encoding ten enzymes were identified as potential genes involved in anthocyanin production in S. miltiorrhiza flowers (Jiang et al., 2020). For most fruit species, the underlying regulatory network and metabolic dynamics that influence overall flavor production during fruit growth and ripening are relatively unexplored. To explore it, Wang et al. developed extensive regulatory networks permitting the discovery of the transcription factor and essential structural genes that control the metabolism of organic acids, soluble sugars, and volatiles in kiwifruit. Furthermore, research discovered a regulatory mechanism including important transcription factors that regulate flavor metabolism. The modulation of flavor metabolism by the identified key transcription factors was confirmed in the different kiwifruit species, demonstrating that dataset is an appropriate tool for clarifying the regulatory factors controlling flavor biosynthetic pathways that have not previously been explored. Overall, the findings shed fresh light on the metabolic regulation of flavor during fruit growth and ripening (Wang et al., 2022). Another study found a complex regulatory network in *Lycopersicon esculentum* after chilling injury using an integrative comparative analysis of metabolomics and transcriptomics data. The DEGs (differentially expressed genes) discovered can serve as indicator genes, reflecting the physiological condition of plants subjected to CI stress. In future investigations, the differentially expressed genes could be utilized as possible genes for genetic modification to increase the fruits' resistance to freezing (Zhang et al., 2019)

20.5.3 Insight into the Biosynthetic Pathway of Medicinal Plants

The huge therapeutic and commercial benefits of medicinal plants have piqued the interest of global investigators. It is acknowledged that the manufacturing of high-quality herbal remedies has become a must for the continued growth of the traditional medicine sector. To that purpose, substantial researchers' hard work has been committed to basic studies to pave the way for standardized plant material verification and the design of medicinal plant metabolic pathways (Ma et al., 2020). Comprehensive metabolite profiling enables the integration of summarized data into biosynthetic pathway manipulation, identification of nutritional quality markers, selection of desired metabolites and genes, agricultural yield improvement, and heritability in current breeding. Furthermore, metabolomics can help anticipate the biological activity of medicinal plants, aid in the bioactivity-guided fractionation process and the discovery of bioactive leads, and serve as a tool for quality monitoring and certification of commercial plant–derived natural products (Marchev et al., 2021). With the development of omics-based approaches, such as metabolomics and genomics, the potential to discover new plant secondary metabolites as well as to give huge data on the regulatory networks of genes involved in secondary metabolite synthesis has been highlighted. Combining gene editing technologies with metabolomics, such as CRISPR-Cas9, provides an effective platform for the identification and detection of identified and unidentified secondary metabolites, as well as the improvement of secondary metabolite synthesis (Fukusaki & Kobayashi, 2005). Modern plant metabolomics, a more advanced form of plant metabolism, has sped up gene discovery and clarified many natural product biosynthesis routes (Sumner et al., 2015). In these investigations, metabolomics plays dual roles: (1) recognizing the spatial and temporal distribution of the target phytochemical as impacted by plant growth and external factors, and (2) identifying similar molecules that can be viewed as alternate products of promiscuous enzymes or as intermediary substances in the production of the target phytochemical. When paired with generalizable concepts of organic synthetic chemistry, such knowledge may pave the way for the discovery of biosynthetic routes. As a result, combining metabolomics and genome-based functional categorizations of gene product allows for the rapid discovery of novel biosynthetic pathways to specialized metabolites (Alami et al., 2022)

20.6 METABOLOMICS-GUIDED BIOTECHNOLOGICAL APPLICATIONS

Metabolomics is an emerging omics technology that quantitatively describes the dynamic changes of metabolites in biological systems. It has been applied in various fields, such as assisted reproductive technology (ART) (Gao & Xiao, 2023), natural product discovery (Choi et al., 2023), microbial metabolomics (de Sá et al., 2023), and the study of pheochromocytoma/paraganglioma (PPGL) (Richter et al., 2023). Metabolomics provides valuable information in evaluating the quality of gametes and embryos in ART (Wu et al., 2022). In natural product discovery, metabolomics looks into biological processes and leads to the discovery of useful biomolecules from natural sources. In microbial metabolomics, careful experimental design and execution are crucial for obtaining genuine results with correct biological interpretation. In the study of PPGL, metabolomics has been used to identify biomarkers and understand the metabolic dysregulation associated with genetic aberrations. Overall, metabolomics plays a significant role in guiding biotechnological applications in various fields.

Metabolomics-guided biotechnological applications represent a cutting-edge approach to harnessing the power of metabolite profiling for diverse fields. This emerging discipline leverages the comprehensive insights provided by metabolomics to drive innovations in agriculture, pharmaceuticals, and environmental sustainability (Floros et al., 2016). By elucidating the metabolic pathways within organisms, it enables precision breeding of crops with enhanced nutritional profiles, improved disease resistance, and higher yields. In the pharmaceutical realm, metabolomics assists in drug discovery, offering a deeper understanding of how compounds interact with biological systems and aiding in the identification of novel therapeutic agents (Parida et al., 2018). Furthermore, in environmental biotechnology, it informs strategies for bioremediation and bioenergy production by uncovering the metabolic potential of microorganisms. As metabolomics continues to evolve and integrate with other omics disciplines, its applications in biotechnology promise to revolutionize industries, leading to a more sustainable agriculture, innovative pharmaceuticals, and cleaner, greener environmental solutions (Wang et al., 2020)

20.6.1 Optimizing Cultivation Conditions for Enhanced Metabolite Yield

Optimizing cultivation conditions for enhanced metabolite yield involves various strategies, such as selection of stable, high-yield variants, manipulating medium components and gas phase composition, metabolic manipulation through gene manipulation, use of inhibitors, use of elicitors, precursor feeding, in situ product removal, and use of immobilization or more differentiated or organized plant material (Su et al., 2007). For example, the optimum cultivation conditions for exopolysaccharide yield by *Cyanothece* sp. 113 cyanobacterium was found as 70.0 g/L of NaCl, 0.9 g/L of MgSO4, 29°C temperature, aeration, and continuous illumination at 86.0 μE/M2/S (Cui & He, 2012) Similarly, the optimum fermentation conditions for *Ganoderma lucidum* included a fermentation time of 6 days, specific pH, rotation speed, inoculation, vaccination days, and incubation temperature as mentioned in Table 20.1 (Manivasagan & Kim, 2014). These studies highlight the importance of understanding and manipulating cultivation conditions to enhance metabolite yield.

The key factors that influence the production of a desired metabolite in a given organism include epigenetic modifications, biodiversity of the microbial community, metabolic engineering, and the morphological type of the organism. Epigenetic modifications, such as chromatin remodeling and DNA methylation, can alter the expression of genes involved in metabolite synthesis (Bind et al., 2022). The biodiversity of the microorganism in contaminated areas and oil reservoirs can affect the efficiency of biodegradation and biosurfactant production (Sharma et al., 2022). *Metabolic engineering* involves genetic manipulations to modulate the metabolism of organisms and optimize the production of desired metabolites (Kulkarni, 2016). The morphological type of the organism, whether it is cultured as free mycelia, aggregated forms, or artificially bound/entrapped cells, can also influence metabolite production (Kenneth et al., 2013). These factors play an important role in enhancing the production of desired metabolites in microorganisms.

20.6.2 Genetic Engineering Strategies Driven by Metabolomics Insights

Genetic engineering strategies driven by metabolomics insights are a powerful approach to optimize the production of desired metabolites in medicinal plants. By gaining a thorough

TABLE 20.1
Various Cultivation Parameters

Temperature	Adjusting temperature for optimal growth and metabolite production	Influences enzyme kinetics and metabolic pathways
pH	Maintaining pH within a specific range to support enzymatic reactions	Affects enzyme activity and stability
Nutrient availability	Optimizing the availability of essential nutrients (e.g., nitrogen, carbon) to meet metabolic demands	Directly impacts the production of precursors and cofactors
Oxygen levels	Controlling oxygen levels through aeration to support aerobic or anaerobic metabolism	Affects energy metabolism and redox balance
Light intensity (for photosynthetic organisms)	Regulating light intensity in phototrophic organisms to optimize photosynthesis	Drives photosynthetic metabolite production
Carbon source	Selecting appropriate carbon sources (e.g., glucose, sucrose) for efficient energy and metabolite production	Directly influences central carbon metabolism
Bioreactor design	Choosing the right bioreactor type (e.g., batch, fed-batch, continuous) for metabolite yield and scalability	Affects culture growth rate and metabolite accumulation
Microbial strain selection	Identifying and using high-yielding strains with desired metabolic pathways	Strain-specific metabolic capabilities
Induction strategies	Employing inducible promoters or triggers to initiate specific metabolic pathways	Controls the timing of metabolite production
Substrate feeding profiles	Implementing controlled substrate feeding strategies for sustained metabolite production	Ensures a continuous supply of precursors
Fermentation time	Optimizing the duration of fermentation to maximize metabolite accumulation	Balances growth and production phases
Cofactor addition	Supplementing with cofactors (e.g., vitamins, coenzymes) required for specific biosynthetic pathways	Enhances enzymatic activity
Downstream processing	Efficiently recovering and purifying metabolites post-cultivation.	Influences the final product yield and purity

understanding of the metabolic pathways involved in the biosynthesis of bioactive compounds, researchers can use genetic manipulation to enhance metabolite yield and quality (Khanijou et al., 2022; Valle et al., 2022; Liu et al., 2023; Barber et al., 2019). Here are some key genetic engineering strategies:

Overexpression of biosynthetic genes. Identify and overexpress genes responsible for key enzymatic steps in the target metabolite's biosynthesis pathway. This can lead to an increased production of the desired metabolite.

Gene silencing. Use techniques like RNA interference (RNAi) to downregulate competing or unwanted pathways that divert precursors away from the desired metabolite production.

Transcription factor manipulation. Modify the expression of transcription factors that regulate the expression of genes involved in metabolite biosynthesis. This can enhance overall pathway activity.

Chimeric genes and synthetic pathways. Create chimeric genes or synthetic pathways to direct precursor molecules toward the desired metabolite, bypassing natural regulatory steps.

Gene editing (CRISPR-Cas9). Employ gene editing technologies like CRISPR-Cas9 to precisely modify the plant genome, enabling the alteration of specific genes related to metabolite production.

Transporter engineering. Enhance the expression or activity of transporters responsible for metabolite export or import within plant cells to improve metabolite accumulation in desired tissues.

Organelle targeting. Direct biosynthetic enzymes to specific organelles (e.g., chloroplasts, mitochondria) to improve the efficiency of metabolite synthesis and reduce metabolic bottlenecks.

Promoter engineering. Design and use strong, tissue-specific, or inducible promoters to control the expression of biosynthetic genes, ensuring metabolite production in specific plant parts.

Metabolic channeling. Engineer protein complexes or organelle structures that facilitate the channeling of intermediates between enzymes in the biosynthetic pathway, reducing substrate competition.

Stacking traits. Combine multiple genetic modifications to simultaneously increase the production of several targeted metabolites or improve resistance against stressors.

Metabolite feedback regulation. Introduce feedback-resistant versions of enzymes to overcome the negative feedback regulation that often limits metabolite biosynthesis.

Metabolic engineering networks. Consider the interconnectedness of metabolic pathways, and engineer broader metabolic networks to optimize the overall yield of multiple related metabolites.

Phloem loading and unloading. Modify phloem loading and unloading mechanisms to facilitate the transport of metabolites to storage tissues or reproductive organs.

Metabolite profiling and iteration. Continuously monitor metabolite levels using metabolomics to fine-tune genetic modifications and cultivation conditions, optimizing metabolite yield over time.

20.6.3 Elicitation Techniques and Metabolic Enhancement

Elicitation techniques have been shown to enhance the synthesis of bioactive compounds in different cell cultures (Du et al., 2020). These techniques involve the application of elicitors, which stimulate metabolite production, at an optimum time to maximize the production of secondary metabolites. The addition of cell viability stabilizers and/or nutrients at the time of elicitation can further enhance metabolite production. Elicitation techniques can be useful in understanding topics that are difficult to discuss in formal interviews and that will reduce power imbalances between interviewers and respondents. They also allow participants to explain their conceptions of the world, rather than limiting them to categories derived from theory or previous research (Matzinger, 1994). Overall, elicitation techniques have the potential to enhance metabolic production and improve understanding in various fields, including bioactive compound synthesis and social education research.

Elicitation techniques have been shown to have various effects on metabolic activity. The application of elicitors can trigger defense responses in plants, leading to the activation of enzymes involved in secondary metabolism (Moreno-Escamilla et al., 2018). In case of *Catharanthus roseus* cell suspension cultures, elicitation resulted in the induction of certain enzyme activities, such as anthranilate synthase and tryptophan decarboxylase, leading to an increase in the production of tryptamine (Moreno et al., 1996). In *Artemisia annua*, different chemotypes showed differential metabolic responses to elicitation, with one chemotype showing an accumulation of artemisinic acid and artemisinin, while the other chemotype showed a decrease in these compounds (Wu et al., 2011). Coronatine treatment in *Lemna paucicostata* led to an increase in the amount of hydroxycinnamic acids and phytosterols, suggesting an activation of the biosynthetic pathways for these metabolites (Kim et al., 2017). Overall, elicitation techniques can modulate metabolic activity by influencing enzyme activities and the production of specific compounds.

20.7 CHALLENGES AND FUTURE DIRECTIONS

Metabolomics has the potential to play a significant role in medicinal plant biotechnology research by addressing various challenges and providing future directions. Metabolomics can be used to identify new active compounds, determine the vegetation period, investigate adulteration, evaluate correlation activity and metabolite levels, and monitor quality control of medicinal plants (Waris et al., 2022). Metabolomics has emerged as a vital tool in the realm of medicinal plant biotechnology research, offering unprecedented insights into the complex chemical landscapes of plants with therapeutic potential. However, this promising field also faces its share of challenges. Accurate metabolite identification and annotation remain a hurdle, hampering our ability to fully decipher the diverse array of plant compounds (Yang et al., 2019b). Standardization, data integration, and quantification issues persist, necessitating concerted efforts to enhance methodological rigor. To unlock metabolomics' full potential, future directions must focus on integrating multiomics data, harnessing artificial intelligence for data analysis, and expanding metabolite databases. High-throughput techniques, particularly in plant engineering, offer exciting prospects, as does the use of metabolomics for pharmacological discovery and environmental adaptation studies (Vahid et al., 2023). These advancements, coupled with ongoing efforts to address existing challenges, promise to further illuminate the intricate chemistry of medicinal plants, facilitating the development of novel drugs, nutraceuticals, and sustainable biotechnological solutions for diverse health and environmental challenges. It allows for the comprehensive analysis of primary and secondary metabolites in complicated natural products, overcoming the difficulties associated with their chemical differences (Shafi & Zahoor, 2021). Metabolomics profiling can provide valuable insights into the secondary metabolites present in MAPs (medicinal and aromatic plants), which are used for the development of herbal-based medicines and pharmacologically active compounds (Sun et al., 2020). Additionally, metabolomics can contribute to understanding the biosynthesis of nutritional and healthy metabolites in plants and their responses to plant-derived biohazards in humans, thus promoting human nutrition and health (Saito, 2018). Furthermore, metabolomics-based functional genomics can be applied to non-model specialized plants, including medicinal plants, for the identification of novel genes, metabolites, and networks involved in the synthesis of specialized plant products (Parray et al., 2019). Overall, metabolomics has the potential to revolutionize medicinal plant biotechnology research by providing valuable insights and directions for future studies.

20.8 CONCLUSION

In this chapter, we have explored the burgeoning field of metabolomics and its significant implications in medicinal plant biotechnology research. Metabolomics offers a comprehensive approach to unraveling the complex biochemical processes occurring within medicinal plants, shedding light on their secondary metabolites, bioactive compounds, and the intricate networks governing their biosynthesis.

In conclusion, *metabolomics* is a powerful tool that has revolutionized the way we study medicinal plants. Its applications span from drug discovery to biotechnological interventions, offering a holistic understanding of plant metabolites and their potential role in medicine and biotechnology. As we move forward, collaboration between researchers, biotechnologists, and pharmacologists will be essential to fully harness the potential of metabolomics in advancing medicinal plant research and improving human health. The future holds exciting opportunities for the integration of metabolomics into mainstream biotechnological and pharmaceutical practices, with the promise of unlocking new therapeutic agents and solutions from the plant kingdom.

REFERENCES

Alami, M. M., Ouyang, Z., Zhang, Y., Shu, S., Yang, G., Mei, Z., & Wang, X. (2022). The current developments in medicinal plant genomics enabled the diversification of secondary metabolites' biosynthesis. *International Journal of Molecular Sciences*, *23*(24), 15932. https://doi.org/10.3390/ijms232415932

Alseekh, S., & Fernie, A. R. (2021). *Using Metabolomics to Assist Plant Breeding* (pp. 33–46). https://doi.org/10.1007/978-1-0716-1201-9_3

Aravindaram, K., & Yang, N.-S. (2013). Metabolomics in medicinal plant research. In *Biotechnology for Medicinal Plants* (pp. 275–294). Springer Berlin Heidelberg. https://doi.org/10.1007/978-3-642-29974-2_12

Awashra, M., & Młynarz, P. (2023). The toxicity of nanoparticles and their interaction with cells: An in vitro metabolomic perspective. *Nanoscale Advances*, *5*(10), 2674–2723. https://doi.org/10.1039/D2NA00534D

Barber, D., Villaseñor, A., & Escribese, M. M. (2019). Metabolomics strategies to discover new biomarkers associated to severe allergic phenotypes. *Asia Pacific Allergy*, *9*(4), e37. https://doi.org/10.5415/apallergy.2019.9.e37

Bind, S., Bind, S., Sharma, A. K., & Chaturvedi, P. (2022). Epigenetic modification: A key tool for secondary metabolite production in microorganisms. *Frontiers in Microbiology*, *13*. https://doi.org/10.3389/fmicb.2022.784109

Böhme, K., Calo-Mata, P., Barros-Velázquez, J., & Ortea, I. (2019). Recent applications of omics-based technologies to main topics in food authentication. *TrAC Trends in Analytical Chemistry*, *110*, 221–232. https://doi.org/10.1016/j.trac.2018.11.005

Castro-Moretti, F. R., Gentzel, I. N., Mackey, D., & Alonso, A. P. (2020). Metabolomics as an emerging tool for the study of plant–pathogen interactions. *Metabolites*, *10*(2), 52. https://doi.org/10.3390/metabo10020052

Choi, Y. H., Jang, Y. P., & Frédérich, M. (2023). Editorial: Applications of metabolomics to the discovery of biomolecules from natural products. *Frontiers in Molecular Biosciences*, *10*. https://doi.org/10.3389/fmolb.2023.1190730

Crispin, M. C., & Wurtele, E. S. (2013). Use of metabolomics and transcriptomics to gain insights into the regulation and biosynthesis of medicinal compounds: Hypericum as a model. In *Biotechnology for Medicinal Plants* (pp. 395–411). Springer Berlin Heidelberg. https://doi.org/10.1007/978-3-642-29974-2_17

Cui, M., & He, G. (2012). Enhanced production of mycelia biomass and bioactive metabolite of the American Ganoderma lucidum in submerged culture. *2012 International Conference on Biomedical Engineering and Biotechnology*, 252–255. https://doi.org/10.1109/iCBEB.2012.198

de Sá, H., dos Santos, E., Carvalho, S., Nunes, R., & Canuto, G. (2023). Fundamentals and analytical strategies for metabolomics workflow: An overview and microbial applications. *Brazilian Journal of Analytical Chemistry*. https://doi.org/10.30744/brjac.2179-3425.RV-136-2022

Du, L., Li, D., Zhang, J., Du, J., Luo, Q., & Xiong, J. (2020). Elicitation of Lonicera japonica Thunb suspension cell for enhancement of secondary metabolites and antioxidant activity. *Industrial Crops and Products*, *156*, 112877. https://doi.org/10.1016/j.indcrop.2020.112877

Fan, R., Xie, S., Wang, S., Yu, Z., Sun, X., Du, Q., Yang, Y., & Han, R. (2023). Identification markers of goat milk adulterated with bovine milk based on proteomics and metabolomics. *Food Chemistry: X*, *17*, 100601. https://doi.org/10.1016/j.fochx.2023.100601

Feng, Z., Ding, C., Li, W., Wang, D., & Cui, D. (2020). Applications of metabolomics in the research of soybean plant under abiotic stress. *Food Chemistry*, *310*, 125914. https://doi.org/10.1016/j.foodchem.2019.125914

Floros, D. J., Jensen, P. R., Dorrestein, P. C., & Koyama, N. (2016). A metabolomics guided exploration of marine natural product chemical space. *Metabolomics*, *12*(9), 145. https://doi.org/10.1007/s11306-016-1087-5

Fukusaki, E., & Kobayashi, A. (2005). Plant metabolomics: Potential for practical operation. *Journal of Bioscience and Bioengineering*, *100*(4), 347–354. https://doi.org/10.1263/jbb.100.347

Gao, J., & Xiao, Y. (2023). Metabolomics and its applications in assisted reproductive technology. *IET Nanobiotechnology*, *17*(5), 399–405. https://doi.org/10.1049/nbt2.12141

Gomez-Casati, D. F., Zanor, M. I., & Busi, M. V. (2013). Metabolomics in plants and humans: Applications in the prevention and diagnosis of diseases. *BioMed Research International*, *2013*, 1–11. https://doi.org/10.1155/2013/792527

Hasanpour, M., Iranshahy, M., & Iranshahi, M. (2020). The application of metabolomics in investigating anti-diabetic activity of medicinal plants. *Biomedicine & Pharmacotherapy*, *128*, 110263. https://doi.org/10.1016/j.biopha.2020.110263

Haug, K., Salek, R. M., & Steinbeck, C. (2017). Global open data management in metabolomics. *Current Opinion in Chemical Biology*, *36*, 58–63. https://doi.org/10.1016/j.cbpa.2016.12.024

Jiang, T., Zhang, M., Wen, C., Xie, X., Tian, W., Wen, S., Lu, R., & Liu, L. (2020). Integrated metabolomic and transcriptomic analysis of the anthocyanin regulatory networks in Salvia miltiorrhiza Bge. flowers. *BMC Plant Biology*, *20*(1), 349. https://doi.org/10.1186/s12870-020-02553-7

Kenneth, D. R. S., Nadine, M. B., Suzanne, S., Eileen, C. K., James, E. H., Sidney, C., Trish, G., & Bevan, H. (2013). Dietary factors influence production of the soy isoflavone metabolite S-(−)Equol in healthy adults. *The Journal of Nutrition*, *143*(12), 1950–1958. https://doi.org/10.3945/jn.113.179564

Khanijou, J. K., Kulyk, H., Bergès, C., Khoo, L. W., Ng, P., Yeo, H. C., Helmy, M., Bellvert, F., Chew, W., & Selvarajoo, K. (2022). Metabolomics and modelling approaches for systems metabolic engineering. *Metabolic Engineering Communications*, *15*, e00209. https://doi.org/10.1016/j.mec.2022.e00209

Kim, J.-Y., Kim, H.-Y., Jeon, J.-Y., Kim, D.-M., Zhou, Y., Lee, J. S., Lee, H., & Choi, H.-K. (2017). Effects of coronatine elicitation on growth and metabolic profiles of Lemna paucicostata culture. *PLoS One*, *12*(11), e0187622. https://doi.org/10.1371/journal.pone.0187622

Kotsanopoulos, K., Martsikalis, P. V., Gkafas, G. A., & Exadactylos, A. (2022). The use of various statistical methods for authenticity and detection of adulteration in fish and seafood. *Critical Reviews in Food Science and Nutrition*, 1–19. https://doi.org/10.1080/10408398.2022.2117786

Kulkarni, R. (2016). Metabolic engineering. *Resonance*, *21*(3), 233–237. https://doi.org/10.1007/s12045-016-0318-4

Liu, S., Zhang, J., Sheng, Y., Feng, T., Shi, W., Lu, Y., Guan, X., Chen, X., Huang, J., & Chen, J. (2023). Metabolomics provides new insights into host manipulation strategies by Asobara japonica (Hymenoptera: Braconidae), a fruit fly parasitoid. *Metabolites*, *13*(3), 336. https://doi.org/10.3390/metabo13030336

Liu, Y., Wu, Z., Armstrong, D. W., Wolosker, H., & Zheng, Y. (2023). Detection and analysis of chiral molecules as disease biomarkers. *Nature Reviews Chemistry*, *7*(5), 355–373. https://doi.org/10.1038/s41570-023-00476-z

Ma, X., Meng, Y., Wang, P., Tang, Z., Wang, H., & Xie, T. (2020). Bioinformatics-assisted, integrated omics studies on medicinal plants. *Briefings in Bioinformatics*, *21*(6), 1857–1874. https://doi.org/10.1093/bib/bbz132

Manivasagan, P., & Kim, S.-K. (2014). *Extracellular Polysaccharides Produced by Marine Bacteria* (pp. 79–94). https://doi.org/10.1016/B978-0-12-800269-8.00005-1

Marchev, A. S., Vasileva, L. V., Amirova, K. M., Savova, M. S., Balcheva-Sivenova, Z. P., & Georgiev, M. I. (2021). Metabolomics and health: From nutritional crops and plant-based pharmaceuticals to profiling of human biofluids. *Cellular and Molecular Life Sciences*, *78*(19–20), 6487–6503. https://doi.org/10.1007/s00018-021-03918-3

Matzinger, P. (1994). Tolerance, danger, and the extended family. *Annual Review of Immunology*, *12*(1), 991–1045. https://doi.org/10.1146/annurev.iy.12.040194.005015

Moreno, P. R. H., Poulsen, C., van der Heijden, R., & Verpoorte, R. (1996). Effects of elicitation on different metabolic pathways in Catharanthus roseus (L.)G.Don cell suspension cultures. *Enzyme and Microbial Technology*, *18*(2), 99–107. https://doi.org/10.1016/0141-0229(95)00078-X

Moreno-Escamilla, J. O., Alvarez-Parrilla, E., de la Rosa, L. A., Núñez-Gastélum, J. A., González-Aguilar, G. A., & Rodrigo-García, J. (2018). Effect of elicitors in the nutritional and sensorial quality of fruits and vegetables. In *Preharvest Modulation of Postharvest Fruit and Vegetable Quality* (pp. 71–91). Elsevier. https://doi.org/10.1016/B978-0-12-809807-3.00003-2

Mumtaz, M. W., Hamid, A. A., Akhtar, M. T., Anwar, F., Rashid, U., & Al-Zuaidy, M. H. (2017). An overview of recent developments in metabolomics and proteomics—Phytotherapic research perspectives. *Frontiers in Life Science*, *10*(1), 1–37. https://doi.org/10.1080/21553769.2017.1279573

Oh, S.-W., Imran, M., Kim, E.-H., Park, S.-Y., Lee, S.-G., Park, H.-M., Jung, J.-W., & Ryu, T.-H. (2023). Approach strategies and application of metabolomics to biotechnology in plants. *Frontiers in Plant Science*, *14*. https://doi.org/10.3389/fpls.2023.1192235

Okazaki, Y., & Saito, K. (2012). Recent advances of metabolomics in plant biotechnology. *Plant Biotechnology Reports*, *6*(1), 1–15. https://doi.org/10.1007/s11816-011-0191-2

Pan, H., Yao, C., Yao, S., Yang, W., Wu, W., & Guo, D. (2020). A metabolomics strategy for authentication of plant medicines with multiple botanical origins, a case study of Uncariae Rammulus Cum Uncis. *Journal of Separation Science*, *43*(6), 1043–1050. https://doi.org/10.1002/jssc.201901064

Pan, Y., Lei, X., & Zhang, Y. (2022). Association predictions of genomics, proteinomics, transcriptomics, microbiome, metabolomics, pathomics, radiomics, drug, symptoms, environment factor, and disease networks: A comprehensive approach. *Medicinal Research Reviews*, *42*(1), 441–461. https://doi.org/10.1002/med.21847

Parida, A. K., Panda, A., & Rangani, J. (2018). Metabolomics-guided elucidation of abiotic stress tolerance mechanisms in plants. In *Plant Metabolites and Regulation under Environmental Stress* (pp. 89–131). Elsevier. https://doi.org/10.1016/B978-0-12-812689-9.00005-4

Parray, J. A., Yaseen Mir, M., & Shameem, N. (2019). Plant metabolomics: Sustainable approach towards crop productivity. In *Sustainable Agriculture: Biotechniques in Plant Biology* (pp. 51–70). Springer. https://doi.org/10.1007/978-981-13-8840-8_2

Rafi, M., Septaningsih, D. A., Karomah, A. H., Lukman, L., Prajogo, B., Amran, M. B., & Rohman, A. (2021). Inhibition of α-glucosidase activity, metals content, and phytochemical profiling of Andrographis paniculata from different geographical origins based on FTIR and UHPLC-Q-Orbitrap HRMS metabolomics. *Biodiversitas Journal of Biological Diversity*, *22*(3). https://doi.org/10.13057/biodiv/d220359

Richter, S., Garrett, T. J., Bechmann, N., Clifton-Bligh, R. J., & Ghayee, H. K. (2023). Metabolomics in paraganglioma: Applications and perspectives from genetics to therapy. *Endocrine-Related Cancer*, *30*(6). https://doi.org/10.1530/ERC-22-0376

Rinschen, M. M., Ivanisevic, J., Giera, M., & Siuzdak, G. (2019). Identification of bioactive metabolites using activity metabolomics. *Nature Reviews Molecular Cell Biology*, *20*(6), 353–367. https://doi.org/10.1038/s41580-019-0108-4

Rivera-Pérez, A., Romero-González, R., & Garrido Frenich, A. (2021). Feasibility of applying untargeted metabolomics with GC-orbitrap-HRMS and chemometrics for authentication of black pepper (Piper nigrum L.) and identification of geographical and processing markers. *Journal of Agricultural and Food Chemistry*, *69*(19), 5547–5558. https://doi.org/10.1021/acs.jafc.1c01515

Saito, K. (2018). Development of plant metabolomics and medicinal plant genomics. *YAKUGAKU ZASSHI*, *138*(1), 1–18. https://doi.org/10.1248/yakushi.17-00193

Saito, K., & Matsuda, F. (2010). Metabolomics for functional genomics, systems biology, and biotechnology. *Annual Review of Plant Biology*, *61*, 463–489. https://doi.org/10.1146/annurev.arplant.043008.092035

Sakurai, N. (2022). Recent applications of metabolomics in plant breeding. *Breeding Science*, *72*(1), 21065. https://doi.org/10.1270/jsbbs.21065

Shafi, A., & Zahoor, I. (2021). Metabolomics of medicinal and aromatic plants: Goldmines of secondary metabolites for herbal medicine research. In *Medicinal and Aromatic Plants* (pp. 261–287). Elsevier. https://doi.org/10.1016/B978-0-12-819590-1.00012-4

Sharma, S., Tiwari, P., & Pandey, L. (2022). *Design of Consortium for the Production of Desired Metabolites* (pp. 179–195). https://doi.org/10.1007/978-981-16-5465-7_8

Simpson, H. L., & Campbell, B. J. (2015). Review article: Dietary fibre-microbiota interactions. *Alimentary Pharmacology & Therapeutics*, *42*(2), 158–179. https://doi.org/10.1111/apt.13248

Su, C., Chi, Z., & Lu, W. (2007). Optimization of medium and cultivation conditions for enhanced exopolysaccharide yield by marine Cyanothece sp. 113. *Chinese Journal of Oceanology and Limnology*, *25*(4), 411–417. https://doi.org/10.1007/s00343-007-0411-3

Sucher, N. J., Hennell, J. R., & Carles, M. C. (2013). Genomic and transcriptomic profiling: Tools for the quality production of plant-based medicines. In *Biotechnology for Medicinal Plants* (pp. 439–455). Springer Berlin Heidelberg. https://doi.org/10.1007/978-3-642-29974-2_19

Sumner, L. W., Lei, Z., Nikolau, B. J., & Saito, K. (2015). Modern plant metabolomics: Advanced natural product gene discoveries, improved technologies, and future prospects. *Natural Product Reports*, *32*(2), 212–229. https://doi.org/10.1039/C4NP00072B

Sun, H., Li, X., Zhang, A., Zhang, K., Yan, G., Han, Y., Wu, F., & Wang, X. (2019). Exploring potential biomarkers of coronary heart disease treated by Jing Zhi Guan Xin Pian using high-throughput metabolomics. *RSC Advances*, *9*(20), 11420–11432. https://doi.org/10.1039/C8RA10557J

Sun, H., Zhang, A., Zou, D., Sun, W., Wu, X., & Wang, X. (2014). Metabolomics coupled with pattern recognition and pathway analysis on potential biomarkers in liver injury and hepatoprotective effects of yinchenhao. *Applied Biochemistry and Biotechnology*, *173*(4), 857–869. https://doi.org/10.1007/s12010-014-0903-5

Sun, W., Chen, Z., Hong, J., & Shi, J. (2020). Promoting human nutrition and health through plant metabolomics: Current status and challenges. *Biology*, *10*(1), 20. https://doi.org/10.3390/biology10010020

Tian, H., Lam, S., & Shui, G. (2016). Metabolomics, a powerful tool for agricultural research. *International Journal of Molecular Sciences*, *17*(11), 1871. https://doi.org/10.3390/ijms17111871

Vahid, F., Hajizadeghan, K., & Khodabakhshi, A. (2023). Nutritional metabolomics in diet–breast cancer relations: Current research, challenges, and future directions—A review. *Biomedicines*, *11*(7), 1845. https://doi.org/10.3390/biomedicines11071845

Valle, A., Soto, Z., Muhamadali, H., Hollywood, K. A., Xu, Y., Lloyd, J. R., Goodacre, R., Cantero, D., Cabrera, G., & Bolivar, J. (2022). Metabolomics for the design of new metabolic engineering strategies for improving aerobic succinic acid production in Escherichia coli. *Metabolomics*, *18*(8), 56. https://doi.org/10.1007/s11306-022-01912-9

Wang, G., Wang, Y., Man, H., Lee, Y.-W., Shi, J., & Xu, J. (2020). Metabolomics-guided analysis reveals a two-step epimerization of deoxynivalenol catalyzed by the bacterial consortium IFSN-C1. *Applied Microbiology and Biotechnology*, *104*(13), 6045–6056. https://doi.org/10.1007/s00253-020-10673-1

Wang, R., Shu, P., Zhang, C., Zhang, J., Chen, Y., Zhang, Y., Du, K., Xie, Y., Li, M., Ma, T., Zhang, Y., Li, Z., Grierson, D., Pirrello, J., Chen, K., Bouzayen, M., Zhang, B., & Liu, M. (2022). Integrative analyses of metabolome and genome-wide transcriptome reveal the regulatory network governing flavor formation in kiwifruit (Actinidia chinensis). *New Phytologist*, *233*(1), 373–389. https://doi.org/10.1111/nph.17618

Wang, X., Ren, J., Zhang, A., Sun, H., Yan, G., Han, Y., & Liu, L. (2019). Novel applications of mass spectrometry-based metabolomics in herbal medicines and its active ingredients: Current evidence. *Mass Spectrometry Reviews*, *38*(4–5), 380–402. https://doi.org/10.1002/mas.21589

Waris, M., Koçak, E., Gonulalan, E. M., Demirezer, L. O., Kır, S., & Nemutlu, E. (2022). Metabolomics analysis insight into medicinal plant science. *TrAC Trends in Analytical Chemistry*, *157*, 116795. https://doi.org/10.1016/j.trac.2022.116795

Wen, W., Zhao, X., Wang, H., & Cao, X. (2023). A real-time metabolomic profiling approach to identify virgin olive oil adulteration by rapid evaporative ionization mass apectrometry. *Food Analytical Methods*, *16*(5), 985–996. https://doi.org/10.1007/s12161-023-02466-9

Wu, W., Yuan, M., Zhang, Q., Zhu, Y., Yong, L., Wang, W., Qi, Y., & Guo, D. (2011). Chemotype-dependent metabolic response to methyl jasmonate elicitation in Artemisia annua. *Planta Medica*, *77*(10), 1048–1053. https://doi.org/10.1055/s-0030-1250744

Wu, W., Zhang, L., Zheng, X., Huang, Q., Farag, M. A., Zhu, R., & Zhao, C. (2022). Emerging applications of metabolomics in food science and future trends. *Food Chemistry: X*, *16*, 100500. https://doi.org/10.1016/j.fochx.2022.100500

Xiao, Q., Mu, X., Liu, J., Li, B., Liu, H., Zhang, B., & Xiao, P. (2022). Plant metabolomics: A new strategy and tool for quality evaluation of Chinese medicinal materials. *Chinese Medicine*, *17*(1), 45. https://doi.org/10.1186/s13020-022-00601-y

Yang, L., Yang, Y., Huang, L., Cui, X., & Liu, Y. (2023). From single- to multi-omics: Future research trends in medicinal plants. *Briefings in Bioinformatics*, *24*(1). https://doi.org/10.1093/bib/bbac485

Yang, M., & Lao, L. (2019). Emerging applications of metabolomics in traditional Chinese medicine treating hypertension: Biomarkers, pathways and more. *Frontiers in Pharmacology*, *10*. https://doi.org/10.3389/fphar.2019.00158

Yang, Q., Zhang, A. H., Miao, J. H., Sun, H., Han, Y., Yan, G. L., Wu, F. F., & Wang, X. J. (2019a). Metabolomics biotechnology, applications, and future trends: A systematic review. In *RSC Advances* (Vol. 9, Issue 64, pp. 37245–37257). Royal Society of Chemistry. https://doi.org/10.1039/c9ra06697g

Yang, Q., Zhang, A. H., Miao, J. H., Sun, H., Han, Y., Yan, G. L., Wu, F. F., & Wang, X. J. (2019b). Metabolomics biotechnology, applications, and future trends: A systematic review. *RSC Advances*, *9*(64), 37245–37257. https://doi.org/10.1039/C9RA06697G

Younis, I. Y., Mohsen, E., Ibrahim, R. M., Fernie, A. R., Alseekh, S., & Salem, M. A. (2023). Non-targeted metabolomics and chemometrics for saffron (Crocus sativus L.) authentication and adulteration detection in relation to its anticholinesterase activity. *Food Chemistry Advances*, *2*, 100217. https://doi.org/10.1016/j.focha.2023.100217

Zhang, W.-F., Gong, Z.-H., Wu, M.-B., Chan, H., Yuan, Y.-J., Tang, N., Zhang, Q., Miao, M.-J., Chang, W., Li, Z., Li, Z.-G., Jin, L., & Deng, W. (2019). Integrative comparative analyses of metabolite and transcript profiles uncovers complex regulatory network in tomato (Solanum lycopersicum L.) fruit undergoing chilling injury. *Scientific Reports*, *9*(1), 4470. https://doi.org/10.1038/s41598-019-41065-9

Zhu, L., Zhou, Y., Li, X., Zhao, J., Guo, N., & Xing, H. (2018). Metabolomics analysis of soybean hypocotyls in response to phytophthora sojae infection. *Frontiers in Plant Science*, *9*. https://doi.org/10.3389/fpls.2018.01530

Index

A

B

C

D

O

P

Q

R

S

T

U

V

W

FSC
www.fsc.org
MIX
Papier | Fördert gute Waldnutzung
FSC® C083411

Kolibri 360 GmbH
Ferdinand-Jühlke-Straße 7
99095 Erfurt, Deutschland
produktsicherheit@kolibri360.de